Data Science in R
A Case Studies Approach to Computational Reasoning and Problem Solving

Chapman & Hall/CRC
The R Series

Series Editors

John M. Chambers
Department of Statistics
Stanford University
Stanford, California, USA

Torsten Hothorn
Division of Biostatistics
University of Zurich
Switzerland

Duncan Temple Lang
Department of Statistics
University of California, Davis
Davis, California, USA

Hadley Wickham
RStudio
Boston, Massachusetts, USA

Aims and Scope

This book series reflects the recent rapid growth in the development and application of R, the programming language and software environment for statistical computing and graphics. R is now widely used in academic research, education, and industry. It is constantly growing, with new versions of the core software released regularly and more than 6,000 packages available. It is difficult for the documentation to keep pace with the expansion of the software, and this vital book series provides a forum for the publication of books covering many aspects of the development and application of R.

The scope of the series is wide, covering three main threads:
- Applications of R to specific disciplines such as biology, epidemiology, genetics, engineering, finance, and the social sciences.
- Using R for the study of topics of statistical methodology, such as linear and mixed modeling, time series, Bayesian methods, and missing data.
- The development of R, including programming, building packages, and graphics.

The books will appeal to programmers and developers of R software, as well as applied statisticians and data analysts in many fields. The books will feature detailed worked examples and R code fully integrated into the text, ensuring their usefulness to researchers, practitioners and students.

Published Titles

Stated Preference Methods Using R, *Hideo Aizaki, Tomoaki Nakatani, and Kazuo Sato*

Using R for Numerical Analysis in Science and Engineering, *Victor A. Bloomfield*

Event History Analysis with R, *Göran Broström*

Computational Actuarial Science with R, *Arthur Charpentier*

Statistical Computing in C++ and R, *Randall L. Eubank and Ana Kupresanin*

Reproducible Research with R and RStudio, *Christopher Gandrud*

Introduction to Scientific Programming and Simulation Using R, Second Edition, *Owen Jones, Robert Maillardet, and Andrew Robinson*

Nonparametric Statistical Methods Using R, *John Kloke and Joseph McKean*

Displaying Time Series, Spatial, and Space-Time Data with R, *Oscar Perpiñán Lamigueiro*

Programming Graphical User Interfaces with R, *Michael F. Lawrence and John Verzani*

Analyzing Sensory Data with R, *Sébastien Lê and Theirry Worch*

Analyzing Baseball Data with R, *Max Marchi and Jim Albert*

Growth Curve Analysis and Visualization Using R, *Daniel Mirman*

R Graphics, Second Edition, *Paul Murrell*

Data Science in R: A Case Studies Approach to Computational Reasoning and Problem Solving, *Deborah Nolan and Duncan Temple Lang*

Multiple Factor Analysis by Example Using R, *Jérôme Pagès*

Customer and Business Analytics: Applied Data Mining for Business Decision Making Using R, *Daniel S. Putler and Robert E. Krider*

Implementing Reproducible Research, *Victoria Stodden, Friedrich Leisch, and Roger D. Peng*

Using R for Introductory Statistics, Second Edition, *John Verzani*

Advanced R, *Hadley Wickham*

Dynamic Documents with R and knitr, *Yihui Xie*

Data Science in R
A Case Studies Approach to Computational Reasoning and Problem Solving

Deborah Nolan
University of California, Berkeley
USA

Duncan Temple Lang
University of California, Davis
USA

CRC Press
Taylor & Francis Group
Boca Raton London New York

CRC Press is an imprint of the
Taylor & Francis Group, an **informa** business

A CHAPMAN & HALL BOOK

CRC Press
Taylor & Francis Group
6000 Broken Sound Parkway NW, Suite 300
Boca Raton, FL 33487-2742

© 2015 by Taylor & Francis Group, LLC
CRC Press is an imprint of Taylor & Francis Group, an Informa business

No claim to original U.S. Government works

Printed on acid-free paper
Version Date: 20150310

International Standard Book Number-13: 978-1-4822-3481-7 (Paperback)

This book contains information obtained from authentic and highly regarded sources. Reasonable efforts have been made to publish reliable data and information, but the author and publisher cannot assume responsibility for the validity of all materials or the consequences of their use. The authors and publishers have attempted to trace the copyright holders of all material reproduced in this publication and apologize to copyright holders if permission to publish in this form has not been obtained. If any copyright material has not been acknowledged please write and let us know so we may rectify in any future reprint.

Except as permitted under U.S. Copyright Law, no part of this book may be reprinted, reproduced, transmitted, or utilized in any form by any electronic, mechanical, or other means, now known or hereafter invented, including photocopying, microfilming, and recording, or in any information storage or retrieval system, without written permission from the publishers. Printed in Canada.

For permission to photocopy or use material electronically from this work, please access www.copyright.com (http://www.copyright.com/) or contact the Copyright Clearance Center, Inc. (CCC), 222 Rosewood Drive, Danvers, MA 01923, 978-750-8400. CCC is a not-for-profit organization that provides licenses and registration for a variety of users. For organizations that have been granted a photocopy license by the CCC, a separate system of payment has been arranged.

Trademark Notice: Product or corporate names may be trademarks or registered trademarks, and are used only for identification and explanation without intent to infringe.

Visit the Taylor & Francis Web site at
http://www.taylorandfrancis.com

and the CRC Press Web site at
http://www.crcpress.com

To our families — Zoë and Suzana, and Dave, Ben, and Sam,

and to our mentors John Chambers and Terry Speed.

Contents

Preface xv

Acknowledgments xix

Authors xxi

Co-Authors xxiii

I Data Manipulation and Modeling   1

1 Predicting Location via Indoor Positioning Systems   3
Deborah Nolan and Duncan Temple Lang

- 1.1 Introduction 3
  - 1.1.1 Computational Topics 4
- 1.2 The Raw Data 4
  - 1.2.1 Processing the Raw Data 8
- 1.3 Cleaning the Data and Building a Representation for Analysis 12
  - 1.3.1 Exploring Orientation 14
  - 1.3.2 Exploring MAC Addresses 16
  - 1.3.3 Exploring the Position of the Hand-Held Device 18
  - 1.3.4 Creating a Function to Prepare the Data 19
- 1.4 Signal Strength Analysis 21
  - 1.4.1 Distribution of Signal Strength 21
  - 1.4.2 The Relationship between Signal and Distance 26
- 1.5 Nearest Neighbor Methods to Predict Location 31
  - 1.5.1 Preparing the Test Data 31
  - 1.5.2 Choice of Orientation 32
  - 1.5.3 Finding the Nearest Neighbors 34
  - 1.5.4 Cross-Validation and Choice of k 36
- 1.6 Exercises 41

2 Modeling Runners' Times in the Cherry Blossom Race   45
Daniel Kaplan and Deborah Nolan

- 2.1 Introduction 45
  - 2.1.1 Computational Topics 47
- 2.2 Reading Tables of Race Results into R 47
- 2.3 Data Cleaning and Reformatting Variables 55
- 2.4 Exploring the Run Time for All Male Runners 63
  - 2.4.1 Making Plots with Many Observations 63
  - 2.4.2 Fitting Models to Average Performance 67
  - 2.4.3 Cross-Sectional Data and Covariates 74
- 2.5 Constructing a Record for an Individual Runner across Years 79
- 2.6 Modeling the Change in Running Time for Individuals 88

		2.7	Scraping Race Results from the Web	93
		2.8	Exercises	100

3 Using Statistics to Identify Spam — 105
Deborah Nolan and Duncan Temple Lang

- 3.1 Introduction ... 105
  - 3.1.1 Computational Topics ... 106
- 3.2 Anatomy of an email Message ... 107
- 3.3 Reading the email Messages ... 110
- 3.4 Text Mining and Naïve Bayes Classification ... 113
- 3.5 Finding the Words in a Message ... 116
  - 3.5.1 Splitting the Message into Its Header and Body ... 116
  - 3.5.2 Removing Attachments from the Message Body ... 117
  - 3.5.3 Extracting Words from the Message Body ... 124
  - 3.5.4 Completing the Data Preparation Process ... 126
- 3.6 Implementing the Naïve Bayes Classifier ... 127
  - 3.6.1 Test and Training Data ... 128
  - 3.6.2 Probability Estimates from Training Data ... 129
  - 3.6.3 Classifying New Messages ... 131
  - 3.6.4 Computational Considerations ... 135
- 3.7 Recursive Partitioning and Classification Trees ... 138
- 3.8 Organizing an email Message into an R Data Structure ... 140
  - 3.8.1 Processing the Header ... 141
  - 3.8.2 Processing Attachments ... 144
  - 3.8.3 Testing Our Code on More email Data ... 146
  - 3.8.4 Completing the Process ... 148
- 3.9 Deriving Variables from the email Message ... 150
  - 3.9.1 Checking Our Code for Errors ... 155
- 3.10 Exploring the email Feature Set ... 158
- 3.11 Fitting the rpart() Model to the email Data ... 160
- 3.12 Exercises ... 164

4 Processing Robot and Sensor Log Files: Seeking a Circular Target — 171
Samuel E. Buttrey, Timothy H. Chung, James N. Eagle, and Duncan Temple Lang

- 4.1 Description ... 171
  - 4.1.1 Computational Topics ... 172
- 4.2 The Data ... 173
  - 4.2.1 Reading an Entire Log File ... 175
  - 4.2.2 Exploring Log Files ... 179
  - 4.2.3 Visualizing the Path ... 184
  - 4.2.4 Exploring a "Look" ... 187
  - 4.2.5 The Error Distribution for Range Values ... 190
- 4.3 Detecting a Circular Target ... 194
  - 4.3.1 Connecting Segments Behind the Robot ... 198
  - 4.3.2 Determining If a Segment Corresponds to a Circle ... 200
- 4.4 Detecting the Target with Streaming Data in Real Time ... 213

5 Strategies for Analyzing a 12-Gigabyte Data Set: Airline Flight Delays 217
Michael Kane

- 5.1 Introduction 217
  - 5.1.1 Computational Topics 218
- 5.2 Acquiring the Airline Data Set 219
- 5.3 Computing with Massive Data: Getting Flight Delay Counts . . . 219
  - 5.3.1 The *R* Programming Environment 219
  - 5.3.2 The *UNIX* Shell 221
  - 5.3.3 An *SQL* Database with *R* 223
  - 5.3.4 The `bigmemory` Package with *R* 227
- 5.4 Explorations Using Parallel Computing: The Distribution of Flight Delays 229
  - 5.4.1 Writing a Parallelizable Loop with `foreach` 230
  - 5.4.2 Using the Split-Apply-Combine Approach for Better Performance . 231
  - 5.4.3 Using Split-Apply-Combine to Find the Best Time to Fly 232
- 5.5 From Exploration to Model: Do Older Planes Suffer Greater Delays? . . . 236

II Simulation Studies 239

6 Pairs Trading 241
Cari Kaufman and Duncan Temple Lang

- 6.1 The Problem 241
  - 6.1.1 Computational Topics 245
- 6.2 The Data Format 246
- 6.3 Reading the Financial Data 247
- 6.4 Visualizing the Time Series 250
- 6.5 Finding Opening and Closing Positions 251
  - 6.5.1 Identifying a Position 251
  - 6.5.2 Displaying Positions 254
  - 6.5.3 Finding All Positions 256
  - 6.5.4 Computing the Profit for a Position 257
  - 6.5.5 Finding the Optimal Value for k 260
- 6.6 Simulation Study 263
  - 6.6.1 Simulating the Stock Price Series 265
  - 6.6.2 Making `stockSim()` Faster 273

7 Simulation Study of a Branching Process 277
Deborah Nolan and Duncan Temple Lang

- 7.1 Introduction 277
  - 7.1.1 The Monte Carlo Method 279
  - 7.1.2 Computational Topics 281
- 7.2 Exploring the Random Process 281
- 7.3 Generating Offspring 284
  - 7.3.1 Checking the Results 286
  - 7.3.2 Considering Alternative Implementations 287
- 7.4 Profiling and Improving Our Code 289
- 7.5 From One Job's Offspring to an Entire Generation 290
- 7.6 Unit Testing 292
- 7.7 A Structure for the Function's Return Value 293
- 7.8 The Family Tree: Simulating the Branching Process 294
- 7.9 Replicating the Simulation 299
  - 7.9.1 Analyzing the Simulation Results 301

8 A Self-Organizing Dynamic System with a Phase Transition — 309
Deborah Nolan and Duncan Temple Lang

- 8.1 Introduction and Motivation — 309
  - 8.1.1 Computational Topics — 310
- 8.2 The Model — 310
  - 8.2.1 The Order Cars Move — 312
- 8.3 Implementing the BML Model — 314
  - 8.3.1 Creating the Initial Grid Configuration — 314
  - 8.3.2 Testing the Grid Creation Function — 318
  - 8.3.3 Displaying the Grid — 321
  - 8.3.4 Visualizing the Grid — 322
  - 8.3.5 Simple and Convenient Object-Oriented Programming — 325
  - 8.3.6 Moving the Cars — 327
- 8.4 Evaluating the Performance of the Code — 334
- 8.5 Implementing the BML Model in C — 346
  - 8.5.1 The Algorithm in C — 348
  - 8.5.2 Compiling, Loading, and Calling the C Code — 355
- 8.6 Running the Simulations — 359
  - 8.6.1 Exploring Car Velocity — 360
- 8.7 Experimental Compilation — 362

9 Simulating Blackjack — 367
Hadley Wickham

- 9.1 Introduction — 367
  - 9.1.1 Computational Topics — 368
- 9.2 Blackjack Basics — 368
  - 9.2.1 Testing Functions — 370
- 9.3 Playing a Hand of Blackjack — 372
  - 9.3.1 Creating Functions for the Player's Actions — 373
- 9.4 Strategies for Playing — 376
  - 9.4.1 Developing the Optimal Strategy — 379
- 9.5 Playing Many Games — 382
- 9.6 A More Accurate Card Dealer Shoe — 384
- 9.7 Counting Cards — 390
- 9.8 Putting It All Together — 393
- 9.9 Exercises — 394

III Data and Web Technologies — 397

10 Baseball: Exploring Data in a Relational Database — 399
Deborah Nolan and Duncan Temple Lang

- 10.1 Introduction — 399
  - 10.1.1 Computational Topics — 400
- 10.2 Sean Lahman's Database — 401
  - 10.2.1 Connecting to the Baseball Database from within R — 401
- 10.3 Aggregating Salaries into Payroll — 403
- 10.4 Merging Payroll Data with Information in Other Tables — 408
  - 10.4.1 Adding Team Names to the Payroll Data — 409
  - 10.4.2 Adding World Series Records to the Payroll Data — 411

(7.10 Exercises — 306)

	10.5	Exploring the Extreme Salaries	412
	10.6	Exercises	415

11 CIA Factbook Mashup — 419
Deborah Nolan and Duncan Temple Lang

	11.1	Introduction	419
		11.1.1 Computational Topics	421
	11.2	Acquiring the Data	421
		11.2.1 Extracting Latitude and Longitude from a CSV File	421
	11.3	Integrating Data from Different Sources	423
	11.4	Preparing the Data for Plotting	424
		11.4.1 Redoing the Merge of the Factbook and Location Data	428
	11.5	Plotting with Google Earth™	430
	11.6	Extracting Demographic Information from the CIA *XML* File	435
	11.7	Generating *KML* Directly	442
	11.8	Additional Computational Tasks	448
		11.8.1 Creating Plotting Symbols	448
		11.8.2 Efficiency in Generating *KML* from Strings	448
		11.8.3 Extracting Latitude and Longitude from an *HTML* File	450
	11.9	Exercises	451

12 Exploring Data Science Jobs with Web Scraping and Text Mining — 457
Deborah Nolan and Duncan Temple Lang

	12.1	Introduction and Motivation	457
		12.1.1 Computational Topics	459
	12.2	Exploring Different Web Sites	459
	12.3	Preliminary/Exploratory Scraping: The Kaggle Job List	465
		12.3.1 Processing the Text	469
		12.3.2 Generalizing to Other Posts	470
		12.3.3 Scraping the Kaggle Post List	473
	12.4	Scraping CyberCoders.com	475
		12.4.1 Getting the Skill List from a Job Post	478
		12.4.2 Finding the Links to Job Postings in the Search Results	482
		12.4.3 Finding the Next Page of Job Post Search Results	487
		12.4.4 Putting It All Together	488
	12.5	A Reusable Generic Framework for Arbitrary Sites	489
	12.6	Scraping Career Builder	492
	12.7	Scraping Monster.com	494
	12.8	Analyzing the Results: The Important Skills	495
	12.9	Note on Web Scraping	503
	12.10	Exercises	503

Index — 507

Colophon — 515

Preface

Our aim in writing this book is two-fold. We want students to be able to read about the computational reasoning and details of real-world data analyses. We also want to provide, hopefully, interesting and useful material to help statistics faculty teach the important aspects of a new, expanded curriculum to a new type of statistics and data science student. This enhanced curriculum includes exposure to data analytic and computational reasoning, rather than focusing primarily on statistical methodology. Our goal is not to provide short, tidy answers and solutions, but to explore the problems, possible solutions, and the thought process involved in addressing data science projects.

Goals of the Book
There are many different languages people commonly use to do data analysis and data science. We focus primarily on *R*, but also use several other domain specific languages (DSLs) and even touch on languages such as the *UNIX* shell and *C*. This book is not intended to teach the syntax or semantics of the *R* language, or any of the other languages we use. Nor is it written to list the large number of packages and functions that data scientists commonly use in *R*. Instead, we wrote the book so that people could experience the *thought process* involved in solving authentic computational problems related to data analysis problems. There are many books that teach programming by introducing the important ideas in a section and illustrating them with one or more examples. These are very useful and essential starting points. However, the code in the examples in these books is the final, polished version written by an expert, as it should be. These do not expose the reader to the actual *process* of writing code, but just the final result. Our aim is to illustrate the process by which programmers approach a problem and reason about different ways of implementing the solution. This process is very dynamic and iterative. We write some code, test it, change it, refine and extend it and generalize it. Often, we "start over," having learned from the first attempt, or prototype, and develop a more succinct, clearer version. Along the way, we make trade-offs between simplicity, efficiency, generality, reuse, correct and approximate results, and so on. We try to find ways to minimize changes to the code while making it faster and more flexible. In this book, we try to illustrate this entire process and the often implicit decisions experienced programmers make. The hope is to complement the textbooks and provide students, researchers (and even faculty) with a glimpse into how professional data scientists think about daily computational tasks.

Using These Case Studies in Statistical Computing Courses
Developing a new course in statistical computing (or any topic) is a very time-consuming task for an instructor. We often have to learn some new topics, or at least their details, and prioritize and order them to identify which ones should be in the course and in what order. We also have to develop assignments and plenty of them so that we can swap them from year to year. We could give synthetic programming assignments to help the students learn, for example, vectorization and loops, or regular expressions. These are terrific introductory exercises when the students are first learning the fundamental concepts, but do not necessarily grow into significant assignments or mini-projects. We strongly favor giving the students authentic real-world data analysis projects in our statistical computing courses that tie the new concepts into the regular data science workflow. We want to expose the students to the

daily activities of data scientists. We also think the context is interesting for the students and helps them to see the wide range of applications of data analysis. Furthermore, we want to introduce statistical methods and concepts, along with the computational topics, that they may not see in other courses. For these reasons, our statistical computing courses act as a catch-all where we cover many "real world" topics necessary for a data scientist to master for everyday work.

With all of these goals in mind, finding pedagogically interesting projects and assignments that the students can actually complete and which interest them and illustrate the particular topics is extremely challenging. When we started developing our computing courses at Berkeley and Davis, we spent many days/weeks developing assignments. We had many ideas for possible data sets and sources. For every one we turned into an assignment, we "interviewed" 4 or 5 other problems. Some were interesting, but were too simple or too complex. Some did not have an interesting statistical/data analysis question at the end of the data processing. Others didn't lend themselves to teaching the particular computational and statistical topics on which we wanted the student to focus. Our hope is that this book and its case studies will reduce the barrier for instructors to incorporate interesting problems into statistical and computing courses that focus on data science skills.

In the current era of data science, we have many rich and interesting data sets to use for research and teaching. The Data Expo competitions that people such as Debby Swayne, Paul Murrell, and Hadley Wickham have organized are excellent sources of interesting, challenging, and manageable problems. Data repositories, such as the one at University of California, Irvine (UCI), have also grown in number and diversity. Sites such as Kaggle.com also provide interesting problems and data. Our focus in this book is slightly different from these. We try to start with the raw data, and identify and explore potential questions of interest, rather than have the question prescribed for us or the data pre-processed. We feel that it is important for students to experience both acquiring and working with unstructured or semi-structured data, and also narrowing down and carefully framing the questions of interest about those data. This motivation comes from our experience in industrial research labs (IBM and Bell Laboratories), summer schools such as Explorations in Statistics Research (ESR), and teaching at UC Berkeley and Davis.

Broad Topics

This book is a collection of somewhat non-traditional assignments and sample solutions and exercises. We chose problems that address various topics, technologies, and characteristics we want the students to be exposed to and learn. These include:

- non-standard data formats (robot logs, email messages);

- text processing and regular expressions;

- newer/less-traditional technologies (Web scraping, Web services, *JSON*, *XML*, *HTML*, *KML* and Google Earth™);

- statistical methods (classification trees, k-nearest neighbors, naïve Bayes);

- visualization and exploratory data analysis;

- relational databases and *SQL*;

- simulation;

- implementing algorithms;

- large data and efficiency;

Preface   xvii

- software design, development, and testing;

- using and interfacing to other languages such as the UNIX shell, *C*, and *Python*.

There are many other computational topics we would like to cover such as modern statistical and machine learning methods, version control, dynamic documents, parallel computing, Hadoop and MapReduce, data technologies, advanced text processing concepts, and so on. Space and time does not permit us to cover these in this book.

The case studies included here are are a subset of those that we have used in our classes. They are not perfect and can be criticized for their different deficiencies. In spite of this, we hope they are valuable to students and instructors alike. We also hope they will catalyze us and other people to distribute more case studies, problems, data sets, and so on to help students learn the computing and statistical reasoning skills they need. We are attempting to collect these at `http://rdatasciencecases.org`. We welcome any suggestions, corrections and contributions. Indeed, we hope to build a larger collection of pedagogical materials, and also a community of collaborating educators.

Target Audience

Since the book does not attempt to teach the "nuts and bolts" of any of the languages used in the case studies, it is not intended to be a standalone textbook. We do think it will serve as a useful compendium for students at all levels who deal with processing data. It can serve as supplementary reading for a statistical computing course at the undergraduate level and first graduate course. We also expect people who are practicing, or emerging, data scientists who haven't had explicit courses in statistical computing will find the book valuable. In these regards, we expect it to be a useful book for self-learners and undergraduate and graduate students (and even faculty) seeking to go beyond the introductory texts. The material in the book is for people who want to explore both the thought process and details of how regular computing for a broad range of data science problems is done.

The Themes of the Three Parts

We have divided the book into 3 parts, with each part having a general theme. All of these focus on computing problems, but also visualization, data technologies, and less-commonly taught statistical/machine learning techniques.

Part I contains case studies that involve reading and transforming raw data, manipulating and visualizing them, and then using statistical techniques to try to solve a problem or understand relationships between variables. The data are typically in non-standard format or source (e.g., in Web pages). The statistical techniques are not very complex, but are different from what students typically see in undergraduate courses.

Part II focuses on using simulation to understand stochastic processes for their own sake and also explore how we can use simulation to model interesting situations. These case studies also explore some advanced computing topics such as reference classes and efficient idioms and computational approaches.

The final set of case studies in Part III explore different data technologies. These include databases, visualization with *KML*, and scraping data from Web pages with *HTTP* requests and text processing.

The division of case studies into the different sections are not precise and absolute. For example, some of the simulation topics involve data manipulation, while some of the data manipulation and modeling chapters involve simulation. The data technologies studies involve a lot of data manipulation. All of the case studies involve visualization – both for understanding and exploring data, and also debugging code.

The primary focus of this book is statistical computing and how to access, transform, manipulate, explore, visualize, and reason about data. However, in addition to the technologies and computing, all of the case studies are based on interesting statistical, mathematical and engineering problems that are worthy of study themselves. The chapters blend computational details with statistical and data analysis concepts. The analyses of the data and results are intentionally short and not comprehensive. Our aim is to whet the reader's interest in the specific application and stop/suspend at a point where interested students can do a great deal more exploration of the data and the statistical approaches to solving the problem. The chapters provide the computational foundation for the problems and we leave further exploration to the students and instructors, but describe many possible exercises and different directions to pursue.

Typographic Conventions

In several case studies we use other languages in addition to R, such as *SQL* and *C*. While the context should clearly indicate that a code block is in a language other than R, we also specify this in the margin of the page. For example, a *UNIX grep* command appears as

```
grep position2d JRSPdata_2010_03_10.log | grep -v ' 004 '
```

In the process of writing code, we introduce errors and mistakes with the idea that these are useful for learning how best to approach and solve computational problems. For this reason, some of the code in this book is purposefully incorrect or ill-advised (i.e., it works but is not a good approach). We identify such code with a no-entry symbol in the page margin, e.g.,

```
createGrid(c(3, 5), .5)

Error in grid[pos] = sample(rep(c("red", "blue"), numCars)) :
  (converted from warning) number of items to replace is not
  a multiple of replacement length
```

We should note that the code we present in this book for creating the figures differs slightly from the code we actually used to create the displayed figures. In typical daily usage, we add titles to our plots when we create them in R, either for interactive viewing or for including in presentations and reports. However, we have not done that for this book because the graphics are displayed in figures which include their own captions and titles. In order to avoid redundancy, we have eliminated the specification of a plot title in our code.

One final convention pertains to the exercises. Some case studies have exercises scattered throughout the chapter while others collect the exercises at the end of each chapter. To help identify and locate an exercise, we have added a question mark in the margin next to the exercise. For example,

Q.1 Write a function to include two series on the same plot. Be sure to add a title to your figure.

Available Materials

The Web site http://rdatasciencecases.org provides data, code, and supplementary material for this book. It also provides ideas and details for additional case studies. We hope others will contribute their case studies so that we can make these available to the community at large.

Acknowledgments

We, of course, want to thank all of the contributors who provided case studies included in the book. They gave a lot of time and thought in preparing their chapters. This was a long process and they were very patient and understanding.

The initial idea for this book of case studies came from an NSF-funded workshop we led in 2007 to explore the role of computing in the statistics curricula. One of the ideas was to share teaching resources, and this is one outcome of that.

We thank the numerous participants at our different NSF-funded workshops focused on computing in the statistics curricula. These workshops were divided into a) developing model curricula, b) creating case studies, and c) facilitating instructors to teach modern statistical computing. The interest, enthusiasm, feedback and input from all the participants was extremely important. We particularly thank the attendees of the 2009 workshop who brought ideas and materials for case studies: Samuel Frame (Cal Poly, San Luis Obispo), Robert Gould (UCLA), Albyn Jones (Reed College), Michael Kane (Yale University), Daniel Kaplan (Macalester College), Cari Kaufman (UC Berkeley), Guy Lebanon (Purdue University, now Amazon), Matt Levinson (UCLA), Thomas Lumley (University of Washington, now University of Auckland), John Monahan (NC State University), Roger Peng (Johns Hopkins University), Andrew Schaffner (Cal Poly, San Luis Obispo), Luke Tierney (University of Iowa), Frances Tong (UC Berkeley, now Becton Dickinson Technologies), John Verzani (CUNY Staten Island), Mark Daniel Ward (Purdue University), Charlotte Wickham (UC Berkeley, now Oregon State University), and Hadley Wickham (Rice University, now RStudio).

The researchers who presented one- or two-day case studies in the NSF-sponsored "Explorations in Statistics Research" (ESR) summer workshop that we have run over many years helped shape our thinking about how to present advanced modern statistics and data science to undergraduates. Specifically, we would like to thank Andreas Buja (Wharton School of Business, University of Pennsylvania), Amanda Cox (New York Times), Francesca Dominici (Johns Hopkins, now Harvard), Chris Genovese (Carnegie Mellon University), Carie Grimes (Google), Mark Hansen (UCLA, now Columbia University), Dave Higdon (Los Alamos National Laboratory), Diane Lambert (Bell Labs, Lucent Technologies, and now Google), Dave Madigan (Columbia University), Doug Nychka (NCAR), Roger Peng (Johns Hopkins), Katie Pollard (Gladstone Institute, UCSF), John Rice (UC Berkeley), Patrick Ryan (Janssen Research and Development), Steve Sein (NCAR), Jas Sekhon (UC Berkeley), Terry Speed (UC Berkeley and Walter and Eliza Hall Institute of Medical Research), Claudia Tebaldi (Climate Central), and Chris Volinsky (AT&T Shannon Labs).

In addition to the presenters, many researchers and faculty helped at the ESR with tutorials, career panels, etc. We thank Joe Blitzstein (Harvard University), Dianne Cook (Iowa State University), Nick Horton (Smith College and now Amherst College), David James (Bell Labs, Lucent Technologies, now Novartis), Cari Kaufman (UC Berkeley), and Debby Swayne (AT&T Shannon Labs).

We also want to thank the numerous teaching assistants who have served in our courses at Berkeley and Davis and also the ESR workshops. They helped to iron out some of the issues with teaching some of these case studies, identifying issues and problems the students encountered, and providing valuable feedback. These include Gabe Becker, Neal Fultz, Tammy Greasby, Brianna Hirst, Wayne Lee, Erin Melcon, Rakhee Patel, Nick Ulle, and Charlotte Wickham. Thanks also go to Ann Cannon who provided helpful feedback on an early version of Chapter 2.

We thank our editor John Kimmel for his continued support and encouragement for our projects.

This material is based in part upon work supported by the National Science Foundation under Grant Numbers DUE-0618865, DMS-0840001, and DUE-1043634.

Authors

Deborah Nolan has led many efforts to improve instruction in mathematics and statistics and to engage undergraduates in educational outreach. She holds the Zaffaroni Family Chair in Undergraduate Education at Berkeley, and received the University's Distinguished Teaching Award at Berkeley and the William R. Kenan, Jr. Visiting Professorship for Distinguished Teaching at Princeton. She is a fellow of the American Statistical Association, and former Chair of both its Computing Section and its Education Section. She is also a Fellow of the Institute of Mathematical Statistics. She co-directs the math and science teacher preparation program, Cal Teach, and master teacher in-service program, Math for America, Berkeley. She is the author of several books, including this one.

Duncan Temple Lang has been involved in the development of R and S for 20 years, and has developed over 100 R packages. He focuses on exploring and developing new possibilities for statistical computing, typically investigating new and ambitious paradigms and technologies from other disciplines and incorporating them, currently, into the R environment. He is currently working on compilation for R using an LLVM-based approach; provenance for R computations; type inference; and a fast, flexible framework for Bayesian and likelihood computations in R (http://r-nimble.org); and graphical processing units (GPUs). He recently became the Director of the UC Davis Data Science Initiative.

Nolan and **Temple Lang** are the authors of the book *XML and Web Technologies for Data Science in R*. They have also organized and led several NSF-funded summer programs aimed at attracting students to graduate studies in statistics, and short workshops in data science topics. Together, they developed a course, Concepts in Computing with Data, on their respective campuses, and they have collaborated on systems for interactive, reproducible, dynamic documents, and Web-based visualization.

Co-Authors

Samuel E. Buttrey
Naval Postgraduate School
Monterey, CA, USA

Timothy H. Chung
Naval Postgraduate School
Monterey, CA, USA

James N. Eagle
Naval Postgraduate School
Monterey, CA, USA

Michael Kane
Yale University
New Haven, CT, USA

Daniel Kaplan
Macalester College
Saint Paul, MN, USA

Cari Kaufman
University of California, Berkeley
Berkeley, CA, USA

Hadley Wickham
RStudio
Houston, TX, USA

Part I

Data Manipulation and Modeling

1
Predicting Location via Indoor Positioning Systems

Deborah Nolan
University of California, Berkeley

Duncan Temple Lang
University of California, Davis

CONTENTS

1.1	Introduction	3
	1.1.1 Computational Topics	4
1.2	The Raw Data	4
	1.2.1 Processing the Raw Data	8
1.3	Cleaning the Data and Building a Representation for Analysis	12
	1.3.1 Exploring Orientation	14
	1.3.2 Exploring MAC Addresses	16
	1.3.3 Exploring the Position of the Hand-Held Device	18
	1.3.4 Creating a Function to Prepare the Data	19
1.4	Signal Strength Analysis	21
	1.4.1 Distribution of Signal Strength	21
	1.4.2 The Relationship between Signal and Distance	26
1.5	Nearest Neighbor Methods to Predict Location	31
	1.5.1 Preparing the Test Data	31
	1.5.2 Choice of Orientation	32
	1.5.3 Finding the Nearest Neighbors	34
	1.5.4 Cross-Validation and Choice of k	36
1.6	Exercises	41
	Bibliography	43

1.1 Introduction

The growth of wireless networking has generated commercial and research interests in statistical methods to reliably track people and things inside stores, hospitals, warehouses, and factories. Global positioning systems (GPS) do not work reliably inside buildings, but with the proliferation of wireless local area networks (LANs), indoor positioning systems (IPS) can utilize WiFi signals detected from network access points to answer questions such as: where is a piece of equipment in a hospital? where am I? and who are my neighbors? Ideally, with minimal training, calibration, and equipment, these questions can be answered well in near real-time.

To build an indoor positioning system requires a reference set of data where the signal strength between a hand-held device such as a cellular phone or laptop and fixed access points (routers) are measured at known locations throughout the building. With these training data, we can build a model for the location of a device as a function of the strength of the signals between the device and each access point. Then we use this model to predict the location of a new unknown device based on the detected signals for the device. In this chapter, we examine nearly one million measurements of signal strength recorded at 6 stationary WiFi access points (routers) within a building at the University of Mannheim and develop a statistical IPS.

Our first step in this process is to understand how the data were collected and formatted. In Section 1.2, we do this by reading documentation provided by the researchers who have recorded the data and by carrying out our own investigations. After we have a sense of the data, we organize them into a structure suitable for analysis. We then clean the data in Section 1.3, and before we begin our modeling, we examine signal strength more closely to better understand its statistical properties (in Section 1.4). Then in Section 1.5 we pursue a nearest neighbor method for predicting location and we test it on a second set of data, also provided by the researchers at Mannheim.

1.1.1 Computational Topics

- string manipulation
- data structures and representation, including variable length observations
- aggregating data in ragged arrays
- exploratory data analysis and visualization
- modular functions
- debugging
- nearest neighbor methods
- cross-validation for parameter selection

1.2 The Raw Data

Two relevant data sets for developing an IPS are available on the CRAWDAD site (A Community Resource for Archiving Wireless Data At Dartmouth) [2]. One is a reference set, termed "offline," that contains signal strengths measured using a hand-held device on a grid of 166 points spaced 1 meter apart in the hallways of one floor of a building at the University of Mannheim. The floor plan, which measures about 15 meters by 36 meters, is displayed in Figure 1.1. The grey circles on the plan mark the locations where the offline measurements were taken and the black squares mark 6 access points. These reference locations give us a calibration set of signal strengths for the building, and we use them to build our model to predict the locations of the hand-held device when its position is unknown.

In addition to the (x, y) coordinates of the hand-held device, the orientation of the device was also provided. Signal strengths were recorded at 8 orientations in 45 degree increments (i.e., 0, 45, 90, and so on). Further, the documentation for the data indicates

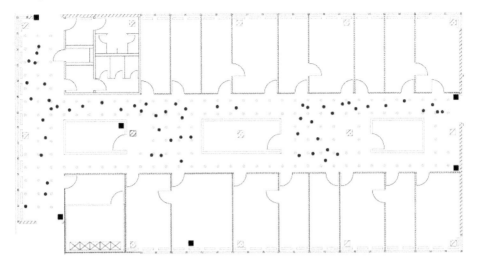

Figure 1.1: Floor Plan of the Test Environment. *In this floor plan, the 6 fixed access points are denoted by black square markers, the offline/training data were collected at the locations marked by grey dots, and the online measurements were recorded at randomly selected points indicated with black dots. The grey dots are spaced one meter apart.*

that 110 signal strength measurements were recorded to each of the 6 access points for every location-orientation combination.

In addition to the offline data, a second set of recordings, called the "online" data, is available for testing models for predicting location. In these data, 60 locations and orientations are chosen at random and 110 signals are measured from them to each access point. The test locations are marked by black dots in Figure 1.1. In both the offline and online data some of these 110 signal strength values were not recorded. Additionally, measurements from other hand-held devices, e.g., phone or laptop, in the vicinity of the experimental unit appear in some offline records.

The documentation for the data [2] describes the format of the data file. Additionally, we can examine the files ourselves with a plain text editor, and we find that each of the two files (offline and online) have the same basic format and start with something similar to

```
# timestamp=2006-02-11 08:31:58
# usec=250
# minReadings=110
t=1139643118358;id=00:02:2D:21:0F:33;pos=0.0,0.0,0.0;degree=0.0;\
00:14:bf:b1:97:8a=-38,2437000000,3;\
00:14:bf:b1:97:90=-56,2427000000,3;\
00:0f:a3:39:e1:c0=-53,2462000000,3;\
00:14:bf:b1:97:8d=-65,2442000000,3;\
00:14:bf:b1:97:81=-65,2422000000,3;\
00:14:bf:3b:c7:c6=-66,2432000000,3;\
00:0f:a3:39:dd:cd=-75,2412000000,3;\
00:0f:a3:39:e0:4b=-78,2462000000,3;\
00:0f:a3:39:e2:10=-87,2437000000,3;\
02:64:fb:68:52:e6=-88,2447000000,1;\
02:00:42:55:31:00=-84,2457000000,1
```

Note that the fourth and subsequent lines displayed here are actually just one line in the text file, but this one line has been formatted here on multiple lines for readability. We have added \ to indicate a continuation of the line.

The available documentation indicates that the format of the data is:

```
t="Timestamp";
id="MACofScanDevice";
pos="RealPosition";
degree="orientation";
MACofResponse1="SignalStrengthValue,Frequency,Mode"; ...
MACofResponseN="SignalStrengthValue,Frequency,Mode"
```

where the units of measurement are shown in Table 1.1. The MAC (media access control) variable refers to the MAC address of a hardware device, which is a unique identifier that allows a network card for a computer, access point, or other piece of equipment to be identified on a network [5]. By convention, this identifier is written in the form mm:mm:¬ mm:ss:ss:ss where mm and ss are 2 hexadecimal digits (0, 1, ..., 9, a, b, c, d, e, f). The first of these 3 sets of pairs of digits, i.e., mm:mm:mm, identifies the manufacturer of the equipment. The second set of 3 pairs (the ss) identifies the particular piece of equipment, both the model and the unique device.

The MACofResponse1 ... MACofResponseN in these data indicate that one line consists of a variable number of MAC address measurements. That is, these records are not of equal length, but form ragged arrays that depend on the number of signals detected. For example, consider another line (the 2,000th) in the input file:

```
t=1139644637174;id=00:02:2D:21:0F:33;pos=2.0,0.0,0.0;degree=45.5;\
00:14:bf:b1:97:8a=-33,2437000000,3;\
00:14:bf:b1:97:8a=-38,2437000000,3;\
00:0f:a3:39:e1:c0=-54,2462000000,3;\
00:14:bf:b1:97:90=-59,2427000000,3;\
00:14:bf:b1:97:8d=-70,2442000000,3;\
00:0f:a3:39:e2:10=-88,2437000000,3;\
00:0f:a3:39:dd:cd=-67,2412000000,3;\
02:00:42:55:31:00=-84,2457000000,1
```

We notice several things: this record has 8 readings; the MAC addresses appear in a different order than in the first record; there are 2 readings from the same access point (the 8a access point); and one of the 8 addresses belongs to an adhoc device because, according to Table 1.1, the mode digit indicates whether the recording is for an adhoc device (1) or access point (3). If we look at the first observation again, we notice that there are more than 6 MAC addresses with a mode of 3. The "extras" are from other floors in the building.

Now that we have a sense of the format of the input file, we can determine how to read the data into a structure that is conducive to analysis. Let's think about how we want to ultimately represent the resulting data in R [8]. There are two reasonably obvious choices. The first is to have a row in a data frame for each row in the input file. In this case, the variables are time, mac-id, x, y, z (for the hand-held device's location), orientation, and then 4 variables for each of the MAC addresses for which we have a signal. These 4 variables are signal, channel, and type of device, as well as the MAC address. Since the raw observations have a different number of recorded signals, our data frame needs enough columns to accommodate the largest number of recordings.

A second approach is to have the same initial variables describing the hand-held device, i.e., time, MAC address, location, and orientation. After this, we have just 4 other variables: the MAC address of the device from which we received a signal, the signal, the channel, and

TABLE 1.1: Units of Measurement

Variable Units
t	timestamp in milliseconds since midnight, January 1, 1970 UTC
id	MAC address of the scanning device
pos	the physical coordinate of the scanning device
degree	orientation of the user carrying the scanning device in degrees
MAC	MAC address of a responding peer (e.g., an access point or a device in `adhoc` mode) with the corresponding values for signal strength in dBm (Decibel-milliwatts), the channel frequency and its mode (access point = 3, device in `adhoc` mode = 1)

This table provides the units of measurement for the variables in the offline and online data.

the type of device. In this scenario, we have a row in our data frame for each signal received. That is, each line in the input file turns into multiple rows in the data frame, corresponding to the number of ';'-separated MAC addresses in that line. For example, the first record in the input file becomes 11 rows and the 2000th observation becomes 8 rows in the data frame.

The first approach yields a natural representation that more directly corresponds to the format of the input file. It also avoids repeating the time and location information and so seems more memory efficient. One of the difficulties it presents, however, is that we have to determine how many columns we need, or more specifically, how many MAC addresses received the signal from the device. Even if we drop the `adhoc` readings, we still have to contend with a different ordering of recordings and multiple measurements from the same MAC address. We most likely need two passes of the data to create our data frame: one to determine the unique MAC addresses and the second to organize the data. While we avoid repeating some information, e.g., timestamp, we need to use `NA` values for the observations that do not have recorded signals from all of the MAC addresses. If there are many of these, then the second approach may actually be more memory efficient. The second approach also appears simpler to create.

With the second approach, we can avoid two passes of the data and read the file into a data frame in just one pass. And, this data structure allows us to use group-by operations on the MAC addresses. For now, we adopt the second approach. Later in our work, we create a data frame with the first format. For this, we do not need to return to the original file, but create the alternative data frame from the existing one.

Another consideration in determining how to read the data into R is whether or not the "comment" lines occur only at the beginning/top of the file. We can search in the file for lines that start with a '#' character. To do this, we read the entire document into R using readLines() with

```
txt = readLines("Data/offline.final.trace.txt")
```

Each line in the offline file has been read into R as a string in the character vector `txt`. We use the substr() function to locate lines/strings that begin with a '#' character and tally them with

```
sum(substr(txt, 1, 1) == "#")

[1] 5312
```

Additionally, we use length() as follows:

```
length(txt)
```

```
[1] 151392
```

to determine that there are 151,392 lines in the offline file. According to the documentation we expect there to be 146,080 lines in the file (166 locations × 8 angles × 110 recordings). The difference between these two (151,392 and 146,080) is 5,312, exactly the number of comments lines.

Generally a good rule of thumb to follow is to check our assumptions about the format of a file and not just look at the first few lines.

1.2.1 Processing the Raw Data

Now that we have determined the desired target representation of the data in R, we can write the code to extract the data from the input file and manipulate it into that form. The data are not in a rectangular form so we cannot simply use a function such as read.table(). However, there is structure in the observations that we can use to process the lines of text. For example, the main data elements are separated by semicolons. Let's see how the semicolon splits the fourth line, i.e., the first line that is not a comment:

```
strsplit(txt[4], ";")[[1]]
```

```
 [1] "t=1139643118358"
 [2] "id=00:02:2D:21:0F:33"
 [3] "pos=0.0,0.0,0.0"
 [4] "degree=0.0"
 [5] "00:14:bf:b1:97:8a=-38,2437000000,3"
 [6] "00:14:bf:b1:97:90=-56,2427000000,3"
 [7] "00:0f:a3:39:e1:c0=-53,2462000000,3"
 [8] "00:14:bf:b1:97:8d=-65,2442000000,3"
 [9] "00:14:bf:b1:97:81=-65,2422000000,3"
[10] "00:14:bf:3b:c7:c6=-66,2432000000,3"
[11] "00:0f:a3:39:dd:cd=-75,2412000000,3"
[12] "00:0f:a3:39:e0:4b=-78,2462000000,3"
[13] "00:0f:a3:39:e2:10=-87,2437000000,3"
[14] "02:64:fb:68:52:e6=-88,2447000000,1"
[15] "02:00:42:55:31:00=-84,2457000000,1"
```

Within each of these shorter strings, the "name" of the variable is separated by an '=' character from the associated value. In some cases this value contains multiple values where the separator is a ','. For example, `"pos=0.0,0.0,0.0"` consists of 3 position variables that are not named.

We can take this vector, which we created by splitting on the semi-colon, and further split each element at the '=' characters. Then we can process the resulting strings by splitting them at the ',' characters. This might look something like

```
unlist(lapply(strsplit(txt[4], ";")[[1]],
              function(x)
                sapply(strsplit(x, "=")[[1]], strsplit, ",")))
```

We end up with a large character vector with the names and data values from the entire first record as individual "tokens." We can then rearrange them into the appropriate form. However, we can do this much more simply and generally using the fact that the *split* parameter for strsplit() can be a regular expression so we can split on any of several characters in a single function call. This means we can use

```
tokens = strsplit(txt[4], "[;=,]")[[1]]
```

to split at a ';', '=' or ',' character.

Before we proceed to write much code to read these data, we ask: Can read.table() take a regular expression as a separator? If so, we can use it instead of readLines(). Unfortunately, it does not. It would slow down the reading of regular text files quite considerably.

Based on the results of the strsplit(), we have all the data elements from the first row. The first 10 elements of tokens give the information about the hand-held device:

```
tokens[1:10]
```

```
[1] "t"         "1139643118358" "id"      "00:02:2D:21:0F:33"
[5] "pos"       "0.0"           "0.0"     "0.0"
[9] "degree"    "0.0"
```

We can extract the values of these variables with

```
tokens[c(2, 4, 6:8, 10)]
```

```
[1] "1139643118358"    "00:02:2D:21:0F:33" "0.0"
[4] "0.0"              "0.0"               "0.0"
```

We know these correspond to the variables time, MAC address, x, y, z, and orientation.

Let's turn our attention to the recorded signals within this observation. These are the remaining values in the split vector, i.e.,

```
tokens[ - ( 1:10 ) ]
```

```
 [1] "00:14:bf:b1:97:8a" "-38"   "2437000000"    "3"
 [5] "00:14:bf:b1:97:90" "-56"   "2427000000"    "3"
 [9] "00:0f:a3:39:e1:c0" "-53"   "2462000000"    "3"
[13] "00:14:bf:b1:97:8d" "-65"   "2442000000"    "3"
[17] "00:14:bf:b1:97:81" "-65"   "2422000000"    "3"
[21] "00:14:bf:3b:c7:c6" "-66"   "2432000000"    "3"
[25] "00:0f:a3:39:dd:cd" "-75"   "2412000000"    "3"
[29] "00:0f:a3:39:e0:4b" "-78"   "2462000000"    "3"
[33] "00:0f:a3:39:e2:10" "-87"   "2437000000"    "3"
[37] "02:64:fb:68:52:e6" "-88"   "2447000000"    "1"
[41] "02:00:42:55:31:00" "-84"   "2457000000"    "1"
```

We can think of these as rows in a 4-column matrix or data frame giving the MAC address, signal, channel, and device type, so let's unravel these and build a matrix from the values. Then we can bind these columns with the values from the first 10 entries. We do this with

```
tmp = matrix(tokens[ - (1:10) ], ncol = 4, byrow = TRUE)
mat = cbind(matrix(tokens[c(2, 4, 6:8, 10)], nrow = nrow(tmp),
            ncol = 6, byrow = TRUE),
        tmp)
```

We confirm that we have 11 rows in the matrix, one for each MAC address, and 10 columns, 6 of which have the same value for each MAC address (e.g., position and orientation):

```
dim(mat)
```

```
[1] 11 10
```

We put all this code into a function so we can repeat this operation for each row in the input file. That is,

```
processLine =
function(x)
{
  tokens = strsplit(x, "[;=,]")[[1]]
  tmp = matrix(tokens[ - (1:10) ], ncol = 4, byrow = TRUE)
  cbind(matrix(tokens[c(2, 4, 6:8, 10)], nrow = nrow(tmp),
         ncol = 6, byrow = TRUE), tmp)
}
```

Let's apply our function to several lines of the input:

```
tmp = lapply(txt[4:20], processLine)
```

Note that we started at the fourth line of the file because the first 3 lines are comments. The result is a list of 17 matrices and we can determine how many signals were detected at each point with

```
sapply(tmp, nrow)
```

```
 [1] 11 10 10 11  9 10  9  9 10 11 11  9  9  9  8 10 14
```

We have done the hard part. Of course, we want to turn these individual matrices into a single data frame. We can stack the matrices together using do.call(). We might be inclined to write a loop to concatenate the second matrix to the first, the third to the earlier result, and so on. This would be very slow. (Try it!) However, do.call() does this stacking efficiently and simply. We call do.call() with the name of the function to call and a list containing the individual arguments that we ordinarily pass to that function separately. For us, this is as simple as

```
offline = as.data.frame(do.call("rbind", tmp))
dim(offline)
```

```
[1] 170  10
```

We are now ready to try this code on the entire dataset. We discard the lines starting with the comment character '#' and then pass each remaining line to processLine().

```
lines = txt[ substr(txt, 1, 1) != "#" ]
tmp = lapply(lines, processLine)
```

When we run this, we get 6 warnings of the form

```
1: In matrix(tokens[c(2, 4, 6:8, 10)], nrow(tmp), 6,
         byrow = TRUE) :
  data length exceeds size of matrix
```

In general, we want to be very cautious about warning messages.

We can try to find the rows to which these warning messages correspond by exploring the result, but it is easier to catch them as they occur. We can ask R to raise an error when a warning is issued and then browse the call stack to examine the state of the computations. We do this by setting an option to handle errors and another to change warnings into errors:

```
options(error = recover, warn = 2)
```

We run the lapply() call again with these new options:

```
tmp = lapply(lines, processLine)
```

When the first warning occurs, we are presented with the message and the call stack

```
Error in matrix(tokens[c(2, 4, 6:8, 10)], nrow(tmp), 6,
    byrow = TRUE) : (converted from warning)
                  data length exceeds size of matrix

Enter a frame number, or 0 to exit

1: lapply(lines, processLine)
2: lapply(lines, processLine)
3: FUN(c("t=1139643118358;id=00:02:2D:21:0F:33;pos=0.0,0.0,0.0...
4: cbind(matrix(tokens[c(2, 4, 6:8, 10)], nrow(tmp), 6,
            byrow = TRUE), tmp)
5: matrix(tokens[c(2, 4, 6:8, 10)], nrow(tmp), 6, byrow = TRUE)
6: .signalSimpleWarning("data length exceeds size of matrix",
                  quote(matrix(tokens[c(2, 4
7: withRestarts({
8: withOneRestart(expr, restarts[[1]])
9: doWithOneRestart(return(expr), restart)

Selection:
```

We select 3, which corresponds to our processLine() function. We can issue R commands in this call frame. For example, we can see what variables are available with ls(). Looking at the variable x, we see its value is

```
[1] "t=1139671993259;id=00:02:2D:21:0F:33;pos=12.0,3.0,0.0;
   degree=315.1"
```

What we notice about this value is that we have no signals detected for this position. This observation is an anomalous case, which our processLine() function needs to handle.

We can modify processLine() to discard such observations or alternatively add a single row to the data frame with the hand-held information and NA values for the MAC, signal, channel, and type. We choose to discard these observations as they do not help us in developing our positioning system. We change our function to return NULL if the tokens vector only has 10 elements. Our revised function is

```
processLine = function(x)
{
  tokens = strsplit(x, "[;=,]")[[1]]
```

```
  if (length(tokens) == 10)
    return(NULL)

  tmp = matrix(tokens[ - (1:10) ], , 4, byrow = TRUE)
  cbind(matrix(tokens[c(2, 4, 6:8, 10)], nrow(tmp), 6,
               byrow = TRUE), tmp)
}
```

We run the updated processLine() and see if the warnings disappear.

```
options(error = recover, warn = 1)
tmp = lapply(lines, processLine)
offline = as.data.frame(do.call("rbind", tmp),
                        stringsAsFactors = FALSE)
```

Indeed, we received no warning messages. Our data frame `offline` has over one million rows:

```
dim(offline)
```

```
[1] 1181628       10
```

Our data frame consists of character-valued variables. A next step is to convert these values into appropriate data types, e.g., convert signal strength to numeric, and to further clean the data as needed. This is the topic of the next section.

1.3   Cleaning the Data and Building a Representation for Analysis

A simple first step to creating a data structure for analysis is to put meaningful names on the variables and convert them to the appropriate data type. We begin by adding names with

```
names(offline) = c("time", "scanMac", "posX", "posY", "posZ",
                   "orientation", "mac", "signal",
                   "channel", "type")
```

Then we convert the position, signal, and time variables to numeric with

```
numVars = c("time", "posX", "posY", "posZ",
            "orientation", "signal")
offline[ numVars ] = lapply(offline[ numVars ], as.numeric)
```

We can also change the type of the device to something more comprehensible than the numbers 1 and 3. To do this, we can make it a factor with the levels, say, `"adhoc"` and `"access point"`. However, in our analysis, we plan to use only the signal strengths measured to the fixed access points to develop and test our model. Given this, we drop all records for `adhoc` measurements and remove the `type` variable from our data frame. We do this with

```
offline = offline[ offline$type == "3", ]
offline = offline[ , "type" != names(offline) ]
dim(offline)
```

```
[1] 978443        9
```

We have removed over 100,000 records from our data frame.

Next we consider the time variable. According to the documentation, time is measured in the number of milliseconds from midnight on January 1st, 1970. This is the origin used for the `POSIXt` format, but with `POSIXt`, it is the number of seconds, not milliseconds. We can scale the value of time to seconds and then simply set the class of the `time` element in order to have the values appear and operate as date-times in R. We keep the more precise time in `rawTime` just in case we need it. We perform the conversion as follows:

```
offline$rawTime = offline$time
offline$time = offline$time/1000
class(offline$time) = c("POSIXt", "POSIXct")
```

Now that we have completed these conversions, we check the types of the variables in the data frame with

```
unlist(lapply(offline, class))
```

and verify that they are what we want:

```
      time1        time2     scanMac        posX         posY
    "POSIXt"    "POSIXct" "character"   "numeric"    "numeric"
       posZ  orientation         mac       signal      channel
   "numeric"    "numeric" "character"   "numeric"  "character"
    rawTime
   "numeric"
```

We have the correct shape for the data and even the correct types. We next verify that the actual values of the data look reasonable. There are many approaches we can take to do this. We start by looking at a summary of each numeric variable with

```
summary(offline[, numVars])
```

```
      time                           posX            posY
 Min.   :2006-02-10 23:31:58   Min.   : 0      Min.   : 0.0
 1st Qu.:2006-02-11 05:21:27   1st Qu.: 2      1st Qu.: 3.0
 Median :2006-02-11 11:57:58   Median :12      Median : 6.0
 Mean   :2006-02-16 06:57:37   Mean   :14      Mean   : 5.9
 3rd Qu.:2006-02-19 06:52:40   3rd Qu.:23      3rd Qu.: 8.0
 Max.   :2006-03-09 12:41:10   Max.   :33      Max.   :13.0

      posZ       orientation        signal
 Min.   :0    Min.   :  0      Min.   :-99
 1st Qu.:0    1st Qu.: 90      1st Qu.:-69
 Median :0    Median :180      Median :-60
 Mean   :0    Mean   :167      Mean   :-62
 3rd Qu.:0    3rd Qu.:270      3rd Qu.:-53
 Max.   :0    Max.   :360      Max.   :-25
```

We also convert the character variables to factors and examine them with

```
summary(sapply(offline[ , c("mac", "channel", "scanMac")],
              as.factor))
```

```
              mac                    channel
00:0f:a3:39:e1:c0:145862     2462000000:189774
00:0f:a3:39:dd:cd:145619     2437000000:152124
00:14:bf:b1:97:8a:132962     2412000000:145619
00:14:bf:3b:c7:c6:126529     2432000000:126529
00:14:bf:b1:97:90:122315     2427000000:122315
00:14:bf:b1:97:8d:121325     2442000000:121325
(Other)            :183831   (Other)   :120757

            scanMac
00:02:2D:21:0F:33:978443
```

After examining these summaries, we find a couple of anomalies:

- There is only one value for `scanMac`, the MAC address for the hand-held device from which the measurements were taken. We might as well discard this variable from our data frame. However, we may want to note this value to compare it with the online data.

- All of the values for `posZ`, the elevation of the hand-held device, are 0. This is because all of the measurements were taken on one floor of the building. We can eliminate this variable also.

We modify our data frame accordingly,

```
offline = offline[ , !(names(offline) %in% c("scanMac", "posZ"))]
```

1.3.1 Exploring Orientation

According to the documentation, we should have only 8 values for orientation, i.e., 0, 45, 90, ..., 315. We can check this with

```
length(unique(offline$orientation))
```

```
[1] 203
```

Clearly, this is not the case. Let's examine the distribution of `orientation`:

```
plot(ecdf(offline$orientation))
```

An annotated version of this plot appears in Figure 1.2. It shows the orientation values are distributed in clusters around the expected angles. Note the values near 0 and near 360 refer to the same direction. That is, an orientation value 1 degree before 0 is reported as 359 and 1 degree past 0 is a 1.

Although the experiment was designed to measure signal strength at 8 orientations – 45 degree intervals from 0 to 315 – these orientations are not exact. However, it may be useful in our analysis to work with values corresponding to the 8 equi-spaced angles. That is, we want to map 47.5 to 45, and 358.2 to 0, and so on. To do this, we take each value and find out to which of the 8 orientations it is closest and we return that orientation. We must handle values such as 358.2 carefully as we want to map them to 0, not to the closer 315. The following function makes this conversion:

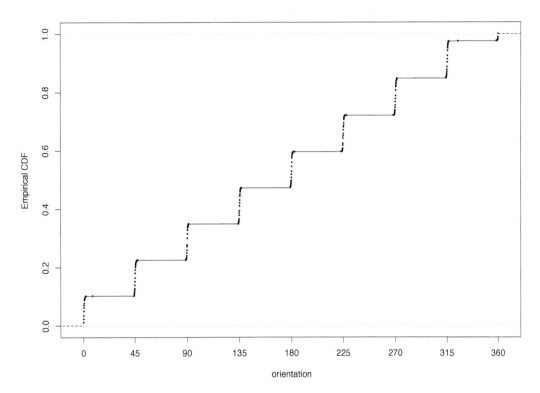

Figure 1.2: Empirical CDF of Orientation for the Hand-Held Device. *This empirical distribution function of orientation shows that there are 8 basic orientations that are 45 degrees apart. We see from the steps in the function that these orientations are not exactly 45, 90, 135, etc. Also, the 0 orientation is split into the two groups, one near 0 and the other near 360.*

```
roundOrientation = function(angles) {
  refs = seq(0, by = 45, length  = 9)
  q = sapply(angles, function(o) which.min(abs(o - refs)))
  c(refs[1:8], 0)[q]
}
```

We use roundOrientation() to create the rounded angles with

```
offline$angle = roundOrientation(offline$orientation)
```

Again, we keep the original variable and augment our data frame with the new angles.

We check that the results are as we expect with boxplots:

```
with(offline, boxplot(orientation ~ angle,
                    xlab = "nearest 45 degree angle",
                    ylab="orientation"))
```

From Figure 1.3 we see that the new values look correct and the original values near 360 degrees are mapped to 0. It also shows the variability in the act of measuring.

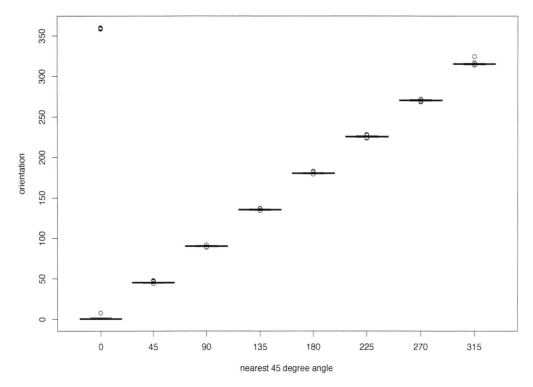

Figure 1.3: Boxplots of Orientation for the Hand-Held Device. *These boxplots of the original orientation against the rounded value confirm that the values have mapped correctly to 0, 45, 90, 135, etc. The "outliers" at the top left corner of the plot are the values near 360 that have been mapped to 0.*

1.3.2 Exploring MAC Addresses

From the summary() information, it seems that there may be a one-to-one mapping between the MAC address of the access points and channel. For example, the summary statistics show there are 126,529 occurrences of the address 00:14:bf:3b:c7:c6 and the same number of occurrences of channel 2432000000. To help us ascertain if we do have a one-to-one mapping, we look at the relationship between the MAC address and channel.

How many unique addresses and channels do we have? There should be the same number, if there is a one-to-one mapping. We find:

```
c(length(unique(offline$mac)), length(unique(offline$channel)))
```

```
[1] 12  8
```

There are 12 MAC addresses and 8 channels. We were given the impression from the building plan (see Figure 1.1) that there are only 6 access points. Why are there 8 channels and 12 MAC addresses? Rereading the documentation we find that there are additional access points that are not part of the testing area and so not seen on the floor plan. Let's check the counts of observations for the various MAC addresses with table():

```
table(offline$mac)
```

```
00:04:0e:5c:23:fc  00:0f:a3:39:dd:cd  00:0f:a3:39:e0:4b
              418             145619              43508
00:0f:a3:39:e1:c0  00:0f:a3:39:e2:10  00:14:bf:3b:c7:c6
           145862              19162             126529
00:14:bf:b1:97:81  00:14:bf:b1:97:8a  00:14:bf:b1:97:8d
           120339             132962             121325
00:14:bf:b1:97:90  00:30:bd:f8:7f:c5  00:e0:63:82:8b:a9
           122315                301                103
```

Clearly the first and the last two MAC addresses are not near the testing area or were only working/active for a short time during the measurement process because their counts are very low. It's probably also the case that the third and fifth addresses are not among the access points displayed on the map because they have much lower counts than the others and these are far lower than the possible 146,080 recordings (recall that there are potentially signals recorded at 166 grid points, 8 orientations, and 110 replications).

According to the documentation, the access points consist of 5 Linksys/Cisco and one Lancom L-54g routers. We look up these MAC addresses at the `http://coffer.com/mac_find/` site to find the vendor addresses that begin with `00:14:bf` belong to Linksys devices, those beginning with `00:0f:a3` belong to Alpha Networks, and Lancom devices start with `00:a0:57` (see Figure 1.4). We do have 5 devices with an address that begins `00:14:bf`, which matches with the Linksys count from the documentation. However, none of our MAC addresses begin with `00:a0:57` so there is a discrepancy with the documentation. Nonetheless, we have discovered valuable information for piecing together a better understanding of the data. For now, let's keep the records from the top 7 devices. We do this with

```
subMacs = names(sort(table(offline$mac), decreasing = TRUE))[1:7]
offline = offline[ offline$mac %in% subMacs, ]
```

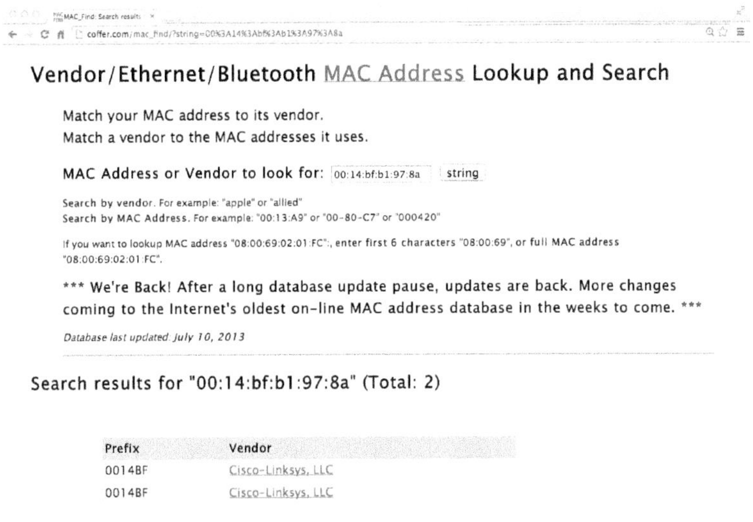

Figure 1.4: Screenshot of the coffer.com Mac Address Lookup Form. *The coffer.com Web site offers lookup services to find the MAC address for a vendor and vice versa.*

Finally, we create a table of counts for the remaining MAC×channel combinations and confirm there is one non-zero entry in each row

```
macChannel = with(offline, table(mac, channel))
apply(macChannel, 1, function(x) sum(x > 0))
```

```
00:0f:a3:39:dd:cd 00:0f:a3:39:e1:c0 00:14:bf:3b:c7:c6
                1                 1                 1
00:14:bf:b1:97:81 00:14:bf:b1:97:8a 00:14:bf:b1:97:8d
                1                 1                 1
00:14:bf:b1:97:90
                1
```

Indeed we see that there is a one-to-one correspondence between MAC address and channel for these 7 devices. This means we can eliminate `channel` from `offline`, i.e.,

```
offline = offline[ , "channel" != names(offline)]
```

1.3.3 Exploring the Position of the Hand-Held Device

Lastly, we consider the position variables, `posX` and `posY`. For how many different locations do we have data? The `by()` function can tally up the numbers of rows in our data frame for each unique (x, y) combination. We begin by creating a list containing a data frame for each location as follows:

```
locDF = with(offline,
             by(offline, list(posX, posY), function(x) x))
length(locDF)
```

```
[1] 476
```

Note that this list is longer than the number of combinations of actual (x, y) locations at which measurements were recorded. Many of these elements are empty:

```
sum(sapply(locDF, is.null))
```

```
[1] 310
```

The null values correspond to the combinations of the xs and ys that were not observed. We drop these unneeded elements as follows:

```
locDF = locDF[ !sapply(locDF, is.null) ]
```

and confirm that we now have only 166 locations with

```
length(locDF)
```

```
[1] 166
```

We can operate on each of these data frames to, e.g., determine the number of observations recorded at each location with

```
locCounts = sapply(locDF, nrow)
```

And, if we want to keep the position information with the location, we do this with

```
locCounts = sapply(locDF,
                   function(df)
                     c(df[1, c("posX", "posY")], count = nrow(df)))
```

We confirm that `locCounts` is a matrix with 3 rows with

```
class(locCounts)
```
```
[1] "matrix"
```

```
dim(locCounts)
```
```
[1]   3 166
```

We examine a few of the counts,

```
locCounts[ , 1:8]
      [,1] [,2] [,3] [,4] [,5] [,6] [,7] [,8]
posX     0    1    2    0    1    2    0    1
posY     0    0    0    1    1    1    2    2
count 5505 5505 5506 5524 5543 5558 5503 5564
```

We see that there are roughly 5,500 recordings at each position. This is in accord with 8 orientations × 110 replications × 7 access points, which is 6,160 signal strength measurements.

We can visualize all 166 counts by adding the counts as text at their respective locations, changing the size and angle of the characters to avoid overlapping text. We first transpose the matrix so that the locations are columns of the matrix and then we make our plot with

```
locCounts = t(locCounts)
plot(locCounts, type = "n", xlab = "", ylab = "")
text(locCounts, labels = locCounts[,3], cex = .8, srt = 45)
```

We see in Figure 1.5 that there are roughly the same number of signals detected at each location.

1.3.4 Creating a Function to Prepare the Data

We have examined all the variables except `time` and `signal`. This process has helped us clean our data and reduce it to those records that are relevant to our analysis. We leave the examination of the signals to the next section where we study its distributional properties. As for `time`, while this variable is not directly related to our model, it indicates the order in which the observations were taken. In an experiment, this can be helpful in uncovering potential sources of bias. For example, the person carrying the hand-held device may have changed how the device was carried as the experiment progressed and this change may lead to a change in the strength of the signal. Plots and analyses of the relationship between time and other variables can help us uncover such potential problems. We leave this investigation as an exercise.

Since we also want to read the online data in R, we turn all of these commands into a function called readData(). Additionally, if we later change our mind as to how we want to handle some of these special cases, e.g., to keep `channel` or `posZ`, then we can make a simple update to our function and rerun it. We might even add a parameter to the function definition to allow us to process the data in different ways. We leave it as an exercise to create readData().

We call readData() to create the offline data frame with

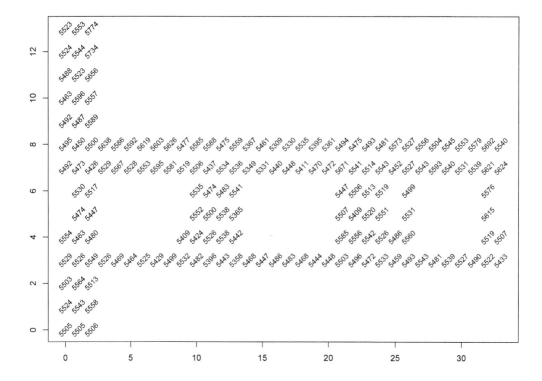

Figure 1.5: Counts of signals detected at each position. *Plotted at each location in the building is the total number of signals detected from all access points for the offline data. Ideally for each location, 110 signals were measured at 8 angles for each of 6 access points, for a total of 5280 recordings. These data include a seventh Mac address and not all signals were detected, so there are about 5500 recordings at each location.*

```
offlineRedo = readData()
```

Then we use the identical() function to check this version of the data frame against the one that we already created:

```
identical(offline, offlineRedo)
```

```
[1] TRUE
```

This confirms that our function behaves as expected.

When we collect code into a function, it is common to forget about some of the variables we need. The code works because they are found in the R session (i.e., globalenv()), but the function does not work in new R sessions or gives the wrong results if we define those global variables differently, by chance. We use the findGlobals() function in the codetools package [10] to identify what variables are global, i.e.,

```
library(codetools)
findGlobals(readData, merge = FALSE)$variables
```

```
[1] "processLine" "subMacs"
```

The processLine() function is a variable since it is referenced in a call to lapply() in readData() so this is not a problem. The variable subMacs is also identified as global. This variable was created in the global environment from the unique values of mac (see Section 1.3.2) and we neglected to include this code in the function. We can update the function to pass it as a parameter with a suitable default value or to create subMacs within the function; then, subMacs is no longer a global variable.

1.4 Signal Strength Analysis

We have used visualization and statistical summaries to help clean and format the data, and now we turn to investigating the properties of the response variable, signal strength. We want to learn more about how the signals behave before designing a model for IPS. The following questions guide us in our investigations.

- We have measured the signal strength to an access point multiple times at each location and orientation. How do these signal strengths behave? That is, what is the distribution of the repeated measurements at each location and orientation? Does signal strength behave similarly at all locations? Or does, the location, orientation, and access point affect this distribution?

- In a laboratory setting, signal strength decays linearly with log distance and a simple triangulation using the signal strength from 3 access points can accurately pinpoint the location of a device [1, 7]. In practice, physical characteristics of a building and human activity can add significant noise to signal strength measurements. How can we characterize the relationship between the signal strength and the distance from the device to the access point? How does the orientation affect this relationship? Is this relationship the same for all access points?

We consider these questions in the next two sections.

1.4.1 Distribution of Signal Strength

We want to compare the distribution of signal strength at different orientations and for different access points, so we need to subdivide our data. We are interested in seeing if these distributions are normal or skewed. We also want to look at their variances.

We consider the impact of orientation on signal strength by fixing a location on the map to see how the signal changes as the experimenter rotates through the 8 angles. We also separately examine the MAC addresses because, for example, at an orientation of 90 degrees the experimenter may be facing toward one access point and away from another. To do this we make simple boxplots with the bwplot() function in the `lattice` package [9] as follows:

```
library(lattice)
bwplot(signal ~ factor(angle) | mac, data = offline,
       subset = posX == 2 & posY == 12
           & mac != "00:0f:a3:39:dd:cd",
       layout = c(2,3))
```

We see in Figure 1.6 that the signal strength varies with the orientation for both close and distant access points. Note we have dropped the records for the MAC address of `00:0f:`¬ `a3:39:dd:cd` because it is identified as the extra address in the next section.

Recall from the summary statistics that signal strengths are measured in negative values. That is,

```
summary(offline$signal)
```

```
   Min. 1st Qu.  Median    Mean 3rd Qu.    Max.
    -98     -67     -59     -60     -53     -25
```

The small values, such as -98, correspond to weak signals and the large values, such as -25, are the strong signals.

When we examine a few other locations, we find a similar dependence of signal strength on angle. For example, we compare the distributions of signal strength for different angles and MAC addresses at the central location of $x = 23$ and $y = 4$; we use the densityplot() function in the `lattice` package because it makes it easy to condition on these variables. We produce 48 density curves for this one location with

```
densityplot( ~ signal | mac + factor(angle), data = offline,
            subset = posX == 24 & posY == 4 &
                    mac != "00:0f:a3:39:dd:cd",
            bw = 0.5, plot.points = FALSE)
```

Many of these distributions look approximately normal, but there are some serious departures with secondary modes and skewness (see Figure 1.7). Also, the center of the distribution varies with angle and MAC address, which indicates that conditioning on angle and MAC address is warranted.

If we want to examine the distribution of signal strength for all 166 locations, 8 angles, and 6 access points, we need to create thousands of boxplots or density curves. We can, instead, examine summary statistics such as the mean and SD or the median and IQR of signal strength for all location–orientation–access point combinations. For each combination, we have roughly 100 observations. To compute summary statistics for these various combinations, we first create a special factor that contains all of the unique combinations of the observed (x, y) pairs for the 166 locations. We can do this with

```
offline$posXY = paste(offline$posX, offline$posY, sep = "-")
```

Next, we create a list of data frames for every combination of (x, y), angle, and access point as follows

```
byLocAngleAP = with(offline,
                    by(offline, list(posXY, angle, mac),
                        function(x) x))
```

Then we can calculate summary statistics on each of these data frames with

```
signalSummary =
  lapply(byLocAngleAP,
         function(oneLoc) {
           ans = oneLoc[1, ]
           ans$medSignal = median(oneLoc$signal)
           ans$avgSignal = mean(oneLoc$signal)
           ans$num = length(oneLoc$signal)
```

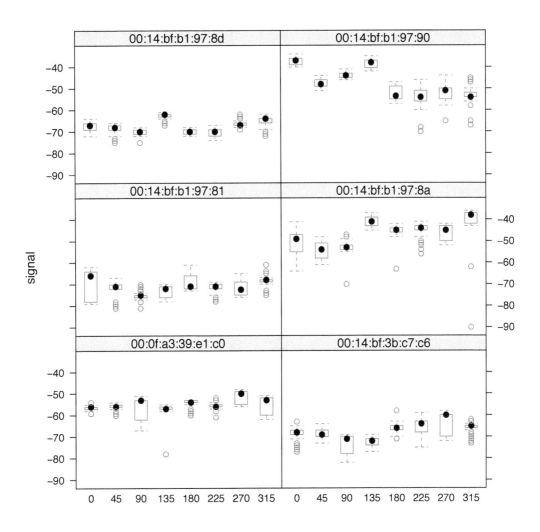

Figure 1.6: Signal Strength by Angle for Each Access Point. *The boxplots in this figure represent signals for one location, which is in the upper left corner of the floor plan, i.e., $x = 2$ and $y = 12$. These boxes are organized by access point and the angle of the hand-held device. The dependence of signal strength on angle is evident at several of the access points, e.g.,* `00:14:bf:97:90` *in the top right panel of the figure.*

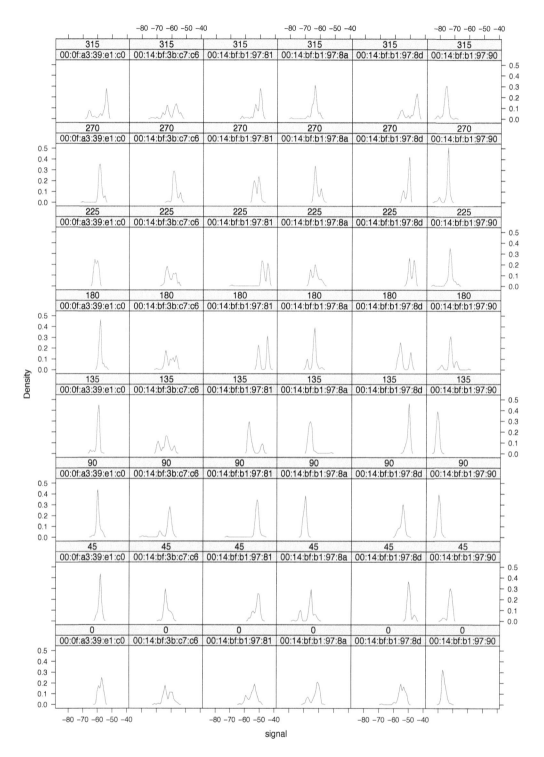

Figure 1.7: Distribution of Signal by Angle for Each Access Point. *The density curves shown here are for the signal strengths measured at the position: $x = 24$ and $y = 4$. These 48 density plots represent each of the access point × angle combinations. There are roughly 110 observations in each panel. Some look roughly normal while many others look skewed left.*

```
            ans$sdSignal = sd(oneLoc$signal)
            ans$iqrSignal = IQR(oneLoc$signal)
            ans
            })

offlineSummary = do.call("rbind", signalSummary)
```

Let's examine the standard deviations and see if they vary with the average signal strength. We can make boxplots of `sdSignal` for subgroups of `avgSignal` by turning `avgSignal` into a categorical variable. We do this with

```
breaks = seq(-90, -30, by = 5)
bwplot(sdSignal ~ cut(avgSignal, breaks = breaks),
       data = offlineSummary,
       subset = mac != "00:0f:a3:39:dd:cd",
       xlab = "Mean Signal", ylab = "SD Signal")
```

We see in Figure 1.8 that the weakest signals have small standard deviations and that it appears that the SD increases with the average signal strength. If we plan to model the behavior of signal strength, then we want to take these features into consideration.

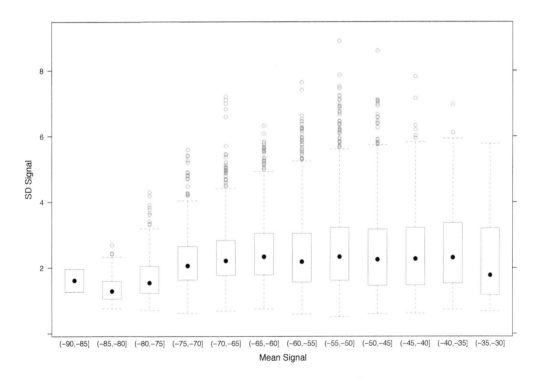

Figure 1.8: SD of Signal Strength by Mean Signal Strength. *The average and SD for the signals detected at each location-angle-access point combination are plotted against each other. The weak signals have low variability and the stronger signals have greater variability.*

We examine the skewness of signal strength by plotting the difference, `avgSignal - medSignal`, against the number of observations. We do this with the smoothScatter()

function so that we avoid problems with over plotting and we also add a local average of the difference between the mean and median to better help us assess its size. We do this with:

```
with(offlineSummary,
     smoothScatter((avgSignal - medSignal) ~ num,
                 xlab = "Number of Observations",
                 ylab = "mean - median"))
abline(h = 0, col = "#984ea3", lwd = 2)
```

We use loess() to locally smooth the differences between the mean and median with

```
lo.obj =
  with(offlineSummary,
       loess(diff ~ num,
             data = data.frame(diff = (avgSignal - medSignal),
                               num = num)))
```

Then we use the fitted model to predict the difference for each value of num and add these predictions to the scatter plot with

```
lo.obj.pr = predict(lo.obj, newdata = data.frame(num = (70:120)))
lines(x = 70:120, y = lo.obj.pr, col = "#4daf4a", lwd = 2)
```

From Figure 1.9 we see that these two measures of centrality are similar to each other; they typically differ by less than 1 to 2 dBm.

1.4.2 The Relationship between Signal and Distance

One way to examine the relationship between distance and signal strength is to smooth the signal strength over the region where it is measured and create a contour plot, similar to a topographical map; that portion of the floor plan where there is strong signal corresponds to the mountainous regions in the contour map. As with our previous analysis of signal strength, we want to control for the access point and orientation. Let's begin by selecting one MAC address and one orientation to examine. We choose the summary statistics for an angle×MAC address combination with, e.g.,

```
oneAPAngle = subset(offline, mac == subMacs[5] & angle == 0)
```

We can make a topographical map using color, i.e., a heat map. The fields package [6] uses the method of thin plate splines to fit a surface to the signal strength values at the observed locations. This package also provides plotting routines for visualizing the surface with a heat map. The Tps() function in fields requires that we provide a unique "z" value for each (x, y) so we must summarize our signal strengths. Rather than use offline, which gives, in oneAPAngle, about 100 recordings of signal strength at each location, we subset offlineSummary with

```
oneAPAngle = subset(offlineSummary,
                    mac == subMacs[5] & angle == 0)
```

Then, after loading fields, we call Tps() to fit a smooth surface to mean signal strength:

```
library(fields)
smoothSS = Tps(oneAPAngle[, c("posX","posY")],
               oneAPAngle$avgSignal)
```

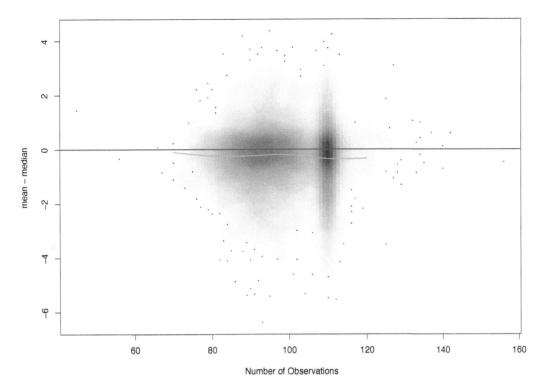

Figure 1.9: Comparison of Mean and Median Signal Strength. *This smoothed scatter plot shows the difference between the mean and median signal strength for each combination of location, access point, and angle against the number of observations. These differences are close to 0 with a typical deviation of 1 to 2 dBm.*

Next, we use predictSurface() to predict the value for the fitted surface at a grid of the observed posX and posY values, i.e.,

```
vizSmooth = predictSurface(smoothSS)
```

Then we plot the predicted signal strength values using plot.surface() as follows:

```
plot.surface(vizSmooth, type = "C")
```

Lastly, we add the locations where the measurements were taken:

```
points(oneAPAngle$posX, oneAPAngle$posY, pch=19, cex = 0.5)
```

We can wrap this plotting routine into its own function so that we can parameterize the MAC address and angle, and if desired, other plotting parameters. Our function, called surfaceSS(), has 3 arguments: *data* for the offline summary data frame, and *mac* and *angle*, which supply the MAC address and angle to select the subset of the data that we want smoothed and plotted. We call surfaceSS() with a couple of MAC addresses and angles to compare them. To do this, we first modify R's plotting parameters so that we can place 4 contour plots on one canvas, and we reduce the size allocated to the margins so more of the canvas is dedicated to the heat maps. We save the current settings for the plotting parameters in parCur with

```
parCur = par(mfrow = c(2,2), mar = rep(1, 4))
```

Then we make 4 calls to our surfaceSS() function using mapply() as follows:

```
mapply(surfaceSS, mac = subMacs[ rep(c(5, 1), each = 2) ],
       angle = rep(c(0, 135), 2),
       data = list(data = offlineSummary))
```

Lastly, we reset the plotting parameters with

```
par(parCur)
```

In Figure 1.10 we see that we can easily identify the location of the access point as the dark red region at the top of the "mountain." We also confirm the effect of the orientation on signal strength. Additionally, a corridor effect emerges. The signal is stronger relative to distance along the corridors where the signals are not blocked by walls.

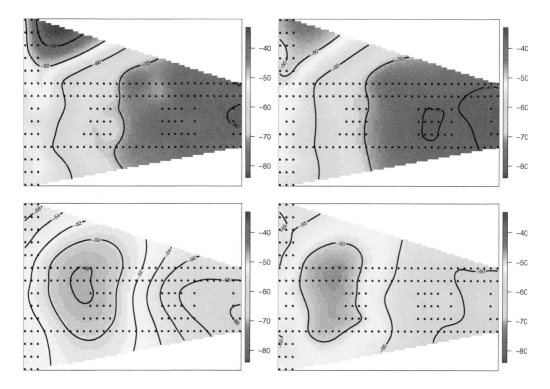

Figure 1.10: Median Signal at Two Access Points and Two Angles. *These four heat maps provide a smooth topographical representation of signal strength. The top two maps are for the access point* 00:14:bf:b1:97:90 *and the angles 0 (left) and 135 (right). The two bottom heat maps represent the signal strength for the* 00:0f:a3:39:e1:c0 *and the same two angles.*

We know the locations of the access points based on the floor plan of the building, but we have not been given their exact location and we do not know the mapping between MAC address and access point. Fortunately, the contour maps that we just created make it easy to connect the MAC address to the access point marked on the floor plan in Figure 1.1.

For example, from Figure 1.10, the signals appearing in the top row of the plot clearly correspond to the access point in the top left corner of the building. Also, according to the documentation, the training data were measured at 1 meter intervals in the building so we can use the grey dots on the plan to estimate the location of the access points. We find that two MAC addresses have similar heat maps and these both correspond to the access point near the center of the building (i.e., $x = 7.5$ and $y = 6.3$). We choose the first of these and leave as an exercise the analysis of the impact of this decision on predicting location.

```
offlineSummary = subset(offlineSummary, mac != subMacs[2])
```

We create a small matrix with the relevant positions for the 6 access points on the floor plan with

```
AP = matrix( c( 7.5, 6.3, 2.5, -.8, 12.8, -2.8,
                1, 14, 33.5, 9.3, 33.5, 2.8),
            ncol = 2, byrow = TRUE,
            dimnames = list(subMacs[ -2 ], c("x", "y") ))
```

Notice that we used the MAC address for the row names. That is,

```
AP
```

```
                   x    y
00:0f:a3:39:e1:c0  7.5  6.3
00:14:bf:b1:97:8a  2.5 -0.8
00:14:bf:3b:c7:c6 12.8 -2.8
00:14:bf:b1:97:90  1.0 14.0
00:14:bf:b1:97:8d 33.5  9.3
00:14:bf:b1:97:81 33.5  2.8
```

These row names are useful when indexing the data.

To examine the relationship between signal strength and distance from the access point, we need to compute the distances from the locations of the device emitting the signal to the access point receiving the signal. We first compute the difference between the x coordinate and access point's x coordinate and the similar difference for the y coordinates. We do this with

```
diffs = offlineSummary[ , c("posX", "posY")] -
        AP[ offlineSummary$mac, ]
```

Then we use these differences to find the Euclidean distance between the position of the hand-held device and the access point with

```
offlineSummary$dist = sqrt(diffs[ , 1]^2 + diffs[ , 2]^2)
```

Finally, we make a series of scatter plots for each access point and device orientation with

```
xyplot(signal ~ dist | factor(mac) + factor(angle),
       data = offlineSummary, pch = 19, cex = 0.3,
       xlab ="distance")
```

The scatter plots appear in Figure 1.11. There appears to be curvature in the plots. A log transformation might improve the relationship. However, the signals are negative values so we need to be careful in taking a log transformation. We leave it to the reader to further investigate this relationship between signal strength and distance.

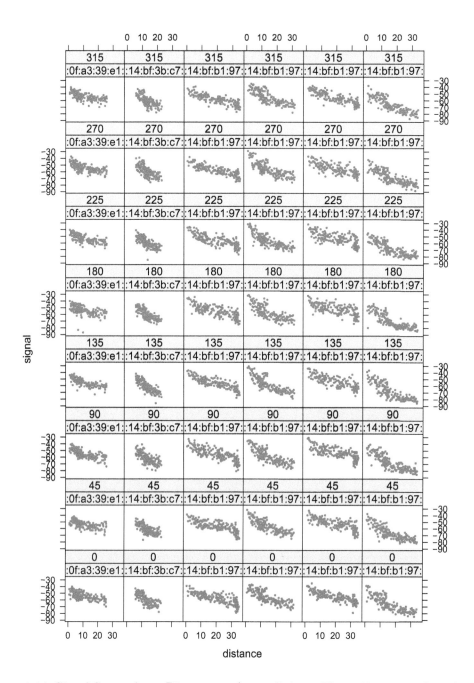

Figure 1.11: Signal Strength vs. Distance to Access Point. *These 48 scatter plots show the relationship between the signal strength and the distance to the access point for each of the 6 access points and 8 orientations of the device. The shape is consistent across panels showing curvature in the relationship.*

1.5 Nearest Neighbor Methods to Predict Location

There are numerous different statistical techniques we can use to estimate the location of a device from the strength of the signal detected between the device and several access points. Here, we use a relatively simple and intuitive approach, called k-nearest neighbors or k-NN for short. The idea behind the nearest neighbor method (for $k = 1$) is as follows: we have training data where the signal is measured to several access points from known positions throughout a building; when we get a new observation, i.e., a new set of signal strengths for an unknown location, we find the observation in our training data that is closest to this new observation. By close we mean the signals recorded between the access points and the new, unobserved location are close to the signal strengths measured between the access points and an observation in the training data. Then, we simply predict the position of the new observation as the position of that closest training observation. For k-nearest neighbors where k is larger than 1, we find the k closest training points (in the signal strength domain) and estimate the new observation's position by an aggregate of the positions of the k training points.

We naturally think of measuring the distance between two sets of signal strengths with Euclidean distance, i.e.,

$$\sqrt{(S_1^* - S_1)^2 + \cdots + (S_6^* - S_6)^2}\,,$$

where S_i is the signal strength measured between the hand-held device and the i-th access point for a training observation taken at some specified location, and S_i^* is the signal measured between the same access point and our new point whose (x, y) values we are trying to predict.

1.5.1 Preparing the Test Data

The online data are in *online.final.trace.txt*, and these observations form our test data. We use the readData() function from Section 1.2 to process the raw data with

```
macs = unique(offlineSummary$mac)
online = readData("Data/online.final.trace.txt", subMacs = macs)
```

We have the locations where these test measurements were taken so that we can assess the accuracy of our predictions. As with the offline data, we create a unique location identifier with

```
online$posXY = paste(online$posX, online$posY, sep = "-")
```

We use this new variable to determine that we have 60 unique test locations, i.e.,

```
length(unique(online$posXY))
```

[1] 60

Also, we tally the number of signal strengths recorded at each location with

```
tabonlineXYA = table(online$posXY, online$angle)
tabonlineXYA[1:6, ]
```

	0	45	90	135	180	225	270	315
0-0.05	0	0	0	593	0	0	0	0
0.15-9.42	0	0	606	0	0	0	0	0
0.31-11.09	0	0	0	0	0	573	0	0
0.47-8.2	590	0	0	0	0	0	0	0
0.78-10.94	586	0	0	0	0	0	0	0
0.93-11.69	0	0	0	0	583	0	0	0

This output indicates that signal strengths were recorded at one orientation for each location.

Given that we are computing distances between vectors of 6 signal strengths, it may be helpful to organize the data in a different structure than we have used so far in this chapter. Specifically, rather than a data frame with one column of signal strengths from all access points, let's organize the data so that we have 6 columns of signal strengths, i.e., one for each of the access points. We summarize the `online` data into this format, providing the average signal strength at each location as follows:

```
keepVars = c("posXY", "posX","posY", "orientation", "angle")
byLoc = with(online,
             by(online, list(posXY),
                function(x) {
                  ans = x[1, keepVars]
                  avgSS = tapply(x$signal, x$mac, mean)
                  y = matrix(avgSS, nrow = 1, ncol = 6,
                        dimnames = list(ans$posXY, names(avgSS)))
                  cbind(ans, y)
                }))

onlineSummary = do.call("rbind", byLoc)
```

We have kept in the data frame only those variables that we use for making and assessing predictions. This new data frame should have 60 rows and 11 variables, including 6 average signal strengths at the corresponding MAC addresses. We confirm this with:

```
dim(onlineSummary)
```

```
[1] 60 11
```

```
names(onlineSummary)
```

```
 [1] "posXY"              "posX"              "posY"
 [4] "orientation"        "angle"             "00:0f:a3:39:e1:c0"
 [7] "00:14:bf:3b:c7:c6"  "00:14:bf:b1:97:81" "00:14:bf:b1:97:8a"
[10] "00:14:bf:b1:97:8d"  "00:14:bf:b1:97:90"
```

1.5.2 Choice of Orientation

In our nearest neighbor model, we want to find records in our offline data, i.e., our training set, that have similar orientations to our new observation because we saw in Section 1.3 that orientation can impact the strength of the signal. To do this, we might consider using all records with an orientation that is within a specified range of the new point's orientation. Since the observations were recorded in 45 degree increments, we can simply specify the number of neighboring angles to include from the training data. For example, if we want only one orientation then we only include training data with angles that match the rounded orientation value of the new observation. If we want two orientations then we pick those two multiples of 45 degrees that flank the new observation's orientation; for three, we choose the closest 45 degree increment and one on either side of it, and so on. That is, for `m` the number of angles and `angleNewObs` the angle of the new observation, we find the angles to include from our training data as follows:

```
refs = seq(0, by = 45, length = 8)
nearestAngle = roundOrientation(angleNewObs)

if (m %% 2 == 1) {
  angles = seq(-45 * (m - 1) /2, 45 * (m - 1) /2, length = m)
} else {
  m = m + 1
  angles = seq(-45 * (m - 1) /2, 45 * (m - 1) /2, length = m)
  if (sign(angleNewObs - nearestAngle) > -1)
    angles = angles[ -1 ]
  else
    angles = angles[ -m ]
}
```

Notice that we handle the case of `m` odd and even separately. Also, we must map the angles to values in `refs`, e.g., -45 maps to 335 and 405 maps to 45, so we adjust `angles` with

```
angles = angles + nearestAngle
angles[angles < 0] = angles[ angles < 0 ] + 360
angles[angles > 360] = angles[ angles > 360 ] - 360
```

After we have the subset of the desired angles, we select the observations from `offlineSummary` to analyze with

```
offlineSubset =
  offlineSummary[ offlineSummary$angle %in% angles, ]
```

Then we aggregate the signal strengths from these angles and create a data structure that is similar to that of `onlineSummary`. Rather than repeat the code again, we turn these computations into a helper function, which we call reshapeSS():

```
reshapeSS = function(data, varSignal = "signal",
                     keepVars = c("posXY", "posX","posY")) {
  byLocation =
    with(data, by(data, list(posXY),
                  function(x) {
                    ans = x[1, keepVars]
                    avgSS = tapply(x[ , varSignal ], x$mac, mean)
                    y = matrix(avgSS, nrow = 1, ncol = 6,
                               dimnames = list(ans$posXY,
                                               names(avgSS)))
                    cbind(ans, y)
```

```
      }))
  newDataSS = do.call("rbind", byLocation)
  return(newDataSS)
}
```

We summarize and reshape `offlineSubset` with

```
trainSS = reshapeSS(offlineSubset, varSignal = "avgSignal")
```

We leave it as an exercise to wrap the code to select the angles and the call to reshapeSS() into a function, called selectTrain(). This function has 3 parameters: *angleNewObs*, the angle of the new observation; *signals*, the training data, i.e., data in the format of `offlineSummary`; and *m*, the number of angles to include from `signals`. The function returns a data frame that matches `trainSS` from above.

We can test our function for an angle of 130 degrees and m of 3, i.e., we aggregate the offline data for angles of 90, 135, and 180. We do this with

```
train130 = selectTrain(130, offlineSummary, m = 3)
```

The results, slightly reformatted for readability, are:

```
head(train130)
```

```
      posXY posX posY  :c0 :c6 :81 :8a :8d :90
0-0   0-0    0    0    -52 -66 -63 -36 -64 -55
0-1   0-1    0    1    -53 -65 -64 -39 -65 -59
0-10  0-10   0    10   -56 -66 -69 -45 -67 -50
0-11  0-11   0    11   -55 -67 -70 -48 -67 -55
0-12  0-12   0    12   -56 -70 -72 -45 -67 -50
0-13  0-13   0    13   -55 -71 -73 -43 -69 -54
```

The selectTrain() function averages the signal strengths for the different angles to produce one set of signal strengths for each of the 166 locations in the training data, i.e.,

```
length(train130[[1]])
```

```
[1] 166
```

However, we may not want to collapse the signal strengths across the m angles, and instead return a set of m×166 signals for each access point. We leave it as an exercise to try this alternative approach.

1.5.3 Finding the Nearest Neighbors

At this point, we have a set of training data that we can use to predict the location of our new point. We want to look at the distance in terms of signal strengths from these training data to the new point. Whether we want the nearest neighbor or the 3 nearest neighbors, we need to calculate the distance from the new point to all observations in the training set. We can do this with the findNN() function:

```
findNN = function(newSignal, trainSubset) {
  diffs = apply(trainSubset[ , 4:9], 1,
                function(x) x - newSignal)
  dists = apply(diffs, 2, function(x) sqrt(sum(x^2)) )
  closest = order(dists)
  return(trainSubset[closest, 1:3 ])
}
```

The parameters to this function are a numeric vector of 6 new signal strengths and the return value from selectTrain(). Our function returns the locations of the training observations in order of closeness to the new observation's signal strength.

We can use some subset of these ordered locations to estimate the location of the new observation. That is, for some value k of nearest neighbors, we can simply average the first k locations. For example, if closeXY contains the x and y values returned from findNN() (these are the ordered training locations), then we estimate the location of the new observation with

```
estXY = lapply(closeXY,
               function(x) sapply(x, function(x) mean(x[1:k])))
```

Of course, we need not take simple averages. For example, we can use weights in the average that are inversely proportional to the distance (in signal strength) from the test observation. In this case, we also need to return the distance values from the findNN() function. This alternative approach allows us to include the k points that are close, but to differentiate between them by how close they actually are from the new observation's signals. The weights might be

$$\frac{1/d_i}{\sum_{i=1}^{k} 1/d_i},$$

for the i-th closest neighboring observation where d_i is the distance from our new point to that nearest reference point (in signal strength space). We may also want to consider different metrics. We have used Euclidean distance, but we may want to try Manhattan distance. We might also be inclined to use medians and not averages when combining neighbors to predict (x, y), if the distribution of the values we are averaging are quite skewed. We leave as exercises these alternatives to the approach presented here.

We have developed two functions, trainSelect() and findNN(), to provide the locations in the training data that have signal strengths close to those of a test observation. We can formalize this approach to make predictions for all of our test data. We do this with predXY() as follows:

```
predXY = function(newSignals, newAngles, trainData,
                  numAngles = 1, k = 3){

  closeXY = list(length = nrow(newSignals))

  for (i in 1:nrow(newSignals)) {
    trainSS = selectTrain(newAngles[i], trainData, m = numAngles)
    closeXY[[i]] =
      findNN(newSignal = as.numeric(newSignals[i, ]), trainSS)
  }

  estXY = lapply(closeXY,
```

```
                function(x) sapply(x[ , 2:3],
                                   function(x) mean(x[1:k])))
  estXY = do.call("rbind", estXY)
  return(estXY)
}
```

We test our functions with the case of 3 nearest neighbors and 3 orientations with:

```
estXYk3 = predXY(newSignals = onlineSummary[ , 6:11],
                 newAngles = onlineSummary[ , 4],
                 offlineSummary, numAngles = 3, k = 3)
```

To assess the fit of the model we can make a map of the actual and predicted locations. Figure 1.12 shows such a map for this model and the 1-NN model. Notice that in general the errors are smaller for 3-NN. Also in the 3-NN model, the large errors seem less problematic as they tend to follow the hallways.

In addition to the visual comparison of the predicted and actual positions, we can compare these fits numerically. For example, we can compute the length of the line segments in each of the figures and sum them to yield a measure of the size of the error. Or, we can find the sum of squared errors with

```
calcError =
function(estXY, actualXY)
   sum( rowSums( (estXY - actualXY)^2) )
```

We apply this function to our two sets of errors to find:

```
actualXY = onlineSummary[ , c("posX", "posY")]
sapply(list(estXYk1, estXYk3), calcError, actualXY)
[1] 659 307
```

This confirms what we saw in the figures, that 3 nearest neighbors do a better job of predicting location than one nearest neighbor. The question remains whether some other value of k makes a better predictor.

1.5.4 Cross-Validation and Choice of k

The choice of k, the number of neighbors to include in the estimate of a new observation's position, is a model selection problem. Ideally, we want to choose the value of k independent of our test data so that we do not overfit our model to the training data. The method of v-fold cross-validation can help us do this. The idea behind it is quite simple: we divide our training data into v non-overlapping subsets of equal size. Then for each subset, we build models with the data that are not in that subset and we assess the predictive ability of the model using the subset that was left out. We repeat this model fitting and assessment for each of the v folds and aggregate the prediction errors across the folds.

In our nearest neighbor scenario, we use all 8 orientations and 6 MAC addresses with each location. This means that we cross-validate on the 166 locations. Suppose that we take $v = 11$; then each fold has `floor(166/v)`, or 15, locations. We can randomly select these locations with

```
v = 11
permuteLocs = sample(unique(offlineSummary$posXY))
permuteLocs = matrix(permuteLocs, ncol = v,
                     nrow = floor(length(permuteLocs)/v))
```

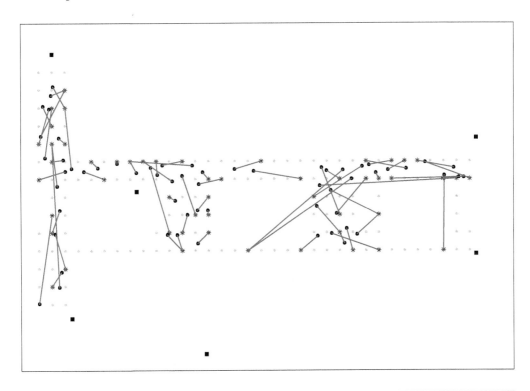

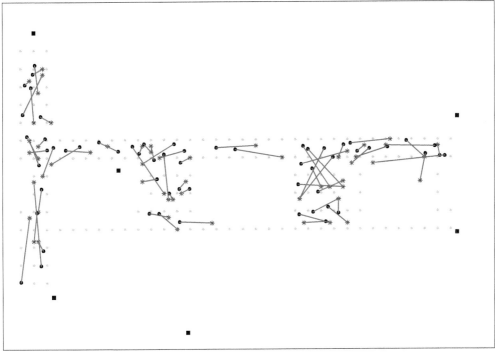

Figure 1.12: Floor Plan with Predicted and Actual Locations. *The red line segments shown in the floor plan connect the test locations (black dots) to their predicted locations (asterisks). The top plot shows the predictions for $k = 1$ and the bottom plot is for $k = 3$ nearest neighbors. In this model, we use as training data the average signal strengths from each of the 166 offline locations (grey dots) to the 6 access points (black squares) for the 3 closest angles to the angle at which the test data was measured.*

We receive a warning message with the call to matrix() because v does not divide evenly into 166, so permuteLocs does not contain all 166 locations. Each subset of 15 locations is used as the "online" or test data so, e.g., the first online fold is

```
onlineFold = subset(offlineSummary, posXY %in% permuteLocs[ , 1])
```

We need to summarize these data so that the data structure matches that of onlineSummary. This includes selecting an orientation at random because each test observation has only one orientation. (Of course, we could find the nearest neighbors for each of the 8 orientations, but we keep things a bit simpler).

Recall from Section 1.5.1 that we summarized the online data into a structure that had 6 columns of signal strength values, one for each access point. It is easier for us to create the test data in its entirety from offline and then divide this data structure into its folds. We can use our function reshapeSS() to do this. However, there is one important difference – we want to select one angle at random for each location. We can augment reshapeSS() to conditionally perform this selection, e.g.,

```
if (sampleAngle)    x = x[x$angle == sample(refs, size = 1), ]
```

After we incorporate this code into reshapeSS() and augment the function definition to include *sampleAngle* with a default value of FALSE, then we can summarize and format offline with

```
keepVars = c("posXY", "posX","posY", "orientation", "angle")

onlineCVSummary = reshapeSS(offline, keepVars = keepVars,
                            sampleAngle = TRUE)
```

Now, our first fold is:

```
onlineFold = subset(onlineCVSummary,
                    posXY %in% permuteLocs[ , 1])
```

This structure makes it easier to use our previous code to find the nearest neighbors. Our training data for the first fold is

```
offlineFold = subset(offlineSummary,
                     posXY %in% permuteLocs[ , -1])
```

This subset is also in the correct format for our earlier application of the nearest neighbor method. That is, we can use our predXY() function with these cross-validated versions of the online and offline data as follows:

```
estFold = predXY(newSignals = onlineFold[ , 6:11],
                 newAngles = onlineFold[ , 4],
                 offlineFold, numAngles = 3, k = 3)
```

Then we find the error in our estimates with

```
actualFold = onlineFold[ , c("posX", "posY")]
calcError(estFold, actualFold)
```

```
[1] 186
```

For each fold, we want to find the k-NN estimates for $k = 1, 2, \ldots, K$, for some suitably large K. And, we want to aggregate the errors over the v folds. We begin simply by wrapping our code from above in loops over the folds and the number of neighbors. We do this as follows, for $K = 20$:

```
K = 20
err = rep(0, K)

for (j in 1:v) {
  onlineFold = subset(onlineCVSummary,
                    posXY %in% permuteLocs[ , j])
  offlineFold = subset(offlineSummary,
                    posXY %in% permuteLocs[ , -j])
  actualFold = onlineFold[ , c("posX", "posY")]
  
  for (k in 1:K) {
    estFold = predXY(newSignals = onlineFold[ , 6:11],
                   newAngles = onlineFold[ , 4],
                   offlineFold, numAngles = 3, k = k)
    err[k] = err[k] + calcError(estFold, actualFold)
  }
}
```

Figure 1.13 shows the sum of squared errors as a function of k. We see that the errors decrease quite a lot at first, e.g., for $k = 1, 2,$ and 3; then the errors level out around values of $k = 5, 6,$ and 7; and after that, the errors begin to increase slowly because the neighbors become too spread out geographically.

We use the value of 5 for the nearest neighbors that we obtained from cross-validation, and we apply it to our original training and test data, i.e.,

```
estXYk5 = predXY(newSignals = onlineSummary[ , 6:11],
               newAngles = onlineSummary[ , 4],
               offlineSummary, numAngles = 3, k = 5)
```

Then we tally the errors in our predictions with

```
calcError(estXYk5, actualXY)
```

```
[1] 276
```

The earlier values for $k = 1$ and $k = 3$ were 659 and 307, respectively. The choice of $k = 5$ may not be the minimizing value for our online data because this value was chosen without reference to the online data. This is the reason we use cross-validation, i.e., we do not use the online data in both the selection of k and the assessment of the predictions.

The code to cross-validate k can take a long time to run. We probably want to examine this code to find ways to speed it up. For example, consider the function predXY(), which we reproduce below:

```
predXY = function(newSignals, newAngles, trainData,
                numAngles = 1, k = 3){
  
  closeXY = list(length = nrow(newSignals))
```

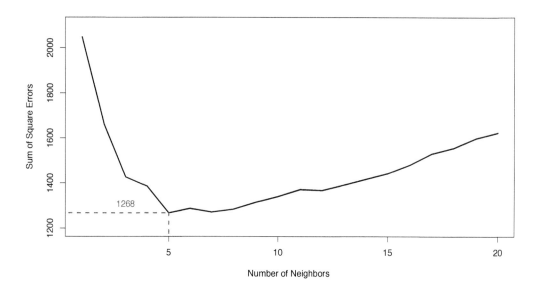

Figure 1.13: Cross Validated Selection of k. *This line plot shows the sum of square errors as a function of the number of neighbors used in predicting the location of a new observation. The sums of squared errors are obtained via cross-validation of the offline data.*

```
  for (i in 1:nrow(newSignals)) {
    trainSS = selectTrain(newAngles[i], trainData, m = numAngles)
    closeXY[[i]] = findNN(newSignal = as.numeric(newSignals[i, ]),
                          trainSS)
  }

  estXY = lapply(closeXY, function(x)
                          sapply(x[ , 2:3],
                                 function(x) mean(x[1:k])))
  estXY = do.call("rbind", estXY)
  return(estXY)
}
```

Recall that findNN() returns all of the positions for the training data, ordered according to their distance from the new observation's signal strengths. We use the first k positions, but since we have all of the locations we can calculate the estimates for all values of k in which we are interested. The cumsum() function is very helpful here, i.e., cumsum(x[1¬:K])/(1:K) provides K means. If we modify predXY() to return all K estimates, then we can eliminate the inner loop over k. This should speed up our code considerably. We leave this to pursue as an exercise.

Finally, we have used the method of cross-validation to select the number of neighbors, but we have another parameter that we have not investigated: the number of angles to include in the training data. This parameter can be selected via cross-validation as well. In fact, the two can be selected jointly via cross-validation. We leave this problem as an exercise for further investigation.

1.6 Exercises

Q.1 Write the code to read the raw training data into the data structure in the first approach described in Section 1.2. That is, the data structure is a data frame with a column for each MAC address that detected a signal. For the column name, use the last two characters of the MAC address, or some other unique identifier.

Q.2 Compare the size of two data structures: the data frame created in Section 1.2 and the data frame created in the previous problem. Which uses less memory? What is the dimension of each? How might this change with different numbers of devices in the building? different number of signals from the less commonly detected devices? Use object.size() and dim() to address these questions.

Q.3 Compare the total time it takes to read the raw data and create the data frame, for the two approaches described in Section 1.2. Do this for different size subsets of the data (chosen at random) and draw a curve of time against input size for each of the approaches. Also, comment on the memory and speed for the two approaches. Use system.time() and Rprof() to make these comparisons.

Q.4 Examine the `time` variable in the offline data. Any change over time in the characteristics of the signal caused by, e.g., reduced battery power in the measuring device as time goes by, or measurements taken on different days may be made by different people with different levels of accuracy. Also, examination of `time` can give insight into how the experiment was carried out. Were the positions close to each other measured at similar times? Do you see any change in the signal strength variability or mean over time? Try controlling for other variables that might affect this relationship.

Q.5 Write the readData() function described in Section 1.3.4. The arguments to this function are the file name, *filename* and the MAC addresses to retain, *subMacs*. Determine whether these parameters should have default values or not. The return value is the data frame described in Section 1.3. Use the findGlobals() function available in `codetools` to check that the function is not relying on any global variables.

Q.6 In Section 1.4.1 we calculated measures of center and location for the signal strengths at each location × angle × access point combination. (See for example Figure 1.9.) Another possible summary statistic we can calculate is the Kolmogorov-Smirnov test-statistic for normality. If the signal strengths are roughly normal, then we expect the p-values to have a uniform distribution. This leads to about 5% of the p-values for the 8000 tests to fall below 0.05.

Q.7 Write the surfaceSS() function that creates plots such as those in Figure 1.10. This function takes 3 arguments: *data* for the offline summary data frame, *mac*, and *angle*. The parameters *mac* and *angle* are used to specify which MAC address and angle are to be selected from the data for smoothing and plotting.

Q.8 Consider the scatter plots in Figure 1.11. There appears to be curvature in the signal strength–distance relationship. Does a log transformation improve this relationship, i.e., make it linear? Note that the signals are negative values so we need to be careful if we want to take the log of signal strength.

Q.9 The floor plan for the building (see Figure 1.1) shows 6 access points. However, the data contain 7 access points with roughly the expected number of signals (166 location

× 8 orientations × 110 replications = 146,080 measurements). With the signal strength seen in the heat maps of Figure 1.10), we matched the access points to the corresponding MAC address. However, two of the MAC addresses seem to be for the same access point. In Section 1.3.2 we decided to keep the measurements from the MAC address `00:0f:a3:39:e1:c0` and to eliminate the `00:0f:a3:39:dd:cd` address. Conduct a more thorough data analysis into these two MAC addresses. Did we make the correct decision? Does swapping out the one we kept for the one we discarded improve the prediction?

Q.10 Write the selectTrain() function described in Section 1.5.2. This function has 3 parameters: $angleNewObs$, the angle of the new observation; $signals$, the training data, i.e., data in the format of `offlineSummary`; and m, the number of angles to include from `signals`. The function returns a data frame that matches `trainSS`, i.e., selectTrain() calls reshapeSS() (see Section 1.5.2 for this function definition).

Q.11 We use Euclidean distance to find the distance between the signal strength vectors. However, Euclidean distance is not robust in that it is sensitive to outliers. Consider other metrics such as the L_1 distance, i.e., the absolute value of the difference. Modify the findNN() function in Section 1.5.3 to use this alternative distance. Does it improve the predictions?

Q.12 To predict location, we use the k nearest neighbors to a set of signal strengths. We average the known (x, y) values for these neighbors. However, a better predictor might be a weighted average, where the weights are inversely proportional to the "distance" (in signal strength) from the test observation. This allows us to include the k points that are close, but to differentiate between them by how close they actually are. The weights might be

$$\frac{1/d_i}{\sum_{i=1}^{k} 1/d_i}$$

for the i-th closest neighboring observation where d_i is the distance from our new test point to this reference point (in signal strength space). Implement this alternative prediction method. Does this improve the predictions? Use calcError() to compare this approach to the simple average.

Q.13 In Section 1.5.4 we used cross-validation to choose k, the number of neighbors. Another parameter to choose is the number of angles at which the signal strength was measured. Use cross-validation to select this value. You might also consider selecting the pair of parameter, i.e., k and the number of angles, simultaneously.

Q.14 The researchers who collected these data implemented a Bayesian approach to predicting location from signal strength. Their work is described in a paper that is available from `http://www.informatik.uni-mannheim.de/pi4/publications/King2006g.pdf`. Consider implementing this approach to building a statistical IPS.

Q.15 Other statistical techniques have been developed to predict indoor positions from wireless local area networks. These include [3, 4, 11]. Consider employing one of their approaches to building and testing a statistical IPS with the CRAWDAD data.

Bibliography

[1] Daniel Faria. Modeling Signal Attenuation in IEEE 802.11 Wireless LANs - Vol. 1. Technical Report TR-KP06-0118, Kiwi Project, Stanford University, 2005.

[2] Thomas King, Stephan Kopf, Thomas Haenselmann, Christian Lubberger, and Wolfgang Effelsberg. CRAWDAD data set mannheim/compass: v. 2008-04-11. http://crawdad.org/mannheim/compass/, 2008.

[3] P. Krishnan, A.S. Krishnakumar, W.H. Ju, C. Mallows, and S. Gani. A system for lease: Location estimation assisted by stationery emitters for indoor rf wireless networks. In *Proceedings IEEE INFOCOM 2004, the 23rd Annual Joint Conference of the IEEE Computer and Communications Societies, Hong-Kong, China, March 7–11.*, 2004.

[4] D. Madigan, W.H. Ju, P. Krishnan, A. Krishnakumar, and I. Zorych. Location estimation in wireless networks: A Bayesian approach. *Statistica Sinica*, 16:495–522, 2006.

[5] Bradley Mitchell. The MAC Address: An Introduction to MAC Addressing. http://compnetworking.about.com/od/networkprotocolsip/l/aa062202a.htm, 2011.

[6] Doug Nychka, Reinhard Furrer, and Stephan Sain. fields: Tools for spatial data. http://cran.r-project.org/web/packages/fields, 2014. R package version 7.1.

[7] Theodore Rappaport. *Wireless Communications: Principles and Practices*. Prentice Hall, New York, 1996.

[8] R Development Core Team. *R: A Language and Environment for Statistical Computing*. Vienna, Austria, 2012. http://www.r-project.org.

[9] Deepayan Sarkar. *Lattice: Multivariate Data Visualization with R*. Springer-Verlag, New York, 2008. http://lmdvr.r-forge.r-project.org/figures/figures.html.

[10] Luke Tierney. codetools: Code Analysis Tools for R. http://cran.r-project.org/web/packages/codetools, 2014. R package version 0.2-9.

[11] M. Youssef and A. Agrawala. On the optimality of WLAN location determination systems. In *Proceedings of the Communication Networks and Distributed Systems Modeling and Simulation Conference.*, 2004.

2

Modeling Runners' Times in the Cherry Blossom Race

Daniel Kaplan
Macalester College

Deborah Nolan
University of California, Berkeley

CONTENTS

2.1	Introduction	45
	2.1.1 Computational Topics	47
2.2	Reading Tables of Race Results into R	47
2.3	Data Cleaning and Reformatting Variables	55
2.4	Exploring the Run Time for All Male Runners	63
	2.4.1 Making Plots with Many Observations	63
	2.4.2 Fitting Models to Average Performance	67
	2.4.3 Cross-Sectional Data and Covariates	74
2.5	Constructing a Record for an Individual Runner across Years	79
2.6	Modeling the Change in Running Time for Individuals	88
2.7	Scraping Race Results from the Web	93
2.8	Exercises	100
	Bibliography	102

2.1 Introduction

In this era of 'free and ubiquitous data,' there is tremendous potential in seeking out data to bring insight to a problem we are working on professionally or to a topic of personal interest. For example, we are interested in understanding how people's physical performance changes as they age. One source of data about this comes from road races. Hundreds of thousands of people participate in road races each year; the race organizers collect information about the runners' times and often publish individual-level data on the Web. These freely accessible data may provide us with insights to our question about performance and age.

One example of the many annual road races is the Cherry Blossom Ten Mile Run held in Washington D.C. in early April when the cherry trees are typically in bloom. The Cherry Blossom started in 1973 as a training run for elite runners who were planning to compete in the Boston Marathon. It has since grown in popularity and in 2012 nearly 17,000 runners ranging in age from 9 to 89 participated. The race has become so popular that entrants are chosen via a lottery or they guarantee a spot by raising $500 for an official race charity. After each year's race, the organizers publish the results at http://www.cherryblossom.org/ (see Figure 2.1). These data offer a tremendous resource for learning about the relationship between age and performance.

Figure 2.1: Screen Shot of Cherry Blossom Run Web site. *This page contains links to each year's race results. The year 1999 is the earliest for which they provide data. Men's and women's results are listed separately.*

The publicly available race results from the Cherry Blossom Ten Mile Run can be scraped from the Web and read into R [3] for analysis. The currently published results include all years from 1999 to 2012. The task of scraping the Web site and formatting the results in a way that can be analyzed in R is a bit challenging because the information reported and the format of this information changes from year to year. Some simple differences in format occur in the format of the table header and the use of footnotes. The tables also include many mistakes, e.g., values that begin in the wrong column, missing headers, and so on. All in all, the acquisition of the data is quite straightforward, but it is an iterative process as we uncover several small errors. We do this statistically, i.e., we examine summary statistics and plots of the data we have read into R, find anomalies, such as all the runners in 2003 being under 9, cross check sample observations with the original tables, modify our code to handle these problem cases in a way that is as general as possible, recreate our data, and repeat. This is the story of "messy" data. It is the focus of Section 2.2 and Section 2.3 of this chapter. Additionally, Section 2.7 covers the topic of scraping the Web for the race results, for those who are interested in the entire process of data acquisition.

After the data have been successfully read into R and cleaned, we study the relationship between run time and age in Section 2.4. Given the popularity of the race, simple tasks such as visualizing the data present challenges, and we consider how to display tens of thousands of observations in an informative manner.

For any one year of race results, we have a cross-sectional view of the performance-age relationship. That is, we are looking at different groups of people of various ages and their run times; we are not viewing an individual's race performance as he or she gets older. However, we do have race results for 14 years and many runners have participated in multiple races. If we can associate run times over years with an individual runner, we can examine how performance changes for an individual as he or she ages. The data include the runner's name, age, and hometown, so we consider how we might use this information to construct longitudinal views of run times for individuals. This is the subject of Section 2.5.

If we study those runners who have competed in multiple years, then we have a longitudinal view of performance. However, we have results for a runner for at most 14 years, so we are unable to view performance for an individual over the full range of participant ages from 18 to 89. Can we piece together these longitudinal data to get estimates for performance as a function of age? We explore approaches for this in Section 2.6.

2.1.1 Computational Topics

- Use regular expressions to extract and clean messy data from pre-formatted text tables and to create unique identifiers for matching records that belong to the same individual.

- Employ statistical techniques to identify bad data and to confirm these problems have been corrected.

- Visualize data that have a large number of observations (~150,000 records).

- Gain experience with the R formula language for plots and modeling.

- Fit piecewise linear models using least squares and non-parametric curves using local averaging.

- Compare data structures, e.g., a data frame and a list of data frames, for holding and working with longitudinal data. This includes the application of 'apply' functions such as tapply(), mapply(), sapply(), and lapply().

- Develop strategies for debugging code with recover() for browsing active function calls after an error.

- Scrape simple Web pages for text content.

2.2 Reading Tables of Race Results into R

Our goal in this section is to transform the raw text tables of race results into data that can be analyzed in R. These tables have been downloaded from the Web and stored in files, named *1999.txt*, ..., *2012.txt* in a directory called *MenTxt* for men and *WomenTxt* for women. The task of downloading the Web pages and extracting the tables is addressed in Section 2.7. If you want to start this project from the "beginning," then skip ahead to that section and return after you have obtained the text files from the Internet.

Let's examine these text tables to get a sense of their format. After that we should have a few ideas about how we might extract information contained in these tables into variables for statistical analysis. Figure 2.2 and Figure 2.3 provide screenshots of two tables as they appear on the Web. By inspection, we see that a call to read.table() will not properly read the text into a data frame because the information, e.g., place and division, are separated by blanks but blanks also appear in the data values, i.e., blanks also occur where they are not being used as variable separators. For example, for the runner's hometown, we see values of Kenya, Tucson AZ, and Blowing Rock NC. The blanks between the different parts of hometown will confuse read.table(). We confirm this when we try to use read.table() to input the data:

```
m2012 = read.table(file="MenTxt/2012.txt", skip = 8)
```

```
Error in scan(file, what, nmax, sep, dec, quote,
    skip, nlines, na.strings) : line 1 did not have 12 elements
```

Note that we skipped the first 8 lines of the file because we observed in Figure 2.2 that these belong to the header of the file.

We need a customized approach to reach this 'table.' From the figures, it appears that the variables are formatted to occupy particular positions in each line of text. That is, the runner's finishing place occupies the first 5 characters, then comes a blank character, the runner's place in his or her division appears in the next 11 spaces, and so on. While the first 2 columns of the 2011 and 2012 male results line up, we see that the columns are not identical across these tables. Given the changes in formats from year to year, we can extract the values from the tables either by programmatically interpreting the format or by using year-dependent fixed-width formats. We take the first approach here and figure out which column is which by programmatically inspecting the table header. We leave the second approach as an exercise. There you examine all 28 tables, determine the start and end position of each column of interest, and use read.fwf() to input the data into R.

```
                Credit Union Cherry Blossom Ten Mile Run
                Washington, DC      Sunday, April 1, 2012

                    Official Male Results (Sorted By Net Time)


     Place Div   /Tot    Num    Name                      Ag Hometown               5 Mile    Time    Pace
     ===== =========== ====== ======================== === ===================== ======= ======= =====
         1     1/347       9 Allan Kiprono              22 Kenya                    22:32   45:15   4:32
         2     2/347      11 Lani Kiplagat              23 Kenya                    22:38   46:28   4:39
         3     1/1093     31 John Korir                 36 Kenya                    23:20   47:33   4:46
         4     1/1457     15 Ian Burrell                27 Tucson AZ                23:50   47:34   4:46
         5     3/347      19 Jesse Cherry               24 Blowing Rock NC          23:50   47:40   4:46
         6     1/1490     37 Ketema Nugusse             31 Ethiopia                 23:42   47:50   4:47
         7     2/1457     13 Josh Moen                  29 Minneapolis MN           24:06   48:38   4:52
         8     3/1457     17 Patrick Rizzo              28 Boulder CO               24:24   49:14   4:56
         9     4/1457     41 Stephen Hallinan           26 Washington DC            25:01   50:18   5:02
        10     2/1490    345 Paolo Natali               31 Washington DC            25:20   50:44   5:05
        11     3/1490    346 David McCollam             32 Bridgeport WV            25:33   50:56   5:06
        12     4/347     299 Frank Devar                23 Washington DC            25:28   50:57   5:06
        13     4/1490    112 Bert Rodriguez             32 Arlington VA             25:31   50:57   5:06
        14     1/931     290 Chris Juarez               41 Alexandria VA            25:28   51:10   5:07
        15     5/1457    108 Darryl Brown               29 Exton PA                 25:28   51:16   5:08
        16     6/1457    119 Jay Luna                   28 Denver CO                25:22   51:17   5:08
        17     7/1457    110 David Burnham              27 Arlington VA             25:27   51:23   5:09
        18     8/1457    296 Karl Dusen                 29 Rockville MD             25:51   51:27   5:09
        19     9/1457    357 Brian Flynn                28 Bridgewater VA           25:34   51:29   5:09
        20    10/1457    114 Carlos Renjifo             29 Columbia MD              25:51   51:43   5:11
        21     5/1490    358 Dustin Meeker              30 Baltimore MD             25:51   51:53   5:12
        22    11/1457    107 Christopher Sloane         28 Rockville MD             25:32   51:57   5:12
        23    12/1457    116 Patrick Reaves             27 Durham NC                25:51   52:16   5:14
        24     6/1490    111 Jake Klim                  31 North Bethesda MD        25:51   52:32   5:16
        25    13/1457    298 Will Viviani               29 Alexandria VA            26:30   52:41   5:17
        26    14/1457    106 Paul Guevara               25 Alexandria VA            26:01   52:54   5:18
        27     7/1490    303 Dickson Mercer             30 Washington DC            26:27   53:04   5:19
```

Figure 2.2: Screen Shot of the 2012 Male Results. *This screenshot shows the results, in race order, for men competing in the 2012 Cherry Blossom 10 Mile Run. Notice that both 5-mile times and net times are provided. We know that the* `Time` *column is net time because it is so indicated in the header of the table.*

Rather than view the Web pages to determine the file format, we can get a better sense of the format if we examine the raw text itself. We use readLines() to read the contents of the file into R, where the return value is a character vector with one string per line of text read. We start by reading the 2012 men's file with

```
els = readLines("MenTxt/2012.txt")
```

The first 10 rows of the 2012 Men's table are

```
els[1:10]
```

Modeling Runners' Times in the Cherry Blossom Race 49

```
          Credit Union Cherry Blossom Ten Mile Run
           Washington, DC     Sunday, April 3, 2011

                      Official Male Results

Place Div  /Tot    Num  Name                  Ag Hometown             5 Mile  Gun Tim Net Tim Pace
===== =========== ====== ===================== == ==================== ======= ======= ======= =====
    1   1/401        3  Lelisa Desisa         21 Ethiopia                        45:36   45:36  4:34
    2   2/401       13  Allan Kiprono         21 Kenya                  23:08    45:41   45:41  4:35
    3   1/1471       5  Ridouane Harroufi     29 Morocco                23:10    46:27   46:27  4:39
    4   3/401       17  Lani Kiplagat         22 Kenya                  23:09    46:30   46:30  4:39
    5   2/1471      27  Macdonard Ondara      26 Kenya                  21:41    46:52   46:52  4:42
    6   3/1471      29  Tesfaye Sendeku       28 Ethiopia               23:15    46:53   46:53  4:42
    7   4/1471      21  Stephen Muange        29 Kenya                  23:24    47:30   47:30  4:45
    8   4/401       23  Simon Cheprot         21 Kenya                  23:14    47:32   47:32  4:46
    9   5/1471      31  Josphat Boit          27 Kenya                  23:24    47:50   47:50  4:47
   10   1/1083      25  Girma Tola            35 Ethiopia               23:27    47:56   47:56  4:48
   11   5/401       47  Ezkyas Sisay          22 Ethiopia               23:34    47:58   47:58  4:48
   12   6/1471      51  Tesfaye Assefa        27 Ethiopia               23:42    48:03   48:03  4:49
   13   7/1471      33  Lucas Meyer           27 Ridgefield CT          24:06    48:26   48:26  4:51
   14   8/1471     296  David Nightingale     25 Washington DC          24:10    48:39   48:39  4:52
   15   9/1471      45  Augustus Maiyo        27 Colorado Springs CA    24:18    49:56   49:56  5:00
   16  10/1471     107  Karl Dusen            28 N Bethesda MD          25:13    50:06   50:06  5:01
   17   1/1332     105  Bert Rodriguez        31 Arlington VA           25:08    50:25   50:25  5:03
   18   6/401      297  Sam Luff              24 Rockville MD           25:22    50:45   50:45  5:05
   19   7/401      106  Jerry Greenlaw        23 Alexandria VA          25:19    50:55   50:55  5:06
   20  11/1471     112  Brian Flynn           27 Weyers Cave VA         25:24    51:08   51:08  5:07
   21  12/1471      49  Birhanu Alemu         28 Ethiopia               25:09    51:10   51:10  5:07
   22   2/1083   20510  Michael Wardian       36 Arlington VA           25:20    51:16   51:16  5:08
   23  13/1471     304  Joe Wiegner           29 Rockville MD           25:25    51:34   51:34  5:10
   24  14/1471     109  Dirk De Heer          29 Silver Spring MD       25:44    51:40   51:40  5:10
   25  15/1471     108  David Burnham         26 Arlington VA           25:37    51:49   51:46  5:11
   26   1/928      114  Fred Kieser           40 Cleveland OH           25:22    51:48   51:48  5:11
   27  16/1471     305  Michael Cassidy       25 Staten Island NY       25:58    52:03   52:03  5:13
```

Figure 2.3: Screen Shot of Men's 2011 Race Results. *This screenshot shows the results, in race order, for men competing in the 2011 Cherry Blossom road race. Notice that in 2011, 3 times are recorded – the time to complete the first 5 miles and the gun and net times for the full run. In contrast, the results from 2012 do not provide gun time.*

```
 [1]  ""
 [2]  "            Credit Union Cherry Blossom Ten Mile Run"
 [3]  "             Washington, DC     Sunday, April 1, 2012"
 [4]  ""
 [5]  "          Official Male Results (Sorted By Net Time)"
 [6]  ""
 [7]  "Place Div    /Tot    Num   Name        ... Time     Pace    "
 [8]  "===== =========== ======  =========... =======  =====   "
 [9]  "    1    1/347           9 Allan Kip...   45:15    4:32   "
[10]  "    2    2/347          11 Lani Kipl...   46:28    4:39   "
```

We also read in and display the first 10 rows of the 2011 male results so we have another table to compare with the 2012 table. We find the following:

```
els2011 = readLines("MenTxt/2011.txt")
els2011[1:10]
```

```
 [1]  ""
 [2]  "            Credit Union Cherry Blossom Ten Mile Run"
 [3]  "             Washington, DC     Sunday, April 3, 2011"
 [4]  ""
 [5]  "                    Official Male Results"
 [6]  ""
 [7]  "Place Div    /Tot    Num   Name        ... Gun Tim Net Tim Pace   "
 [8]  "===== =========== ======  ======...  ======= ======= =====   "
 [9]  "    1    1/401           3 Lelisa...   45:36   45:36   4:34  "
[10]  "    2    2/401          13 Allan ...   45:41   45:41   4:35  "
```

What do we find with this simple inspection?

- Both of the tables have a header.
- The last line of the header is a row of '=' characters, i.e., a separation line.
- There are blanks inserted in the row of '=' characters that mark the start and end of a column of information, e.g., the Pace column occupies 5 spaces.
- The row above the '=' characters gives column names.
- There are two times reported in 2011 (called `Gun Tim` and `Net Tim`) and only one time reported in 2012 (`Time`). The header of the 2012 file tells us that `Time` is net time.

If we examine a few more years of race results, we find other differences between how the data are organized. Some years have column names that are all capitalized; do not include the time at 5 miles; contain a rightmost column that holds some sort of annotation, e.g., #; have headers consisting of 3 lines instead of 8, etc.

Let's use the 2012 men's results as our test case for developing the code to read in all the files. However, we will try to write the code in a general way so that it can potentially be used for all 28 files. Our first step is to find the row with the equal signs. The rows below it contain the data, the row above it holds the column headers, and the row itself supplies the spacings for the columns. We saw earlier that the '=' characters are in the eighth row of the 2012 table. Since the organization of the tables differs a bit from year to year, we use a programmatic search for the equal signs. We use grep() to search through the character strings in els for one that begins with, say, 3 equal signs as follows:

```
eqIndex = grep("^===", els)
eqIndex
```

```
[1] 8
```

Note that an alternative to regular expressions and the grep() function is to use substr() to extract the first 3 characters from each row and compare them to the string "===". That is,

```
first3 = substr(els, 1, 3)
which(first3 == "===")
```

```
[1] 8
```

The choice of 3 '=' characters is somewhat arbitrary. We could have used just one as the equal sign does not appear elsewhere in the document.

Now that we have located this key row in the table, we extract it and the row above it and discard earlier rows with

```
spacerRow = els[eqIndex]
headerRow = els[eqIndex - 1]
body = els[ -(1:eqIndex) ]
```

Our next task is to extract the various pieces of information from each string in body, i.e., the content of the table. How might we extract the runner's age? From inspection, a runner's age appears in the column labeled Ag or AG so we first convert the column names to lower case so we need not search separately for Ag and AG. We use tolower() to do this with

```
headerRow = tolower(headerRow)
```

We can search through `headerRow` for this two letter sequence as follows:

```
ageStart = regexpr("ag", headerRow)
ageStart
[1] 49
attr(,"match.length")
[1] 2
attr(,"useBytes")
[1] TRUE
```

The return value from regexpr() tells us a match was found in position 49 of the character string. If no match is found, then regexpr() returns -1. Now we have the information about the location of runner's age: it begins in position 49 and ends at the 50th position in each row of the table. We use this information to extract each runner's age using substr() as follows:

```
age = substr(body, start = ageStart, stop = ageStart + 1)
head(age)
[1] "22" "23" "36" "27" "24" "31"

summary(as.numeric(age))
   Min. 1st Qu.  Median    Mean 3rd Qu.    Max.    NA's
   9.00   29.00   35.00   37.75   45.00   89.00       1
```

It appears that we have located the runner's age correctly. The youngest male runner in 2012 was 9 and the oldest 89, and there was one runner who did not have an age reported.

We can extract all of our variables in this manner, but the width of a column might change from one year to the next so we generalize our code to search the row of equal signs for blank spaces and use the position of these to determine the locations of the columns. This approach is better than searching for variable names to find the starting position of the column of values. We find the locations of all of the blanks in the line of '=' characters with

```
blankLocs = gregexpr(" ", spacerRow)
blankLocs
[[1]]
[1]  6 18 25 48 51 72 80 88 94
attr(,"match.length")
[1] 1 1 1 1 1 1 1 1 1
attr(,"useBytes")
[1] TRUE
```

Here the g in gregexpr() stands for "global," which means that the function searches for multiple matches in the string, not just the first match. Blank spaces are found at the 6th, 18th, 25th, 48th, 51st, etc. positions.

In general, we want to write our code so that it does not depend on a variable name starting or ending in a particular column. We can extract all the columns of values using `blankLocs` to determine the start and end positions of the columns. The starting position of a column is one character past a blank and the ending position is one character before a blank. In order to properly handle the first column, we can augment `blankLocs` with 0 so the first column starts one character after 0, i.e.,

```
searchLocs = c(0, blankLocs[[1]])
```

We can extract all the columns using substr() with

```
Values = mapply(substr, list(body),
                start = searchLocs[ -length(searchLocs)] + 1,
                stop = searchLocs[ -1 ] - 1)
```

We encapsulate the task of finding the starting and ending positions of the columns into a function, which we call findColLocs(). In the function, we safeguard against the last character in the row of '=' characters not being a blank, we add an additional element to the end of the vector of locations that is one character more than the length of the string. Our function appears as:

```
findColLocs = function(spacerRow) {

  spaceLocs = gregexpr(" ", spacerRow)[[1]]
  rowLength = nchar(spacerRow)

  if (substring(spacerRow, rowLength, rowLength) != " ")
    return( c(0, spaceLocs, rowLength + 1))
  else return(c(0, spaceLocs))
}
```

We can extract all 10 columns of data from the 2012 file, but do we want to keep all of these variables? Do we want to keep the union of all variables across the 14 years? Or, use only a subset? For now, we extract name, age, hometown, and all 3 times, i.e., gun time, net time, and time, and ignore the rest, e.g., place, div, and the 5-mile run time. We encapsulate into a function the code to extract the locations of the desired columns. We need, as inputs to this function, the names of the desired columns, the header row that contains the column names, and the locations of the blanks in the separator row. Our function follows:

```
selectCols =
function(colNames, headerRow, searchLocs)
{
  sapply(colNames,
         function(name, headerRow, searchLocs)
         {
           startPos = regexpr(name, headerRow)[[1]]
           if (startPos == -1)
             return( c(NA, NA) )

           index = sum(startPos >= searchLocs)
           c(searchLocs[index] + 1, searchLocs[index + 1] - 1)
         },
         headerRow = headerRow, searchLocs = searchLocs )
}
```

Notice that the function is simply a wrapper to a call to sapply(). However, this encapsulation makes it easy for us to test our code on individual column names and to extract a subset of columns. For example, we can find the age variable with

```
searchLocs = findColLocs(spacerRow)
ageLoc = selectCols("ag", headerRow, searchLocs)
ages = mapply(substr, list(body),
              start = ageLoc[1,], stop = ageLoc[2, ])

summary(as.numeric(ages))
```

```
   Min. 1st Qu.  Median    Mean 3rd Qu.    Max.    NA's
   9.00   29.00   35.00   37.75   45.00   89.00       1
```

Our more general extraction matches the earlier one.

Another advantage to creating selectCols() and findColLocs() is that these functions make our code modular so our code is easier to follow and to modify in the context of the larger data extraction and cleaning process.

Since the column names vary somewhat from year to year, we use only the first few characters that uniquely identify the desired columns, e.g.,

```
shortColNames = c("name", "home", "ag", "gun", "net", "time")
```

Also, if a file does not have one of the desired variables, then we want the values for that variable to be NA. We can anticipate this situation because we have seen that the 2011 file has gun time and net time, but not time, and the 2012 file has a column labeled time, but no gun or net times.

```
locCols = selectCols(shortColNames, headerRow, searchLocs)

Values = mapply(substr, list(body), start = locCols[1, ],
                stop = locCols[2, ])
```

Let's examine the return value. First we check the type of the return value with

```
class(Values)
```

```
[1] "matrix"
```

The results form a matrix of character strings. (We have not yet converted any values such as age to numeric.) We see that the first few rows of the matrix are

```
colnames(Values) = shortColNames
head(Values)
```

```
     name                home               ag    gun net time
[1,] "Allan Kiprono  "  "Kenya           "  "22"  NA  NA  " 45:15"
[2,] "Lani Kiplagat  "  "Kenya           "  "23"  NA  NA  " 46:28"
[3,] "John Korir     "  "Kenya           "  "36"  NA  NA  " 47:33"
[4,] "Ian Burrell    "  "Tucson AZ       "  "27"  NA  NA  " 47:34"
[5,] "Jesse Cherry   "  "Blowing Rock NC "  "24"  NA  NA  " 47:40"
[6,] "Ketema Nugusse "  "Ethiopia        "  "31"  NA  NA  " 47:50"
```

The 2012 table has a column for time and not gun and net times so the gun and net values are NA. We also check the last few lines with

```
tail(Values)[ , 1:3]
```

```
              name                       home                  ag
[7188,]  "Dana Brown"           " "Randallstown MD"  " "41"
[7189,]  "Jurek Grabowski"      " "Fairfax VA"      " "39"
[7190,]  "Larry Hume"           " "Arlington VA"    " "56"
[7191,]  "Sean-Patrick Alexander" "Alexandria VA"   " "35"
[7192,]  "Joseph White"         " "Forestville MD"  " "  "
[7193,]  "Lee Jordan"           " "Herndon VA"      " "48"
```

Here we see the one runner who did not report an age. It appears that we have successfully captured the information from the table in *MenTxt/2012.txt*.

We wrap up the process of extracting the columns into a function so we can apply it to each year's data. This function calls our helper functions findColLocs() and selectCols(). Our function might look like

```
extractVariables =
  function(file, varNames =c("name", "home", "ag", "gun",
                             "net", "time"))
{
       # Find the index of the row with =s
  eqIndex = grep("^===", file)
       # Extract the two key rows and the data
  spacerRow = file[eqIndex]
  headerRow = tolower(file[ eqIndex - 1 ])
  body = file[ -(1 : eqIndex) ]

       # Obtain the starting and ending positions of variables
  searchLocs = findColLocs(spacerRow)
  locCols = selectCols(varNames, headerRow, searchLocs)

  Values = mapply(substr, list(body), start = locCols[1, ],
                  stop = locCols[2, ])
  colnames(Values) = varNames

  invisible(Values)
}
```

We are ready to create the data frames for each year, but the extractVariables() function expects the file passed to it for the extraction to be a character vector. We first must read the lines of the tables into R. We do this with:

```
mfilenames = paste("MenTxt/", 1999:2012, ".txt", sep = "")
menFiles = lapply(mfilenames, readLines)
names(menFiles) = 1999:2012
```

Similarly, we can read the women's results into womenFiles. These two objects, menFiles and womenFiles, are lists where each list contains 14 character vectors, one for each year. Each of these character vectors contains one string for every row in the corresponding file.

We can now apply the extractVariables() function to menFiles and womenFiles to obtain a list of character matrices. We do this for the men's list with

```
menResMat = lapply(menFiles, extractVariables)
length(menResMat)
```

```
[1] 14
```

```
sapply(menResMat, nrow)
```

```
1999 2000 2001 2002 2003 2004 2005 2006 2007
3190 3017 3622 3724 3948 4156 4327 5237 5276
2008 2009 2010 2011 2012
5905 6651 6911 7011 7193
```

We see that we get reasonable values for the number of rows in our matrices. Our next task is then to convert these character matrices into a format that we can readily analyze. As we do this, we use statistics to check the results and find that additional data cleaning is necessary. This is the topic of the next section.

2.3 Data Cleaning and Reformatting Variables

In this section, we consider how to convert the list of character matrices, `menResMat`, into an appropriate format for analysis. Currently, the data values are all character, which is not conducive to, e.g., finding the median age of the runners. However, we can easily reformat age into numeric values with the as.numeric() function. Do we want to turn the entire matrix into a numeric matrix? Not really. It doesn't make sense to try to convert the runner's name into a numeric value. For this reason, we want to create a data frame because it allows our variables to be different types. We have 6 variables: the runner's name, home town, age, and 3 versions of time. As just mentioned, we want to convert age to a numeric and leave name as character. What about the other variables? We probably want to also keep hometown as character.

Time is stored as a string in the format: `hh:mm:ss`. We want time in a numeric format so it can be more easily summarized and modeled. One possibility is to convert it to minutes, i.e., `hh * 60 + mm + ss/60`. To carry out this computation, we must split the time field up into its constituent pieces and convert each to numeric values. The strsplit() function can be very helpful for splitting strings at, e.g., colons. We also need to reconcile the 3 different recorded times (gun, net, and plain time). Net time is considered more accurate than gun time so we can simply use net time when available and otherwise use gun time or time, whichever is reported. Of course, we can keep all 3 versions of time around and let the analyst explore relationships between them and decide which to use, but we keep things simple for now and just report one time for each runner.

Before we begin converting our character strings into numeric values, we also consider whether there are any new variables we might want to create. If we are to combine all the data from the 14 years of records into one data frame, then we should keep track of the year. Likewise, if we are to combine the men's and women's results then we also want a variable that indicates the sex of the runner. These can be simply made using rep().

We begin with the task of creating the numeric variable age with as.numeric(), e.g., for the 2012 males,

```
age = as.numeric(menResMat[['2012']][ , 'ag'])
```

Note that we subsetted the list to work with the 2012 matrix and then subsetted this matrix to work with the column named 'ag'. We check a few age values with

```
tail(age)
```

```
[1] 41 39 56 35 NA 48
```

These values look reasonable, but let's check more thoroughly that our data extraction works as expected by summarizing each year's ages with

```
age = sapply(menResMat,
             function(x) as.numeric(x[ , 'ag']))
```

```
Warning messages:
1: In FUN(X[[14L]], ...) : NAs introduced by coercion
2: In FUN(X[[14L]], ...) : NAs introduced by coercion
3: In FUN(X[[14L]], ...) : NAs introduced by coercion
```

We received a warning message that our conversion of the character values for age into numeric resulted in NA values, meaning that some of the values do not correspond to numbers. We want to look into the specific cause of these messages, but first we examine age.

We create side-by-side boxplots of the yearly distribution of the age of the runners to give a quick check on the reasonableness of the values.

```
boxplot(age, ylab = "Age", xlab = "Year")
```

Figure 2.4 reveals problems with 2 years. All of the runners in 2003 were under 10 and more than 1 in 4 runners in 2006 were under 10! Clearly something has gone wrong.

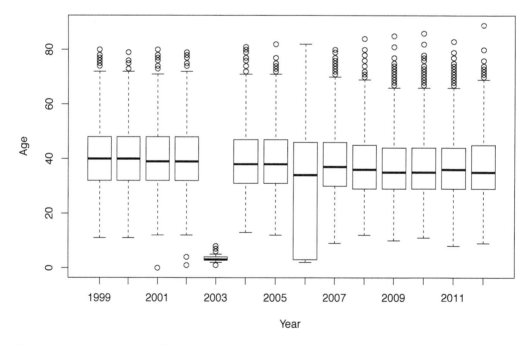

Figure 2.4: Box Plot of Age by Year. *These side-by-side boxplots of age for each race year show a few problems with the data for 2003 and 2006. The runners in these years are unusually young.*

Let's examine the original text for 2003 and 2006:

```
head(menFiles[['2003']])
```

```
[1] ""
[2] "Place Div /Tot    Num    Name                   Ag Homet..."
[3] "===== ========== =====  ====================...== == ======..."
[4] "    1    1/1999      6  John Korir                27 KEN ..."
[5] "    2    2/1999      1  Reuben Cheruiyot          28 KEN ..."
[6] "    3    3/1999      8  Gilbert Okari             24 KEN ..."
```

```
menFiles[['2006']][2200:2205]
```

```
[1] " 2192 1263/2892   1475 Matt Curtis             39 Vienna ..."
[2] " 2193   94/279    1437 Joe McCloskey           59 Columbia..."
[3] " 2194  257/590    7062 Donald Hofmann          48 Princeto..."
[4] " 2195 1264/2892   7049 Claudio Petruzziello    23 Princet..."
[5] " 2196  339/746    3319 Robert Morrison         40 South Bo..."
[6] " 2197 1265/2892   9345 Larry Cooper            32 Arlingt..."
```

We see that in 2003, the age values are shifted to the right one space in comparison to the location of the '=' characters. This means that we are picking up only the digit in the tens place. In 2006, some but not all of the rows have values that are off by one character.

We can easily solve both of these problems by including the value in the "blank" space between columns. We can do this by changing the index for the end of each variable when we perform the extraction. That is, we modify the line in selectCols() that locates the end of a column to include the blank position, i.e.,

```
c(searchLocs[index] + 1, searchLocs[index + 1])
```

When we use this revised calculation in selectCols(), we pick up the blank character after each field. This should not matter when we convert our text data to numeric and if we don't want trailing blanks in our character-valued variables, we can easily remove them with regular expressions.

In the process of confirming our conversion of age from character to numeric, we uncovered problems with our extraction process. We need to modify our helper function selectCols() from Section 2.2 to address the problem. This process is iterative as we continue to check that our data make sense. When we uncover nonsensical results, we investigate them further, which possibly leads to retracing our steps to clean up messy data.

After we modify this one line of code in selectCols() and reapply this updated version of the function to the tables of race results, we check again the summary statistics with boxplots. We find that the problem with too many young runners has cleared up (see Figure 2.5).

We now turn to the warning messages that occurred when we converted the character strings for age to numeric values. We were given several messages 'NAs introduced by coercion'. We count the number of NA values in each year with

```
sapply(age, function(x) sum(is.na(x)))
```

```
1999 2000 2001 2002 2003 2004 2005 2006 2007 2008
   1    1   61    3    2    0   13    2    5    0
2009 2010 2011 2012
   2    6    0    1
```

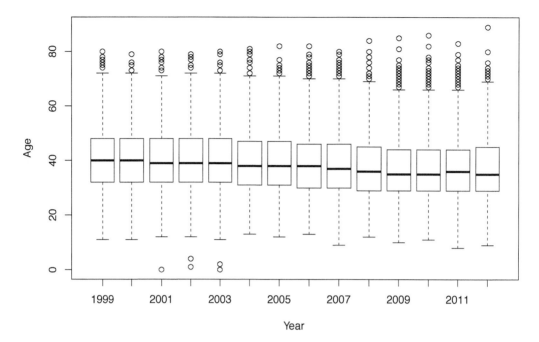

Figure 2.5: Box Plot of Age by Year. *These side-by-side boxplots of age for each race year show a reasonable age distribution. For example, the lower quartile for all years range between 29 and 32. The problems identified earlier for 2003 and 2006 have been addressed.*

In 2001, we have 61 NAs for age. We need to investigate. To make our work simpler, let's assign the 2001 ages to a vector called age2001. We do this with

```
age2001 = age[["2001"]]
```

Let's examine the original rows in the file that correspond to an NA in age2001. Recall that we dropped the header of the file before extracting the variables so we need to add an offset to the location of the NAs in age2001 in order to pick out the correct rows in the original table. We find the offset with

```
grep("^===", menFiles[['2001']])
[1] 5
```

We then find the lines in the original file that have the bad age values with

```
badAgeIndex = which(is.na(age2001)) + 5
menFiles[['2001']][ badAgeIndex ]
   [1] "                    "
   [2] "                              "
   [3] "                       "
   [4] ""
   [5] "                            "
   [6] ""
 ...
  [60] ""
  [61] "# Under USATF OPEN guideline..."
```

With one exception, all of these rows are blank/empty. The one exception is the row that corresponds to the footnote that defines the meaning of the '#' annotation. Where in the table are these rows located?

```
badAgeIndex
```

```
 [1] 1756 1757 1758 1759 1760 1761 1762 1763 1814 ...
[22] 1877 1878 1879 1930 1931 1932 1933 1934 1935 ...
[43] 2898 2899 2900 2901 2902 2903 2904 2955 2956 ...
```

These blank lines are scattered throughout the file. We can modify the extraction by checking for blank rows and removing them. The regular expression,

```
blanks = grep("^[[:blank:]]*$", menFiles[['2001']])
```

locates all rows that are entirely blank. The first argument to `grep` uses several meta characters to specify the pattern to search for. The `^` is an anchor for denoting the start of the string, the `$` anchors to the end of the string, the `[[:blank:]]` denotes the equivalence class of space and tab characters, and the `*` indicates that the blank character can appear 0 or more times. All together the pattern `^[[:blank:]]*$` matches a string that contains any number of blanks from start to end, i.e., only blank lines.

A simpler expression locates the footnote rows, i.e., rows that begin with # or *. We leave as an exercise the task of modifying extractVariables() to remove these unwanted rows. After adding this code to carry out the additional cleaning of the tables, the 61 `NA`s in 2001 are gone as well as many but not all of the other `NA`s in other the years.

Continued inspection of Figure 2.5 uncovers another problem – the minimum values for age in 2001, 2002, and 2003 remain small, i.e., close to 0. That's clearly not possible! Let's find which runners have an age under 5 and look at their records in the original table. For 2001, we have

```
which(age2001 < 5)
```

```
[1] 1377 3063 3112
```

```
menFiles[['2001']][ which(age2001 < 5) + 5 ]
```

```
[1] " 1377   5629 Steve PINKOS            0 Wash..."
[2] " 3003   5033 Jeff LAKE               0 Clar..."
[3] " 3052   5637 Greg RHODE              0 Wash..."
```

Apparently there are runners with an age entered as 0! Since these are the actual values in the table, we leave the decision as to what to do with these runners for later when we analyze the data. At this point, it appears we have successfully taken care of the creation of the age variable. However, we typically clean the variables simultaneously as an error in one variable often leads to errors in others based on position. As we clean the other variables, we may need to re-examine age to ensure that the values for age remain valid.

Next, we turn to the creation of the time variable. As mentioned at the beginning of this section, the time appears as `hh:mm:ss` and we wish to convert it to minutes. However, to carry out this computation, we must split the time field up into its constituent pieces. Also, some runners completed the race in under one hour so their times appear in a slightly different format, i.e., `mm:ss`, and we need to be able to handle both formats in our processing. For simplicity, we again start with converting the time variable for one year, say 2012. We create a vector to develop our code as follows:

```
charTime = menResMat[['2012']][, 'time']
head(charTime, 5)
```

```
[1] "   45:15 " "   46:28 " "   47:33 " "   47:34 " "   47:40 "
```

```
tail(charTime, 5)
```

```
[1] "2:27:11 " "2:27:20 " "2:27:30 " "2:28:58 " "2:30:59 "
```

We split each character string up into its parts using strsplit() with

```
timePieces = strsplit(charTime, ":")
```

The : characters are discarded in the process, and the return value from strsplit() is a list of character vectors. We have one vector for each input string, where the elements of the vector contain the pieces of the string separated by each : character. We confirm that the splitting worked properly by examining the first and last times, i.e.,

```
timePieces[[1]]
```

```
[1] "   45" "15 "
```

```
tail(timePieces, 1)
```

```
[[1]]
[1] "2" "  30" "59 "
```

We convert these elements to numeric values and combine them into one value that reports time in minutes with

```
timePieces = sapply(timePieces, as.numeric)

runTime = sapply(timePieces,
                 function(x) {
                   if (length(x) == 2) x[1] + x[2]/60
                   else 60*x[1] + x[2] + x[3]/60
                 })
```

We check our conversion with

```
summary(runTime)
```

```
   Min. 1st Qu.  Median    Mean 3rd Qu.    Max.
   45.2    77.6    87.5    88.4    97.8   151.0
```

It appears that our time conversion works. We saw earlier that the fastest runner completed the 2012 race in 45 minutes and 15 seconds, which is 45.25 minutes, and the slowest completed it in 2 hours 30 minutes and 59 seconds, which is nearly 151 minutes. We leave it as an exercise to encapsulate this conversion into a function called convertTime().

Let's wrap these conversions into a function to apply to the character matrices in menResMat and return a data frame with variables for analysis. We call this function createDF(). In addition to the conversion of character strings to numeric, we also create two new variables, year and sex. To do this, we must have input arguments to tell us which year we are cleaning and whether the results are for men or women. Lastly, we also choose which time variable to include in the data frame from among the 3 available, with a preference for net time. The function appears as

```
createDF =
function(Res, year, sex)
{
        # Determine which time to use
  useTime = if( !is.na(Res[1, 'net']) )
              Res[ , 'net']
            else if( !is.na(Res[1, 'gun']) )
              Res[ , 'gun']
            else
              Res[ , 'time']

  runTime = convertTime(useTime)

  Results = data.frame(year = rep(year, nrow(Res)),
                       sex = rep(sex, nrow(Res)),
                       name = Res[ , 'name'],
                       home = Res[ , 'home'],
                       age = as.numeric(Res[, 'ag']),
                       runTime = runTime,
                       stringsAsFactors = FALSE)
  invisible(Results)
}
```

We apply our new function, createDF(), to all of the male results as follows:

```
menDF = mapply(createDF, menResMat, year = 1999:2012,
               sex = rep("M", 14), SIMPLIFY = FALSE)

There were 50 or more warnings
(use warnings() to see the first 50)
```

We check the warnings we received with

```
warnings()[ c(1:2, 49:50) ]

Warning messages:
1: In lapply(X = X, FUN = FUN, ...) : NAs introduced by coercion
2: In lapply(X = X, FUN = FUN, ...) : NAs introduced by coercion
3: In lapply(X = X, FUN = FUN, ...) : NAs introduced by coercion
4: In lapply(X = X, FUN = FUN, ...) : NAs introduced by coercion
```

It is likely that the conversion problems are coming from the conversion of time from a character string into minutes because we have already handled the conversion of age. We can check the number of NA values for runTime with

```
sapply(menDF, function(x) sum(is.na(x$runTime)))

1999 2000 2001 2002 2003 2004 2005 2006 2007 2008 2009
   0    0    0    0    0    0    0 5232   83    0  164
2010 2011 2012
  68    0    0
```

There are a large number of NAs in 2007, 2009, and 2010, and it appears that all of the run time values for 2006 are NA.

Let's begin by examining a few of the records in 2007, 2009, and 2010 that have an NA in run time. We find that these are caused by runners who completed half the race but have no final times and by runners who have a footnote after their time, e.g.,

```
"    1       1/54        13 Tadesse Tola                19
          Ethiopia               46:01#  4:37       28:47   "
" 5273    309/309     16370 Stephen Peterson             57
          Washington DC                 #           1:36:29 "
```

We can easily modify createDF() to eliminate the footnote symbols (# and *) from the times and drop records of runners who do not complete the race. These revisions are

```
         # Remove # and * and blanks from time
  useTime = gsub("[#\\*[:blank:]]", "", useTime)

         # Drop rows with no time
  Res = Res[ useTime != "", ]
  runTime = convertTime(useTime[ useTime != "" ])
```

After we apply this revised function to menResMat to create our data frame, all missing values in time are gone, except for 2006.

```
sapply(menDF, function(x) sum(is.na(x$runTime)))
```

```
1999 2000 2001 2002 2003 2004 2005 2006 2007 2008 2009
   0    0    0    0    0    0    0 5232    0    0    0
2010 2011 2012
   0    0    0
```

Close inspection of the header for the 2006 file reveals the problem, but for brevity's sake, we leave that problem to the exercises as well.

At last, we combine the race results for all years and men into one data frame using the do.call() function to call rbind() with the list of data frames as input. That is,

```
cbMen = do.call(rbind, menDF)
save(cbMen, file = "cbMen.rda")
```

The do.call() function is very convenient when we have the individual arguments to a function as elements of a list. For example, the rbind() function's first argument is ..., i.e.,

```
args(rbind)
function (..., deparse.level = 1)
```

The ... argument allows the caller to provide an arbitrary number of arguments that, in the case of rbind(), are bound together into one object. We could call rbind() as follows:

```
rbind(menDF[[1]], menDF[[2]], menDF[[3]], menDF[[4]],
      menDF[[5]], menDF[[6]], menDF[[7]], menDF[[8]],
      menDF[[9]], menDF[[10]], menDF[[11]], menDF[[12]],
      menDF[[13]], menDF[[14]])
```

That's a lot of typing and it requires us to know that there are 14 data frames in `menDF`. With do.call() we supply the inputs to rbind() as a list and do.call() puts together the rbind() function call for us.

We check the dimension of our amalgamated data frame with

```
dim(cbMen)
```

```
[1] 70070     6
```

We also examine a summary of the variables in `cbMen` to check whether any problems arose, e.g., with coercion, in the binding together of the data frames.

Over these 14 years, 70,070 male runners completed the Cherry Blossom race. In addition, more than 70,000 female runners completed the race. We leave it as an exercise to handle the women's race results. In the next section we take a closer look at the race results.

2.4 Exploring the Run Time for All Male Runners

Now that we have completed the extraction of our data from the tables published on the Cherry Blossom Web site, we can begin to study the relationship between age and run time. Typically, we first examine our data graphically in a scatter plot with run time on the y-axis and age on the x-axis. We can make such a scatter plot for the male runners with the following call to plot()

```
plot(runTime ~ age, data = cbMen, ylim = c(40, 180),
     xlab = "Age (years)", ylab = "Run Time (minutes)")
```

Here we included *ylim* to screen out the runner with a run time of 1.5 minutes.

The first argument in this call to plot() is an *R* formula. The formula language is very powerful as it can be used to succinctly express complex relationships and a variety of *R* functions can interpret a formula and carry out an analysis appropriate for the data. In our case, the formula is very simple, `runTime ~ age`, and it indicates that we are interested in how `runTime` depends on, or varies with, `age`. The plot() function builds the visual model based on the representation of the data. Since `runTime` and `age` are both numeric variables, plot() makes a scatter plot with `runTime` on the y-axis and `age` on the x-axis. Later in this section, we see other formulas containing more variables including categorical variables, and we see formulas used with other functions such as lm() and loess().

The resulting plot appears in Figure 2.6. Most of the points appear as a black blob in the scatter plot because so many points have been plotted on top of each other. The shape of the distribution is obscured because we cannot see which regions of the (age, run time) space are more densely populated. Notice also the vertical stripes in the plot. These are the result of runner's age being reported to the nearest year, which results in more over plotting. In the next section, we consider a few alterations to this default scatter plot that address the problem of over plotting.

2.4.1 Making Plots with Many Observations

There are several modifications we can make to the plot in Figure 2.6 to ameliorate the effect of over plotting. We can reduce the size of the plotting symbol, use transparent colors for the plotting symbol, and add a small amount of random noise to the age variable.

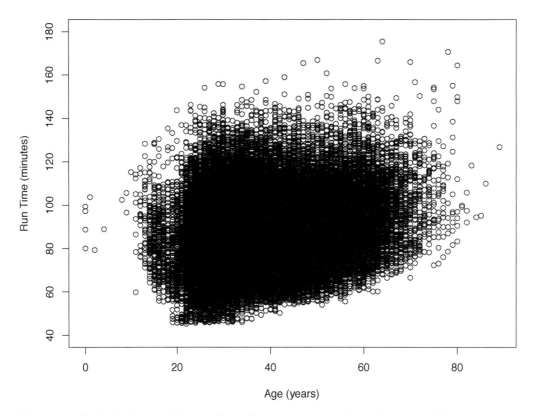

Figure 2.6: Default Scatter Plot for Run Time vs. Age for Male Runners. *This plot demonstrates that a simple scatter plot of run time by age for the 70,000 male runners leads to such severe over plotting that the shape of the data is not discernible.*

Alternatively, we can create a plot that reveals a smoothed version of the density of the points in each region. We can also make a series of boxplots instead of a scatter plot. We demonstrate each of these approaches in this section.

We first modify the call to plot() to change the plotting symbol from a circle to a disk, and we shrink the size of the disk as well. We also use a transparent blue as the color for the disk. If we use a transparent color for the plotting symbol, then when two symbols overlap, their color appears darker. This way, regions with a higher density of observations appear darker than low density areas.

Colors can be specified in many ways in R and other systems too. The RGBA specification provides a triple of red-blue-green components that combine to make a color. The fourth component in the RGBA specification provides the amount of transparency. For our color, we choose one from Cindy Brewer's color palettes that are available in the RColorBrewer package [2]. We load the package and display the objects in the package with

```
library(RColorBrewer)
ls("package:RColorBrewer")

[1] "brewer.pal"         "brewer.pal.info"
[3] "display.brewer.all" "display.brewer.pal"
```

These are the names of the 4 functions available in the package. After reading the help information on display.brewer.all(), we see that it is a good starting place because it displays all of the palettes available in the package. We call it with

```
display.brewer.all()
```

and choose the blue in the Set3 palette as follows

```
Purples8 = brewer.pal(9, "Purples")[8]
Purples8
```

```
[1] "#54278F"
```

The color is stored in hexadecimal format, where red is 54, blue is 27, and green is 8F. This color does not include an alpha transparency, which means that it is an opaque color. However, we can create a transparent version of this color by pasting a transparency value to the end of Purples8 with

```
Purples8A = paste(Purples8, "14", sep = "")
```

We use this color for our plotting symbol.

Additionally, we change the ages of the runners by a small random amount between -0.5 and 0.5. This operation is called jittering, and we jitter the age values with jitter(age, amount = 0.5).

Our resulting plot appears in Figure 2.7. This plot is much improved from the initial one in Figure 2.6. We can see where the bulk of the runners are, including what appears to be a slight upward curvature in run time as age increases and a skew distribution of run time given age. We can also see the small group of runners with very fast run times. We leave the creation of this plot as an exercise.

The smoothScatter() function provides a more formal approach to jittering and using transparency for visualizing the density of runner's time-age distribution. This function produces a smooth density representation of the scatter plot using color, much like in Figure 2.7, but with a more statistical approach to building regions that vary by color intensity. With smoothScatter(), the color at an (x, y) location is determined by the density of points in a small region around that point. This averaging process yields a smoother plot with dark shades corresponding to high density regions. We call smoothScatter() with cbMen as follows:

```
smoothScatter(y = cbMen$runTime, x = cbMen$age,
              ylim = c(40, 165), xlim = c(15, 85),
              xlab = "Age (years)", ylab = "Run Time (minutes)")
```

The resulting plot in Figure 2.8 shows a very similar shape to our plot in Figure 2.7. It has the addition of small black dots to indicate individual points that are far from the main point cloud.

A very different approach to these scatter plots is to graphically display summary statistics of run time for subgroups of runners with roughly the same age. Here, we group the runners into 10-year age intervals and plot the summaries for each subgroup in the form of a boxplot (see Figure 2.9). With these side-by-side boxplots, the size of the data does not obscure the main features, e.g., the quartiles and tails for an age group. To make these boxplots, we categorize age using the cut() function. We first remove those runners under 15 or who have unrealistic run times with

```
cbMenSub = cbMen[cbMen$runTime > 30 &
                !is.na(cbMen$age) & cbMen$age > 15, ]
```

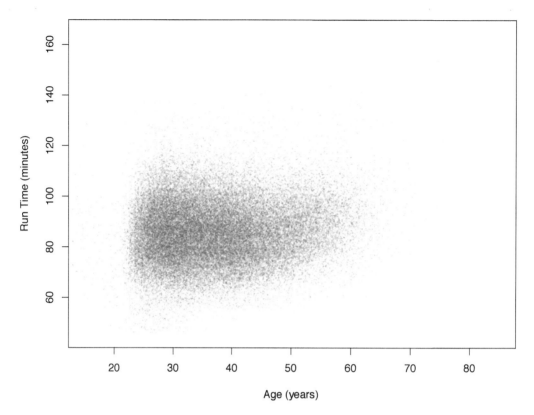

Figure 2.7: Revised Scatter Plot of Male Runners. *This plot revises the simple scatter plot of Figure 2.6 by changing the plotting symbol from a circle to a disk, reducing the size of the plotting symbol, using a transparent color for the disk, and adding a small amount of random noise to age. Now we see the shape of the high density region containing most of the runners and the slight upward trend of time with increasing age.*

The we categorize age with

```
ageCat = cut(cbMenSub$age, breaks = c(seq(15, 75, 10), 90))
table(ageCat)
```

```
ageCat
(15,25] (25,35] (35,45] (45,55] (55,65] (65,75] (75,90]
   5804   25434   20535   12212    5001     751      69
```

This new variable, `ageCat`, is a factor that categorizes age into 10-year intervals with the exception of all of those over 75 being lumped together into one interval.

We use the formula `runTime ~ ageCat` in the call to plot() as follows:

```
plot(cbMenSub$runTime ~ ageCat,
     xlab = "Age (years)", ylab = "Run Time (minutes)")
```

We see in Figure 2.9 that the plot() function has created a series of boxplots rather than a scatter plot. The difference between this function call and the earlier one that produced Figure 2.6 is in the formula provided. Since `ageCat` is a factor, the default plot for the

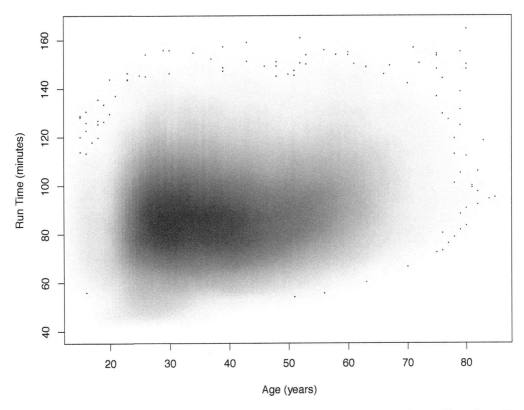

Figure 2.8: *Smoothed Scatter Plot of Male Runners Race Times vs. Age.* This plot offers an alternative to the scatter plot of Figure 2.7 that uses jittering and transparent color to ameliorate the over plotting. Here there is no need to jitter age because the smoothing action essentially does that for us by spreading an individual runner's (age, run time) pair over a small region. The shape of the high density region has a very similar shape to the earlier plot.

formula `time ~ ageCat` is a series of side-by-side boxplots with one boxplot of run time per level of the age factor. We observe in this plot that the upper quartile increases faster with age than the median and lower quartile. In the next section, we try summarizing this relationship between age and run time more formally.

2.4.2 Fitting Models to Average Performance

As seen in Figure 2.9, the average performance seems to curve upward with age. A simple linear model may be inadequate to describe this relationship. To see how well the simple linear model captures the relationship (or not) between run time and age, we fit the model with

```
lmAge = lm(runTime ~ age, data = cbMenSub)
```

Here again we use R's formula language to express the relationship we want fitted to the data. Our formula `runTime ~ age` indicates we want to fit time as a function of age. The lm() function performs least squares to find the best fitting line to our data, which we see has the following intercept and slope:

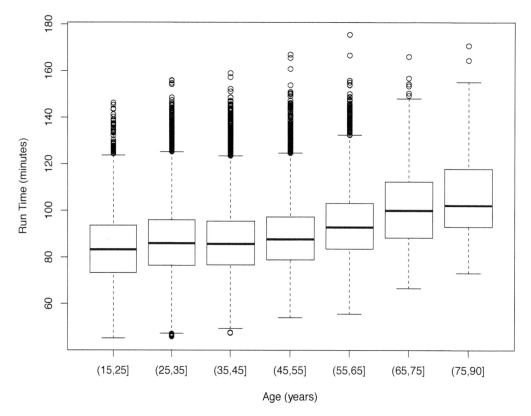

Figure 2.9: Side-by-Side Boxplots of Male Runners' Run Time vs. Age. *This sequence of boxplots shows the quartiles of time for men grouped into 10-year age intervals. As age increases, all the quartiles increase. However, the box becomes asymmetrical with age, which indicates that the upper quartile increases faster than the median and lower quartile.*

```
lmAge$coefficients
```

```
(Intercept)         age
      78.76        0.23
```

We have assigned the return value from lm() to lmAge. This object contains the coefficients from the fit, predicted values, residuals, and other information about the linear least squares fit of run time to age. We can retrieve a brief summary of the fit with a call to summary() as follows:

```
summary(lmAge)
```

```
Call:
lm(formula = runTime ~ age, data = cbMenSub)

Residuals:
   Min     1Q Median     3Q    Max
-40.33 -10.22  -0.95   9.10  82.42

Coefficients:
```

```
              Estimate Std. Error t value Pr(>|t|)
(Intercept)  78.75672    0.20770    379.2   <2e-16 ***
age           0.22529    0.00517     43.6   <2e-16 ***
---
Signif. codes:  0 '***' 0.001 '**' 0.01 '*' 0.05 '.' 0.1 ' ' 1

Residual standard error: 15 on 69804 degrees of freedom
Multiple R-squared:  0.0265,      Adjusted R-squared:  0.0265
F-statistic: 1.9e+03 on 1 and 69804 DF,  p-value: <2e-16
```

Note that the summary() function does not produce the typical quantiles, extreme values, etc. This is because we have passed it an lm object, i.e.,

```
class(lmAge)
```

```
[1] "lm"
```

The summary method for class lm provides a different set of summary statistics that are more appropriate for a fitted linear model.

To help us assess how well the simple linear model fits the data we plot the residuals against age. As with the original scatter plot of run time against age, we need to address the issue of over plotting. We use smoothScatter() to do this. Further, to help us see any curvature in the residuals, we add to the plot a horizontal line at 0. We do this with

```
smoothScatter(x = cbMenSub$age, y = lmAge$residuals,
              xlab = "Age (years)", ylab = "Residuals")
abline(h = 0, col = "purple", lwd = 3)
```

To help us further discern any pattern in the residuals, we augment this residual plot with a smooth curve of local averages of the residuals from the fit. That is, for a particular age, say 37, we take a weighted average of the residuals for those runners with an age in a small neighborhood of 37. Such a locally fitted curve allows us to better see deviations in the pattern of residuals. We fit the curve using loess() with

```
resid.lo = loess(resids ~ age,
                 data = data.frame(resids = residuals(lmAge),
                                   age = cbMenSub$age))
```

Notice that the loess() function also accepts a formula object to describe the relationship to fit to the data. Here we request a fit of the resids variable to age. The data frame provided via the parameter *data* contains these two variables; it may contain others as well, but we have created this data frame specially so that it has only the residuals from lmAge and the ages of runners in cbMenSub. Similar to lm(), the return value from loess() is a special object that contains fitted values as well as other relevant information about the curve fitted to the data.

To add the fitted curve to the smooth scatter of the residuals, we can predict the average residual for each year of age and then use lines() to "connect the dots" between these predictions to form an approximation of the fitted curve. We start by making a vector of age values from 20 to 80 with

```
age20to80 = 20:80
```

Now, if we have the predicted average residual for each of these ages in a vector called, say, resid.lo.pr, then we can add the curve to the smooth scatter with

```
lines(x = age20to80, y = resid.lo.pr,
      col = "green", lwd = 3, lty = 2)
```

We can obtain these predicted values from the predict.loess() function. This function takes the loess object from a fit, e.g., resid.lo and a data frame with variables matching those used in the loess curve fitting, in this case age. That is, we create resid.lo.pr with

```
resid.lo.pr =
  predict(resid.lo, newdata = data.frame(age = age20to80))
```

Notice that we called predict() rather than predict.loess(). The predict() function is a wrapper that allows us to write code that is not dependent on the form of the fit. It takes an object returned from a fit, such as that returned from lm() or loess(), and depending on which class of object it is, predict() calls the relevant function, i.e., predict.lm() for lm objects and predict.loess() for loess objects.

The augmented smoothed scatter plot appears in Figure 2.10. We see that the simple linear model tends to underestimate the run time for men over 60. This confirms our observations from the boxplot and smooth scatter plot of the nonlinear trend in run time. The simple linear model is not able to capture the change in performance with age.

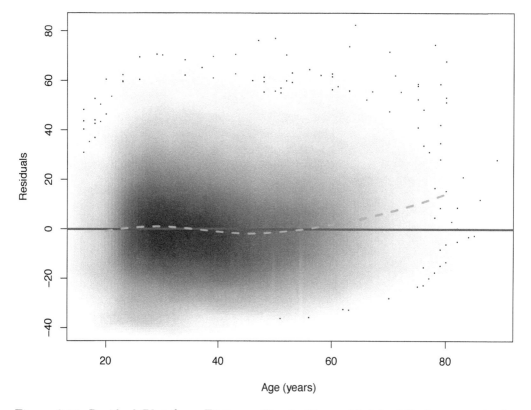

Figure 2.10: Residual Plot from Fitting a Simple Linear Model of Performance to Age. Shown here is a smoothed scatter plot of the residuals from the fit of the simple linear model of run time to age for male runners who are 15 to 80 years old. Overlaid on the scatter plot are two curves. The "curve" in purple is a solid horizontal line at $y = 0$. The green dashed curve is a local smooth of the residuals.

We consider two approaches to a more complex fit: a piecewise linear model and a nonparametric smooth curve. For the latter, we simply take local weighted averages of time as age varies, just as we smoothed the residuals from the linear fit. We use loess() again to do this with

```
menRes.lo = loess(runTime ~ age, cbMenSub)
```

and we make predictions for all ages ranging from 20 to 80 with

```
menRes.lo.pr = predict(menRes.lo, data.frame(age = age20to80))
```

The curve appears in Figure 2.11.

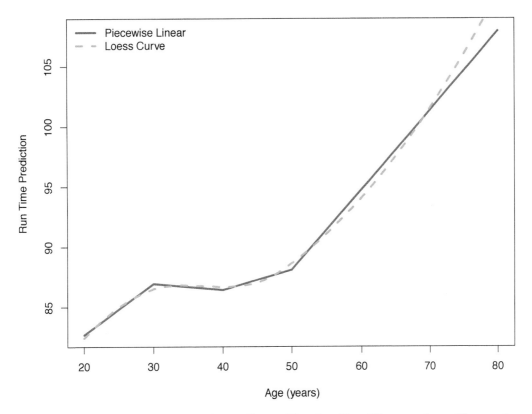

Figure 2.11: Piecewise Linear and Loess Curves Fitted to Run Time vs. Age. *Here we have plotted the fitted curves from loess() and a piecewise linear model with hinges at 30, 40, 50, and 60. These curves follow each other quite closely. However, there appears to be more curvature in the over 50 loess fit that is not captured in the piecewise linear fit.*

Next we fit a piecewise linear model, which consists of several connected line segments. This is similar to the idea of the locally smoothed curve from loess() in that it allows us to bend the line at certain points to better fit the data. The difference is that the fit must be linear between the hinges. We place hinges at 30, 40, 50, and 60 and thus allow the slope of the line to change at these decade markers. The fitted "curve" appears in Figure 2.11.

How do we fit such a model to our data? Before we fit the full piecewise model, we consider a simpler model with one hinge at 50. We first create an over50 variable that takes on the value 0 for ages 50 and under and otherwise holds the number of years over 50, e.g., 1 for someone who is 51, 2 for someone who is 52, and so on. If our fit is $a + b \times age + c \times over50$ then for an age below 50 this is simply $a + b \times age$ and for an age over 50 it is equivalent to $(a - 50c) + (b + c)age$. We see that the coefficient c is the change in the slope from below 50 to above 50, and the intercept makes the line segments connect.

Our first task then is to create this over50 variable. We use the pmax() function, which performs an element-wise or "parallel" maximum. We find the maximum of each element of menRes$age - 50 and 0 with

```
over50 = pmax(0, cbMenSub$age - 50)
```

We then fit this augmented model as follows

```
lmOver50 = lm(runTime ~ age + over50, data = cbMenSub)

summary(lmOver50)

Call:
lm(formula = runTime ~ age + over50, data = cbMenSub)

Residuals:
   Min     1Q  Median     3Q    Max
-40.27 -10.10  -0.88   9.06  79.04

Coefficients:
            Estimate Std. Error t value Pr(>|t|)
(Intercept) 82.75489    0.26504   312.2   <2e-16 ***
age          0.10569    0.00715    14.8   <2e-16 ***
over50       0.56387    0.02337    24.1   <2e-16 ***
---
Signif. codes:  0 '***' 0.001 '**' 0.01 '*' 0.05 '.' 0.1 ' ' 1

Residual standard error: 15 on 69803 degrees of freedom
Multiple R-squared:  0.0345,      Adjusted R-squared:  0.0345
F-statistic: 1.25e+03 on 2 and 69803 DF,  p-value: <2e-16
```

Now the slope of the line for those under 50 is less steep than in our original simple linear model, and for ages over 50, the model indicates the average man slows by 0.67 minutes over the entire race, which is an additional 0.56 minutes a year compared to those under fifty.

We can create the over30, over40, etc. variables as follows:

```
decades = seq(30, 60, by = 10)
overAge = lapply(decades,
                 function(x) pmax(0, (cbMenSub$age - x)))
names(overAge) = paste("over", decades, sep = "")
overAge = as.data.frame(overAge)
tail(overAge)

      over30 over40 over50 over60
69801     36     26     16      6
69802     11      1      0      0
```

```
69803      9     0    0    0
69804     26    16    6    0
69805      5     0    0    0
69806     18     8    0    0
```

Now that we have each of these variables, we can create the model,

```
runTime ~ age + over30 + over40 + over50 + over60
```

This model has an interpretation similar to the model with just `age` and `over50`. That is, the coefficient for, say, `over40` is the change in the slope for ages in (30, 40] to ages in (40, 50]. We find the least squares fit with

```
lmPiecewise = lm(runTime ~ . ,
                 data = cbind(cbMenSub[, c("runTime", "age")],
                              overAge))
```

Here we have used the . in the formula to indicate that the model should include all of the variables in the data frame (other than `runTime`) as covariates.

When we call summary() with the `lm` object `lmPiecewise`, we obtain the coefficients, their standard errors, and other summary statistics for the fit:

```
summary(lmPiecewise)
Call:
lm(formula = time ~ .,
   data = cbind(menRes[, c("time", "age")], overAge))

Residuals:
   Min     1Q Median     3Q    Max
-40.92 -10.12  -0.89   9.02  78.96

Coefficients:
            Estimate Std. Error t value Pr(>|t|)
(Intercept) 74.2286     0.9153   81.10  < 2e-16 ***
age          0.4243     0.0332   12.78  < 2e-16 ***
over30      -0.4770     0.0478   -9.98  < 2e-16 ***
over40       0.2216     0.0407    5.45  5.1e-08 ***
over50       0.4944     0.0529    9.34  < 2e-16 ***
over60      -0.0036     0.0776   -0.05     0.96
---
Signif. codes:  0 '***' 0.001 '**' 0.01 '*' 0.05 '.' 0.1 ' ' 1

Residual standard error: 15 on 69800 degrees of freedom
Multiple R-squared:  0.0359,     Adjusted R-squared:  0.0359
F-statistic:  520 on 5 and 69800 DF,  p-value: <2e-16
```

Notice that the coefficient for `over60` is essentially 0, meaning that those over 60 do not slow down any faster than those in their fifties, i.e., about 0.494 minutes more per year for each year over 50 for a total of about 0.66 minutes for the 10-mile race per year.

How do we plot this piecewise linear function that we have fitted? As with the loess curve, we can use predict() to provide fitted values for each age value from 20 to 80. However, we need to provide predict() with all of the covariates used in making the fit, i.e., `age`, `over30`, `over40`, `over50`, and `over60`. We can create a data frame of these covariates just as we did for the full data set as follows:

```
overAge20 = lapply(decades, function(x) pmax(0, (age20to80 - x)))
names(overAge20) = paste("over", decades, sep = "")
overAgeDF = cbind(age = data.frame(age = age20to80), overAge20)

tail(overAgeDF)
```

```
   age over30 over40 over50 over60
75  75     45     35     25     15
76  76     46     36     26     16
77  77     47     37     27     17
78  78     48     38     28     18
79  79     49     39     29     19
80  80     50     40     30     20
```

Then we call predict() passing it the lm object, i.e., lmPiecewise, with the details of the fit and also the covariates to use to make the predictions, i.e., overAgeDF. That is, we call predict() with

```
predPiecewise = predict(lmPiecewise, overAgeDF)
```

We plot this fitted piecewise linear function with

```
plot(predPiecewise ~ age20to80,
     type = "l", col = "purple", lwd = 3,
     xlab = "Age (years)", ylab = "Run Time Prediction")
```

And we add the loess curve with

```
lines(x = age20to80, y = menRes.lo.pr,
      col = "green", lty = 2, lwd = 3)
legend("topleft", col = c("purple", "green"),
       lty = c(1, 2), lwd= 3,
       legend = c("Piecewise Linear", "Loess Curve"), bty = "n")
```

The two fitted curves appear in Figure 2.11. We see that they follow each other quite closely. The main deviation is in the over 70 group. We did not include a hinge at 70 so our fitted model is unable to capture the sharper increase for those over 70. We may want to consider adding this additional hinge to our model to see if it improves the fit. It may seem that we have made great progress in modeling the average performance, but we must interpret these results with care. For example, suppose, as seems likely, that younger runners who are slow tend to drop out of racing as they age so older runners who do participate are those who tend to be faster. This can bias our estimate of how running speed changes with age. Additionally, these data consist of 14 cross-sectional snapshots of runners. We might ask ourselves whether or not the composition of the participants has changed over this time period. These concerns are the topics of the next two sections.

2.4.3 Cross-Sectional Data and Covariates

In our earlier analysis, we examined the average performance for runners of different ages. That is, we looked at average performance for, e.g., 30-39 year olds and 40-49 year olds in the Cherry Blossom road race. However, we have not seen how a runner's performance changes as he or she ages. These two groups (30-39 and 40-49 year olds) are composed of different people and if these groups of people differ from each other in some significant ways, e.g., those in their 30s are more likely to be world class runners and those in their 40s are more likely to be local amateur athletes, then we might be misled by comparing these two group's average performances. To further complicate the matter, we have data from 14 different races so we are also averaging across the participants in these different races. We expect the average performances to be the same across the years. However, each year we have a self-selected group of participants, and we might wonder whether the composition of the participants has changed over the years. If it has, that could further complicate inference.

We know that the Cherry Blossom 10-mile run has been increasingly popular. Figure 2.12 indicates that the number of male runners has more than doubled over the 14 years. It seems reasonable to question if the demographics of the participants have changed over this time period.

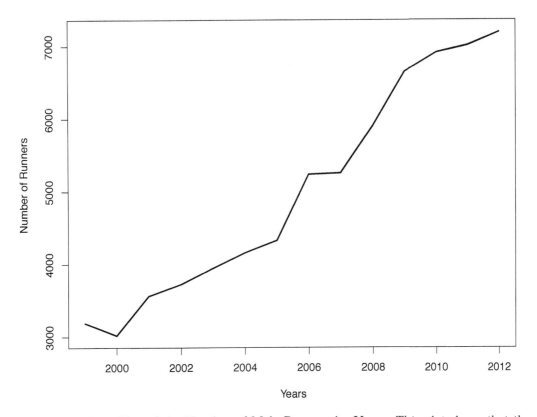

Figure 2.12: Line Plot of the Number of Male Runners by Year. *This plot shows that the number of male runners in the Cherry Blossom 10-mile race has more than doubled from 1999 to 2012.*

Historically, the race was used as a preparation for the Boston Marathon. The fastest runners in the Cherry Blossom primarily come from Ethiopia, Kenya, and Tanzania. And, their times are within a minute or two of the world record of 44:24 set in 2005 by Haile Gebrselassie from Ethiopia, who was 32 at the time (see http://inglog.com/tools/world-records/). Professional runners continue to compete in the Cherry Blossom road race.

Let's compare the distribution of performance for the earliest and latest years, i.e., the 1999 and 2012 races. We see below that while the fastest man has gotten faster from 1999 to 2012, the quartiles of the 2012 distribution are each about 3 minutes slower compared to 1999:

```
summary(cbMenSub$runTime[cbMenSub$year == 1999])
```

```
   Min. 1st Qu.  Median    Mean 3rd Qu.    Max.
   47.0    74.8    84.3    84.3    93.1   171.0
```

```
summary(cbMenSub$runTime[cbMenSub$year == 2012])
```

```
   Min. 1st Qu.  Median    Mean 3rd Qu.    Max.
   45.2    77.6    87.5    88.4    97.8   151.0
```

Could it be that the men competing in 2012 are older and therefore slower than their counterparts in 1999? We can compare the age distributions of the runners in the two races.

For simplicity, we make two vectors of age for the 1999 and 2012 runners with

```
age1999 = cbMenSub[ cbMenSub$year == 1999, "age" ]
age2012 = cbMenSub[ cbMenSub$year == 2012, "age" ]
```

We next superpose the density curves for the two sets of ages. We do this as follows:

```
plot(density(age1999, na.rm = TRUE),
     ylim = c(0, 0.05), col = "purple",
     lwd = 3,  xlab = "Age (years)",  main = "")
lines(density(age2012, na.rm = TRUE),
      lwd = 3, lty = 2, col="green")
legend("topleft", col = c("purple", "green"), lty= 1:2, lwd = 3,
       legend = c("1999", "2012"), bty = "n")
```

Note that the first time we made this plot we used the default horizontal and vertical axis limits in the call to plot(). We found that the vertical axis was not large enough to include the peak of the second density when we added it to the first density plot so we remade this plot and specified `ylim = c(0, 0.05)` to accommodate the higher peak of the 2012 density. We can also use the densityplot() function in the lattice package [5], and this function automatically scales the axes correctly for all the density curves plotted. Nonetheless, typically the visualization process is iterative. We make plots using the default settings for most of the arguments, and as we uncover interesting structure, we remake the plots to adjust the scale and to add information, e.g., axis labels, titles, legends, color, line type and thickness, etc.

The density curves in Figure 2.13 are surprising. The males in 2012 are not older. In fact, the opposite is the case. There are many more younger men in 2012 in comparison to 1999, as evidenced by the sharp peak in the 2012 distribution at about 30. We can also compare these two distributions with a quantile-quantile plot. We leave this as an exercise. The difference in performance between 1999 and 2012 is more subtle than having an aging population of runners. We need to control the covariates, age and year, simultaneously when we analyze race performance.

In the previous section, we saw how the average performance was flat for runners in their 30s and rose slightly in the 40s and more sharply in the 50s and 60s. We make separate smooth curves of time versus age for the 1999 and 2012 runners and plot them together as follows:

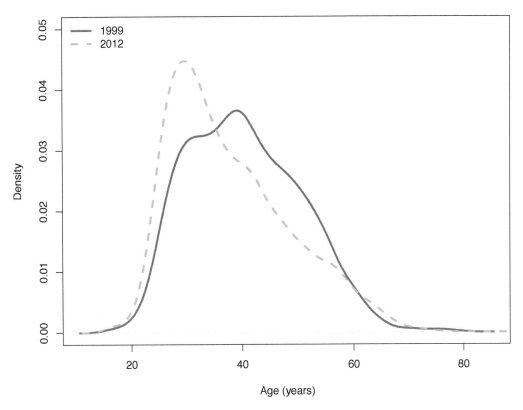

Figure 2.13: Density Curves for the Age of Male Runners in 1999 and 2012. *These two density curves have quite different shapes. The 1999 male runners have a broad, nearly flat mode where they are roughly evenly distributed in age from 28 to 45. In contrast, the 2012 runners are younger with a sharper peak just under 30 years and a skew right distribution.*

```
mR.lo99 = loess(runTime ~ age, cbMenSub[ cbMenSub$year == 1999,])
mR.lo.pr99 = predict(mR.lo99, data.frame(age = age20to80))

mR.lo12 = loess(runTime ~ age, cbMenSub[ cbMenSub$year == 2012,])
mR.lo.pr12 = predict(mR.lo12, data.frame(age = age20to80))

plot(mR.lo.pr99 ~ age20to80,
     type = "l", col = "purple", lwd = 3,
     xlab = "Age (years)", ylab = "Fitted Run Time (minutes)")

lines(x = age20to80, y = mR.lo.pr12,
      col = "green", lty = 2, lwd = 3)

legend("topleft", col = c("purple", "green"), lty = 1:2, lwd = 3,
       legend = c("1999", "2012"), bty = "n")
```

We see in Figure 2.14 that the two curves are similar in shape but the curve for 2012 is higher than the 1999 curve. There appears to be a consistent difference between these two groups of runners. Figure 2.15 shows the difference in predicted run times for these two

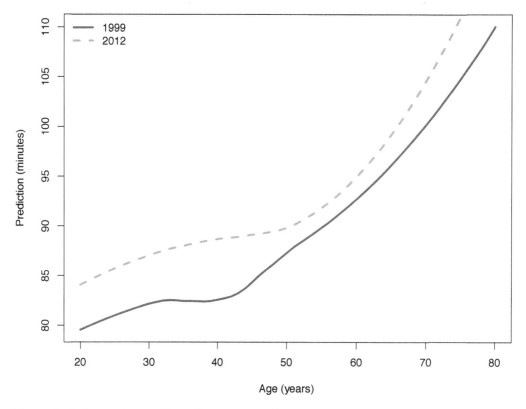

Figure 2.14: Loess Curves Fit to Performance for 1999 and 2012 Male Runners. *This loess fit of run time to age for 2012 male runners sits above the fit for 1999 male runners. The gap between these curves is about 5 minutes for most years. The exception is in the late 40s to early 60s where the curves are within 2–3 minutes of each other. Both curves have a similar shape.*

curves. This difference narrows to 2 minutes for men in their 50s and gradually widens for men in their 60s, 70s, and 80s from 2.5 to 8.5 minutes. We leave it as an exercise to compare the run time age relationship for all 14 years of data.

We mention one last idea for comparing these two distributions of runners, and we leave it to the exercises to carry out this comparison. In track, there is a performance standard called age grading that measures an individual's performance based on his or her age. It normalizes the individual's run time by the world record for that distance for that age group [1]. Since the fastest runners in the Cherry Blossom road race perform close to the world record, we might use the fastest runner in each age category to normalize the times. To minimize the year-to-year fluctuations, we can smooth the fastest times and use these smoothed times to normalize each runner's time. When we do this, we find the age graded performances roughly follow the Normal distribution. However, the 1999 runners tend to be better than their 2012 counterparts as evidenced by the peak at 1.4 rather than 1.5 and a smaller IQR.

The run time distribution appears to have changed over the years, and this points out the main issue with cross-sectional studies. However, there is an advantage to having 14 years' worth of race results. It is possible that some runners have participated in the race

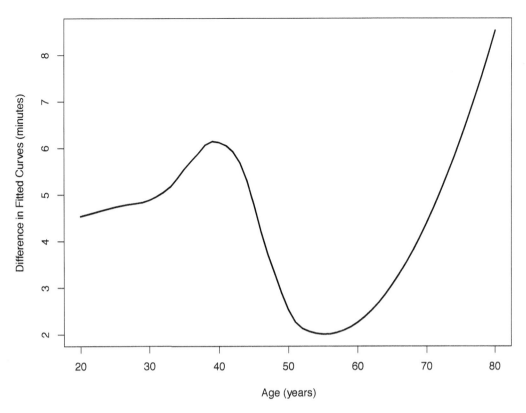

Figure 2.15: *Difference between Loess Curves. This line plot shows the difference between the predicted run time for 2012 and 1999 male runners.*

over several years and we can study how each runner's performance changes as he or she grows older. In order to do this, we need to connect runners across the years.

2.5 Constructing a Record for an Individual Runner across Years

We want to match records from runners who have participated in more than one Cherry Blossom run. The race results do not have unique identifiers for each person so we need to construct these from the information we have on each race entrant. Ideally we use all of the information, i.e., the runner's name, home, age, race time, and the year of the race. However, if this information is reported inconsistently from one year to the next, then this can reduce the number of matches. On the other hand, even using all of this information we may be incorrectly matching records from two different athletes. Whatever approach we devise will not be completely accurate, and the purpose of this section is to investigate several possibilities and settle on one that we think is reasonable.

We consider the following questions:

- How many entrants are there over the 14 years?
- How many unique names are there among these entrants?

- How many names appear twice, 3 times, 4 times, etc. and what name occurs most often?
- How often does a name appear more than once in a year?

Answering these questions gives us a sense of the magnitude of the matching problem. Additionally, we consider how to improve the matching by cleaning the name and home values. For example, recall that we picked up some trailing blanks when we parsed the text tables. Now might be a good time to eliminate them. We also noted earlier that capitalization was inconsistent. This can prove problematic for matching records. Other issues with cleaning the character strings crop up as we begin to examine the records more carefully. Before we answer these questions let's clean the names.

Any blanks appearing before or after a name can be dropped. Also, if there are multiple blanks between, e.g., the first and last name, we can convert them to one blank. The gsub() function is helpful here. We create a helper function, trimBlanks() to do this as follows:

```
trimBlanks = function(charVector) {
  nameClean = gsub("^[[:blank:]]+", "", charVector)
  nameClean = gsub("[[:blank:]]+$", "", nameClean)
  nameClean = gsub("[[:blank:]]+", " ", nameClean)
}
```

The first substitution eliminates all beginning blanks, the second all trailing blanks, and the third substitutes multiple contiguous blanks with a single blank. Notice that we use the meta character [:blank:] so that we find all forms of space including tabs. We clean the names with

```
nameClean = trimBlanks(cbMenSub$name)
```

Now we can begin to answer the questions about the uniqueness of the names. We do this by examining summary statistics and sets of records.

How many entrants are there over the 14 races? We use length() to find out:

```
length(nameClean)
```

[1] 69806

Recall, we have dropped those records with a run time under 30 minutes, and age under 16.

How many unique names are there?

```
length(unique(nameClean))
```

[1] 42884

How many names appear once, twice, etc.? We can determine this by two calls to table(), i.e.,

```
table(table(nameClean))
```

1	2	3	4	5	6	7	8	9	10
29293	7716	2736	1386	712	417	249	149	92	56

11	12	13	14	15	17	18	19	30
44	19	7	3	1	1	1	1	1

What does this table tell us? We see that over 7000 names appear 2 times throughout the 14 races. One name appears 30 times, and we know this name must correspond to at least 3 people because we have only 14 years of race results.

Which name appears 30 times? We can find that with

```
head( sort(table(nameClean), decreasing = TRUE), 1)
```

```
Michael Smith
          30
```

Let's examine other information about these 30 Michael Smiths. We extract them from our data frame with

```
mSmith = cbMenSub[nameClean == "Michael Smith", ]
```

The home towns include:

```
head(unique(mSmith$home))
```

```
[1] "Annapolis MD      "  "Bethesda MD        "
[3] " Annapolis MD     "  " Chevy Chase MD    "
[5] " Annandale VA     "  "Annapolis MD       "
```

There are several version of Annapolis MD that differ due to extra blanks. Clearly we need to clean the home field as well.

Thinking ahead, we might ask: can we do more cleaning to potentially improve the matching? We have seen that the column headers have inconsistent capitalization. The same is undoubtedly the case for the name. We can check this, but we can also simply proceed to make all characters lower case letters with

```
nameClean = tolower(nameClean)
```

We check the most common name again:

```
head( sort(table(nameClean), decreasing = TRUE), 1)
```

```
michael smith
          33
```

This additional cleaning picked up 3 more Michael Smiths.

Additionally, we can remove punctuation such as a period after someone's middle initial and any stray commas. We do this in one call to gsub() with

```
nameClean = gsub("[,.]", "", nameClean)
```

With so many duplicate names, let's figure out how many times a name appears in the same year. We can create a table of year-name combinations with

```
tabNameYr = table(cbMenSub$year, nameClean)
```

and then call max() to find the cell in the table with the greatest count, i.e.,

```
max(tabNameYr)
```

```
[1] 5
```

Is this Michael Smith again? It takes a bit of work to find the name associated with this maximum. The table saved in `tabNameYr` is of class `table`, which we see is a numeric vector with 3 attributes, `dim`, `dimnames`, and `class`. Calls to class(), mode(), and attributes(), help us figure this out, i.e.,

```
class(tabNameYr)
```

```
[1] "table"
```

```
mode(tabNameYr)
```

```
[1] "numeric"
```

```
names(attributes(tabNameYr))
```

```
[1] "dim"      "dimnames" "class"
```

There are several implications of this data structure. First, some matrix functions work on a `table`, e.g., we can call dim() and colnames() and find

```
dim(tabNameYr)
```

```
[1]    14 39077
```

```
head(colnames(tabNameYr), 3)
```

```
[1] "8illiam maury"   "a gudu memon"   "a miles simmons"
```

Notice we have uncovered another piece of messy data! To find out which cell has a count of 5, we can use which(), but to find the row and column location, we need to include the *arr.ind* argument in our call. That is,

```
which( tabNameYr == max(tabNameYr) )
```

```
[1] 356034
```

```
which( tabNameYr == max(tabNameYr), arr.ind = TRUE )
```

```
     row   col
2012  14 25431
```

Finally, we locate the name(s) with

```
indMax = which( tabNameYr == max(tabNameYr), arr.ind = TRUE )
colnames(tabNameYr)[indMax[2]]
```

```
[1] "michael brown"
```

It's Michael Brown, not Michael Smith!

Now that we have a cleaned version of runner's name, we add it to our data frame with

```
cbMenSub$nameClean = nameClean
```

We use this format of the name to create our unique person identifier.

We can also derive an approximation to year of birth because we have the runner's age and the year of the race. The difference between these two is an approximation to age because the race is held on the first Sunday in April every year. Those runners who have a birthday in the first 7 days of April may have their age reported inconsistently from one race year to the next. What fraction of the records can we expect to have this problem? We create a new variable `yob` in our data frame with

```
cbMenSub$yob = cbMenSub$year - cbMenSub$age
```

Also, we uncovered an issue with blanks and capitalization in names of the hometowns. We leave it as an exercise to clean the values for `home` and add the cleaned version of `home` to `cbMenSub` as `homeClean`.

Let's look closer at the values for these new and cleaned variables for the Michael Browns in our data frame. We do this with

```
vars = c("year", "homeClean", "nameClean", "yob", "runTime")
mb = which(nameClean == "michael brown")
birthOrder = order(cbMenSub$yob[mb])
cbMenSub[mb[birthOrder], vars]
```

```
year       homeClean     nameClean  yob  runTime
2000         tucson az michael brown 1939   96.88
2010      north east md michael brown 1953   92.27
2011      north east md michael brown 1953   85.95
2012      north east md michael brown 1953   88.43
2009         oakton va michael brown 1957   99.73
2008        ashburn va michael brown 1958   93.73
2009        ashburn va michael brown 1958   88.57
2010        ashburn va michael brown 1958   99.75
2012         reston va michael brown 1958   89.95
2006      chevy chase michael brown 1966   84.57
2010   chevy chase md michael brown 1966   79.35
2012   chevy chase md michael brown 1966   95.82
2004      berryville va michael brown 1978   76.32
2008       arlington va michael brown 1984   84.68
2010       new york ny michael brown 1984  110.88
2011       arlington va michael brown 1984   81.70
2012       arlington va michael brown 1984   70.93
2012         clifton va michael brown 1988   84.88
```

What observations can we make about these various `michael brown` rows?

- The 3 entries for `michael brown` born in 1953 seem to be the same person because all have a hometown of "north east md". Additionally, the 3 race times are within 7 minutes of each other.

- The 4 entries for `michael brown` born in 1958 have race years of 2008, 2009, 2010, and 2012. The most recent entry lists Reston, VA for a hometown while the other 3 show Ashburn, VA. Do we have 1, 2, 3, or 4 different `michael browns` here? The 2010 entrant ran the slowest of the 4 races by about 11 minutes and the other 3 times are closer. An Internet search reveals that Reston and Ashburn are within 22 km of each other. It is conceivable that these 4 records belong to the same individual who moved from Ashburn to Reston between April 2010 and 2012. We can't know for sure.

- Another 3 `michael brown` entries have 1966 for a birth year. All 3 list Chevy Chase as the hometown, except that for 2006, the state (MD) is not provided. When we examine more locations for other runners in 2006, we find that none of them list a state. These 3 `michael brown` records also have a range of 11 minutes for time with the middle year (2010) being the fastest. These records are likely for the same person, but we have uncovered an inconsistency in how home is reported for 2006 compared to the other years.

- Next, we have 4 records for `michael brown` born in 1984, with races in 2008, 2010, 2011, and 2012. Of these, the 2010 record seems to be a different person as his home is listed as New York, NY and his race time is 25 to 40 minutes slower than the other 3 records. The other 3 all have the same hometown of Arlington, VA. They also have increasingly better times with a 2008 time of 84 and a 2012 time of 71 minutes. It is not unreasonable to think that these 3 records belong to the same person who has been training and running faster as he ages. Again, we cannot know this for sure.

- Lastly, notice that the 5 `michael browns` who registered for the 2012 race have different years of birth (1953, 1958, 1966, 1984, and 1988) and 5 different home towns. These are 5 different people.

We summarize our various observations to make a first attempt to create an identifier for individuals. We might paste together the cleaned name and the derived year of birth. We do this with

```
cbMenSub$ID = paste(nameClean, cbMenSub$yob, sep = "_")
```

We have ignored the information provided by the hometown and the run times and so have created the least restrictive identifier.

Since our goal is to study how an athlete's time changes with age, let's focus on those IDs that appear in at least 8 races. To do this, we first determine how many times each ID appears in `cbMenSub` with

```
races = tapply(cbMenSub$year, cbMenSub$ID, length)
```

Then we select those IDs that appear at least 8 times with

```
races8 = names(races)[which(races >= 8)]
```

and we subset `menRes` to select the entries belonging to these identifiers with

```
men8 = cbMenSub[ cbMenSub$ID %in% races8, ]
```

Finally, we organize the data frame so that entries with the same ID are contiguous. This makes it easier to manually examine records, etc. We can do this with

```
orderByRunner = order(men8$ID, men8$year)
men8 = men8[orderByRunner, ]
```

An alternative organization for the data is to store them as a list with an element for each ID in `races8`. In this list, each element is a data frame containing only those results for the records with the same ID. We can create this list with

```
men8L = split(men8, men8$ID)
names(men8L) = races8
```

Which data structure is preferable? That depends on what we want to do with the data. In the following we show how to accomplish a task using both approaches to help make a comparison between the two structures. In the next section, we find it easiest to work with the list of data frames as we often need to apply a function of multiple arguments to each runner's entries.

How many IDs do we have left?

```
length(unique(men8$ID))
```

```
[1] 480
```

This is the same as `length(men8L)`. We might also want to discard matches if the performance varies too much from year to year. How large a fluctuation would make us think that we have mistakenly connected two different people? Of course, we don't want to bias our results by eliminating an individual whose run times vary a lot. Let's look at a few records where the year-to-year difference in time exceeds, say, 20 minutes. We determine which satisfy this constraint with

```
gapTime = tapply(men8$runTime, men8$ID,
                 function(t) any(abs(diff(t)) > 20))
```

or with

```
gapTime = sapply(men8L, function(df)
                         any(abs(diff(df$runTime)) > 20))
```

How many of these runners have gaps of more than 20 minutes?

```
sum(gapTime)
```

```
[1] 49
```

Slightly reformatted displays of the first two of these athletes are

```
lapply(men8L[ gapTime ][1:2], function(df) df[, vars])
```

```
$`abiy zewde_1967`
     year         homeClean    nameClean  yob runTime
     1999      gaithersburg md abiy zewde 1967   96.52
     2000   montgomery vill md abiy zewde 1967   96.63
     2001   montgomery vill md abiy zewde 1967   89.10
     2002   montgomery vill md abiy zewde 1967  123.00
     2003      gaithersburg md abiy zewde 1967   97.68
     2004   montgomery vill md abiy zewde 1967  100.37
     2006         gaithersburg abiy zewde 1967  108.40
     2008   montgomery vill md abiy zewde 1967   98.78
     2009  montgomery villag md abiy zewde 1967   98.50
     2010  montgomery villag md abiy zewde 1967   99.92
     2011  montgomery villag md abiy zewde 1967  113.10
     2012  montgomery villag md abiy zewde 1967   84.88

$`adam hughes_1978`
     year       homeClean    nameClean  yob runTime
     2005 washington dc adam hughes 1978   80.38
```

```
2006      washington    adam hughes 1978    85.17
2007   washington dc    adam hughes 1978    77.78
2008   washington dc    adam hughes 1978    74.23
2009   washington dc    adam hughes 1978   108.07
2010   washington dc    adam hughes 1978   103.07
2011   washington dc    adam hughes 1978    77.12
2012   washington dc    adam hughes 1978    77.77
```

The name `abiy zewde` seems unusual enough to most likely be the same person participating in 12 of the 14 races even though the hometown has changed over the years and the race results differ by nearly 40 minutes with one of the fastest times being the most recent when he was 45 years old. In fact, a Google search locates a Web page at http://storage.athlinks.com/racer/results/65866776 with his published race times from several different runs. A screenshot of this page appears in Figure 2.16. Clearly these entries all belong to the same person.

Do we want to further restrict our matching to those with the same hometown? This eliminates, e.g., the `abiy zewde` records, even though we're quite certain the records all belong to the same individual. We could identify the mismatches and manually examine them for potentially false matches. We need to eliminate the state abbreviation from the end of those records that have one because the 2006 records do not have it. We can substitute a blank followed by 2 letters occurring at the end of the string with an empty string, i.e.,

```
gsub("[[:blank:]][a-z]{2}$", "", home)
```

This may result in matches that are too liberal, e.g., matching Springfield IL and Springfield MA. We leave it as an exercise to determine how to limit the matches to those where the entries have the same hometown and to assess whether this additional restriction should be added to the matching process.

Here, we consider a less strict matching where we match only those records that have the same values for the state of residence. To do this, we create a new variable that holds the 2 letter abbreviation for the state. We return to work with `cbMenSub` because the data structure is simpler and we maintain consistency. We extract the last 2 characters from each home string. This is the state, if it is present. We know that in 2006, the state was not present so we set these to NA. For athletes who come from outside the US, we pick up the last two letters of either the country or province, but these should not dramatically affect our matches.

We first determine how many characters are in each value for `home` with

```
homeLen = nchar(cbMenSub$homeClean)
```

Then we use it to extract the last two characters and add them back to our data frame with

```
cbMenSub$state = substr(cbMenSub$homeClean,
                       start = homeLen - 1, stop = homeLen)
```

And, we set the 2006 values to NA:

```
cbMenSub$state[cbMenSub$year == 2006] = NA
```

Next, we recreate the new ID so that it includes state. We do this with

```
cbMenSub$ID = paste(cbMenSub$nameClean, cbMenSub$yob,
                   cbMenSub$state, sep = "_")
```

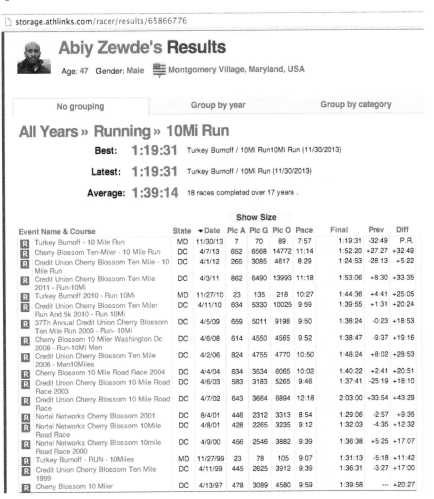

Figure 2.16: Screen Shot of One Runner's Web Page of Race Results. *This Web page at* http://storage.athlinks.com *contains the race results of one runner who participated in the Cherry Blossom run for 12 of the 14 years for which we have data. Notice that his fastest time was from his most recent run in 2012 where he completed the race in under 85 minutes. He was 45 at that time. Also, his slowest time was 123 minutes in 2002 at the age of 35.*

Then, we again select those IDs that occur at least 8 times with

```
numRaces = tapply(cbMenSub$year, cbMenSub$ID, length)
races8 = names(numRaces)[which(numRaces >= 8)]
men8 = cbMenSub[ cbMenSub$ID %in% races8, ]
orderByRunner = order(men8$ID, men8$year)
men8 = men8[orderByRunner, ]

men8L = split(men8, men8$ID)
names(men8L) = races8
```

In the next section we work solely with the list structure.

This addition to the runner id further reduces the number of runners who have completed 8 races, i.e.,

```
length(races8)
[1] 306
```

We now have 306 athletes who have the same name, year of birth, and state and who have run in 8 of the 14 races. We carry on with this set of matches we have obtained thus far, and in the next section, we examine how each runner's performance changes as he grows older.

2.6 Modeling the Change in Running Time for Individuals

The Cherry Blossom race results include recordings for athletes from 20 to 80 years old. However, we don't have records for any one person that covers this 60-year span. That's not possible because we have only 14 years of race results so we can at most observe a 20 year old until he turns 33 or an 80 year old when he was 67. This means when we examine the performance of an individual over time, we are looking at short time series that are at most 14 years long. To examine performance from 20 to 80 necessarily means that we rely on the cross-sectional aspect of the data, but there is information to be gleaned in these short time series.

It's reasonable to imagine that over a short period of time, say 8 to 10 years, a runner's performance is roughly linear with age. (We saw this with the piecewise linear model for the cross-sectional data). We can make plots to ascertain if this is the case. We have over 300 runners to plot so to limit the effect of over plotting, we make several plots of different subsets of the data. We begin by dividing the runners into 9 groups to make 9 plots in a 3-by-3 grid. We assign roughly the same number of runners to each group with

```
groups = 1 + (1:length(men8L) %% 9)
```

To make each plot, we create a blank canvas with, e.g.,

```
plot( x = 40, y = 60, type = "n",
      xlim = c(20, 80), ylim = c(40, 160),
      xlab = "Age (years)", ylab = "Run Time (minutes)")
```

Then we add the lines for each runner in the group. We make the lines different colors and line types to help distinguish between the runners. The addRunners() function below adds a line for each runner:

```
addRunners = function(listRunners, colors, numLty)
{
  numRunners = length(listRunners)
  colIndx = 1 + (1:numRunners) %% length(colors)
  ltys = rep(1:numLty, each = length(colors), length = numRunners)

  mapply(function(df, i) {
          lines(df$runTime ~ df$age,
                col = colors[colIndx[i]], lwd = 2, lty = ltys[i])
        }, listRunners, i = 1:numRunners)
}
```

We can create the 9 blank canvases and add the runners' lines with

```
colors = c("#e41a1c", "#377eb8","#4daf4a", "#984ea3",
           "#ff7f00", "#a65628")
par(mfrow = c(3, 3), mar = c(2, 2, 1, 1))
invisible(
  sapply(1:9, function(grpId){
    plot( x = 0, y = 0, type = "n",
          xlim = c(20, 80), ylim = c(50, 130),
          xlab = "Age (years)", ylab = "Run Time (minutes)")
    
    addRunners(men8L[ groups == grpId ], colors, numLty = 6)
  }) )
```

The invisible() function hides the return value from sapply(). Since our function adds lines to the canvas, it returns NULL for each iteration, which we can safely ignore.

In Figure 2.17 we see 9 line plots, each containing about 30 athletes. Some of the athletes fluctuate quite a bit, and we might want to revisit the notion of dropping runners that may be the combination of two different people's records because they fluctuate more than expected. We leave this as an exercise. Otherwise, fitting a line to each individual's performance seems a reasonable approach.

A longitudinal analysis of each individual runner implicitly controls for covariates that may influence performance, e.g., gender. One exception is the race condition in any given year – some years might be slow and some fast due to changes in the course or weather. However, it seems plausible that such an effect is uncorrelated with age and so amounts to measurement noise.

Now that we have our list of runners, we wish to fit a line to each runner's performance. If we write a function that carries out this work for one runner, then we can apply it to all of the runners in our list. What do we need this function to do? We can have it fit a line via lm(). What do we want the function to return? We are interested in the coefficient for age, but we need to be able to interpret it in the context of age. Since we have multiple ages for each runner, let's return a middle value for age. And, while we are at it, let's also return a predicted value for the runner's performance at that age. What inputs do we need for our function? Really just the runner's run time and age. We can pass these into our function as separate parameters or we can pass in our data frame. If we do the latter, then we need to know the names of the variables to fit. Let's do this. Let's also have our function add the fitted line segment to a plot. We can make this an optional operation by adding a parameter that has a default value of FALSE so the function only adds the line when the call specifies this parameter as TRUE. This function appears below:

```
fitOne = function(oneRunner, addLine = FALSE, col = "grey") {
  lmOne = lm(runTime ~ age, data = oneRunner)
  if (addLine)
    lines(x = oneRunner$age, y = predict(lmOne),
          col = col, lwd = 2, lty = 2)
  
  ind = floor( (nrow(oneRunner) + 1) / 2)
  res = c(coefficients(lmOne)[2], oneRunner$age[ind],
          predict(lmOne)[ind])
  names(res) = c("ageCoeff", "medAge", "predRunTime")
  return(res)
}
```

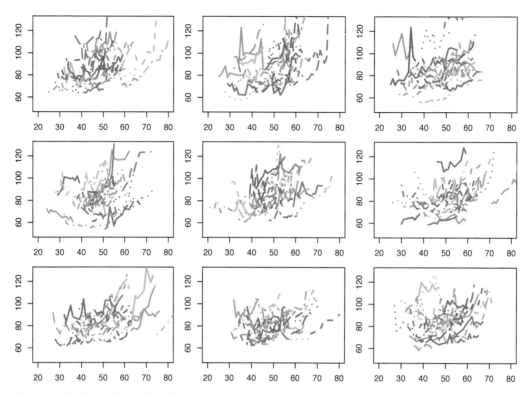

Figure 2.17: Run Times for Multiple Races. *These line plots show the times for male runners who completed at least 8 Cherry Blossom races. Each set of connected segments corresponds to the run times for one athlete. Looking at all line plots, we see a similar shape to the scatter plot in Figure 2.7, i.e., an upward curve with age. However, we can also see how an individual's performance changes. For example, many middle-aged runners show a sharp increase in run time with age but that is not the case for all. Some of them improve and others change more slowly.*

We leave it as an exercise to augment this function to also return the SD of the errors and the SE of the coefficient for age. We also leave as an exercise an alternative approach that fits all of the athletes' lines together so that the noise is pooled across athletes.

We call fitOne() to add the fitted lines to one of the line plots for the runners in Figure 2.17:

```
plot( x = 0, y = 0, type = "n",
      xlim = c(20, 80), ylim = c(50, 130),
      xlab = "Age (years)", ylab = "Run Time (minutes)")

addRunners(men8L[ groups == 9 ], colors, numLty = 6)
lapply(men8L[groups == 9], fitOne, addLine = TRUE, col = "black")
```

See the black dashed line segments in Figure 2.18. These line segments seem to capture each runner's performance.

Next, to examine the runner-to-runner variability, we fit lines to all 306 athletes with the following call:

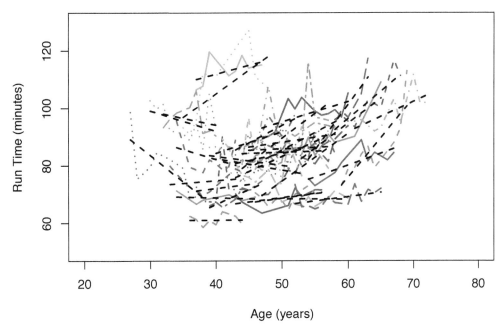

Figure 2.18: Linear Fits of Run Time to Age for Individual Runners. *Here we have augmented the bottom-right line plot from Figure 2.17 with the least squares fit of run time for each of the athletes. These are the 30 or so black dashed line segments plotted on each of the individual runner's times series.*

```
men8LongFit = lapply(men8L, fitOne)
```

We can extract the 306 coefficients for age and each runner's representative age with

```
coeffs = sapply(men8LongFit, "[", "ageCoeff" )
ages = sapply(men8LongFit, "[", "medAge")
```

Now we have a single coefficient that represents the relationship between run time and age for each runner who ran at least 8 times (and who resided in the same state over those race years). This coefficient has units of minutes in run time for the 10-mile race per year. A positive coefficient means that the runner is slowing down by that number of minutes a year.

We see in Figure 2.19 how these coefficients vary with age. There is plenty of individual variation in performance with a few in their 50s and 60s getting faster and many in their 30s slowing down. However, we also see a relationship between age and run time. There appears to be a positive linear trend in the coefficients. We fit this with

```
longCoeffs = lm(coeffs ~ ages)
```

The summary from the fit appears below

```
summary(longCoeffs)
```

```
Call:
lm(formula = coeffs ~ ages)

Residuals:
   Min      1Q  Median      3Q     Max
-4.403  -0.638  -0.025   0.565   3.354

Coefficients:
            Estimate Std. Error t value Pr(>|t|)
(Intercept) -1.95844    0.30549   -6.41  5.5e-10 ***
ages         0.05526    0.00618    8.95  < 2e-16 ***
---
Signif. codes:  0 '***' 0.001 '**' 0.01 '*' 0.05 '.' 0.1 ' ' 1

Residual standard error: 1 on 304 degrees of freedom
Multiple R-squared:  0.209,     Adjusted R-squared:  0.206
F-statistic: 80.1 on 1 and 304 DF,  p-value: <2e-16
```

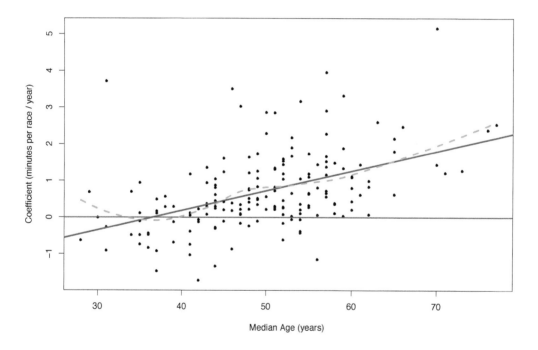

Figure 2.19: Coefficients from Longitudinal Analysis of Athletes. *This scatter plot displays the slope of the fitted line to each of the 300+ runners who competed in at least 8 Cherry Blossom road races. A negative coefficient indicates the runner is getting faster as he ages. The plot includes a least squares fitted line and a loess fitted curve. Notice that nearly all of the coefficients for those over 50 are positive. The typical size of this coefficient for a 50-year old is about one minute per year.*

We have added to the plot this fitted line along with a horizontal reference line at 0 and a smooth curve fit to the coefficients using loess(). This graph suggests that, on average, performance improves for people who are younger than about 35. That is, the age coefficient is negative for ages under 35. The hypothetical "average" runner who is older than 35 slows down. By age 60, the typical runner slows by about 1.3 minutes per year, about twice as fast as indicated by the cross-sectional analysis.

2.7 Scraping Race Results from the Web

The race results for the Cherry Blossom Ten Mile Run are available at http://www.cherryblossom.org. Figure 2.1 shows a screen shot of the site's main Web page with links leading to each year's results. The results for men in, e.g., 2012, are displayed in the screen shot in Figure 2.2 and again in Figure 2.20. We see that the data are simply formatted in what appears to be a block of plain text arranged in fixed-width columns. We can examine the source code for the Web page to check if this is the case. We do this in, e.g., a Google Chrome browser by clicking on View -> Developer -> View Source. When we do this we see that the table itself contains no *HTML* markup, and it has been inserted into a `<pre>` node within the document. (The *HTML* source for the page shown in Figure 2.2 is shown in Figure 2.20.) It should be very easy to extract this "table" from the *HTML* for further processing.

```
<!-- saved from url=(0022)http://internet.e-mail -->
<html>
<pre>
                     Credit Union Cherry Blossom Ten Mile Run
                     Washington, DC      Sunday, April 1, 2012

                        Official Male Results (Sorted By Net Time)

Place  Div      /Tot    Num    Name                        Ag Hometown               5 Mile   Time     Pace
=====  =========== ======   ==================== == ==================  =======  =======  =====
   1     1/347              9 Allan Kiprono              22 Kenya                 22:32    45:15    4:32
   2     2/347             11 Lani Kiplagat              23 Kenya                 22:38    46:28    4:39
   3     1/1093            31 John Korir                 36 Kenya                 23:20    47:33    4:46
   4     1/1457            15 Ian Burrell                27 Tucson AZ             23:50    47:34    4:46
and so on
7190   375/375          18780 Larry Hume                 56 Arlington VA          1:07:17  2:27:20 14:44
7191  1093/1093         19104 Sean-Patrick Alexander 35 Alexandria VA             1:08:44  2:27:30 14:45
7192                    22280 Joseph White                  Forestville MD        1:10:04  2:28:58 14:54
7193   648/648           6555 Lee Jordan                 48 Herndon VA            1:09:06  2:30:59 15:06
</pre>
```

Figure 2.20: Screen Shot of the Source for Men's 2012 Cherry Blossom Results. *This screen shot is of the HTML source for the male results for the 2012 Cherry Blossom road race. While the format is not quite the same as the female results for 2011 (see Figure 2.21), both are plain text tables within `<pre>` nodes in an HTML document.*

We examine one more year to ascertain if the format is the same. When we view the source for the page of 2011 women's results, we see that the basic format is the same. A screen shot of the source for 2011 female results appears in Figure 2.21. However, the columns are not identical. In 2011, a net time is reported as well as a time. And, following the pace column there is a column labeled S, which has an exclamation mark for the first few runners and nothing for the rest. Our task here is simply to extract the text table so we need only locate the table and extract it as a block of text. The functions in Section 2.3 take care of turning the columns of information into variables.

We use the htmlParse() function in the **XML** package [6] to scrape the 2012 male's page from the site.

```
 1 <!-- saved from url=(0022)http://internet.e-mail -->
 2 <html>
 3 <pre>
 4                 Credit Union Cherry Blossom Ten Mile Run
 5                 Washington, DC      Sunday, April 3, 2011
 6
 7                 Female Official Results (Sorted By Net Time)
 8
 9 Place Div   /Tot    Num   Name                  Ag Hometown              5 Mile  Time    Net Tim Pace  S
10 ===== =========== ====== ===================== == ===================== ======= ======= ======= ===== =
11   1   1/2706       14 Julliah Tinega           25 Kenya                          54:02   54:02   5:25 1
12   2   1/937        16 Risper Gesabwa           22 Kenya                 27:17    54:03   54:03   5:25 1
13   3   1/1866       48 Tgist Tufa               30 Ethiopia              27:17    54:13   54:13   5:26 1
14   4   2/1866       44 Alemtsehay Misganaw      30 Ethiopia              27:17    55:17   55:17   5:32 1
15   5   2/2706       24 Claire Hallissey         28 United Kingdom        28:01    56:17   56:17   5:38 1
16   6   3/1866       40 Kelly Jaske              34 Portland OR           27:58    57:06   57:06   5:43 1
17   7   4/1866      156 Michelle Miller          30 Damascus MD           29:33    59:20   59:20   5:56
18   8   1/1265      148 Sharon Lemberger         37 Stamford CT           29:50    59:40   59:40   5:58
19   9   3/2706      167 Katie Howery             25 Verona WI             29:38    59:52   59:52   6:00
20  10   2/1265      157 Kara Waters              36 Ellicott City MD              1:00:03 1:00:03  6:01
```

Figure 2.21: Screen Shot of the Source for Women's 2011 Cherry Blossom Results. *This screen shot is of the HTML source for the female results for the 2011 Cherry Blossom road race. Notice that times given are for the midpoint of the race (*5 Mile*) and for two finish times (*Time *and* Net Tim*). Also notice the leftmost column labeled* S*. While the format is different than the male results for 2012, both are plain text tables within* <pre> *nodes in an HTML document.*

```
library(XML)
ubase = "http://www.cherryblossom.org/"
url = paste(ubase, "results/2012/2012cucb10m-m.htm", sep = "")
doc = htmlParse(url)
```

We saw from the *HTML* source that we want to extract the text content of the <pre> node. We can access all <pre> nodes in the document with the simple *XPath* expression, //pre. We do this with

```
preNode = getNodeSet(doc, "//pre")
```

The getNodeSet() function returns a list where each element corresponds to one of the <pre> nodes in the document. In our case, there is only one such node. Next, we use the xmlValue() function to extract the text content from this node as follows:

```
txt = xmlValue(preNode[[1]])
```

Let's examine the contents of txt. We first determine how many characters it contains and then examine the start and end. We do this with

```
nchar(txt)
```

[1] 690904

```
substr(txt, 1, 50)
```

[1] "\r\n   Credit Union Cherry Blossom Ten "

```
substr(txt, nchar(txt) - 50, nchar(txt))
```

[1] "   48 Herndon VA   1:09:06 2:30:59 15:06 \r\n"

It appears that we have successfully extracted the information from the Web page. We also see that the individual lines end with \r\n. We can use these characters to split up the 690904 characters into separate strings corresponding to lines in the table. That is,

```
els = strsplit(txt, "\\r\\n")[[1]]
```

Now we have 7201 lines of information, i.e.,

```
length(els)
```

```
[1] 7201
```

The first few of these contain the header information, i.e.,

```
els[1:3]
```

```
[1] ""
[2] "                    Credit Union Cherry Blossom Ten Mile Run"
[3] "                    Washington, DC        Sunday, April 1, 2012"
```

and the last line contains information for one of the runners:

```
els[ length(els) ]
```

```
[1] " 7193    648/648      6555 Lee Jordan                
        48 Herndon VA                1:09:06 2:30:59 15:06 "
```

We have succeeded in extracting the rows of the table as elements of a character vector.

Let's formalize our code into a function that we can apply to each of the 28 Web pages (a page for each of the men's and women's races from 1999 to 2012). We want our function to take as input the URL for the Web page and return a character vector with one element per line, including the header lines and the rows in the table of results. We arrange our previous code into a function as

```
extractResTable =
       # Retrieve data from web site, find preformatted text,
       # return as a character vector.
function(url)
{
  doc = htmlParse(url)
  preNode = getNodeSet(doc, "//pre")
  txt = xmlValue(preNode[[1]])
  els = strsplit(txt, "\r\n")[[1]]

  return(els)
}
```

Let's try our function with the 2012 men's results.

```
m2012 = extractResTable(url)
```

```
identical(m2012, els)
```

```
[1] TRUE
```

Our function has extracted the same results as before. Let's now apply it to all of the men's results across the years.

If we have a vector of all the URLs then we can simply apply our function to the vector. We make this vector by pasting together the base URL to the year-specific information as follows:

```
ubase = "http://www.cherryblossom.org/"
urls = paste(ubase, "results/", 1999:2012, "/",
             1999:2012, "cucb10m-m.htm", sep = "")
```

Now we can apply extractResTable() to urls with

```
menTables = lapply(urls, extractResTable)

Error in preNode[[1]] : subscript out of bounds
```

We have an error that indicates there is a problem with preNode.

To find more information about what is causing this error, we turn on the error handling by setting the error option to the recover() function object so that when an error occurs, recover() is called. This function gives us access to the active call frames so that we can examine the objects and see if they are what we expect. We set the options() and call the extractResTable() again:

```
options(error = recover)
menTables = lapply(urls, extractResTable)

Error in preNode[[1]] : subscript out of bounds

Enter a frame number, or 0 to exit

1: lapply(urls, extractResTable)
2: FUN(c("http://www.cherryblossom.org/results/1999/...
3: #13: xmlValue(preNode[[1]])

Selection:
```

This presents us with a simple view of the current set of function calls in effect when the error occurred. This is the "call stack". The first is our top-level call to lapply(). The second entry is the actual call to our extractResTable() function. lapply() calls this for us but uses the name FUN since that is the name of the parameter in lapply() containing the function we want to call for each element of the first argument to lapply(). extractRestTable() calls several functions, but the error occurred in the third expression xmlValue(preNode[[¬ 1]]). This is the third and final element in the current call stack.

The recover() function allows us to select which of these calls we want to examine. At the Selection: prompt, we enter that number, e.g., 2 and the return key. This will put is directly in call frame of this particular call to the function. We will be able to examine (and modify) the current values of arguments and local variables and execute arbitrary code there.

We choose the second element of the call stack as this is the function call to extractResTable(). We do this with

```
Selection: 2
Called from: lapply(urls, extractResTable)
Browse[1]>
```

After selecting this frame, we use R's browser capabilities to examine the objects in this environment. We find:

```
Browse[1]> ls()
```

```
[1] "doc"     "preNode" "url"
```

```
Browse[1]> url
```

```
[1] "http://www.cherryblossom.org/results/1999/1999cucb10m-m.htm"
```

```
Browse[1]> length(preNode)
```

```
[1] 0
```

It appears that there is no `<pre>` node in the 1999 race results Web page.
Let's check this out by visiting the site. When we paste the URL

```
http://www.cherryblossom.org/results/1999/1999cucb10m-m.htm
```

into the Web browser, we find that it takes us to the main page shown in Figure 2.1, not to the page we were expecting. When we use the navigation system on the main Web page to go to the 1999 men's results we see the problem. The URL is not as we expected. Instead, it is

```
http://www.cherryblossom.org/cb99m.htm
```

This is a quite different format of the URL from what we created based on the 2011 and 2012 URLs. It tells us that we need to determine all 28 URLs by using the Web site's navigation system. We can do this programmatically, but here we simply gather the URLs for the male results into a character vector called menURLs, i.e.,

```
menURLs =
  c("cb99m.htm", "cb003m.htm", "results/2001/oof_m.html",
    "results/2002/oofm.htm", "results/2003/CB03-M.HTM",
    "results/2004/men.htm", "results/2005/CB05-M.htm",
    "results/2006/men.htm", "results/2007/men.htm",
    "results/2008/men.htm", "results/2009/09cucb-M.htm",
    "results/2010/2010cucb10m-m.htm",
    "results/2011/2011cucb10m-m.htm",
    "results/2012/2012cucb10m-m.htm")
```

We include only the portion of the URL that follows the base: `http://www.cherryblossom.org/`.
We reconstruct the urls vector so that it contains the proper Web addresses. We paste the URLs together with

```
urls = paste(ubase, menURLs, sep = "")
```

```
urls[1:3]
```

```
[1] "http://www.cherryblossom.org/cb99m.htm"
[2] "http://www.cherryblossom.org/cb003m.htm"
[3] "http://www.cherryblossom.org/results/2001/oof_m.html"
```

We again try to read the results into R with

```
menTables = lapply(urls, extractResTable)
names(menTables) = 1999:2012
```

This time, we receive no error messages. Of course, simply because we didn't run into any errors, does not mean that we have properly extracted the data. We need to check the results to see if they contain the information expected.

Let's first check the length of each of the character vectors. From the Web site we have seen that several thousand runners compete each year so we expect several thousand elements in our vectors.

```
sapply(menTables, length)
```

```
1999 2000 2001 2002 2003 2004 2005 2006 2007 2008
3193    1 3627 3727 3951 4164 4335 5245 5283 5913
2009 2010 2011 2012
   1 6919 7019 7201
```

Hmmm, the 2000 and 2009 extractions resulted in vectors with a single element.

The file names for these two years are correct so this requires digging deeper. We view the source of the 2000 Web page to see if it is formatted as expected. Below are the first few lines of the 2000 document:

```
<html>
<body bgcolor="#CCFFFF">
<p align="center"><font color="#800000" size="4" face="Arial">
<strong>Nortel Networks Cherry Blossom 10mile Road Race<br>
Washington, DC *** April 9, 2000
</strong></font><strong><font color="#800000" face="Arial"><br>
<h3 align="center"><font color="#800000" face="Arial">
Official Results, MEN *** Gun Time Is The Official Time</font>
</h3>
<BR>
<PRE><Strong>
</font><font color="#800000" face="Courier New">
PLACE DIV /TOT   NUM   NAME                      AG ...
===== ========= ===== ====================== == ...
```

Let's rearrange the *HTML* tags and use indentation to see if there is a problem with the format of the document. Below is the same content displayed in a more readable format:

```
<html>
<body bgcolor="#CCFFFF">
 <p align="center">
  <font color="#800000" size="4" face="Arial">
    <strong>
Nortel Networks Cherry Blossom 10mile Road Race<br>
Washington, DC *** April 9, 2000
    </strong>
  </font>
  <strong>
    <font color="#800000" face="Arial">
      <br>
      <h3 align="center"><font color="#800000" face="Arial">
Official Results, MEN *** Gun Time Is The Official Time
       </font>
      </h3>
```

```
    <BR>
    <PRE>
      <Strong>
        </font>
        <font color="#800000" face="Courier New">
PLACE DIV /TOT   NUM   NAME                    AG ...
===== =========  =====  =====================  == ...
```

This document is not well-formed *HTML*. The htmlParse() function can fix many problems with malformed documents, e.g., closing a `<br>` tag and matching case for tag names. However, this function can only do so much. Notice that the `<font>` and `<h3>` tags are not properly nested, and similarly the closing `</font>` tag that appears after the `<pre>` tag is problematic. If htmlParse() closes the `<pre>` tag so that the tags in the document are properly nested, then the `<pre>` node does not contain the table of race results.

We can programmatically edit the *HTML* so that it is well formed. Alternatively, we can try another *XPath* expression for locating the content for this particular file. We proceed with the second of these options and leave the first as an exercise.

If we want to handle one of the years differently than the others, then we need a way to distinguish between the two approaches. One way to do this might be to add a second argument to the function definition that indicates with which year we are working. Then our code can check the year, and if it is 2000, we can extract the table of results differently. We supply a default value to *year* so that if we don't specify this argument then the function carries out the default extraction. We provide a modified extractResTable() to do this:

```
extractResTable =
  # Retrieve data from web site,
  # find the preformatted text,
  # and return as a character vector.
function(url, year = 1999)
{
  doc = htmlParse(url)

  if (year == 2000) {
    # Get text from 4th font element
    # File is ill-formed so <pre> search doesn't work.
    ff = getNodeSet(doc, "//font")
    txt = xmlValue(ff[[4]])
  }
  else {
    preNode = getNodeSet(doc, "//pre")
    txt = xmlValue(preNode[[1]])
  }

  els = strsplit(txt, "\r\n")[[1]]
  return(els)
}
```

Since we now have two arguments to our function, we use mapply() to call extractResTable():

```
years = 1999:2012
menTables = mapply(extractResTable, url = urls, year = years)
```

```
names(menTables) = years
sapply(menTables, length)
1999 2000 2001 2002 2003 2004 2005 2006 2007 2008
3193 3019 3627 3727 3951 4164 4335 5245 5283 5913
2009 2010 2011 2012
   1 6919 7019 7201
```

We have cleared up the problem with 2000, but the problem with 2009 remains. We leave it as an exercise to modify extractResTable() to handle this special case. Once modified, we find that there are 6659 rows in the 2009 table.

Now that we have the function working for the Web pages of men's results, we can try it on the women's pages. When we do, we find that all works fine except for the year 2009. As it happens, we don't need any special handling for the women's results for that year. We leave it as an exercise to modify the function again so that the 2009 women's results use the default processing, rather than the special 2009 processing needed for the men's results.

We save the data for further processing.

```
save(menTables, file = "CBMenTextTables.rda")
```

Lastly, an alternative to saving the list of character vectors in an R data format is to write the character vectors out as plain text files. We can use writeLines() to do this. In fact, we can modify extractResTable() to accept a *file* argument. If supplied, the function writes the results to a file with that name, and if NULL then the function returns the character vector. Again, we leave this enhancement as an exercise.

2.8 Exercises

Q.1 Write a function that uses read.fwf() to read the 28 text tables in *MenTxt/* and *WomenTxt/* into R. These are called *1999.txt*, *2000.txt*, etc., and are described in greater detail in Section 2.2. Examine the tables in a plain text editor to determine the start and end position of each column of interest (name, hometown, age, and gun and net time). Use statistics to explore the results and confirm that you have extracted the information from the correct positions in the text.

Q.2 Revise the extractVariables() function (see Section 2.2) to remove the rows in menTables that are blank. In addition, eliminate the rows that begin with a '*' or a '#'. You may find the following regular expression helpful for locating blank rows in a table

```
grep("^[[:blank:]]*$", body)
```

The pattern uses several meta characters. The ^ is an anchor for the start of the string, the $ anchors to the end of the string, the [[:blank:]] denotes the equivalence class of any space or tab character, and * indicates that the blank character can appear 0 or more times. All together the pattern ^[[:blank:]]*$ matches a string that contains any number of blanks from start to end.

Q.3 Find the record where the time is only 1.5. What happened? Determine how to handle the problem and which function needs to be modified: extractResTable(), extractVariables(), or cleanUp(). In your modification, include code to provide a warning message about the rows that are being dropped for having a time that is too small.

Q.4 Examine the head and tail of the 2006 men's file. Look at both the character matrix in the list called menResMat and the character vector in the list called menFiles (see Section 2.2). (Recall that the desired character matrix in menResMat and the character vector in menFiles both correspond to the element named "2006"). What is wrong with the hometown? Examine the header closely to figure out how this error came about. Modify the extractVariables() function to fix the problem.

Q.5 Write the convertTime() function described in Section 2.3. This function takes a string where time is in either the format hh:mm:ss or mm:ss. The return value is the time as a numeric value of the number of minutes. Design this function to take a character vector with multiple strings and return a numeric vector.

Q.6 Modify the createDF() function in Section 2.3 to handle the formatting problem with the 2006 male file. You will need to carefully inspect the raw text file in order to determine the problem.

Q.7 Follow the approach developed in Section 2.2 to read the files for the female runners and then process them using the functions in Section 2.3 to create a data frame for analysis. You may need to generalize the createDF() and extractVariables() functions to handle additional oddities in the raw text files.

Q.8 Modify the call to the plot() function that created Figure 2.6 to create Figure 2.7. To do this, read the documentation for plot() to determine which parameters could be helpful; that is, ?plot.default contains helpful information about the commonly used graphical parameters.

Q.9 Modify the piecewise linear fit from Section 2.4.2 to include a hing at 70. Examine the coefficients from the fit and compare the fitted curve to the loess curve. Does the additional hing improve the fit? Is the piecewise linear model closer to the loess curve?

Q.10 We have seen that the 1999 runners were typically older than the 2012 runners. Compare the age distribution of the runners across all 14 years of the races. Use quantile–quantile plots, boxplots, and density curves to make your comparisons. How do the distributions change over the years? Was it a gradual change?

Q.11 Normalize each male runner's time by the fastest time for the runner of the same age. To do this, find the fastest runner for each year of age from 20 to 80. The tapply() function may be helpful here. Smooth these times using loess(), and find the smoothed time using predict(). Use these smoothed times to normalize each run time. Use density plots, quantile–quantile plots, and summary statistics to compare the distribution of the age-normalized times for the runners in 1999 and 2012. What do you find? Repeat the process for the women. Compare the women in 1999 to the women in 2012 and to the men in 1999 and 2012.

Q.12 Clean the strings in home in menRes to remove all leading and trailing blanks and multiple contiguous blanks. Also make all letters lower case and remove any punctuation such as '.' or ',' or ''' characters from the string. Assign the cleaned version of home into menRes. Call it homeClean.

Q.13 In Section 2.5 we created an id for a runner by pasting together name, year of birth, and state. Consider using the home town instead of the state. What is the impact on the matching? How many runners have competed in at least 8 races using this new id? What if you reduced the number of races to 6? Should this additional restriction be used in the matching process?

Q.14 Further refine the set of athletes in the longitudinal analysis by dropping those IDs (see Section 2.5) who have a large jump in time in consecutive races and who did not compete for two or more years in a row. How many unique IDs do you have when you include these additional restrictions? Does the longitudinal analysis in Section 2.6 change?

Q.15 Follow the procedures developed in Section 2.5 to clean the female runners' names and hometowns, and create longitudinal records for the females. Then follow the modeling of Section 2.6 to investigate the run time-age relationship for women who competed in multiple races.

Q.16 Consider adapting a non-parametric curve fitting approach to the longitudinal analysis. Rice [4] suggests modeling an individual's behavior as a combination of an average curve plus an individual curve. That is, the predicted performance for an individual comes from the sum of the average curve and the individual's curve: $Y_i(t) = \mu(t) + f_i(t) + \epsilon$, where $Y_i(t)$ is the performance of individual i at age t. He suggests a "two-step" process to do this: (a) take a robust average of all of the smoothed curves for the individuals; (b) subtract this average smoothed curve from the individual data points and smooth the residuals.

Rather than using only the individual's run times to produce the individual's curve, Rice also suggests smoothing over a set of nearest neighbors' times. Here a nearest neighbor is a runner with similar times for similar age.

Q.17 In Section 2.7, we discovered that the *HTML* file for the male 2000 results was so poorly formatted that htmlParse() was unable to fix it to allow us to extract the text table from the `<pre>` tag. In this exercise, we programmatically edit this *HTML* file so that we can use htmlParse() as desired. To do this, begin by reading the *HTML* file located at http://www.cherryblossom.org/cb003m.htm using readLines(). Carefully examine the *HTML* displayed in Section 2.7 and come up with a plan for correcting it. Consider whether you want to drop `<font>`s or close them properly. Once you have fixed the problem so that the `<pre>` tag contains the text table, pass your corrected *HTML* to htmlParse(). You may want to use a text connection to do this rather than writing the file to disk and reading it in.

Q.18 Revise the extractResTable() function in Section 2.7 so that it can read the male 2009 results. Carefully examine the raw *HTML* to determine how to find the information about the runners. Work with XPath to locate `<div>` and `<pre>` tags and extract the text value. The female 2009 results do not need this special handling. Modify the extractResTable() function to determine whether to perform the special processing of the `<div>` and `<pre>` tags.

Q.19 Revise the extractResTable() function in Section 2.7 so that it takes an additional parameter: *file*. Give the *file* parameter a default value of NULL. When NULL, the parsed results are returned from extractResTable() as a character vector. If not NULL, the results are written to the file named in *file*. The writeLines() function should be helpful here.

Bibliography

[1] Owen Barder. *Running for Fitness.* Owen Barder, Addis Ababa, 2010.

[2] Erich Neuwirth. RColorBrewer: ColorBrewer palettes. http://cran.r-project.org/web/packages/RColorBrewer, 2011. R package version 1.0-5.

[3] R Development Core Team. *R: A Language and Environment for Statistical Computing.* Vienna, Austria, 2012. http://www.r-project.org.

[4] John Rice. Functional and Longitudinal Data Analysis: Perspectives on smoothing. *Statistica Sinica*, 14:631–647, 2004.

[5] Deepayan Sarkar. *Lattice: Multivariate Data Visualization with R.* Springer-Verlag, New York, 2008. http://lmdvr.r-forge.r-project.org/figures/figures.html.

[6] Duncan Temple Lang. XML: Tools for parsing and generating *XML* within *R* and *S-PLUS*. http://www.omegahat.org/RSXML, 2011. R package version 3.4.

3

Using Statistics to Identify Spam

Deborah Nolan
University of California, Berkeley

Duncan Temple Lang
University of California, Davis

CONTENTS

3.1	Introduction	105
	3.1.1 Computational Topics	106
3.2	Anatomy of an email Message	107
3.3	Reading the email Messages	110
3.4	Text Mining and Naïve Bayes Classification	113
3.5	Finding the Words in a Message	116
	3.5.1 Splitting the Message into Its Header and Body	116
	3.5.2 Removing Attachments from the Message Body	117
	3.5.3 Extracting Words from the Message Body	124
	3.5.4 Completing the Data Preparation Process	126
3.6	Implementing the Naïve Bayes Classifier	127
	3.6.1 Test and Training Data	128
	3.6.2 Probability Estimates from Training Data	129
	3.6.3 Classifying New Messages	131
	3.6.4 Computational Considerations	135
3.7	Recursive Partitioning and Classification Trees	138
3.8	Organizing an email Message into an R Data Structure	140
	3.8.1 Processing the Header	141
	3.8.2 Processing Attachments	144
	3.8.3 Testing Our Code on More email Data	146
	3.8.4 Completing the Process	148
3.9	Deriving Variables from the email Message	150
	3.9.1 Checking Our Code for Errors	155
3.10	Exploring the email Feature Set	158
3.11	Fitting the rpart() Model to the email Data	160
3.12	Exercises	164
	Bibliography	169

3.1 Introduction

People are terrific at spotting spam in their mail reader with a quick glance at the subject line and sender, and when that approach is not conclusive, a glimpse at the contents of the message is usually enough to classify the message. But how do we design an automated procedure to classify and eliminate these unwanted messages to save us the time and irritation of having to sort through them in our inbox? Spam filters used by mail readers examine various characteristics of an email before deciding whether to place it in your inbox or spam folder. This decision is in part based on a statistical analysis of a large amount of email that has been hand classified as spam (unwanted) or ham (wanted). In this chapter, we examine over 9000 messages that have been classified by SpamAssassin (http://spamassassin.apache.org) for the purpose of developing and testing spam filters.

Before we can begin to analyze the information present in the SpamAssassin corpus, we need to process the messages into a form conducive to statistical analysis. The content of the message itself might be useful for analysis, but how do we organize and quantify this information? We can take a text mining approach where we tally up all the words occurring in a message and compare the frequencies of these words in ham and spam. Alternatively, or additionally, we can derive variables from characteristics of the message and use these to classify email. For example, the amount of capitalization in the subject line may be useful in ascertaining whether or not the message is spam.

Clearly we need to do a lot of text processing to get these thousands of messages into shape for either type of analysis. A first task is to bring the email into R [4] for processing. To do this we need to read thousands of files as each message is in its own file. This is the topic of Section 3.3. Then for each message, we locate its various parts: a) the header, which contains information about, e.g., the sender and subject; b) the message itself; and c) any attachments. However, before we can design the data extraction, we need to know more about the organization and format of a general email message. We describe the structure of email messages in Section 3.2.

We also need to know more about the analysis we want to perform. For text mining, we use the naïve Bayes method to approximate the likelihood a message is spam given the message content. This approach requires processing the message content to locate the words within the message, clean them up, e.g., handle punctuation and capitalization, and tally them. We describe the naïve Bayes technique in Section 3.4, then we prepare the email for analysis in Section 3.5, and carry out the analysis in Section 3.6. Alternatively, we employ a decision tree that uses derived variables, which represent characteristics of a message, to classify the messages. In order to derive these variables, we need to process the header and body of the email message to extract information. For example, we can count the number of characters in the message or look for excess capitalization and punctuation in the subject line. We follow the same sequence of tasks for this approach as we do with text mining. That is, in Section 3.7, we describe decision trees; then we process the email in Section 3.8 and derive the variables and explore them in Section 3.9 and Section 3.10, respectively; and lastly, we apply a recursive partitioning method in Section 3.11 to build a decision tree from the derived set of features. The presentations of these two approaches do not depend on one another so the reader may choose to focus on one approach only.

3.1.1 Computational Topics

The computational work in this chapter involves extensive text processing, which includes the following types of computations.

- Locate and process thousands of files by programmatically finding the names of the files and reading them into R in a general and automated manner.

- Manipulate strings and apply regular expressions to strings to turn the unstructured text in a message into structured information for analysis.

- Use cross-validation to select among competing models, e.g., by varying parameters in fitting a model, and evaluate the selected model using independent test data. This includes finding the Type I and II errors incurred when applying different model fits to test data.

- Efficiently and accurately compute with vectors containing large numbers of small values.

- Explore the email to design variables and develop intuition for what may be useful features in predicting spam and ham.

- Classification trees and recursive partitioning.

- Debugging techniques.

- Complex data structures, e.g., lists of lists of vectors and data frames.

- Writing modular functions to process the email files.

3.2 Anatomy of an email Message

An electronic mail message has two parts, a header and body. Analogous to surface mail, the header acts as an envelope, and the body is the letter that contains the contents for the recipient. In addition to the date, sender, and subject, the header has many other pieces of information, such as the message id, the carbon-copy (cc) recipients, and routing information as the message is relayed. These pieces of information are provided in a `key: value` format. That is, the key is the name of the kind of information being provided, and the value for that key follows the colon. For example, `From: debnolan@gmail.com` has a key of `From` and the value is the address of the sender. Sometimes the value contains more than one piece of information, and when this occurs, the pieces are separated by semicolons, e.g., `Content-Type: TEXT/PLAIN; charset=US-ASCII`.

The body of the message is separated from the header by a single empty line. When an attachment is added to a message, the attachment is included in the body of the message. That is, even with attachments, email messages still consist only of these two basic parts – header and body. They may not appear that way when you see them in your email reader, but that is because your email application displays the message in a way that makes it easy for you to read. To figure out what portion of the body is the message and what is an attachment, mail readers use an Internet standard called MIME (Multipurpose Internet Mail Extensions). When attachments are present, the `Content-Type` key in the header has a MIME type value of `multipart`, which indicates there are several documents in the body of the message. In addition, a boundary string is provided in the value. The boundary is a unique string not otherwise in the message, and it marks the start and end of the attachments. The receiving email reader looks for this string in the message body and uses it to divide the message into its various pieces.

We provide two sample messages as examples. The first is a plain text message with no attachments. It consists of an instructor's response to an inquiry sent by a student. The header includes 14 key: value pairs. Note the `Date` key includes a time-zone offset, the `Message-ID` key gives the unique ID to track the message from the `stat.berkeley.edu` email server, and the `Content-Type` key indicates a MIME type of `TEXT/PLAIN`; that is, the message is plain text with no attachments. Notice also that there is an additional line at the start of the message that is not in the key: value format, i.e.,

```
From nolan@stat.Berkeley.EDU Sun Feb  2 22:16:19 2014 -0800
Date: Sun, 2 Feb 2014 22:16:19 -0800 (PST)
From: nolan@stat.Berkeley.EDU
X-X-Sender: nolan@kestrel.Berkeley.EDU
To: Txxxx Uxxx <txxxx@uclink.berkeley.edu>
Subject: Re: prof: did you receive my hw?
In-Reply-To: <web-569552@calmail-st.berkeley.edu>
Message-ID:
  <Pine.SOL.4.50.040202216120.2296-100000@kestrel.Berkeley.EDU>
References: <web-569552@calmail-st.berkeley.edu>
MIME-Version: 1.0
Content-Type: TEXT/PLAIN; charset=US-ASCII
Status: O
X-Status:
X-Keywords:
X-UID: 9079

Yes it was received.

------------------------------------

On Sun, 2 Feb 2014, txxxx wrote:

> hey prof .nolan,
>
> i sent out my hw on sunday night. i just wonder did you receive
> it because i am kinda scared thatyou didnt' receive it.
> like i just wonder how do i know if you got it or not, since
> the cal mail system is kinda weird sometimes.  thanks
>
> txxxx
>
```

The second message consists of a text message and 2 attachments. It was sent by a student to the instructor and then forwarded by the instructor to the teaching assistant. We display only a small part of each attachment. The first attachment is a *PDF* file and the second is an *HTML* file.

```
From nolan@stat.Berkeley.EDU Sun Feb  2 22:18:56 2014 -0800
Date: Sun, 2 Feb 2014 22:18:55 -0800 (PST)
From: nolan@stat.Berkeley.EDU
X-X-Sender: nolan@kestrel.Berkeley.EDU
To: Gxxx   <lxxx@stat.Berkeley.EDU>
Subject: Assignment 1 sorry (fwd)
Message-ID:
  <Pine.SOL.4.50.040202218470.2296-201000@kestrel.Berkeley.EDU>
MIME-Version: 1.0
Content-Type: MULTIPART/Mixed;
   BOUNDARY="_===669732====calmail-me.berkeley.edu===_"
```

```
Content-ID: <Pine.SOL.4.50.040202218471.2296@kestrel.Berkeley.EDU>
Status: RO
X-Status:
X-Keywords:
X-UID: 9080

--_===669732====calmail-me.berkeley.edu===_
Content-Type: TEXT/PLAIN; CHARSET=US-ASCII; FORMAT=flowed
Content-ID: <Pine.SOL.4.50.040202218472.2296@kestrel.Berkeley.EDU>


---------- Forwarded message ----------
Date: Sun, 02 Feb 2014 21:50:47 -0800
From: Yyyy Zzz <Zzz@uclink.berkeley.edu>
To: nolan@stat.Berkeley.EDU
Subject: Assignment 1 sorry

I am sorry to send this email again, but my outbox told
me that the last email only send 1 attached file.
I am sending this again to make sure you recieve all
the necessary files.
Thank You and sorry for the inconvenience.

--_===669732====calmail-me.berkeley.edu===_
Content-Type: APPLICATION/PDF; CHARSET=US-ASCII
Content-Transfer-Encoding: BASE64
Content-ID: <Pine.SOL.4.50.040202218473.2296@kestrel.Berkeley.EDU>
Content-Description:
Content-Disposition: ATTACHMENT; FILENAME="PLOTS.pdf"

JVBERi0xLjEKJYHigeOBz4HTDQoxIDAgb2JqCjw8Ci9DcmVhdGlvbkRhdGUgKEQ6Mj
MDIxMTIwMTEpCi9Nb2REYXRlIChEOjIwMDQwMjAyMTEyMDExKQovVGl0bGUgKFI...

--_===669732====calmail-me.berkeley.edu===_
Content-Type: TEXT/HTML; CHARSET=US-ASCII
Content-Transfer-Encoding: BASE64
Content-ID: <Pine.SOL.4.50.040202218474.2296@kestrel.Berkeley.EDU>
Content-Description:
Content-Disposition: ATTACHMENT; FILENAME="Stat133HW1.htm"

PGh0bWwgeG1sbnM6bz0idXJuOnNjaGVtYXMtbWljcm9zb2Z0LWNvbTpvZmZpY2U6b2
Ig0KeG1sbnM6dz0idXJuOnNjaGVtYXMtbWljcm9zb2Z0LWNvbTpvZmZpY2U6d29...

--_===669732====calmail-me.berkeley.edu===_--
```

The first part of the body is the forwarded message, which is a regular part of the body. Note that it has its own short header indicating the content type is plain text. Next comes the *PDF* attachment, which the owner has named *PLOTS.pdf*, and the third part of the email is an *HTML* attachment. According to the header for the attachments, they are both encoded in base64, which is an encoding designed for representing binary data in an ASCII string format.

The Content-Type key in the header provides the boundary string for separating the attachments. This string is _===669732====calmail-me.berkeley.edu===_. We find 4 occurrences of the boundary in the email: at the start of the message, between the message and first attachment, between the first and second attachment, and at the end of the last attachment. Notice that the actual separator is the boundary string preceded by 2 hyphens, i.e., --_===669732====..., except at the end of the message where the boundary is both preceded and followed with 2 hyphens.

Now that we have a more complete understanding of the anatomy of an email message, we can better determine how to convert email into data that can be analyzed. In the next section, we tackle the job of reading email into R, and in Section 3.5 and Section 3.8 we use the information about a general email message to store the email in data structures that are appropriate for further processing and analysis.

3.3 Reading the email Messages

In order to read the raw text messages into R, we need to know where they are located and how they are organized in the file system on our computer. As mentioned already, these messages are made available by SpamAssassin (http://spamassassin.apache.org), but for convenience they also have been placed in the R package RSpamData [5]. After we know how the information is organized in the package, we can develop a function to read the email into R.

We can use a file finder, such as Spotlight, to check the organization of the messages or use simple command-line shell tools to examine the contents of the RSpamData package. Alternatively, we can use R to perform these operations. This is the approach that we take. In order to find the files in a machine-independent manner, we can use the system.file() function. We begin by finding the full path name to the RSpamData/ directory with

```
spamPath = system.file(package = "RSpamData")
```

We can list the files in *RSpamData* with list.files() and the directories with list.dirs(). We list the directories as follows:

```
list.dirs(spamPath, full.names = FALSE)
 [1]  ""                      "help"
 [3]  "html"                  "latex"
 [5]  "man"                   "messages"
 [7]  "messages/easy_ham"     "messages/easy_ham_2"
 [9]  "messages/hard_ham"     "messages/spam"
[11]  "messages/spam_2"       "Meta"
[13]  "R"                     "R-ex"
```

Note this listing is recursive, unless we turn off recursion with recursive = FALSE. The directory named messages/ has 5 subdirectories. We list only those files and directories in messages/ with

```
list.files(path = paste(spamPath, "messages",
                 sep = .Platform$file.sep))
```

Using Statistics to Identify Spam

```
[1] "easy_ham"   "easy_ham_2" "hard_ham"
[3] "spam"       "spam_2"
```

Notice that we specify the file separator in a machine-independent manner The names of these directories suggest that the ham messages have been organized into those that are easy or difficult to detect. The file names are a bit more inscrutable. We examine a few file names in `easy_ham/` with

```
head(list.files(path = paste(spamPath, "messages", "easy_ham",
                             sep = .Platform$file.sep)))
```

```
[1] "00001.7c53336b37003a9286aba55d2945844c"
[2] "00002.9c4069e25e1ef370c078db7ee85ff9ac"
[3] "00003.860e3c3cee1b42ead714c5c874fe25f7"
[4] "00004.864220c5b6930b209cc287c361c99af1"
[5] "00005.bf27cdeaf0b8c4647ecd61b1d09da613"
[6] "00006.253ea2f9a9cc36fa0b1129b04b806608"
```

We check that the names of the messages in the spam directories look the same as those in the ham directories with

```
head(list.files(path = paste(spamPath, "messages", "spam_2",
                             sep = .Platform$file.sep)))
```

```
[1] "00001.317e78fa8ee2f54cd4890fdc09ba8176"
[2] "00002.9438920e9a55591b18e60d1ed37d992b"
[3] "00003.590eff932f8704d8b0fcbe69d023b54d"
[4] "00004.bdcc075fa4beb5157b5dd6cd41d8887b"
[5] "00005.ed0aba4d386c5e62bc737cf3f0ed9589"
[6] "00006.3ca1f399ccda5d897fecb8c57669a283"
```

The SpamAssassin Web page at http://spamassassin.org/publiccorpus/ gives a description of the naming convention for these files. According to the Web site, the messages are named by a message number and their MD5 checksum. The MD5 checksum is a unique identifier derived from the contents of the file.

How many files are there all together? We use length() and list.files() to find out with

```
dirNames = list.files(path = paste(spamPath, "messages",
                      sep = .Platform$file.sep))
length(list.files(paste(spamPath, "messages", dirNames,
                        sep = .Platform$file.sep)))
```

```
[1] 9353
```

There are over 9000 messages in the 5 directories combined. These are not equally divided between the 5 directories, i.e.,

```
sapply(paste(spamPath, "messages", dirNames,
             sep = .Platform$file.sep),
       function(dir) length(list.files(dir)) )
```

```
  messages/easy_ham  messages/easy_ham_2   messages/hard_ham
               5052                 1401                 501
      messages/spam       messages/spam_2
               1001                 1398
```

There are only 501 messages in *hard_ham* and about one quarter of the email is spam.

Given the organization and volume of files, we cannot simply read the files into *R* by writing calls such as

```
readLines("messages/easy_ham/00006.3ca1f399ccda5d897fecb8c57...")
```

How do we read the contents of the files into *R* in a more general, automated, machine-independent manner? And, is readLines() the function we should use?

Let's address the second question first. The manual, *R Data Import/Export* [3] describes several functions available to us in *R* for reading input from files. Depending on what we want to end up with, different functions are easier to use; they all have different purposes and some provide greater control at the expense of additional complexity. Since our email is free-formatted text, it is probably easiest to import the contents of each file as a sequence of lines. In other words, when we read a message, we want to obtain a character vector with one string per line in the file. The readLines() function offers this capability so we use it.

Now to automate the process, we want to avoid typing the file name ourselves. If a file's name has been assigned to a string, say, in the variable fileName, then we can pass the file name to readLines() with

```
readLines(fileName)
```

In our case this string might be

```
~/RPackages/RSpamData/messages/easy_ham/
    00006.3ca1f399ccda5d897fecb8c57669a283
```

In order to find the file names in a machine-independent manner, we again use the list.files() function. Recall that spamPath contains the full path name to RSpamData and dirNames contains the names of the 5 subdirectories in messages/ that contain the message files. We construct the full name for these directories by pasting together these strings with

```
fullDirNames = paste(spamPath, "messages", dirNames,
                sep = .Platform$file.sep)
```

We obtain the full names of the files within the first directory with

```
fileNames = list.files(fullDirNames[1], full.names = TRUE)
fileNames[1]
```

```
[1] "/Users/nolan/RPackages/RSpamData/messages/easy_ham/
    00001.7c53336b37003a9286aba55d2945844c"
```

Then, we can read the first message in *easy_ham* with

```
msg = readLines(fileNames[1])
head(msg)
```

```
[1] "From exmh-workers-admin@redhat.com   Thu Aug 22 12:36:23..."
[2] "Return-Path: <exmh-workers-admin@spamassassin.taint.org>"
[3] "Delivered-To: zzzz@localhost.netnoteinc.com"
[4] "Received: from localhost (localhost [127.0.0.1])"
[5] "\tby phobos.labs.netnoteinc.com (Postfix) with ESMTP id..."
[6] "\tfor <zzzz@localhost>; Thu, 22 Aug 2002 07:36:16 -0400..."
```

Using Statistics to Identify Spam 113

We have successfully located the email files and determined how to programmatically read them into R as character vectors, but before we attempt to wrap this code into a function and read all the email into R, let's consider how to prepare each message for analysis. To do this, we want to think about what we want to end up with. What parts of the message do we want to keep? How do we want to analyze the email? If we just start writing code, there is a danger that we will get confused and the code will become intertwined with doing several different things. The answers to these questions depend on the kind of analysis we want to perform. In the next section, we provide a brief summary of the naïve Bayes approach to classifying email that uses only the content of the message. After we understand how to carry out this statistical analysis, we will be able to answer these questions about how to process and store the email.

Lastly to assist us, we select a small set of email messages to use as test cases as we develop our code. We have chosen, by manual inspection, 15 ham message files from the first directory that exhibit different characteristics of email. We read them into R as follows:

```
indx = c(1:5, 15, 27, 68, 69, 329, 404, 427, 516, 852, 971)
fn = list.files(fullDirNames[1], full.names = TRUE)[indx]
sampleEmail = sapply(fn, readLines)
```

Of course, we have no spam or hard ham in our sample. We may want to revisit this selection later to ensure that we have email that are representative of the different cases our code needs to be able to handle.

3.4 Text Mining and Naïve Bayes Classification

Naïve Bayes is a probability-based approach to classification. This approach begins by studying the content of a large collection of email messages that have already been read and classified as spam or ham. Then when a new message comes to us, we use the information gleaned from our "training" set to compute the probability that the new message is spam. For example, suppose we receive a new message that says, "Are your taxes too high?". We use the probability:

$$\mathbb{P}(\text{message is spam} \,|\, \text{message content: "Are your taxes too high?"})$$

to determine how likely it is that a message with this content is spam. We can also include in this probability computation other information about the email such as the number of attachments and the percentage of letters that are capitals, but our focus in this section is only on the text of the message. To compute this probability, we re-express it using Bayes' Rule as follows:

$$\mathbb{P}(\text{message is spam} \,|\, \text{message content}) = \frac{\mathbb{P}(\text{message content} \,|\, \text{spam})\mathbb{P}(\text{spam})}{\mathbb{P}(\text{message content})}$$

At first glance, it doesn't look as though Bayes' Rule has simplified the probability calculations at all. Now we have 3 probabilities to compute instead of one. However, the probability of spam is easily estimated by the proportion of spam messages in the training set of messages. And, we soon see that we do not need to calculate $\mathbb{P}(\text{message content})$.

We do need to compute the probability that a spam message has the new message's content. This is where the "naïve" simplification of Bayes' Rule comes into play. We assume that the chance a particular word is in the message is independent of all other words in the message. That is, the probability that `high` appears in a spam message is independent of the probability that `taxes` appears in the message, i.e., $\mathbb{P}(\text{high}|\text{taxes}) = \mathbb{P}(\text{high})$. Clearly this is not the case, but making this assumption greatly simplifies the computations and turns out to still be effective in identifying spam.

Suppose the set of unique words in all of the training messages ranges from `apple` to `zebra`. Then this naïve assumption says that the likelihood of our message content is the following product of probabilities for all words:

$$\mathbb{P}(\text{message content} \mid \text{spam}) \approx \mathbb{P}(\textit{not } \text{apple}|\text{spam}) \times \cdots \times$$
$$\mathbb{P}(\text{high} \mid \text{spam}) \times \cdots \times \mathbb{P}(\text{taxes} \mid \text{spam}) \times$$
$$\cdots \times \mathbb{P}(\textit{not } \text{zebra}|\text{spam}).$$

That is, each probability in the product on the right-hand side of the above approximation is either the probability that an individual word is present given the message is spam or the probability that the word is absent given the message is spam. To estimate these probabilities we have only to find the frequencies of these words in the spam portion of the training set. For example, we approximate $\mathbb{P}(\text{high} \mid \text{spam})$ by the empirical fraction of spam messages containing the word `high` in the training set.

Similarly, we can estimate $\mathbb{P}(\text{message content} \mid \text{ham})$ using Bayes' Rule and the naïve simplification. That is,

$$\mathbb{P}(\text{message content} \mid \text{ham}) \approx \mathbb{P}(\textit{not } \text{apple}|\text{ham}) \times \cdots \times$$
$$\mathbb{P}(\text{high} \mid \text{ham}) \times \cdots \times \mathbb{P}(\text{taxes} \mid \text{ham}) \times$$
$$\cdots \times \mathbb{P}(\textit{not } \text{zebra}|\text{ham}).$$

And, we estimate each of the probabilities on the right-hand side of the approximation with the empirical proportion, e.g., we approximate $\mathbb{P}(\text{high} \mid \text{ham})$ by the fraction of ham messages that contain the word `high`.

The classification of a message depends on the likelihood ratio:

$$\frac{\mathbb{P}(\text{spam} \mid \text{message content})}{\mathbb{P}(\text{ham} \mid \text{message content})}$$

The possible values of this ratio range between 0 and ∞. The ratio is 1 when spam and ham are equally likely, greater than 1 when the probability the message is spam is more likely than the probability it is ham, and less than 1 when it is less likely.

It is often easier to work with this ratio in part because it eliminates the need to compute $P(\text{message content})$. Applying the naïve Bayes approximation to the numerator and denominator yields:

$$\frac{\mathbb{P}(\text{spam}|\text{message content})}{\mathbb{P}(\text{ham}|\text{message content})}$$
$$\approx \frac{\mathbb{P}(\textit{not } \text{apple}| \text{spam})}{\mathbb{P}(\textit{not } \text{apple}| \text{ham})} \times \cdots \times \frac{\mathbb{P}(\text{high}| \text{spam})}{\mathbb{P}(\text{high}| \text{ham})} \times \cdots \times$$
$$\frac{\mathbb{P}(\textit{not } \text{zebra}|\text{spam})}{\mathbb{P}(\textit{not } \text{zebra} |\text{ham})} \times \frac{\mathbb{P}(\text{spam})}{\mathbb{P}(\text{ham})}$$

Our calculation has been reduced to a product of ratios of probabilities that are simple to estimate from the training data.

Using Statistics to Identify Spam

One further mathematical convenience we use is to take the log of the ratio of the conditional probabilities above. Two benefits to taking the log are that the product becomes a sum, and values between 0 and 1 get "stretched out" between $-\infty$ and 0. This latter fact often means that the logged values have good statistical properties. The log odds ratio appears as:

$$\log\left(\frac{P(\text{ spam } | \text{ message content})}{P(\text{ ham } | \text{ message content})}\right)$$
$$\approx \log\left(\frac{P(not\ \text{apple}|\text{spam})}{P(\ not\ \text{apple}|\text{ham})}\right) + \cdots + \log\left(\frac{P(\text{high}|\text{spam})}{P(\text{high}|\text{ham})}\right) +$$
$$\cdots + \log\left(\frac{P(not\ \text{zebra}|\text{spam})}{P(not\ \text{zebra}|\text{ham})}\right) + \log\left(\frac{P(\text{spam})}{P(\text{ham})}\right)$$

One last computational consideration arises when a word appears solely in ham. In this situation, our estimate is 0 for the probability of this word given a message is spam, which is problematic when we take logs. To remedy this, we "smooth" all of the word counts by adding 0.5 to them, i.e.,

$$\mathbb{P}(\text{high} \mid \text{spam}) \approx \frac{\#\text{ of spam messages with high } + 1/2}{\#\text{ of spam messages } + 1/2}$$

We use the approximate log odds to classify messages. Our simple classification rule uses the training data to estimate probabilities like the one above; then we compute the log odds on test messages. That is, we include, e.g., either $\log P(\text{high} \mid \text{spam})$ or $\log \mathbb{P}(not\ \text{high} \mid \text{spam})$ in the sum depending on whether or not the test message contains the word high. When the log likelihood ratio for a new message exceeds some threshold, it is classified as spam. Otherwise, it is classified as ham.

Now that we understand the naïve Bayes approach to text mining and classification, we can determine how to process the email for this kind of analysis. From each message, we need the set of words it contains, and we need the collection of unique words across all the messages. This all-encompassing set is called a "bag of words" (BOW). The probabilities that we compute are based on the presence and absence in a message of each word in the bag of words. Our tasks then are to:

- Transform a message body into a set of the words.

- Combine the words across messages into a bag of words.

After we have prepared each message in this way, we can carry out our analysis. For this analysis, we need to:

- Tally the frequencies in the training data of words in spam and ham separately to estimate the probability a word appears in a message given it is spam (or ham) from the proportion of spam (or ham) messages containing that word.

- Estimate the likelihood that a new test message is spam (or ham) given its contents, i.e., given the message's words compute the naïve Bayes version of the log likelihood ratio.

- Find a threshold for the log likelihood ratio, where a message with a value above the threshold is classified as spam. We choose this threshold by examining the error rates for the test data.

Since we do not have test messages, we divide our email corpus into 2 parts and use one part as our pseudo test set and the other as our training set. We do this after we have simplified all of the email into their word sets.

3.5 Finding the Words in a Message

We need to access the body of the message in order to extract its words. Also, we need to eliminate the attachments from the body as we are not interested in this portion of the body. (We leave it as an exercise to extract the words from any text attachments and include them in the set of a message's words.) Once we have located the relevant portion of the message body, our task is to extract its words. We tackle each of these 3 steps in turn and place the code in functions, splitMessage(), dropAttach(), and findMsgWords(), respectively.

3.5.1 Splitting the Message into Its Header and Body

Recall from Section 3.2 that the header and body of the message are separated by an empty line. This should be the first empty line in the email. We can find this line by finding all the empty lines and then choosing the first of these. We work with the first message in sampleEmail and find the index of the first empty line with

```
msg = sampleEmail[[1]]
which(msg == "")[1]
[1] 63
```

An alternative way to do this is to use the match() function. It returns the position of the first matching element in the specified object. If we look for "" in our lines, we get the location of the line separating the header and body:

```
match("", msg)
[1] 63
```

Let's assign this location to splitPoint, i.e.,

```
splitPoint = match("", msg)
```

To confirm that we have correctly found the division between the header and body, we examine a few lines in msg on either side of splitPoint with

```
msg[ (splitPoint - 2):(splitPoint + 6) ]
[1] "List-Archive: <https://listman.spamassassin.taint.org/..."
[2] "Date: Thu, 22 Aug 2002 18:26:25 +0700"
[3] ""
[4] "    Date:        Wed, 21 Aug 2002 10:54:46 -0500"
[5] "    From:        Chris Garrigues <cwg-dated-1030377287..."
[6] "    Message-ID:  <1029945287.4797.TMDA@deepeddy.vircio.com>"
[7] ""
[8] ""
[9] "  | I can't reproduce this error."
```

This may be a bit confusing because we have an indented header from another message within the body of this message, but it appears we have correctly located the empty line that marks the beginning of the body.

Simple subsetting gives us the header and the body. The header is the first, second, third, etc. lines up to but not including splitPoint, and the body includes all lines past splitPoint. That is,

Using Statistics to Identify Spam

```
header = msg[1:(splitPoint-1)]
body = msg[ -(1:splitPoint) ]
```

We close this section by collecting the code we have written into the splitMessage() function. The input to this function is the character vector returned from readLines() (see Section 3.3), and the output is a list of 2 character vectors comprising the header and body. Our simple function is

```
splitMessage = function(msg) {
  splitPoint = match("", msg)
  header = msg[1:(splitPoint-1)]
  body = msg[ -(1:splitPoint) ]
  return(list(header = header, body = body))
}
```

We apply this function to our sample messages with

```
sampleSplit = lapply(sampleEmail, splitMessage)
```

We have found the body of the message, and we next tackle the removal of any attachments.

3.5.2 Removing Attachments from the Message Body

We saw in Section 3.2 that when an email message has attachments, the MIME type is multipart and the Content-Type field provides a boundary string that can be used to locate the attachments. In the example provided there, the Content-Type field is

```
Content-Type: MULTIPART/Mixed;
   BOUNDARY="_===669732====calmail-me.berkeley.edu===_"
```

It seems our first step is to find the Content-Type key and use its value to determine whether or not an attachment is present. If so, then we find the boundary string and use this string to locate the attachments.

We work with the first message in our sample and use the grep() function to locate Content-Type in the header with

```
header = sampleSplit[[1]]$header
grep("Content-Type", header)
```

```
[1] 46
```

We have successfully found the Content-Type key in the 46th element of header. We next use this Content-Type's value to determine whether the message has any attachments, i.e., we check whether or not the Content-Type is "multipart". When we examine the messages in sampleEmail we see that the MIME type is not consistently capitalized so we convert header[46] to lower case before searching for the term multipart. We again use grep() to do this, i.e.,

```
grep("multi", tolower(header[46]))
```

```
integer(0)
```

It appears this message has no attachments. We double check with

```
header[46]

[1] "Content-Type: text/plain; charset=us-ascii"
```

Indeed, it has only a plain text body.

We can apply this call to grep() to all of the headers in the list of sample messages with

```
headerList = lapply(sampleSplit, function(msg) msg$header)
CTloc = sapply(headerList, grep, pattern = "Content-Type")
CTloc

[[1]]
[1] 46
...
[[6]]
[1] 54

[[7]]
integer(0)
...
```

The sapply() did not return a vector as expected because the seventh element has no Content-Type key. To remedy this, we can check for a missing Content-Type field and return 0 or NA in this case so that we have a numeric vector to work with, i.e.,

```
sapply(headerList, function(header) {
                 CTloc = grep("Content-Type", header)
                 if (length(CTloc) == 0) return(NA)
                 CTloc
              })

 [1] 46 45 42 30 44 54 NA 21 17 52 31 52 52 27 31
```

Finally, we add the check for a multipart MIME type with

```
hasAttach = sapply(headerList, function(header) {
  CTloc = grep("Content-Type", header)
  if (length(CTloc) == 0) return(FALSE)
  grepl("multi", tolower(header[CTloc]))
})

hasAttach

 [1] FALSE FALSE FALSE FALSE FALSE  TRUE FALSE  TRUE  TRUE
[10]  TRUE  TRUE  TRUE  TRUE  TRUE  TRUE
```

Note that grepl() returns a logical indicating whether there was a match or not. Several of the messages in our sample have attachments.

We have used grep() and grepl() to search for specific literals in our header strings. For example, grepl("multi", header) searches in each element of the character vector header for an m followed by a u then by an l and so on. This sequence of 5 literals can appear anywhere in the string, and if it does, grep() returns the indices of the elements where it found a match. This is a very simple example of pattern matching using regular expressions. The first argument to grep() is a regular expression and the second is the vector of strings in which to search. Regular expression matching is far more powerful and flexible than this simple example demonstrates. We use more of the features of regular expressions next as we search for the boundary string.

We need to extract the boundary string from those messages that have attachments in order to locate and remove the attachments. There are several ways to extract the boundary string from the Content-Type value. We leave a string manipulation approach to the exercises and use regular expressions and the sub() function here. Essentially we want to discard all of the string except for the boundary so our goal is to create a regular expression that identifies that part of the string which is the boundary. We locate the boundary string in our sixth message as follows:

```
header = sampleSplit[[6]]$header
boundaryIdx = grep("boundary=", header)
header[boundaryIdx]

[1] "    boundary=\"==_Exmh_-1317289252P\";"
```

The boundary string begins after 'boundary="' and ends before the ';' character.

In pseudocode, we want to create a pattern like the following:

```
any characters followed by
  boundary="(string we are looking for)"; any characters
```

The actual pattern we use is '.*boundary="(.*)";.*'. This pattern uses many of the special characters available in the regular expression language. The pattern begins with a '.; character, which stands for any literal. It is followed by the '*; quantifier, meaning any number of times so the pattern '.*' matches any number of arbitrary literals. However, these must be followed by the literals 'boundary=' and a quotation mark. The boundary value is the string that follows, up to a quotation mark, followed by a semicolon and then any characters. That part of the pattern within the parentheses is our boundary string. That is, the pattern (.*) does not match the literal parentheses, but uses them to group together the characters that match the pattern within them. Note that it matches any characters any number of times, but it must be followed by a quotation mark and a semicolon. The use of the parentheses to identify a sub-pattern gives us access to these matching characters later. They can be referred to using a variable, specifically \\1. In the following call to sub(), we use the contents of this variable as a substitute for the entire string. That is,

```
sub(".*boundary=\"(.*)\";.*", "\\1", header[boundaryIdx])

[1] "==_Exmh_1547759024P"
```

The first argument to sub() is the pattern that we search for in header[boundaryIdx], and the second argument contains the substitution for the matching substring. The sub() function allows us to modify part of the input string. Here we are processing all of it to remove the pieces we do not want. That is, if we have written our first pattern correctly, we match the entire string and replace it with the piece that contains the boundary string.

Although our first application of our pattern successfully extracted the boundary string from the header, pattern matching can be tricky and we want to try it on other strings. For example, we apply the call to sub() to the ninth message in our sample, i.e.,

```
header2 = headerList[[9]]
boundaryIdx2 = grep("boundary=", header2)
header2[boundaryIdx2]
```

```
[1] "Content-Type: multipart/alternative;
     boundary=Apple-Mail-2-874629474"
```

Notice that the boundary string does not appear in quotes and there is no semicolon at the end. Our pattern matching fails, i.e.,

```
sub('.*boundary="(.*)";.*', "\\1", header2[boundaryIdx2])
```

```
[1] "Content-Type: multipart/alternative;
     boundary=Apple-Mail-2-874629474"
```

We have not successfully located the boundary string because of the missing quotation marks and semicolon. Searching for quotation marks is potentially problematic as not all boundaries appear in quotes. If we eliminate quotation marks from the string then we can drop them from our pattern as well. This is a simpler approach than searching for optional quotation marks. We eliminate them with

```
boundary2 = gsub('"', "", header2[boundaryIdx2])
```

The substitution string is empty so this is equivalent to eliminating the quotation marks from `header2[boundaryIdx2]`. Notice that we use the gsub() function, rather than sub(). The "g" stands for global, which means that all occurrences of a quotation mark in the string are found and substituted, rather than only the first occurrence.

We have not yet solved the problem of correctly identifying that portion of the string that contains the boundary information because we have not addressed the case of a Content-Type value that has no semicolon. Let's change our pattern to make the semicolon optional by adding a '?' after the semicolon in the pattern. Let's also allow any number of blanks (0 or more) between boundary= and the boundary string, i.e.,

```
sub(".*boundary= *(.*);?.*", "\\1", boundary2)
```

```
[1] "Apple-Mail-2-874629474"
```

That seems to have done it!

Let's check that this revised pattern successfully finds the boundary string in our first example. When we do, we find that the pattern no longer finds the boundary string in that message's Content-Type value. It worked before, but we have broken the pattern matching, i.e.,

```
boundary = gsub('"', "", header[boundaryIdx])
sub(".*boundary= *(.*);?.*", "\\1", boundary)
```

```
[1] "==_Exmh_-1317289252P;"
```

Our pattern no longer correctly finds the end of the boundary string, but instead includes the semicolon. This is a case of greedy matching. We are allowing any character within our parentheses, including the semicolon, and the semicolon at the end of the string is now optional. We can exclude the semicolon from matching by using [^;] in the expression, which matches all characters except the semicolon. Our revised pattern is

```
sub(".*boundary= *([^;]*);?.*", "\\1", boundary)
```

```
[1] "==_Exmh_-1317289252P"
```

Now we have again successfully located the boundary string from the sixth message, and when we try our revised regular expression on the ninth message, we find that it still works.

Although we did not initially identify the task of finding the boundary string as a separate function, we can wrap this code into its own function, which we call getBoundary(). The only input required is the header and the function returns the boundary string. We do this with

```
getBoundary = function(header) {
  boundaryIdx = grep("boundary=", header)
  boundary = gsub('"', "", header[boundaryIdx])
  gsub(".*boundary= *([^;]*);?.*", "\\1", boundary)
}
```

We are now ready to search through the body of the message for attachments. To get a better sense of the format of these bodies and attachments, we examine a few more messages, e.g.,

```
sampleSplit[[6]]$body
 [1] "--==_Exmh_-1317289252P"
 [2] "Content-Type: text/plain; charset=us-ascii"
 [3] ""
 [4] "> From:    Chris Garrigues <cwg-exmh@DeepEddy.Com>"
 [5] "> Date:    Wed, 21 Aug 2002 10:40:39 -0500"
 [6] ">"
...
[43] "  World War III:  The Wrong-Doers Vs. the Evil-Doers."
[44] ""
[45] ""
[46] ""
[47] ""
[48] "--==_Exmh_-1317289252P"
[49] "Content-Type: application/pgp-signature"
[50] ""
[51] "-----BEGIN PGP SIGNATURE-----"
[52] "Version: GnuPG v1.0.6 (GNU/Linux)"
[53] "Comment: Exmh version 2.2_20000822 06/23/2000"
[54] ""
[55] "iD8DBQE9ZQJ/K9b4h5R0IUIRAiPuAJwL4mUus5whLNQZC8MsDlGpEdK..."
[56] "PcGgN9frLIM+C5Z3vagi2wE="
[57] "=qJoJ"
[58] "-----END PGP SIGNATURE-----"
[59] ""
[60] "--==_Exmh_-1317289252P--"
[61] ""
[62] ""
[63] ""
[64] "_____"
[65] "Exmh-workers mailing list"
[66] "Exmh-workers@redhat.com"
[67] "https://listman.redhat.com/mailman/listinfo/exmh-workers"
[68] ""
```

We see that this body contains one attachment, which is a PGP signature, and each body part has its own short header. Also, there are 8 lines following the end of the attachment.

Another message body in our sample appears as

```
 [1]  "This is a multi-part message in MIME format."
 [2]  ""
 [3]  "------=_NextPart_000_0005_01C26412.7545C1D0"
 [4]  "Content-Type: text/plain;"
 [5]  "\tcharset=\"iso-8859-1\""
 [6]  "Content-Transfer-Encoding: 7bit"
 [7]  ""
 [8]  "liberalism"
...
[27]  " http://www.english.upenn.edu/~afilreis/50s/schleslib.html"
[28]  ""
[29]  "------=_NextPart_000_0005_01C26412.7545C1D0"
[30]  "Content-Type: application/octet-stream;"
[31]  "\tname=\"Liberalism in America.url\""
[32]  "Content-Transfer-Encoding: 7bit"
[33]  "Content-Disposition: attachment;"
[34]  "\tfilename=\"Liberalism in America.url\""
[35]  ""
[36]  "[DEFAULT]"
[37]  "BASEURL=http://www.english.upenn.edu/~afilreis/50s/sch..."
[38]  "[InternetShortcut]"
[39]  "URL=http://www.english.upenn.edu/~afilreis/50s/schlesl---"
[40]  "Modified=E0824ED43364C201DE"
[41]  ""
[42]  "------=_NextPart_000_0005_01C26412.7545C1D0--"
[43]  ""
[44]  ""
[45]  ""
```

Here we find that there are a few lines in the body preceding the first boundary string and a few lines after the closing string. Lines 4, 5, and 6 contain header information for the first part of the body, i.e., the message. Lines 30 through 34 are header lines for the attachment. Also note that there is an empty line between the header information for each portion of the body and the content itself, e.g., lines 7 and 35 are empty. That is, each body part has a structure that mimics the structure of the message with header information separated from the content with a blank line.

We examine one more message, the 11th in our sample:

```
 [1]  ""
 [2]  "---------------090602010909000705010009"
 [3]  "Content-Type: text/plain; charset=ISO-8859-1; format=flowed"
 [4]  "Content-Transfer-Encoding: 8bit"
 [5]  ""
 [6]  "Geege wrote:"
...
[63]  "Check out the pictures."
[64]  ""
[65]  ""
```

```
[66] ""
[67] ""
[68] "--------------09060201090900070501009--"
[69] ""
[70] ""
```

Note that this body contains no attachment. There are two occurrences of the boundary string — one at the start of the body and one at the end. That is, there is no boundary string to separate the message text from the attachment. The header information within the body indicates that the format is flawed.

With these examples in hand, we can begin to design a way to extract the attachments from the body of the message. Our investigation has shown that some messages do not have an attachment even though their header indicates that they are supposed to. We also must decide what to do with the lines that appear before the first boundary string and after the closing boundary string. Additionally, we might want to address the situation when the last boundary string is not found. We have not come across such a case yet, but it seems like a reasonable precaution to take. Let's write our function to do the following.

- Drop the blank lines before the first boundary string.

- Keep the lines following the closing string as part of the first portion of the body and not the attachments.

- Use the last line of the email as the end of the attachment if we find no closing boundary string.

We prepare the last message in our sample with

```
boundary = getBoundary(headerList[[15]])
body = sampleSplit[[15]]$body
```

We search in the body for the boundary string preceded by 2 hyphens with

```
bString = paste("--", boundary, sep = "")
bStringLocs = which(bString == body)
bStringLocs
```

```
[1]    2 35
```

These lines in the body mark the start of each portion of the email. Next, we find the closing boundary with

```
eString = paste("--", boundary, "--", sep = "")
eStringLoc = which(eString == body)
eStringLoc
```

```
[1] 77
```

We can locate the first part of the message from the body, excluding the attachments, with

```
msg = body[ (bStringLocs[1] + 1) : (bStringLocs[2] - 1)]
tail(msg)
```

```
[1] ">"        ">Yuck" ">    "    ">"        ""         ""
```

To add the lines that appear after the last attachment to this part of the message, we do the following

```
msg = c(msg, body[ (eStringLoc + 1) : length(body) ])
tail(msg)
```

```
[1] "" ""    ""    ""    ""
```

It appears we have added several empty lines.

We leave as an exercise the creation of the dropAttach() function. It follows the basic operations explored in this section. However, the special cases described earlier need to be addressed, e.g., when there is no attachment despite the header supplying a MIME type of multipart and a boundary string.

Next we explore how to extract the words from a message.

3.5.3 Extracting Words from the Message Body

We begin the task of separating the message into words by looking over the first few lines of a couple of messages in sampleSplit with

```
head(sampleSplit[[1]]$body)
```

```
[1] "     Date:         Wed, 21 Aug 2002 10:54:46 -0500"
[2] "     From:
       Chris Garrigues <cwg-dated-1030377287.06fa6d@DeepEddy.Com>"
[3] "     Message-ID: <1029945287.4797.TMDA@deepeddy.vircio.com>"
[4] ""
[5] ""
[6] "  | I can't reproduce this error."
```

and

```
msg = sampleSplit[[3]]$body
head(msg)
```

```
[1] "Man Threatens Explosion In Moscow "
[2] ""
[3] "Thursday August 22, 2002 1:40 PM"
[4] "MOSCOW (AP) - Security officers on Thursday
        seized an unidentified man who"
[5] "said he was armed with explosives and threatened
        to blow up his truck in"
[6] "front of Russia's Federal Security Services
        headquarters in Moscow, NTV"
```

What are some issues that we can see from these few lines that we need to handle in our code? Clearly, we need to address capitalization because we want to count Federal and federal as the same word. We also want to discard periods, commas, semicolons, etc., but do we keep haven't and Russia's as words? Do we eliminate numbers? What about URLs and email addresses? And how do we handle plural words and past tense? That is, do we treat services as different from service and are armed and arm distinct words?

These few lines have uncovered many issues that we need to make decisions about. In text mining, words are often "stemmed," meaning that the plural is converted to the singular and past tense to present, etc. There are stemming packages in R to perform this conversion, which we leave as an exercise to investigate. Also, depending on the purpose of the analysis many short common words are ignored in text analysis. That is, words such as: as, to, of, so, etc. are dropped from the document's content. These words are called stop words and we leave it as an exercise to create a vector of English stop words from the functionality provided in the tm package [1]. (Of course, each language has its own stop words and stemming procedures.)

To process the text, we can use regular expressions. For example, we can substitute all punctuation and digits with a blank so, e.g., don't becomes don t and Russia's is changed to Russia s. This may not be the ideal transformation, but it is simple and a good first attempt. The outcome creates "words" of one letter, which we want to eliminate. The conversion appears as

```
tolower(gsub("[[:punct:]0-9[:blank:]]+", " ", msg))
```

Here the + converts multiple occurrences of these unwanted characters into a single blank. The term [:punct:] is a named character class that matches any punctuation symbol, the term 0-9 stands for the digits 0, 1, 2, ..., 9, and the term [:blank:] is a named class that matches a space or a tab character. All 3 of these terms are collected within one set of [] meaning that any punctuation mark, digit, or type of blank is treated as an equivalent character and matched.

We check this transformation for a few lines in one message. The original lines are

```
msg[ c(1, 3, 26, 27) ]
```

[1] "Man Threatens Explosion In Moscow "
[2] "Thursday August 22, 2002 1:40 PM"
[3] "4 DVDs Free +s&p Join Now"
[4] "http://us.click.yahoo.com/pt6YBB/NXiEAA/mG3HAA/7gSolB/TM"

We convert the strings as described above and find:

```
cleanMsg = tolower(gsub("[[:punct:]0-9[:blank:]]+", " ", msg))
cleanMsg[ c(1, 3, 26, 27) ]
```

[1] "man threatens explosion in moscow "
[2] "thursday august pm"
[3] " dvds free s p join now"
[4] "http us click yahoo com pt ybb nxieaa mg haa gsolb tm"

Notice that much of the *URL* in the original string becomes several gibberish words. We leave it as an exercise to improve this translation process.

After the translation, we can divide the strings into words by splitting the string on blanks, i.e.,

```
words = unlist(strsplit(cleanMsg, "[[:blank:]]+"))
```

We can drop one letter words with

```
words = words[ nchar(words) > 1 ]
```

and we can remove the stop words with a vector of stop words called, e.g., stopWords, with

```
words = words[ !( words %in% stopWords) ]
head(words)

[1] "man"      "threatens" "explosion" "moscow"
[5] "thursday" "august"
```

Notice the element `"in"` was dropped from the collection of words in this message. We leave it as an exercise to wrap this code up into a function called findMsgWords(). This function takes as input the message content with all attachments eliminated. The return value from findMsgWords() is a vector of the unique words in the message, i.e., we only track which words are in the message, not the number of times they appear. Additionally, we suggest considering whether it is simpler to split the string by blanks first and then process the punctuation, digits, etc.

3.5.4 Completing the Data Preparation Process

We have completed all of the tasks to process one message. Before we can move on to compute the log likelihood ratio (Section 3.4), we need to process all of the email in the SpamAssassin corpus. Then we can determine the bag of words and estimate the probability, e.g., that a spam message has the word `federal` in it.

Below is our function to carry out the processing of all the email in one of the directories. Note that the input is the full path name of the directory. The function contains calls to our functions, splitMessage(), getBoundary(), dropAttach(), and findMsgWords(). The code uses many of the computations from the preparatory code for processing the set of sample messages.

```
processAllWords = function(dirName, stopWords)
{
     # read all files in the directory
  fileNames = list.files(dirName, full.names = TRUE)
     # drop files that are not email, i.e., cmds
  notEmail = grep("cmds$", fileNames)
  if ( length(notEmail) > 0) fileNames = fileNames[ - notEmail ]

  messages = lapply(fileNames, readLines, encoding = "latin1")

     # split header and body
  emailSplit = lapply(messages, splitMessage)
     # put body and header in own lists
  bodyList = lapply(emailSplit, function(msg) msg$body)
  headerList = lapply(emailSplit, function(msg) msg$header)
  rm(emailSplit)

     # determine which messages have attachments
  hasAttach = sapply(headerList, function(header) {
    CTloc = grep("Content-Type", header)
    if (length(CTloc) == 0) return(0)
    multi = grep("multi", tolower(header[CTloc]))
    if (length(multi) == 0) return(0)
    multi
  })
```

Using Statistics to Identify Spam

```
    hasAttach = which(hasAttach > 0)
    
         # find boundary strings for messages with attachments
    boundaries = sapply(headerList[hasAttach], getBoundary)
    
         # drop attachments from message body
    bodyList[hasAttach] = mapply(dropAttach, bodyList[hasAttach],
                                boundaries, SIMPLIFY = FALSE)
    
         # extract words from body
    msgWordsList = lapply(bodyList, findMsgWords, stopWords)
    
    invisible(msgWordsList)
}
```

Finally, we apply processAllWords() to each directory with

```
msgWordsList = lapply(fullDirNames, processAllWords, 
                      stopWords = stopWords)
```

In addition to the collections of words in the messages, we also need to know which messages are spam or ham. We can create a logical vector based on the number of elements in each list.

```
numMsgs = sapply(msgWordsList, length)
numMsgs
```

```
[1] 5051 1400  500 1000 1397
```

The first 3 directories are ham and the last 2 are spam so we create the logical with

```
isSpam = rep(c(FALSE, FALSE, FALSE, TRUE, TRUE), numMsgs)
```

Of course, we could use the file names to ascertain whether a message is spam or not.

Finally, we flatten the 5 lists in msgWordsList into one list with

```
msgWordsList = unlist(msgWordsList, recursive = FALSE)
```

Now that we have cleaned and extracted the words from all the email in the Spam Assassin corpus, we can proceed with the text analysis.

3.6 Implementing the Naïve Bayes Classifier

Recall from Section 3.4 that our goal is to estimate from the training data the following probabilities:

$$\mathbb{P}(\text{a word is present} \mid \text{spam}) \approx \frac{\#\text{ of spam messages with this word } + 1/2}{\#\text{ of spam messages } + 1/2}$$

and

$$\mathbb{P}(\text{a word is absent} \mid \text{spam}) \approx \frac{\text{\# of spam messages without this word} + 1/2}{\text{\# of spam messages} + 1/2}$$

And we must also make similar estimates for ham messages. We also noted there that we work with the log of ratios of these probabilities because taking logs reduces products to sums and tends to have better statistical properties. Thus, for a new message we compute the log likelihood ratio as

$$\sum_{\text{words in message}} \log \mathbb{P}(\text{word present} \mid \text{spam}) - \log \mathbb{P}(\text{word present} \mid \text{ham})$$
$$+ \sum_{\text{words not in message}} \log \mathbb{P}(\text{word absent} \mid \text{spam}) - \log \mathbb{P}(\text{word absent} \mid \text{ham})$$
$$+ \log \mathbb{P}(\text{spam}) - \log \mathbb{P}(\text{ham})$$

Additionally, we note that we can drop the last term, $\log \mathbb{P}(\text{spam}) - \log \mathbb{P}(\text{ham})$ as it is constant across messages.

Before we can proceed with these calculations, we need to divide the corpus into test and training sets. We do this next.

3.6.1 Test and Training Data

We want both the training and test data to be representative of the email corpus. One way to achieve this is to randomly select email for the two subsets. We may also want to control the sampling process so that the proportion of spam in the test and training sets matches the corpus proportion. Also, we typically want at least half of the data to be in the training set. In our case, we choose to use 2/3 of the data for training and 1/3 for testing.

We tally the number of spam and ham messages so that we know how many to sample from each subset. We do this with

```
numEmail = length(isSpam)
numSpam = sum(isSpam)
numHam = numEmail - numSpam
```

The sample() function can help us select which of the messages are in the test set. We start by setting the random seed so that if we need to repeat the selection process as we debug our code, we get the same subset each time.

```
set.seed(418910)
```

Then we determine the indices of test spam and ham messages with

```
testSpamIdx = sample(numSpam, size = floor(numSpam/3))
testHamIdx = sample(numHam, size = floor(numHam/3))
```

We use these indices to select the word vectors from msgWordsList with

```
testMsgWords = c((msgWordsList[isSpam])[testSpamIdx],
                 (msgWordsList[!isSpam])[testHamIdx] )
trainMsgWords = c((msgWordsList[isSpam])[ - testSpamIdx],
                  (msgWordsList[!isSpam])[ - testHamIdx])
```

Using Statistics to Identify Spam

Notice that the way we have organized the test and training collections, all spam messages are first, followed by ham. Rather than subset `isSpam`, we can create the test and train versions of `isSpam` using `rep()` with

```
testIsSpam = rep(c(TRUE, FALSE),
                 c(length(testSpamIdx), length(testHamIdx)))
trainIsSpam = rep(c(TRUE, FALSE),
                  c(numSpam - length(testSpamIdx),
                    numHam - length(testHamIdx)))
```

Now that we have created our test and training sets, we use the training data to develop our probability estimates and then apply these to our test messages and create the log likelihood ratio for the test messages.

3.6.2 Probability Estimates from Training Data

To estimate the probability a word is present or absent in a message given the message is spam or ham, we need a complete listing of all words, i.e., the bag of words. We can use an external source for this dictionary or create our own from our training data. We do the latter with

```
bow = unique(unlist(trainMsgWords))
```

In our new training subset there are over 80,000 unique words:

```
length(bow)
```

```
[1] 80481
```

We know that some of these are garbage words from *URL*s that we haven't cleaned properly. Also, we have not stemmed words, which means there are "duplicate" words in the bag of words. Addressing these issues is left to the exercises. We do find that even without this additional cleaning, our classifier performs well.

For each word in `bow`, we compute the number of spam messages in our training set that contain that word. We can work our way through the spam messages updating the counts as we encounter additional occurrences of words. Those familiar with other programming languages might think it is natural to write two nested loops that cycle over each message and each word in that message, but this does not take advantage of the vectorized nature of R. We consider an alternative approach that uses subsetting by name. We start by creating a vector to hold the counts for all the words with

```
spamWordCounts = rep(0, length(bow))
```

We use `bow` to add names to the elements of `spamWordCounts` with

```
names(spamWordCounts) = bow
```

Consider the following code and consider what it does:

```
tmp = lapply(trainMsgWords[trainIsSpam], unique)
tt = table( unlist(tmp) )
spamWordCounts[ names(tt) ] = tt
```

We process each message and retrieve only the unique words. This avoids any repeated words from any one email. Then we collapse these words from across all the spam messages and calculate the frequency table. Because no word is repeated in any message, the frequency of a given word in tt is the number of spam messages in which it occurred. We can then update the elements of spamWordCounts that occur in any spam message using the names on the frequency table.

The code above is ostensibly similar to how we computed the bag of words (bow). We used unlist() and unique(). However, we used these in a very different way, namely applying unique() on each message and then unlist()ing the resulting elements. An important aspect of this is that we can update only the elements of bow with the counts from tt and leave the counts for words that did not appear in any spam message unchanged.

Once we have our counts, we can estimate the probabilities with

```
spamWordProbs = (spamWordCounts + 0.5) / (sum(trainIsSpam) + 0.5)
```

We can do the same for the ham probabilities and then take the log of the ratios of these spam and ham probabilities with

```
log(spamWordProbs) - log(hamWordProbs)
```

Similarly, we need to estimate the probabilities for a word not being in the message, i.e.,

```
log(1 - spamWordProbs) - log(1 - hamWordProbs)
```

We collect all of these operations into a function called computeFreqs(). For the function signature, we have

```
function(wordsList, spam, bow = unique(unlist(wordsList)))
```

There are 3 parameters for computeFreqs(): *wordsList*, *spam*, and *bow*. Notice that the default value for *bow* is computed from the words supplied in the *wordsList* argument. This means that there is no need to compute the bag of words in advance if we wish to use only the words in *wordsList* for our bag of words. On the other hand, a different bag of words can be supplied if desired.

This function returns the building blocks for the log likelihood ratio, i.e., for each word we find

$$\log \mathbb{P}(\text{word present}|\text{ spam}) - \log \mathbb{P}(\text{word present}|\text{ ham})$$

and

$$\log \mathbb{P}(\text{word absent}|\text{ spam}) - \log \mathbb{P}(\text{word absent}|\text{ ham})$$

These values are returned in a matrix that has a column for each word in the bag of words. The matrix has a row for each of the above log likelihood ratios, and a row for each word's observed proportion in spam and ham. The computeFreqs() function appears as:

```
computeFreqs =
function(wordsList, spam, bow = unique(unlist(wordsList)))
{
  # create a matrix for spam, ham, and log odds
  wordTable = matrix(0.5, nrow = 4, ncol = length(bow),
                     dimnames = list(c("spam", "ham",
                                       "presentLogOdds",
                                       "absentLogOdds"), bow))
```

Using Statistics to Identify Spam

```
  # For each spam message, add 1 to counts for words in message
  counts.spam = table(unlist(lapply(wordsList[spam], unique)))
  wordTable["spam", names(counts.spam)] = counts.spam + .5

  # Similarly for ham messages
  counts.ham = table(unlist(lapply(wordsList[!spam], unique)))
  wordTable["ham", names(counts.ham)] = counts.ham + .5

  # Find the total number of spam and ham
  numSpam = sum(spam)
  numHam = length(spam) - numSpam

  # Prob(word|spam) and Prob(word | ham)
  wordTable["spam", ] = wordTable["spam", ]/(numSpam + .5)
  wordTable["ham", ] = wordTable["ham", ]/(numHam + .5)

  # log odds
  wordTable["presentLogOdds", ] =
     log(wordTable["spam",]) - log(wordTable["ham", ])
  wordTable["absentLogOdds", ] =
     log((1 - wordTable["spam", ])) - log((1 -wordTable["ham", ]))

  invisible(wordTable)
}
```

We apply this function to our training data with

```
trainTable = computeFreqs(trainMsgWords, trainIsSpam)
```

Now, `trainTable` can be used to construct the log likelihood ratio for a new message. That is, we select values from the matrix `trainTable` that correspond to the words that appear in a new message and the words that are absent from the message, and we use these to compute the log likelihood ratio for the message. This value is then used to classify the message as spam or ham. We do this next.

3.6.3 Classifying New Messages

The `trainTable` object has all of the individual word probabilities needed to construct the log likelihood ratio for a message. To do this we need to combine these estimated probabilities where we take the log odds from the "present" row of `trainTable` for each word appearing in the message and similarly take the log odds from the "absent" row of the table for all those words in the bag of words that do not appear in the message. We combine these to create the likelihood that the message is spam versus ham using

$$\sum_{\text{words in message}} \log \mathbb{P}(\text{word present}|\text{ spam}) - \log \mathbb{P}(\text{word present}|\text{ ham})$$
$$+ \sum_{\text{words not in message}} \log \mathbb{P}(\text{word absent}|\text{ spam}) - \log \mathbb{P}(\text{word absent}|\text{ ham})$$

For example, consider the set of words in the first message in `testMsgWords`,

```
newMsg = testMsgWords[[1]]
```

There is the possibility that a test message contains a word that is not in the bag of words. When this happens we do not include it in our calculation as we have no information about the likelihood a message with this word is spam or ham. We drop these new words from newMsg with

```
newMsg = newMsg[!is.na(match(newMsg, colnames(trainTable)))]
```

For the remaining words that are in newMsg, we locate the columns in the frequency table that contain them with the logical vector:

```
present = colnames(trainTable) %in% newMsg
```

Then we compute the log of the ratio of the probability a message is spam versus ham with

```
sum(trainTable["presentLogOdds", present]) +
   sum(trainTable["absentLogOdds", !present])
```

```
[1] 255
```

We know the first message in testMsgWords is spam, and we see the log likelihood ratio computed for it is large and positive, indicating spam. We can try a test ham message as well, e.g.,

```
newMsg = testMsgWords[[ which(!testIsSpam)[1] ]]
newMsg = newMsg[!is.na(match(newMsg, colnames(trainTable)))]
present = (colnames(trainTable) %in% newMsg)
sum(trainTable["presentLogOdds", present]) +
    sum(trainTable["absentLogOdds", !present])
```

```
[1] -125
```

This message has a large negative value, which indicates it is ham.

We place this simple code into a function so that we can calculate the log likelihood ratio (LLR) for all of the test messages. Our function, computeMsgOdds() appears as

```
computeMsgLLR = function(words, freqTable)
{
       # Discards words not in training data.
   words = words[!is.na(match(words, colnames(freqTable)))]

      # Find which words are present
   present = colnames(freqTable) %in% words

   sum(freqTable["presentLogOdds", present]) +
      sum(freqTable["absentLogOdds", !present])
}
```

We apply this function to each of the messages in our test set with

```
testLLR = sapply(testMsgWords, computeMsgLLR, trainTable)
```

Using Statistics to Identify Spam

We want to use these values to classify the test messages as spam or ham. A value that is positive indicates spam is more likely and a negative value indicates ham is more likely, but we are free to choose some other value as a threshold for classification.

We compare the summary statistics of the LLR values for the ham and spam in the test data with

```
tapply(testLLR, testIsSpam, summary)
```

```
$`FALSE`
   Min.  1st Qu.  Median    Mean  3rd Qu.    Max.
  -1360    -127    -102     -117    -82      700

$`TRUE`
   Min.  1st Qu.  Median    Mean  3rd Qu.    Max.
    -61       7      50      138    131    23600
```

We see from these statistics and the boxplots in Figure 3.1 that there is a good deal of separation of the ham and spam.

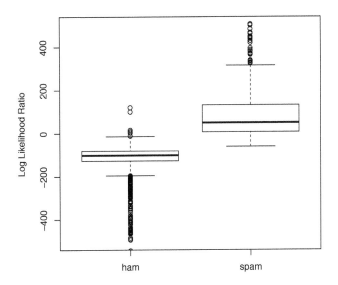

Figure 3.1: Boxplot of Log Likelihood Ratio for Spam and Ham. *The log likelihood ratio, $\log(P(\ spam\ |\ message\ content)/P(\ ham\ |\ message\ content))$, for 3116 test messages was computed using a naïve Bayes approximation based on word frequencies found in manually classified training data. The test messages are grouped according to whether they are spam or ham. Notice most ham messages have values well below 0 and nearly all spam values are above 0.*

We have 3116 LLR values corresponding to each test message, and we need to decide on a cut-off τ, where we classify a message as spam or ham according to whether or not the LLR exceeds this threshold. We assess the choice of τ using our test data. That is, we find the proportion of ham messages in the test set with LLR values that exceed the threshold and so are misclassified as spam. This is the Type I error rate for the test data. Likewise, we find the proportion of LLR values for spam messages in the test set that are below the threshold and so misclassified as ham, which is the Type II error rate.

We can write a simple R function to compute the rate of misclassification of ham as spam for a particular value of τ. This function takes 3 inputs: the value of τ, the vector of LLR values for the test messages, and the hand-classified type of each message (spam or ham). This function appears as

```
typeIErrorRate =
function(tau, llrVals, spam)
{
  classify = llrVals > tau
  sum(classify & !spam)/sum(!spam)
}
```

Note that we do not divide by the total number of messages, but only by the number of ham messages. It is important to divide by the right number here, which is the total number of ham messages as these are the only ones that can contribute to a Type I error.

The typeIErrorRate() function is not vectorized in its argument, *tau*. For example, in order to find τ that yields a 0.5% Type I error rate, we examine the boxplots in Figure 3.1. From the plot we make an initial guess that $\tau = 0$. We use typeIErrorRate() to calculate the Type I error with this threshold for the test messages, and find it is 0.3%. Then, we calculate the error for a few τ values below 0 and, and find that for $\tau = -20$ we get an error rate of 0.5%, i.e.,

```
typeIErrorRate(0, testLLR,testIsSpam)
```

[1] 0.0035

```
typeIErrorRate(-20, testLLR,testIsSpam)
```

[1] 0.0056

Typically, we want to find the error rate for a vector of τs because we want to find one that provides an acceptable Type I error. In its current form, if we want to use typeIErrorRate() to calculate the Type I error for a vector of values, we need a loop in the form of an sapply() call.

In theory, to select a threshold, we need to search over all possible values of τ. However, it should be clear after a little thought that we can at least restrict the interval. Any value of τ less than the minimum of the LLR values means that we classify all messages as spam and the Type I error rate is 1. Similarly, any value of τ greater than the maximum of the LLR values implies that we classify every message in our sample as ham so our Type I error rate is 0. Additionally, we need to keep in mind that there are also errors in misclassifying spam as ham. The Type II error is 1 when we use the largest observed LLR value in our test set because all spam is classified as ham, which is clearly not acceptable either.

We also note that the Type I error rate only changes at values of τ that match one of the observed LLR values in our set of messages. That is, for 2 values of τ, say τ_1 and τ_2, if there are no LLR values from the test set between them, then their associated Type I errors must be the same. Likewise, the Type II error rates for τ_1 and τ_2 are the same. This means that we only need to compute the error rate at the 3116 LLR values for the test messages.

Our estimate of the Type I error rate is a step function and only changes at each of the observed LLR values. We can do even better than this to reduce the set of possible τs that we search over. It is not all LLR values that potentially cause a change in the Type I error. Only the values corresponding to ham messages will affect the Type I error because messages that are spam do not contribute to the Type I error.

These observations about the Type I and II error rates for our test messages imply that we can determine the error rates as a function of τ more conveniently and efficiently. The following function does this by looking only at the *llrVals* values for ham messages and recognizing that the number of Type I errors decreases by 1 at each of these values and so is $i/$(number of ham messages). Note that the function ignores ties for the ratios, but these are unlikely since they should be unique. Our function is defined as

```
typeIErrorRates =
function(llrVals, isSpam)
{
  o = order(llrVals)
  llrVals = llrVals[o]
  isSpam = isSpam[o]

  idx = which(!isSpam)
  N = length(idx)
  list(error = (N:1)/N, values = llrVals[idx])
}
```

In essence, we have found a vectorized way to compute the Type I errors. We can compute the Type II errors similarly. We leave this as an exercise.

The plot in Figure 3.2 shows that a threshold of -43 looks reasonable. A Type I error rate of 0.01 coincides with $\tau = -43$, and our Type II error rate is 0.02. If we want a smaller Type I error, say 0.001, then we need to set the threshold at $\tau = 120$ and that leads to a very high Type II error of 0.73, i.e. 73% of the spam is misclassified as ham.

We have used the test set here to both select the threshold τ and evaluate the Type I and II errors for that threshold. The implication of this is that the threshold we have chosen may work well with this particular test set but not others, and it may underestimate the size of the errors. Ideally we select τ from other data, independent of our training and test data. To address this problem, we can apply the method of cross-validation. With cross-validation, we partition the training data into k parts at random. Then we use each of these parts to act as a test set and compute the LLR values for the messages in this subset using the remaining data as a training set. We pool all of these LLR values from all k validation sets to select the threshold τ. In this case, when we use $k = 5$ we find that $\tau = -33$ corresponds to a 1% Type I error. Finally, we apply this threshold to our original test set and find for $\tau = -33$, the Type I error is 0.8% and the Type II error is 4%. We leave it as an exercise to carry out this cross-validation.

This completes the naïve Bayes approach to spam classification using word vectors. Before we turn to the second approach where we derive characteristics of email as variables to predict spam and ham, we briefly examine some of the computational considerations in calculating the LLR.

3.6.4 Computational Considerations

In computing the log likelihood ratio for a message, we used the following representation of this quantity to guide how we wrote the code

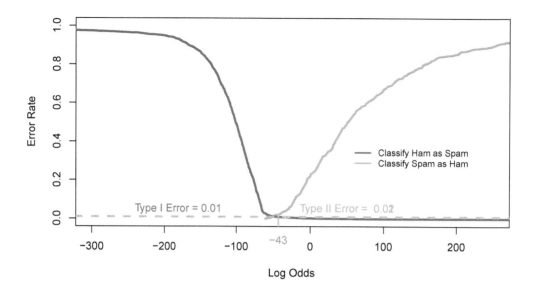

Figure 3.2: Comparison of Type I and II Error Rates. *The Type I and II error rates for the 3116 test messages are shown as a function of the threshold τ. For example, with a threshold of $\tau = -43$, all messages with an LLR value above -43 are classified as spam and those below as ham. In this case, 1% of ham is misclassified as spam and 2% of spam is misclassified as ham.*

$$\sum_{\text{words in message}} \log \mathbb{P}(\text{word present}|\text{ spam}) - \log \mathbb{P}(\text{word present}|\text{ ham})$$
$$+ \sum_{\text{words not in message}} \log \mathbb{P}(\text{word absent}|\text{ spam}) - \log \mathbb{P}(\text{word absent}|\text{ ham})$$

In other words, we first computed from the observed proportions in our training set the estimates to $\mathbb{P}(\text{word in message}|\text{ spam})$, $\mathbb{P}(\text{word not in message}|\text{ spam})$, $\mathbb{P}(\text{word}|\text{ ham})$, and $\mathbb{P}(\text{not word}|\text{ ham})$. Then, we took logs of these estimated probabilities and combined them to calculate the LLR for a particular message. That is, we selected which of these terms to include in the above sum, according to whether each word in the bag of words was present or absent from that message. Given our bag of words consists of more than 80,000 words, we want to consider whether there are faster or more accurate ways to carry out these computations.

The following are equivalent representations of the log likelihood ratio:

$$LLR = \log\left(\prod_{\text{words in message}} \frac{\mathbb{P}(\text{word present}|\text{ spam})}{\mathbb{P}(\text{word present}|\text{ ham})}\right) +$$

$$\log\left(\prod_{\text{words not in message}} \frac{\mathbb{P}(\text{word absent}|\text{ spam})}{\mathbb{P}(\text{word absent}|\text{ ham})}\right)$$

$$= \log\left(\frac{\prod_{\text{in msg}} \mathbb{P}(\text{word present}|\text{ spam})}{\prod_{\text{in msg}} \mathbb{P}(\text{word present}|\text{ ham})}\right) + \log\left(\frac{\prod_{\text{not in msg}} \mathbb{P}(\text{word absent}|\text{ spam})}{\prod_{\text{not in msg}} \mathbb{P}(\text{word absent}|\text{ ham})}\right)$$

$$\propto \log \frac{\prod_{\text{in msg}} \#\text{spam with word}}{\prod_{\text{in msg}} \#\text{ham with word}} \times \frac{\prod_{\text{not in msg}} \#\text{spam without word}}{\prod_{\text{not in msg}} \#\text{ham without word}}$$

These alternative mathematical expressions each suggest a different approach to carrying out the computation of the log odds. We leave it as an exercise to write code for them and compare the results to our approach.

Why might these various alternatives not give us the same answer? A computer is a finite state machine, meaning that it has only a fixed amount of space to store a number so some numbers can only be approximated, e.g., irrational numbers. Additionally, the order of operations can matter. For example, if we have one large number and many small numbers, then adding up all of the small numbers first and then adding this total to the large number can produce a more accurate result than adding the small numbers one at a time to the large number. Below is an artificial example that makes this point:

```
smallNums = rep((1/2)^40, 2000000)
largeNum = 10000

print(sum(smallNums), digits = 20)

[1] 1.8189894035458564758e-06

print(largeNum + sum(smallNums), digits = 20)

[1] 10000.000001818989404

for (i in 1:length(smallNums)) {
  largeNum = largeNum + smallNums[i]
}
print(largeNum, digits = 20)

[1] 10000
```

In our case, we are working with thousands of small numbers such as the proportion of spam that contains a particular work. It might matter quite a bit how we compute the LLR.

3.7 Recursive Partitioning and Classification Trees

As with the naïve Bayes approach, we first describe the kind of analysis we want to do and this leads us to determining how to prepare the raw email. With the word vectors, we examined only the words in the message to classify it as ham or spam. We ignored all the other information available in the message, such as information contained in the header or that can be derived from the content. From experience we know that this other information is often enough for us to determine whether a messages is spam or ham. This determination is almost an unconscious process, but let's try to examine that process and uncover the thoughts that induce our reactions to classify a message one way or the other.

- Is it that we don't recognize the sender's name?

- Is the punctuation in the subject line unusual?

- Are there special words in the subject line that make it look like generic email about a marketing topic?

- Is the mail in reply to a message that we initially sent, i.e., with a subject line that we recognize?

- Was the mail sent at an odd time of the day?

For example, a message with the subject line that begins "CLAIM YOUR LUCKY WIN-NING...", we easily recognize as spam for two reasons: the subject is about winning a prize and the letters are all capitalized. If we transfer features such as these into quantitative measures on the messages, then we can analyze and compare spam and ham using these characteristics. That is, we can represent each message as a collection of features, such as whether or not the subject line begins Re: or the quantity of exclamation marks in the body of the message, and we use these characteristics to differentiate between spam and ham. The determination of which features to code into variables is itself a statistical process.

Our goal is to create a classifier based on these characteristics. One popular method that works well in a variety of settings is recursive partitioning. The basic idea is quite intuitive: the method splits the data into two groups according to the value of a particular variable. For example, it may split the messages according to whether the percentage of capital letters in a message exceeds 10% or not. Once the data are split into the two groups, a subsequent split divides one of these sub-groups into two groups, again according to the value of some variable. This splitting of sub-groups continues until the messages are partitioned into subsets that are nearly all spam or ham. At each stage, the same criteria for selecting a variable for the split is applied to each subgroup recursively, and thus the name recursive partitioning.

For example, Figure 3.3 shows a recursive partition that begins by splitting the email into two groups according to whether the percentage of capital letters falls below 13% (left branch) or not (right branch). For those messages with fewer than 13% capitals, the next split is determined by whether the message characters contain fewer than 3.9% *HTML* tags (left branch) or not (right branch). The email messages that have fewer than 13% capitals and fewer than 3.9% *HTML* tags are classified as ham. For these messages, the Type II error rate is quite high because 653 of the 1598 (653 + 8 + 13 + 281 + 643) spam messages meet these criteria and are classified incorrectly as ham. The other messages in other parts of the tree are also partitioned according to other yes–no questions. The resulting tree is called a classification tree, or a decision tree, which are other names for recursive partitioning.

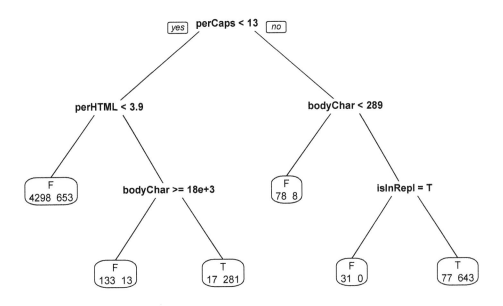

Figure 3.3: Example Tree from a Recursive Partition. *This tree is a simple example of a recursive partition fitted model. It was fitted using the rpart() function and restricting the tree depth to 3 levels. The first yes–no question is whether the percentage of capitals in the message is less than 13. If not, the second question is whether there are fewer than 289 characters in the message. If the answer to this question is also no, then the next question is whether the message header contains an* InReplyTo *key. If the answer is again no, then the message is classified as spam. Of the 6232 messages in the training set, 77 ham and 643 spam fall into this leaf. The spam have been correctly classified and the 77 ham have been misclassified.*

The partitioning process attempts to make the observations in the resulting subgroups as similar as possible, i.e., all spam or all ham. The observations in a subgroup that form a leaf at the bottom of the tree are given the same classification. If the messages in a leaf are as homogeneous as possible, then we reduce misclassification errors.

Now that we have a general idea of this approach, we can identify the tasks we need to accomplish to carry it out. We must:

- Process the email into a format that gives easy access to the information in the header, body, and attachments.

- Develop functions for the features of interest that transform the email messages into variables for analysis.

- Apply to these derived variables the recursive partitioning method and assess how well it predicts spam and ham.

We tackle these tasks over the next 4 sections.

3.8 Organizing an email Message into an *R* Data Structure

Now that we have a sense of the kind of information that we want to turn into feature sets for our email, we consider how to represent the messages in *R*. That is, how do we store the raw email message in a data structure that is conducive to converting it into measurable characteristics for analysis? What parts of the message do we want to keep and how do we want to organize them? A reasonable structure to create might be a list of *R* objects with one object per message. Then for each message we have the following considerations:

- The header elements tell us about the sender, recipient(s), date sent, routing information of the message, mail program used to compose the message, etc. Since we work with the header in terms of its key-value pairs, it seems easiest to represent the contents of the header as a named character vector where the name of an element corresponds to the key and the string itself holds the value for that key.

- The message text does not require any special processing. Simply maintaining it as a collection of text lines at this point seems most expedient.

- We want to extract the attachments from the body of the email. Since each attachment has its own header, we may want the header and body of the attachment handled in a manner parallel to the header and body of the message. Alternatively, we may want to keep only summary information about each attachment, such as its MIME type and length. We take the latter approach and leave the former as an exercise.

- We want the information as to whether the message is spam or ham to be part of the message-element. A logical seems an appropriate data type. This information comes from the name of the directory that contains the message, not from the message itself.

- While the file names of the messages are not relevant to processing their contents, it is often useful when debugging to know the name of a message's source file. This allows us to easily identify and view the original source of the message and compare it with the resulting *R* object. We can use the file name as the name of the element in the list that corresponds to the individual email message.

Given these considerations, it makes sense to store each message as a named list that contains: a named character vector for the header; a character vector with the message content; a data frame with the summary information about the attachments; and a logical indicating if the message is spam or not.

We want to develop our code as separate tasks, where these tasks are encapsulated in separate functions. For example, we can write a function, processHeader(), to format the header of a single message into a named vector and another function, processAttach(), to create the data frame that summarizes a message's attachments. Then we gradually work up or outwards to process all of the messages. These top-level steps are merely loops over the directories and the individual messages within each of the directories. The function for reading an individual message should be the same for all directories and messages. We focus on preparing one message and handle special cases as we encounter them in our sample email.

A first step might be to split the message into two parts — the header and body. This task was accomplished in Section 3.5.1 where we prepared the email for the naïve Bayes analysis of the message text. We should be able to reuse the splitMessage() function that we have developed already. Recall that we split the messages into their headers and bodies with

Using Statistics to Identify Spam

```
sampleSplit = lapply(sampleEmail, splitMessage)
```

We address the task of creating the named vector from the headers in `sampleSplit` in the next subsection.

The task of processing the attachments is similar to the task of removing the attachments from the message body, which we carried out in Section 3.5.2. However, it is not directly applicable because we discarded the attachments. Yet we may be able to use some of that code here. For example, the getBoundary() function that finds the boundary string marking the location of the attachments is useful here as well. In Section 3.8.2, we examine how we can adapt the code from Section 3.5.2 for summarizing the attachments.

3.8.1 Processing the Header

Our plan is to process the header by converting the key: value pairs into a named vector, where the name of each element in the vector is taken as the key. Having a separate function to carry out this task allows us to test it without having to repeatedly read the entire message, and if desired, we can use this function to process the header in attachments within the message body.

To determine how to process the header, we examine a few lines of the header in one of our sample messages with

```
header = sampleSplit[[1]]$header
header[1:12]
```

```
 [1] "From exmh-workers-admin@redhat.com  Thu Aug 22 12:36:23..."
 [2] "Return-Path: <exmh-workers-admin@spamassassin.taint.org>"
 [3] "Delivered-To: zzzz@localhost.netnoteinc.com"
 [4] "Received: from localhost (localhost [127.0.0.1])"
 [5] "\tby phobos.labs.netnoteinc.com (Postfix) with ESMTP id..."
 [6] "\tfor <zzzz@localhost>; Thu, 22 Aug 2002 07:36:16 -0400..."
 [7] "Received: from phobos [127.0.0.1]"
 [8] "\tby localhost with IMAP (fetchmail-5.9.0)"
 [9] "\tfor zzzz@localhost (single-drop); Thu, 22 Aug 2002 12..."
[10] "Received: from listman.spamassassin.taint.org (listman...."
[11] "    dogma.slashnull.org (8.11.6/8.11.6) with ESMTP id g..."
[12] "    <zzzz-exmh@spamassassin.taint.org>; Thu, 22 Aug 200..."
```

We make the following observations about these 12 header lines that are relevant to our extraction process:

- Some key: value pairs appear on multiple lines so there is not a one-to-one correspondence between lines in the `header` variable and key: value pairs. For example, lines 4, 5, and 6 are all part of the same `Received` value, and the same is true for lines 7, 8, and 9 and for lines 10–12. We also note that when the value continues over multiple lines, these additional lines begin with blanks as in lines 11 and 12 or with a tab character, i.e., \t, as with lines 5 and 6.

- The first line in the header is not in the key: value format. The information in this line also appears elsewhere in the header. This identifies the start of a new message in general.

- Colons can appear in the value portion of the key: value pair, e.g., line 6 contains a time in the format 07:36:16.

Our function needs to handle these various situations as it transforms the header. In what order do we address them? Also, to address the second issue, we need to decide what we want to do with the first line of the header. Do we discard it because the information appears elsewhere in the header or do we keep it? Let's be conservative and keep this line, but change it so that it follows the key: value format of the other lines.

To handle the situation where a value appears on multiple lines in the header, we can collapse these extra lines into one. Then, each key: value pair occupies one line in the revised header vector. It makes sense to fix these two issues first because then all of the lines have the same format when we go about splitting the lines up into their keys and values.

We can address the problem with the first line of the header not being in the key: value format by simply substituting the string "From" that appears at the start of the line with, say, "Top-From:". By choosing a special key that is not used in typical email headers, it will not be confused with other key: value pairs. We do this with the following call to sub():

```
header[1] = sub("^From", "Top-From:", header[1])
header[1]
```

```
[1] "Top-From: exmh-workers-admin@redhat.com   Thu Aug 22..."
```

Here we have used the regular expression '^From' to locate 'From' at the start of the string. That is, the character ^ anchors the pattern we are searching for to the start of the string, and any From later in the string is not matched. The second argument to sub() is the substring that is substituted for the initial 'From'. Alternatively, we can use simple string manipulation functions available in R for finding and modifying this first header line. For example, we can remove the first 4 characters from the string and paste 'Top-From:' to the front of this shortened string.

Now all of the information appears in a key: value format. We might ask: does R provide a function to read files in this format? Our initial searches on the Internet and in the documentation for input/export of files in R do not turn up anything useful. We can write code to handle the continuation lines, and, e.g., catenate them to the previous line, and code to identify the key and value. In fact the first time that we processed the header, this is what we did, and we make this approach an exercise. Later, we discovered the read.dcf() function that handles this format. According to the documentation for the function, read.dcf() reads the format:

- Regular lines are of the form key: value and start with a non-whitespace character.
- Lines starting with whitespace are continuation lines (to the preceding field).
- Fields may appear more than once in a record.
- Records are separated by one or more empty (i.e., whitespace only) lines.

We can try read.dcf() on our sample header. Since it is already in R as a character vector, we use a text connection to read it, i.e.,

```
headerPieces = read.dcf(textConnection(header), all = TRUE)
```

The return value is a data frame with one row, where, e.g., the Delivered-To element is a list that contains the values for the 2 Delivered-To keys in the header. That is,

```
headerPieces[, "Delivered-To"]
```

Using Statistics to Identify Spam

```
[[1]]
[1] "zzzz@localhost.netnoteinc.com"
[2] "exmh-workers@listman.spamassassin.taint.org"
```

We can convert `headerPieces` into a character vector and use the key for the name of each of these values. We have duplicate names when there are duplicate fields in the header. We do this with

```
headerVec = unlist(headerPieces)
dupKeys = sapply(headerPieces, function(x) length(unlist(x)))
names(headerVec) = rep(colnames(headerPieces), dupKeys)
```

We confirm that we have 2 elements in `headerVec` named "Delivered-To" with

```
headerVec[ which(names(headerVec) == "Delivered-To") ]
```

```
                      Delivered-To
            "zzzz@localhost.netnoteinc.com"
                      Delivered-To
 "exmh-workers@listman.spamassassin.taint.org"
```

The header vector has 36 elements, i.e.,

```
length(headerVec)
```

```
[1] 36
```

The raw header was originally 62 lines, but apparently, 26 of these lines were continuation lines. Moreover, these 36 elements include 10 duplicate names,

```
length(unique(names(headerVec)))
```

```
[1] 26
```

We can put this code into our processHeader() function. What are the inputs and outputs of this function? We only need the original header vector as input, and the return value from processHeader() is the named character vector. Our function follows:

```
processHeader = function(header)
{
      # modify the first line to create a key:value pair
  header[1] = sub("^From", "Top-From:", header[1])

  headerMat = read.dcf(textConnection(header), all = TRUE)
  headerVec = unlist(headerMat)

  dupKeys = sapply(headerMat, function(x) length(unlist(x)))
  names(headerVec) = rep(colnames(headerMat), dupKeys)

  return(headerVec)
}
```

Let's call processHeader() on the rest of our sample messages. Recall the headers and bodies of the messages in `sampleEmail` have already been separated and assigned to `sampleSplit`. We apply processHeader() to them with

```
headerList = lapply(sampleSplit,
                    function(msg) {
                      processHeader(msg$header)} )
```

We can access the value of, e.g., the Content-Type key with subsetting by name, i.e.,

```
contentTypes = sapply(headerList, function(header)
                                       header["Content-Type"])
names(contentTypes) = NULL
contentTypes
```

```
 [1] " text/plain; charset=us-ascii"
 [2] " text/plain; charset=US-ASCII"
 [3] " text/plain; charset=US-ASCII"
 [4] " text/plain; charset=\"us-ascii\""
 [5] " text/plain; charset=US-ASCII"
 [6] " multipart/signed;\n     boundary=..."
 [7] NA
 [8] " multipart/alternative;\n ...       "
 [9] " multipart/alternative;  boundary=Apple-Mail-2-874629474"
[10] " multipart/signed;\n     boundary=..."
[11] " multipart/related;\n    boundary=..."
[12] " multipart/signed;\n     boundary=..."
[13] " multipart/signed;\n     boundary=..."
[14] " multipart/mixed;\n     boundary=..."
[15] " multipart/alternative;\n    boundary=..."
```

We see that in our sample one of the 15 messages has no Content-Type specified, i.e., it yields NA. When we examine the raw element, we confirm that we properly processed it, i.e., the Content-Type key is not present in the original header.

We next tackle processing the body of the message; in particular, we extract the attachments from the body and summarize them.

3.8.2 Processing Attachments

We saw in Section 3.5.2 that we can determine which messages have attachments by examining the Content-Type, e.g.,

```
hasAttach = grep("^ *multi", tolower(contentTypes))
hasAttach
```

```
[1]   6  8  9 10 11 12 13 14 15
```

The hasAttach variable contains the indices of those test messages with attachments. In Section 3.5.2 we also determined how to find the boundary string that separates the attachments in the body. We can use the function we developed there to extract the boundary strings, e.g.,

```
boundaries = getBoundary(contentTypes[ hasAttach ])
boundaries
```

Using Statistics to Identify Spam

```
[1] "==_Exmh_-1317289252P"
[2] "------=_NextPart_000_00C1_01C25017.F2F04E20"
[3] "Apple-Mail-2-874629474"
[4] "==_Exmh_-518574644P"
[5] "------------0906020109090007050100009"
[6] "==_Exmh_-451422450P"
[7] "==_Exmh_267413022P"
[8] "------=_NextPart_000_0005_01C26412.7545C1D0"
[9] "------------08020906070003030908805"
```

We can use the boundary string to locate the attachments in the body. Before we design our function, we first need to decide what sort of information we want to store about the attachments. Some possibilities include the attachment's MIME type and its length. We can find the MIME type in a manner similar to how we found the value of the boundary string in the header. We can determine the length of the attachment from the locations of the boundary string in the body, i.e., it is the number of lines between the boundary strings. If we do not want to include the header information in the length then that is a bit more work.

Our investigation of the sample email in Section 3.5.2 revealed that we need to be aware that some messages may not have an attachment, even though their header indicates that they are supposed to. We also must decide what to do with the lines before the first boundary string and after the closing boundary string. As a precaution, we might want to address the situation where the last boundary string is not found. Let's write our function to do the following.

- Drop the lines before the first boundary string, which marks the first part of the body.

- Keep the lines following the closing string as part of the body without attachments.

- Include the header information in the line count for the attachments.

- Use the last line of the email as the end of the attachment if we find no closing boundary string.

We follow the ideas of Section 3.5.2 to process one of the messages in our sample. For simplicity, we begin by assigning the boundary string and the body to `boundary` and `body`, respectively, e.g.,

```
boundary = boundaries[9]
body = sampleSplit[[15]]$body
```

We saw earlier that we can locate the attachments by searching for the boundary string preceded by 2 hyphens in the body, i.e.,

```
bString = paste("--", boundary, sep = "")
bStringLocs = which(bString == body)
bStringLocs
```

```
[1]    2 35
```

The lines (2 and 35) mark the start of the body and the start of the single attachment. Next, we find the closing boundary with

```
eString = paste("--", boundary, "--", sep = "")
eStringLoc = which(eString == body)
eStringLoc
```

```
[1] 77
```

In addition to extracting the content that is not part of an attachment, which the dropAttach() function from Section 3.5.2 does, we also want to process the attachments to find their MIME types and lengths. For example, we find the number of lines in the attachments in our current sample message with

```
diff(c(bStringLocs[-1], eStringLoc))
```

```
[1] 42
```

We leave as an exercise the creation of the processAttach() function. The inputs to this function are one message's body and Content-Type value. The function returns a list with 2 elements: the first part of the body, which contains the message and no attachments; and a data frame called attachDF, which has 2 variables, aLen and aType, that provide the length and type of each attachment, respectively. It follows the basic operations explored in this section. However, the special cases described earlier need to be addressed in the function, e.g., when there is no attachment despite the header giving a MIME type of multipart and a boundary string.

3.8.3 Testing Our Code on More email Data

We have developed our functions using a simple message so we want to check them more carefully on a larger set of messages. These checks typically lead to code refinements to handle cases we have not seen yet. Additionally, we may anticipate potential problems, e.g., a missing Content-Type, and test whether the code handles this problem, even though we have not encountered the problem.

To begin, we can test our processAttach() function on all of the messages in sampleEmail that have attachments. We have their corresponding Content-Types in contentTypes, and we apply processAttach() with

```
bodyList = lapply(sampleSplit, function(msg) msg$body)
attList = mapply(processAttach, bodyList[hasAttach],
                 contentTypes[hasAttach],
                 SIMPLIFY = FALSE)
```

Let's examine the attachment lengths. We do this with

```
lens = sapply(attList, function(processedA)
                          processedA$attachDF$aLen)
head(lens)
```

```
[[1]]
[1] 12

[[2]]
[1] 44 44

[[3]]
[1] 83

[[4]]
[1] 12
```

```
[[5]]
NULL

[[6]]
[1] 12
```

We see a few curious results: the fifth message has NULL for its attachment lengths and the second message supposedly has 2 attachments of the same length. Recall that some of the email with a multipart Content-Type do not have attachments. A brief examination of the body of the fifth message confirms that is the case so this is not an error.

When we examine the data frame `attachDF` for the second email, we find the following:

```
attList[[2]]$attachDF
```

```
  aLen                                                  aType
1   44                                              text/html
2   44 <META http-equiv=3DContent-Type content=3Dtext/html; =
```

Notice that the MIME type of the second "attachment" looks a bit suspicious. The literal term `Content-Type` appears to be located within an *HTML* tag. We confirm that this is the case by examining the body of the message more closely. We find the number of lines in the body with

```
body = bodyList[hasAttach][[2]]
length(body)
```

```
[1] 86
```

Since the attachment is 44 lines long, we look at the body in the neighborhood of the 40th line with

```
body[35:45]
```

```
 [1] ""
 [2] "------=_NextPart_000_00C1_01C25017.F2F04E20"
 [3] "Content-Type: text/html;"
 [4] "\tcharset=\"Windows-1252\""
 [5] "Content-Transfer-Encoding: quoted-printable"
 [6] ""
 [7] "<!DOCTYPE HTML PUBLIC \"-//W3C//DTD HTML 4.0 Transition..."
 [8] "<HTML><HEAD>"
 [9] "<META http-equiv=3DContent-Type content=3D\"text/html; ="
[10] "charset=3Dwindows-1252\">"
[11] "<META content=3D\"MSHTML 6.00.2716.2200\" name=3DGENERA..."
```

We see the actual Content-Type and its MIME type in the third line and the mistaken one in the ninth line. Clearly we need to improve our pattern matching. Once we do this, then the duplicates are gone.

Now that we have cleared up the problem with the curious message with 2 attachments of the same length, we see that we have no messages with more than one attachment in our sample. We may want to add a few more messages to our sample, ones with multiple attachments, so that we can test our code for these cases.

As another example of testing code, consider the getBoundary() function. When we first designed it in Section 3.5.2, the input to the function was all of the lines in the header. In the current situation, we know exactly where to search for the string and we pass to getBoundary() only one line. This means that the code in this function could be simplified for our current purposes. We could make a second copy of getBoundary() that eliminates the unnecessary search for the line with the `boundary=` string. However, we prefer to keep one version of this function and design the code to be general and flexible so we can reuse it without having to copy it or modify it and potentially break it. Having copies of functions means that we have to update all copies if we find a bug or add a feature. Good software design and development minimizes code duplication and rewards generality when it doesn't add to the complexity.

3.8.4 Completing the Process

We have completed and tested all of the tasks to transform one message, and we are ready to apply these tasks to all of the email in the SpamAssassin corpus. Below is our function to carry out the processing of all the email in one of the directories. Note that the inputs are the full path name of the directory and a logical indicating whether all of the email in the directory is spam or not. The function contains calls to our functions, splitMessage(), processHeader(), and processAttach(). The code is quite similar to the preparatory code for processing the sample messages with the addition of some post-processing to create the desired data structure and some preprocessing to read in the email.

We realize that the task of reading the email messages from the 9000 files is the same for both analyses so we abstract this code from processAllWords() in Section 3.5.4 to use for both functions. This function, readEmail() appears below

```
readEmail = function(dirName) {
     # retrieve the names of files in directory
  fileNames = list.files(dirName, full.names = TRUE)
     # drop files that are not email
  notEmail = grep("cmds$", fileNames)
  if ( length(notEmail) > 0) fileNames = fileNames[ - notEmail ]

     # read all files in the directory
  lapply(fileNames, readLines, encoding = "latin1")
}
```

The processAllEmail() function pulls together these various tasks:

```
processAllEmail = function(dirName, isSpam = FALSE)
{
     # read all files in the directory
  messages = readEmail(dirName)
  fileNames = names(messages)
  n = length(messages)

     # split header from body
  eSplit = lapply(messages, splitMessage)
  rm(messages)

     # process header as named character vector
  headerList = lapply(eSplit, function(msg)
```

```
                              processHeader(msg$header))

      # extract content-type key
  contentTypes = sapply(headerList, function(header)
                                       header["Content-Type"])

      # extract the body
  bodyList = lapply(eSplit, function(msg) msg$body)
  rm(eSplit)

      # which email have attachments
  hasAttach = grep("^ *multi", tolower(contentTypes))

      # get summary stats for attachments and the shorter body
  attList = mapply(processAttach, bodyList[hasAttach],
                   contentTypes[hasAttach], SIMPLIFY = FALSE)

  bodyList[hasAttach] = lapply(attList, function(attEl)
                                          attEl$body)

  attachInfo = vector("list", length = n )
  attachInfo[ hasAttach ] = lapply(attList,
                               function(attEl) attEl$attachDF)

      # prepare return structure
  emailList = mapply(function(header, body, attach, isSpam) {
                       list(isSpam = isSpam, header = header,
                            body = body, attach = attach)
                     },
                     headerList, bodyList, attachInfo,
                     rep(isSpam, n), SIMPLIFY = FALSE )
  names(emailList) = fileNames

  invisible(emailList)
}
```

Finally, we apply this over-arching function processAllEmail() to each directory with

```
emailStruct = mapply(processAllEmail, fullDirNames,
                     isSpam = rep( c(FALSE, TRUE), 3:2))
emailStruct = unlist(emailStruct, recursive = FALSE)
```

We extract from emailStruct the same set of sample email messages that we used in developing processAllEmail() with

```
sampleStruct = emailStruct[ indx ]
```

We use these in the next section to help us develop the functions to create the feature set.

3.9 Deriving Variables from the email Message

We begin by "dreaming up" a set of features for using to classify spam. Although we are focusing on identifying spam, don't forget that it can be much easier to identify spam if we can also classify ham messages. That is, if we can determine that a message is ham, then there is less doubt about it being spam! Table 3.1 provides a list of 29 variables that may prove useful in distinguishing between spam and ham.

We focus our attention on the problem of how to create a few of these variables. Let's consider the use of capitalization. The over use of capitalization in email is referred to as yelling, and from experience, we have seen that messages that yell a lot tend to be spam. There are several approaches to quantifying the amount of yelling in a message. We may, for example, look at the subject line in the header, and determine whether it is all capitals or not; we may wish to report the percentage of capital letters among all letters used in the body; or, we may count the number of lines in the body that are entirely capitalized. In the first case, it is natural to create a logical to indicate whether or not a subject is all capitals. Recall that the subject is an element of the header vector in each message element of emailStruct, and it is named Subject.

A simple approach to determining if the subject is all caps is to split the subject character string into substrings with one character per substring, and determine which are upper case letters. The R built-in vector LETTERS is a 26-element character vector containing the capital letters. We can use it as follows,

```
header = sampleStruct[[1]]$header
subject = header["Subject"]
els = strsplit(subject, "")
all(els %in% LETTERS)
```

[1] FALSE

We can make up some of our own test cases to test that our code performs as expected. Below are some subject lines that we want to be sure we handle properly:

```
testSubject = c("DEAR MADAME", "WINNER!", "")
```

When we try this code on the test cases, we immediately identify several problems:

```
els = strsplit(testSubject, "")
sapply(els, function(subject) all(subject %in% LETTERS))
```

[1] FALSE FALSE TRUE

The subject "DEAR MADAME" returns FALSE because it contains a blank character, which is not an upper case letter. The "WINNER!" returns FALSE because it contains an exclamation mark, which is also not a capital. The blank subject line returns TRUE because the return value from the strsplit() is character(0); consequently, the return value from the %in%() function is logical(0), and when all() is applied to logical(0), the result is TRUE.

We clearly need to improve our code to discount punctuation and blanks in the subject and to handle headers with no subject. To address the issue with punctuation and blanks, we can strip these characters from the subject before looking for upper case letters. The following call to gsub() "substitutes" a blank or punctuation mark with nothing, meaning it drops them from the character string:

TABLE 3.1: Variable Definition Table

Variable	Type	Definition
isRe	logical	TRUE if Re: appears at the start of the subject.
numLines	integer	Number of lines in the body of the message.
bodyCharCt	integer	Number of characters in the body of the message.
underscore	logical	TRUE if email address in the From field of the header contains an underscore.
subExcCt	integer	Number of exclamation marks in the subject.
subQuesCt	integer	Number of question marks in the subject.
numAtt	integer	Number of attachments in the message.
priority	logical	TRUE if a Priority key is present in the header.
numRec	numeric	Number of recipients of the message, including CCs.
perCaps	numeric	Percentage of capitals among all letters in the message body, excluding attachments.
isInReplyTo	logical	TRUE if the In-Reply-To key is present in the header.
sortedRec	logical	TRUE if the recipients' email addresses are sorted.
subPunc	logical	TRUE if words in the subject have punctuation or numbers embedded in them, e.g., w!se.
hour	numeric	Hour of the day in the Date field.
multipartText	logical	TRUE if the MIME type is multipart/text.
hasImages	logical	TRUE if the message contains images.
isPGPsigned	logical	TRUE if the message contains a PGP signature.
perHTML	numeric	Percentage of characters in *HTML* tags in the message body in comparison to all characters.
subSpamWords	logical	TRUE if the subject contains one of the words in a spam word vector.
subBlanks	numeric	Percentage of blanks in the subject.
noHost	logical	TRUE if there is no hostname in the Message-Id key in the header.
numEnd	logical	TRUE if the email sender's address (before the @) ends in a number.
isYelling	logical	TRUE if the subject is all capital letters.
forwards	numeric	Number of forward symbols in a line of the body, e.g., >>> xxx contains 3 forwards.
isOrigMsg	logical	TRUE if the message body contains the phrase original message.
isDear	logical	TRUE if the message body contains the word dear.
isWrote	logical	TRUE if the message contains the phrase wrote:.
avgWordLen	numeric	The average length of the words in a message.
numDlr	numeric	Number of dollar signs in the message body.

This table provides several possible variables that can be derived from an email message and used for classifying spam.

```
gsub("[[:punct:] ]", "", testSubject)

[1] "DEARMADAME" "WINNER"        ""
```

We probably want to eliminate numbers too. This may be more cleanly and simply accomplished by eliminating the complement, i.e., all non-alpha characters. That is,

```
gsub("[^[:alpha:]]", "", testSubject)

[1] "DEARMADAME" "WINNER"        ""
```

Further testing of our code uncovers yet another peculiarity: a message with a header that consists entirely of non-alpha characters returns an empty character string when we eliminate non-alpha characters. This raises the question of how to handle missing data. In the case where there is no subject line, a value of NA seems an accurate and informative return value because it indicates the message does not have a subject. However, when a subject has only special characters, then a value of FALSE is an appropriate return value.

Our set of expressions can be gathered into an R function called isYelling(). The argument to the function is a message of the format we have created in Section 3.8. Recall that each message is converted into a list where the element called header contains the header as a named vector. The return value of isYelling() is a logical indicating whether all of the alpha characters in the subject line are upper case, with NA indicating the message has no subject. Note that in our function, shown below, we take an alternative approach to determining whether the subject line is all yelling than that discussed:

```
isYelling = function(msg) {
  if ( "Subject" %in% names(msg$header) ) {
     el = gsub("[^[:alpha:]]", "", msg$header["Subject"])
     if (nchar(el) > 0)
        nchar(gsub("[A-Z]", "", el)) < 1
     else
        FALSE
  } else
     NA
}
```

Another way to measure the presence of yelling is by the percentage of capitalized letters in the message body. The denominator of this proportion could be the number of characters in the email message, the number of non-blank characters, the number of alpha characters, etc. There is no right answer here, but several reasonable ones. For example, it seems reasonable to eliminate all non-alpha characters and compute: 100×#upper case letters /#upper and lower case letters. The following function perCaps() does just that.

```
perCaps =
function(msg)
{
  body = paste(msg$body, collapse = "")

     # Return NA if the body of the message is "empty"
  if(length(body) == 0 || nchar(body) == 0) return(NA)

     # Eliminate non-alpha characters
```

Using Statistics to Identify Spam

```
    body = gsub("[^[:alpha:]]", "", body)
    capText = gsub("[^A-Z]", "", body)
    100 * nchar(capText)/nchar(body)
}
```

We apply perCaps() to our sample messages with

```
sapply(sampleStruct, perCaps)
```

```
 [1]  4.5  7.5  7.4  5.1  6.1  7.7  5.5 10.1 10.9  6.5
[11]  9.6 12.0  9.2  1.7  6.4
```

We have many different variables that we can create to quantify the difference between spam and ham. To help us keep track of the corresponding functions we create a list of functions. Then, we can apply this list of functions to our email structure to make a data frame of variables. Each column in the data frame corresponds to the output from one of these functions, and each row corresponds to a message. The list shown below contains functions to create 4 of the 29 variables shown in Table 3.1.

```
funcList = list(
  isRe = function(msg) {
      "Subject" %in% names(msg$header) &&
        length(grep("^[ ]*Re:", msg$header[["Subject"]])) > 0
  },
  numLines = function(msg)
                length(msg$body),
  isYelling = function(msg) {
    if ( "Subject" %in% names(msg$header) ) {
      el = gsub("[^[:alpha:]]", "", msg$header["Subject"])
      if (nchar(el) > 0)
        nchar(gsub("[A-Z]", "", el)) < 1
      else
        FALSE
    }
    else NA
  },
  perCaps = function(msg) {
    body = paste(msg$body, collapse = "")

        # Return NA if the body of the message is "empty"
    if(length(body) == 0 || nchar(body) == 0) return(NA)

        # Eliminate non-alpha characters
    body = gsub("[^[:alpha:]]", "", body)
    capText = gsub("[^A-Z]", "", body)
    100 * nchar(capText)/nchar(body)
  }
)
```

Notice the two logical expressions on either side of && in the isRe() function. The expression on the right-hand side yields an error if the header has no element named Subject, but fortunately, the expression on the left-hand side evaluates to FALSE in this situation so the right-hand side is not evaluated. We rely on R's lazy evaluation in this expression to avoid an error. Other versions of isRe() might include checking for Fwd: Re:, or it might examine the In-Reply-To field instead of the Subject line. We leave these alternatives as exercises.

Why do we need to put these functions into a list? If they are in a list, then we can more easily apply them to our email corpus. That is, rather than writing explicit code, such as

```
isRe(msg)
perCaps(msg)
numLines(msg)
...
```

we can apply the functions with generic code such as

```
lapply(funcList, function(func)
                   sapply(sampleStruct, function(msg) func(msg)))

$isRe
 [1]  TRUE FALSE FALSE FALSE  TRUE  TRUE  TRUE FALSE
 [9]  TRUE  TRUE  TRUE  TRUE  TRUE FALSE  TRUE

$numLines
 [1] 50 26 38 32 31 54 35 36 65 58 70 31 38 28 34

$isYelling
Subject Subject Subject Subject Subject Subject Subject Subject
  FALSE   FALSE   FALSE   FALSE   FALSE   FALSE   FALSE   FALSE
Subject Subject Subject Subject Subject Subject Subject
  FALSE   FALSE   FALSE   FALSE   FALSE   FALSE   FALSE

$perCaps
 [1]  4.5  7.5  7.4  5.1  6.1  7.7  5.5 10.1 10.9  6.5
[11]  9.6 12.0  9.2  1.7  6.4
```

If we add or remove functions in funcList our code to create the derived variables does not change because it does not depend on the name or number of functions.

The following function, createDerivedDF(), implements this approach. We allow the list provided in the *operations* argument to include expressions as well as functions. The operation is applied to the email slightly differently, depending on whether it is a function or expression. For an expression, we evaluate the expression in an environment which contains the variable msg. This is the individual message object to be processed. The expression must refer to that variable directly. The benefit of an expression is that it is very slightly simpler to write than a function. However, it is harder to reason about how it will be evaluated and where the variables to which it refers are located. A function is self-contained and we can write its code in terms of our parameter names, not those defined by others. There is a trade-off. Our function to create the derived variables from the list of functions and expressions is

```
createDerivedDF =
function(email = emailStruct, operations = funcList,
         verbose = FALSE)
{
  els = lapply(names(operations),
```

Using Statistics to Identify Spam

```
                function(id) {
                  if(verbose) print(id)
                  e = operations[[id]]
                  v = if(is.function(e))
                         sapply(email, e)
                       else
                         sapply(email, function(msg) eval(e))
                  v
           })

    df = as.data.frame(els)
    names(df) = names(operations)
    invisible(df)
}
```

Notice that one variable is created at a time using the code in the list of operations, and the name of an element in `operations` becomes the corresponding variable/column name in the data frame. We try createDerivedDF() on our sample messages with

```
sampleDF = createDerivedDF(sampleStruct)
head(sampleDF)
```

```
    isRe  numLines isYelling   perCaps
1   TRUE        50     FALSE       4.5
2  FALSE        26     FALSE       7.5
3  FALSE        38     FALSE       7.4
4  FALSE        32     FALSE       5.1
5   TRUE        31     FALSE       6.1
6   TRUE        54     FALSE       7.7
```

We leave as exercises the tasks of deriving the remaining variables and any others that you think might be promising as possible predictors of spam or ham. The data frame `emailDF` in `RSpamData` contains the conversion of the complete set of 9348 email into the 29 variables shown in Table 3.1 plus the `isSpam` indicator for spam. We can load and query it with

```
load("Data/spamAssassinDerivedDF.rda")
dim(emailDF)
```

```
[1] 9348   30
```

3.9.1 Checking Our Code for Errors

How do we know if our code to convert the email into variables performs correctly? The code might not be giving us syntax errors, but is it giving us the values we want? To answer this question we can try any of the following approaches:

- Create a second way to accomplish the same task and compare the 2 sets of results. If they are not identical then the differences point to problems with the code.

- Examine the cases where NAs or large values occur, and determine if they are caused by an error in the code or if they are legitimate.

- Perform exploratory data analysis on the results to confirm basic characteristics of the data. If we find surprising or inconsistent results, then we investigate how they arose.

For an example of the first approach, consider again the function perCaps() that we wrote to compute the percentage of letters in the body that are capitalized. As described earlier, we can check the upper case alpha characters in the body against those in LETTERS. The following function takes this alternative approach.

```
perCaps2 =
function(msg)
{
  body = paste(msg$body, collapse = "")

      # Return NA if the body of the message is "empty"
  if(length(body) == 0 || nchar(body) == 0) return(NA)

      # Eliminate non-alpha characters and empty lines
  body = gsub("[^[:alpha:]]", "", body)
  els = unlist(strsplit(body, ""))
  ctCap = sum(els %in% LETTERS)
  100 * ctCap / length(els)
}
```

We apply perCaps() and perCaps2() to all the email and compare the results. We see that they produce the same values for each message.

```
pC = sapply(emailStruct, perCaps)
pC2 = sapply(emailStruct, perCaps2)
identical(pC, pC2)
```

[1] TRUE

We next consider an example of the second approach to error checking; that is, we examine unusual values to see if they indicate problems with our code. We find that a few messages have a value of NA for the number of exclamation marks in the subject line of the message,

```
indNA = which(is.na(emailDF$subExcCt))
```

Then we find which of the raw email messages have no subject line,

```
indNoSubject = which(sapply(emailStruct,
                            function(msg)
                              !("Subject" %in% names(msg$header))))
```

Finally, we compare the 2 sets of indices to see if they are the same, i.e., if the NAs in the subExcCt correspond to those messages that have no subject line.

```
all(indNA == indNoSubject)
```

[1] TRUE

Using Statistics to Identify Spam

This is indeed the case and is one indication that our code is correct.

Lastly, we consider an example of the third approach to testing our code: we compare the values in 2 variables to see if they are consistent with one another. Let's take the number of lines in the body of a message, `numLines`, and the number of characters in the body of the message, `bodyCharCt`. We picked these two because there are constraints on their relationship. Namely, the number of lines should not exceed the number of characters in the body of the message. (Messages with many empty lines may violate this condition.) The following code makes this comparison, and determines which messages, if any, violate this inequality.

```
all(emailDF$bodyCharCt > emailDF$numLines)
```

```
[1] TRUE
```

We also examine the relationship between these 2 variables in Figure 3.4. As might be expected, they appear to have a highly linear association.

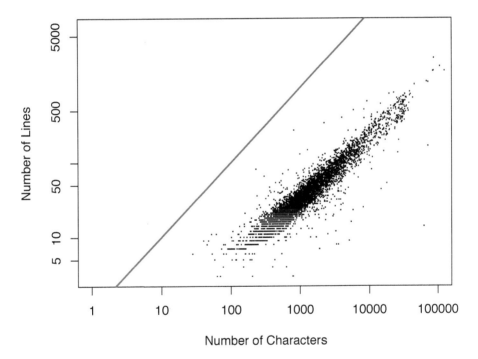

Figure 3.4: Comparison of Two Measures of Length for a Message. *This scatter plot shows the relationship between the number of lines and the number of characters in the body of a message. The plot is on log scale, and 1 is added to all of the values before taking logs to address issues with empty bodies. The line $y = x$ is added for comparison purposes.*

3.10 Exploring the email Feature Set

We have derived a few dozen variables from the header, body, and attachments of the messages. Before conducting a formal analysis, we may ask ourselves whether or not these derived variables are useful for predicting spam. Before embarking on a complex statistical analysis it is often a good idea to perform some simple analyses to get a better understanding of the data and its possibilities. This exploratory analysis may indicate that we need to do a better job of deriving variables from the messages. As seen already, this not only helps us in our future statistical analysis, it also helps us check that the variables have been correctly coded.

Let's examine the percentage of capitals in the message body, comparing this percentage between spam and ham messages. Side-by-side boxplots in Figure 3.5 help us compare the distributions of these two groups.

```
percent = emailDF$perCaps
isSpamLabs = factor(emailDF$isSpam, labels = c("ham", "spam"))
boxplot(log(1 + percent) ~ isSpamLabs,
        ylab = "Percent Capitals (log)")
```

We see from the boxplots that about 75% of the regular email have values below the lower quartile for spam. This variable may be useful for classification.

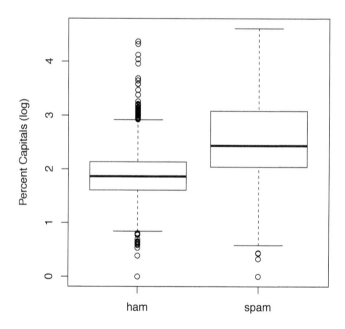

Figure 3.5: Use of Capitalization in email. *These boxplots compare the percentage of capital letters among all letters in a message body for spam and ham. The use of a log scale makes it easier to see that nearly 3/4 of the spam have more capital letters than nearly all of the ham.*

Another way to see this is with a quantile–quantile plot. If the 2 distributions have roughly the same shape then their paired quantiles fall on a line. We see that this is indeed the case for the percentage of capital letters in the email, although the spam messages have a larger average number of capital letters and a greater spread than non-spam messages. A slope other than 1 indicates the distributions have different spreads, and an intercept other than 0 indicates a shift in the mean of the distributions. We leave it as an exercise to create this quantile–quantile plot.

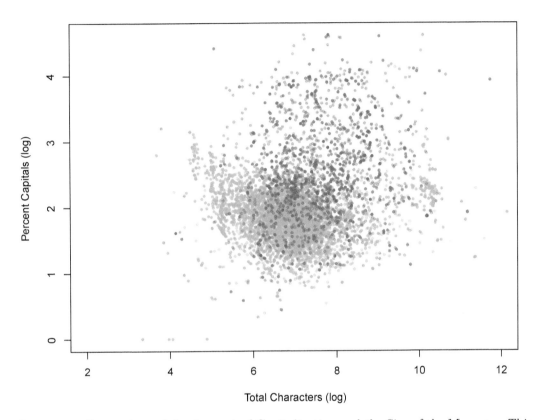

Figure 3.6: Comparison of the Amount of Capitalization and the Size of the Message. *This scatter plot examines the relationship between the percentage of capital letters among all letters in a message and the total number of characters in the message. Spam is marked by purple dots and ham by green. The darker color indicates overplotting. We see here that the spam tends to be longer and have more capital letters than ham.*

In addition, we can compare the joint distribution of the percentage of capital letters in the email and the total number of characters in the body of the message for spam vs. ham messages. We make a scatter plot of these 2 variables

```
colI = c("#4DAF4A80", "#984EA380")
logBodyCharCt = log(1 + emailDF$bodyCharCt)
logPerCaps = log(1 + emailDF$perCaps)
plot(logPerCaps ~ logBodyCharCt, xlab = "Total Characters (log)",
     ylab = "Percent Capitals (log)",
     col = colI[1 + emailDF$isSpam],
     xlim = c(2,12), pch = 19, cex = 0.5)
```

In Figure 3.6, the ham messages are denoted by green dots and the spam messages are purple dots. We see a lot of overlap between ham and spam but the spam messages tend to be longer and have a greater percentage of capitals.

A simple tabulation of the number of attachments in a message against whether it is spam shows that there is very little difference between spam and ham because most messages have no attachments.

```
table(emailDF$numAtt, isSpamLabs)
      ham spam
   0 6624 2158
   1  314  230
   2   11    6
   4    0    1
   5    1    2
  18    1    0
```

It is doubtful that this variable will be useful for classification.

Finally, we compare spam and ham messages by examining some of the logical variables. The variable `isRe` indicates whether or not the subject line has an `Re:` in it, and the variable `numEnd` indicates whether or not the sender's address has a number at the end of it, e.g., `beng3000@gmail.com` has a numeric end. We compare these against whether or not the message is spam with

```
colM = c("#E41A1C80", "#377EB880")
isRe = factor(emailDF$isRe, labels = c("no Re:", "Re:"))
mosaicplot(table(isSpamLabs, isRe), main = "",
           xlab = "", ylab = "", color = colM)

fromNE = factor(emailDF$numEnd, labels = c("No #", "#"))
mosaicplot(table(isSpamLabs, fromNE), color = colM,
           main = "", xlab="", ylab = "")
```

In the mosaic plots in Figure 3.7 we see that the spam messages are less likely to contain an `Re:` but more likely to have a numeric end to the sender's address.

3.11 Fitting the rpart() Model to the email Data

We have examined some of the variables that we have derived from the email and uncovered a few that may prove useful in partitioning the email. This exploratory analysis may have suggested other variables to consider. We also may want to look more deeply into relationships between the variables. For now, we proceed to apply the rpart() method to the 29 variables in `emailDF` and assess how well they can classify email.

Let's first learn more about the `rpart` package [6]. We can read the documentation provided for `rpart` and available at http://cran.r-project.org/web/packages/rpart/rpart.pdf. We also can search for additional documentation and tutorials on the Internet. In this process, we discover that the variables in the data frame must all be either factors or numeric. For a factor, a split is made according to a particular level of the factor; that is, if an observation has that particular factor level then it branches to the left, otherwise it branches to the right. Our variables are either numeric or logical so we must convert the logicals to factors. We use the following function to perform the conversion:

Using Statistics to Identify Spam

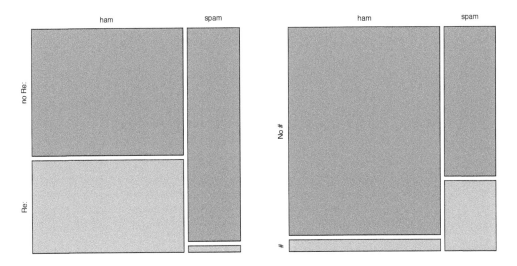

Figure 3.7: Exploring Categorical Measures Derived from email. *These two mosaic plots use area to denote the proportion of messages that fall in each category. The plot on the top shows those messages that have an* Re: *in the subject line tend not to be spam. The bottom plot shows that those messages that are from a user with a number at the end of their email address tend to be spam. However, few messages are sent from such users so it is not clear how helpful this distinction will be in our classification problem.*

```
setupRpart = function(data) {
  logicalVars = which(sapply(data, is.logical))
  facVars = lapply(data[ , logicalVars],
                   function(x) {
                     x = as.factor(x)
                     levels(x) = c("F", "T")
                     x
                   })
  cbind(facVars, data[ , - logicalVars])
}
```

We apply setupRpart() to our data frame

```
emailDFrp = setupRpart(emailDF)
```

Now that our data are properly formatted, we split them into training and test sets, as we did in the naïve Bayes analysis. We use the same subsets as with the text mining of Section 3.6.1 so we can more accurately compare these 2 approaches. To do this, we reset the random seed to the value used earlier and generate the indices for the test set as follows:

```
set.seed(418910)
testSpamIdx = sample(numSpam, size = floor(numSpam/3))
testHamIdx = sample(numHam, size = floor(numHam/3))
```

We use these indices to select the rows of the data frame with

```
testDF =
  rbind( emailDFrp[ emailDFrp$isSpam == "T", ][testSpamIdx, ],
         emailDFrp[emailDFrp$isSpam == "F", ][testHamIdx, ] )
trainDF =
  rbind( emailDFrp[emailDFrp$isSpam == "T", ][-testSpamIdx, ],
         emailDFrp[emailDFrp$isSpam == "F", ][-testHamIdx, ])
```

Then we fit the classification tree with the following call to rpart()

```
rpartFit = rpart(isSpam ~ ., data = trainDF, method = "class")
```

The documentation for the rpart() function informs us that rpart() returns an object of class rpart and that there is a specialized plotting function that can plot a tree representation of the object. To find out more about this plotting function, we can read the help for plot.rpart(). We can access this specialized plotting function with the call plot(rpartFit), and we find that the plot has problems with over plotting. With another Internet search, we discover an alternative plotting library called rpart.plot [2]. We plot the fitted tree with

```
library(rpart.plot)
prp(rpartFit, extra = 1)
```

The resulting tree appears in Figure 3.8. Other functions that can handle rpart objects include predict() to classify data using the tree, and print() to produce text summaries of the rpart object.

Let's see how well the classifier does at predicting whether the test messages are spam or ham. We use predict() on testDF to obtain the predictions with

```
predictions = predict(rpartFit,
       newdata = testDF[, names(testDF) != "isSpam"],
       type = "class")
```

To find out how well our tree has performed in classifying these test messages, we compare the predictions from the fitted rpart object to the hand classifications. First we find the Type I error, the proportion of ham messages that have been misclassified as spam. We do this with

```
predsForHam = predictions[ testDF$isSpam == "F" ]
summary(predsForHam)

FALSE   TRUE
 2192    125


sum(predsForHam == "T") / length(predsForHam)

[1] 0.054
```

And the Type II error rate is

```
predsForSpam = predictions[ testDF$isSpam == "T" ]
sum(predsForSpam == "F") / length(predsForSpam)

[1] 0.16
```

Using Statistics to Identify Spam

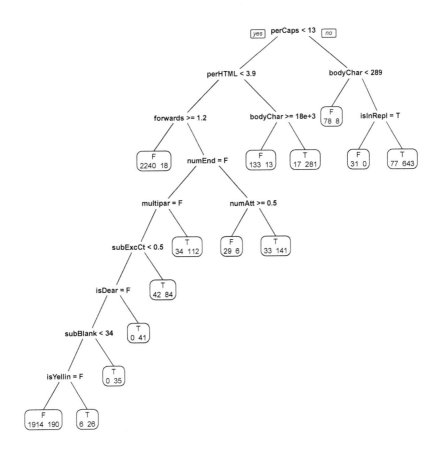

Figure 3.8: Tree for Partitioning email to Predict Spam. *This tree was fitted using rpart() on 6232 messages. The default values for all of the arguments to rpart() were used. Notice the leftmost leaf classifies as ham those messages with fewer than 13% capitals, fewer than 4% HTML tags, and at least 1 forward. Eighteen spam messages fall into this leaf and so are misclassified, but 2240 of the ham is properly classified using these 3 yes–no questions.*

We see that our classifier did reasonably well with Type I errors, but the Type II error rate is 16%. Can we do better? We used the default settings for the partitioning algorithm, and it may be that changing them could improve the results.

The rpart.control() function allows the user to control various aspects of the tree-fitting procedure. It takes the following arguments,

```
args(rpart.control)
```

```
function (minsplit = 20L, minbucket = round(minsplit/3),
    cp = 0.01, maxcompete = 4L, maxsurrogate = 5L,
    usesurrogate = 2L, xval = 10L, surrogatestyle = 0L,
    maxdepth = 30L, ...)
```

The documentation for rpart.control() explains more about how these various parameters are used in fitting the tree to the data. Let's explore the complexity parameter, which is used as a threshold where any split that does not decrease the overall lack of fit by cp is not considered. The default value is 0.01 so let's examine how the fit changes when we try different values for cp, e.g.,

```
complexityVals = c(seq(0.00001, 0.0001, length=19),
                   seq(0.0001, 0.001, length=19),
                   seq(0.001, 0.005, length=9),
                   seq(0.005, 0.01, length=9))
```

We call rpart() with each of these values for cp and use the resulting model to classify the test data with

```
fits = lapply(complexityVals, function(x) {
         rpartObj = rpart(isSpam ~ ., data = trainDF,
                          method="class",
                          control = rpart.control(cp=x) )

         predict(rpartObj,
                 newdata = testDF[ , names(testDF) != "isSpam"],
                 type = "class")
       })
```

We assess the Type I and II errors for these fitted models applied to our test data with

```
spam = testDF$isSpam == "T"
numSpam = sum(spam)
numHam = sum(!spam)
errs = sapply(fits, function(preds) {
                typeI = sum(preds[ !spam ] == "T") / numHam
                typeII = sum(preds[ spam ] == "F") / numSpam
                c(typeI = typeI, typeII = typeII)
              })
```

The errors are displayed in Figure 3.9. We see there that the smallest Type I error that we are able to achieve is about 0.04, which occurs for a complexity value of about 0.001. The Type II error for this complexity value is 10.5%. The text mining approach has smaller Type I and Type II errors. The poorer performance of recursive partitioning may be due to the variables that were used in the model. Or, it may be due to the parameter settings in rpart(). We leave as an exercise the task of developing a better feature set for predicting spam using rpart(). Additionally, we leave as an exercise the creation of a combined approach that employs both word vectors and derived variables in predicting spam.

3.12 Exercises

Q.1 We hand-selected email to belong to the sample set in sampleEmail. Instead of this approach, use the sample function to choose messages at random for the sample. Be sure to take files from all 5 directories of email. Read these files into R and test several functions with these new messages, e.g., getBoundary() and dropAttach() from Section 3.5.2, to make sure that they work properly.

Using Statistics to Identify Spam 165

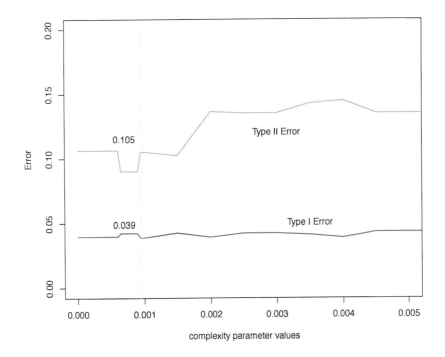

Figure 3.9: Type I and II Errors for Recursive Partitioning. *This plot displays the Type I and II errors for predicting spam as a function of the size of the complexity parameter in the rpart() function. The complexity parameter is a mechanism for specifying the threshold for choosing a split for a subgroup. Splits that do not achieve a gain in fit of at least the size of the parameter value provided are not made. The Type I error is minimized at a complexity parameter value of 0.001 for an error rate of 3.9%. The Type II error rate for this complexity parameter value is 10.5%.*

Q.2 In the text mining approach to detecting spam we ignored all attachments in creating the set of words belonging to a message (see Section 3.5.2). Write a function to extract words from any plain text or *HTML* attachment and include these words in the set of a message's words. Try to reuse the findMsg() function and modify the dropAttach() function to accept an additional parameter that indicates whether or not the words in attachments are to be extracted. Does this change improve the classification?

Q.3 The string manipulation functions in R can be used instead of regular expression functions for finding, changing, and extracting substrings from strings. These functions include: strsplit() to divide a string up into pieces, substr() to extract a portion of a string, paste() to glue together multiple strings, and nchar(), which returns the number of characters in a string. Write your own version of getBoundary() (see Section 3.5.2) using these functions to extract the boundary string from the Content-Type. Debug your function with the messages in `sampleEmail`.

Q.4 Write the dropAttach() function for Section 3.5.2. This function has two inputs, the body of a message and the boundary string that marks the location of the attachments. It returns the body without its attachments. Include in the return value the lines of the body that follow the first boundary string up to the string marking the first attachment and the lines following the ending boundary string. Be sure to consider the idiosyncratic cases of no attachments and a missing ending boundary string.

Q.5 Write the function findMsgWords() of Section 3.5.3. This function takes as input the message body (with no attachments) and the return value is a vector of the unique words in the message. That is, we only track which words are in the message, not the number of times these words appear in the message. This function should eliminate punctuation, digits, and blanks from the message. Consider whether it is simpler to split the string by blanks first and then process the punctuation, digits, etc. The function should convert capital letters to lower case, and drop all stop words and words that are only one letter long. A vector of stop words is available in the tm package. Use the stopwords() function in this package to create the vector.

Q.6 Try to improve the text cleaning in findMsgWords() of Section 3.5.3 by stemming the words in the messages. That is, make plural words singular and reduce present and past tenses to their root words, e.g., run, ran, runs, running all have the same "stem." To do this, use the stemming functions available in the text mining package tm. Incorporate this stemming process into the findMsgWords() function. Then recreate the vectors of words for all the email and see if the classification improves.

Q.7 Consider the treatment of *URLs* in the text cleaning in findMsgWords() of Section 3.5.3. Notice that this function often turns a *URL* into gibberish. Should we drop *URLs* altogether from the messages, or should we try to keep the *URL* as one whole "word"? Why might these alternatives be better or worse than the approach taken in Section 3.5.3? Try one of these alternatives and compare it to the approach of that section to see if it improves the classification.

Q.8 In Section 3.6.4 we saw a few alternative mathematical expressions for the naïve Bayes approximation to the log likelihood ratio of the chance a message is spam or ham. Each suggests a different approach to carrying out the computation. Create alternative versions of computeFreqs() and computeMsgOdds() in Section 3.6 to calculate the log odds. Compare the accuracy of these alternatives to the approach used in Section 3.6. Also consider timing the computation with a call to system.time() to determine if one approach is much faster or slower than the others.

Q.9 The function computeFreqs() in Section 3.6 uses a default bag of words constructed directly from the words passed to it in the argument *wordsList*. However, it is possible to supply a different bag of words via the *bow* parameter. When this happens, it may be the case that some words in *wordsList* are not found in the bag of words. This causes an error when running the function. Determine which lines of code in computeFreqs() are problematic and update the function to handle this situation. Be sure to test your code.

Q.10 Use the typeIErrorRates() function in Section 3.6.3 as a guide to write a function that computes the Type II error rates, i.e., the proportion of spam messages that are misclassified as ham. As with the Type I error rates, convince yourself that the error rate is monotonic in τ, that it changes only at the values of the LLR in the provided messages, and that you only need to consider these values for the spam messages. This function, called typeIIErrorRates(), has the same inputs as the typeIErrorRates() function and returns the same structure. The only difference is that the rates returned are based on Type II errors.

Using Statistics to Identify Spam

Q.11 In Section 3.8.1, we used the read.dcf() function to read the key: value data in the email headers. In this exercise, we use regular expressions to extract the keys and their values from the header.

The first step in the process is to find the continuation lines for a value, and then collapse them with the first line of the value. These continuation lines start with either a blank space or a tab character. Use regular expressions to locate them. Then paste them to the first line of the value.

Next break the revised set of key:value strings into the keys and values. Again use regular expressions to do this. Then create the names vector from these keys and values.

Q.12 Write the processAttach() function for Section 3.8.2. This function has two inputs, the body of a message and the Content-Type value. It returns a list of 2 elements. The first is called body and it contains the message body without its attachments. Be sure to consider the idiosyncratic cases of no attachments and a missing ending boundary string. The second element is a data frame called attachDF. This data frame has one row for each attachment in the message. There are 2 variables in this data frame, aLen and aType, which hold the number of lines and the MIME type of each attachment, respectively.

Q.13 Write a function to handle an alternative approach to measure yelling: count the number of lines in the email message text that consist entirely of capital letters. Carefully consider the case where the message body is empty. How do you modify your code to report a percentage rather than a count? In considering this modification, be sure to make clear how you handle empty lines, lines with no alpha-characters, and messages with no text.

Q.14 Write alternative implementations for the isRe() function in Section 3.9. For one alternative, instead of only checking whether the subject of a message begins Re:, look also for Fwd: Re:. For a second alternative, check for Re: anywhere in the subject, not just at the beginning of the string. Analyze the output from these 3 functions, including the original isRe() function in Section 3.9. How many more messages have a return value of TRUE for these alternatives, and are they all ham? Which do you think is most useful in predicting spam?

Q.15 Choose several of the ideas listed in Table 3.1 for deriving features from the email and write functions for them. Be sure to check your code against what you expect. Try writing one of these functions in two different ways and compare the output from each. Use exploratory data analysis techniques to check that your code works as expected. Does the output of your function match the values in the corresponding columns of emailDF? If not, why do you think this might be the case? Does it appear that this derived variable will be useful in identifying spam or ham?

Q.16 Consider other variables that are not listed in Table 3.1 that might by useful in the classification problem. Write functions to derive them from the messages in emailStruct and add them to emailDF. Refit the classification tree with the enhanced data frame. Were these new variables chosen to partition the messages? Is the error in classification improved?

Q.17 Carry out additional exploratory analysis as described in Section 3.9.1. Include in your analysis a quantile–quantile plot of perCaps for the ham and spam.

Q.18 Write code to handle the attachments in the message separately from the text in the body of the message. Since each attachment has its own header, try processing the header and body of the attachment in a manner similar to the message header and body. Use the processHeader() function to do this. You may need to revise processHeader() to handle cases that arise in the attachments and not in the main headers of the messages.

Q.19 Consider the other parameters that can be used to control the recursive partitioning process. Read the documentation for them in the rpart.control() documentation. Also, carry out an Internet search for more information on how to tweak the rpart() tuning parameters. Experiment with values for these parameters. Do the trees that result make sense with your understanding of how the parameters are used? Can you improve the prediction using them?

Q.20 In Section 3.6.3 we used the test set that we had put aside to both select τ, the threshold for the log odds, and to evaluate the Type I and II errors incurred when we use this threshold. Ideally, we choose τ from another set of messages that is both independent of our training data and our test data. The method of cross-validation is designed to use the training set for training and validating the model. Implement 5-fold cross-validation to choose τ and assess the error rate with our training data. To do this, follow the steps:

  (a) Use the sample() function to permute the indices of the training set, and organize these permuted indices into 5 equal-size sets, called folds.

  (b) For each fold, take the corresponding subset from the training data to use as a 'test' set. Use the remaining messages in the training data as the training set. Apply the functions developed in Section 3.6 to estimate the probabilities that a word occurs in a message given it is spam or ham, and use these probabilities to compute the log likelihood ratio for the messages in the training set.

  (c) Pool all of the LLR values from the messages in all of the folds, i.e., from all of the training data, and use these values and the typeIErrorRate() function to select a threshold that achieves a 1% Type I error.

  (d) Apply this threshold to our original/real test set and find its Type I and Type II errors.

Q.21 Often in statistics, when we combine two methods, the resulting hybrid outperforms the two pure methods. For example, consider a naïve Bayes approach that incorporates the derived variables into the probability calculation. For simplicity, you might try a few variables that result from the important splits in the recursive partitioning tree from Figure 3.8, e.g., whether or not the percentage of capitals in the message body exceeds 13%. These variables have only 2 values as the splits are based on yes–no questions. Develop a hybrid classifier that uses both the word vectors and these additional features.

Q.22 An alternative to recursive partitioning is the kth nearest neighbor method. This method computes the distance between 2 messages using the values of the derived variables from Section 3.9, or some subset of them. Use the email in trainDF to find the k closest messages to each email in testDF. Then use these k neighbors to classify the test message as spam or ham. That is, use the neighbors' classifications to vote on the classification for the test message. Compare this method for predicting spam to the recursive partitioning approach. Use both Type I and Type II errors in making your comparison. Include a comparison of the two approaches from a computational perspective. Is one much faster or easier to implement?

Bibliography

[1] Ingo Feinerer and Kurt Hornik. tm: Text Mining Package. http://cran.r-project.org/web/packages/tm, 2014. R package version 0.5-10.

[2] Stephen Milborrow. rpart.plot: Plot rpart models. http://cran.r-project.org/web/packages/rpart.plot, 2014. R package version 1.4-4.

[3] R Core Team. *R Data Import/Export*, 2012. http://cran.r-project.org/doc/manuals/R-data.html.

[4] R Development Core Team. *R: A Language and Environment for Statistical Computing*. Vienna, Austria, 2012. http://www.r-project.org.

[5] Duncan Temple Lang and Deborah Nolan. RSpamData: SpamAssassin Public Data. http://omegahat.org/RSpamData, 2004. R package version 1.0.

[6] Terry Therneau, Beth Atkinson, and Brian Ripley. rpart: Recursive Partitioning and Regression Trees. http://cran.r-project.org/web/packages/rpart, 2014. R package version 4.1-8.

4

Processing Robot and Sensor Log Files: Seeking a Circular Target

Samuel E. Buttrey
Naval Postgraduate School

Timothy H. Chung
Naval Postgraduate School

James N. Eagle
Naval Postgraduate School

Duncan Temple Lang
University of California, Davis

CONTENTS

4.1	Description	171
	4.1.1 Computational Topics	172
4.2	The Data	173
	4.2.1 Reading an Entire Log File	175
	4.2.2 Exploring Log Files	179
	4.2.3 Visualizing the Path	184
	4.2.4 Exploring a "Look"	187
	4.2.5 The Error Distribution for Range Values	190
4.3	Detecting a Circular Target	194
	4.3.1 Connecting Segments Behind the Robot	198
	4.3.2 Determining If a Segment Corresponds to a Circle	200
4.4	Detecting the Target with Streaming Data in Real Time	213
	Bibliography	215

4.1 Description

In this case study, we explore robots searching for a circular target in a rectangular course that contains numerous obstacles (see Figure 4.1). The robots use a search strategy to move around the course, avoiding the obstacles and searching for the target in the shortest time possible. The robot continuously reports its location and also what it "sees" all around it. It searches for the target and ends when it determines it has found it, or after 30 minutes of searching. The robot can detect objects up to a distance of 2 meters away. In this chapter, we focus on processing these location and sight records and developing a classifier to detect if the robot is "looking at" the target. We use a statistical approach to determine if the shape the robot currently "sees" is consistent with the circular shape of the target (with known radius).

We look at log files for 100 different experiments (or runs), each log file containing the entire path information for that robot and its search for the target. The data include the location of the robot as it moves and what it "sees" at each of these positions. We explore the characteristics of each of these experiments, e.g., whether they found the target, how long the experiment lasted (up to the 30-minute time limit), how fast the robot moved, the locations of the obstacles, and the variability in the measurements. We develop the classifier for detecting the target and explore its operating characteristics, e.g. type I and type II error rates. We then discuss how to use the functionality to read lines in the log file to do classification from this streaming data in real time.

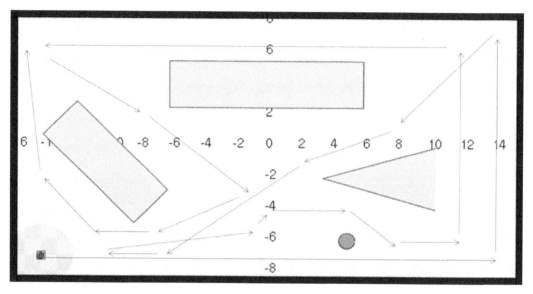

Figure 4.1: Example of the Course. *This shows a sample path through the course. The robot starts in the lower left corner. The circular target can be seen at approximately (4.5, -6.5). There are two rectangular obstacles and one triangular obstacle. The horizontal dimensions range from -15 to +15, and the vertical from -8 to +8.*

4.1.1 Computational Topics

- Text processing of log files
- Visualization
- Non-linear least squares
- Numerical optimization
- Goodness-of-fit criteria
- Streaming data

4.2 The Data

We have numerous data files in the `logs/` directory. We can find their names with the list.files() function:

```
ff <- list.files("logs", full.names = TRUE)
```

We look at these file names and see

```
  [1] "logs/01groundTruth.log"
  [2] "logs/JRSPdata_2010_03_10_12_12_31.log"
  [3] "logs/JRSPdata_2010_03_10_12_12_50.log"
     ....
[102] "logs/LASER"
[103] "logs/README"
```

The files corresponding to the experiments, that is, runs, start with `JRSPdata` and end with **log**. So we can specify this pattern to get only these file names with

```
ff <- list.files("logs", full.names = TRUE,
                 pattern = "JRSPdata.*\\.log")
```

The pattern is a regular expression. This call returns a vector of 100 file names.

How large are these files? We can use file.info() to get the size of each file:

```
info <- file.info(ff)
```

`info` contains information about who created the files, who can modify them, etc. However, we are interested in the `size` element of this data frame. This gives the number of bytes in each file. We can convert this to megabytes[1] and look at the distribution of this with

```
summary(info$size/1024^2)
   Min. 1st Qu.  Median    Mean 3rd Qu.    Max.
 0.6832  5.4060  7.1640  8.7560  9.3240 31.0500
```

We see that the smallest file is less than a megabyte and many of the files are between 7 and 9 megabytes. The largest file is about 31 megabytes. We can plot the distribution of the file size, shown in Figure 4.2, with

```
plot(density(info$size/1024^2), xlab = "megabytes")
```

and we can also compute the upper quantiles with

```
quantile(info$size/1024^2, seq(.9, 1, by = .01))
   90%   91%   92%   93%   94%   95%   96%   97%   98%   99%  100%
 15.68 16.84 23.92 25.19 25.41 26.19 27.25 29.88 29.93 30.48 31.05
```

So there are some reasonably large files. In total, there are `sum(info$size)/1024^2`, i.e., 875.6 megabytes of data. This isn't enormous, but it is significant so that we have to consider processing it efficiently. This is especially true considering we want to develop code that can process many more log files and also we ultimately want to process the data in

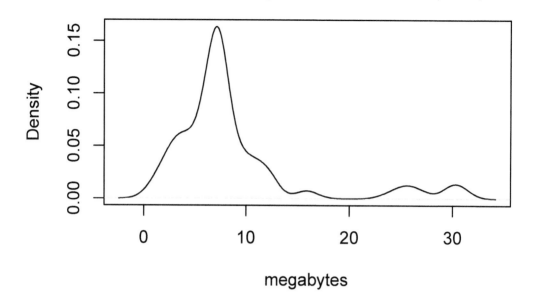

Figure 4.2: Log File Size. *This shows the distribution of the size of the 100 log files.*

real-time, i.e., as the robot is delivering the data to us and needs to know if it has located the target.

The log files are text files. They are not in a simple rectangular format such as comma-separated values CSV. Instead, they are structured in a standardized format defined and used by the Player Project. This project develops and distributes software for robot and sensor applications. The file format is documented at http://playerstage.sourceforge.net/doc/Player-svn/player/group_ _tutorial__datalog.html. The important idea is that each line contains a record, but there are different types of records. Each record type contains different information and has a different structure for the values it contains. This is why the data are non-rectangular, as each record type has a different number of values measuring different characteristics.

The first 12 lines of a particular log file (the fourth) are

```
## Player version 2.1.3
## File version 0.3.0
## Format:
## - Messages are newline-separated
## - Common header to each message is:
##    time     host    robot  interface index  type    subtype
##   (double) (uint)  (uint)  (string)  (uint) (uint)  (uint)
## - Following the common header is the message payload
0000000000.100 16777343 6668 laser 00 004 001 +0.000 +0.000
 ↻  0.000   0.156   0.155
0000000000.200 16777343 6668 position2d 00 004 001 -00.040
 ↻  +00.000 +0.000 +00.440 +00.380
```

---

[1] A megabyte contains 1024^2 bytes.

```
0000000000.200 16777343 6668 position2d 00 001 001 -14.000
↻ -07.000 +0.785 +00.000 +00.000 +00.000 0
0000000000.200 16777343 6668 laser 00 001 001 0001
↻ -3.1416 +3.1416 +0.01740495 +2.0000 0361 1.838  0 1.807
↻    0 1.778   0 1.749   0 1.723   0 1.697   0 1.673   0 1.650
```

(We have reformatted this to appear on the page. A line that starts with ↻ is actually a continuation of the previous line.)

Except for comment lines, which start with the pound sign (#) and can be ignored, every line of a log file starts with 7 common/shared fields of meta-data, separated by space characters. The names of these 7 fields are listed in the sixth line of the data file above. Of these 7 fields, the first, fourth, and sixth are of interest for our purposes: these give, respectively, the time, the "interface" (which describes the purpose of the record), and the type of the message. A number of different kinds of message are possible, but for our purposes only two combinations of interface and type are important. Lines with interface `position2d` and type value `001` give the current position, orientation, and yaw of the robot (this last measuring the angle clockwise from East to the direction the robot's head is facing, in radians). Lines with interface `laser` and type `001` give the measurements made during the robot's data collection, which we will call a "look." The `laser` line is associated with the previous `position2d` line and so these form a natural pair.

Data are collected every few seconds. There are two steps. During a look, the robot records its position and heading via a `position2d` record, and then looks all around itself, starting from the direction immediately behind it and continuing in one-degree increments in a `laser` record. The last look is, like the first, immediately behind the robot, so each data acquisition consists of 361 readings.

During a look, each reading produces a (Range, Intensity) pair. We ignore the Intensity for our purposes. The range gives the extent of the robot's view, i.e., the distance to something it can detect. The robot's vision is limited to 2 meters, so if there is no object visible within that distance, the observed value for range will be 2m. Otherwise, of course, the Range will be smaller.

Distance readings potentially contain measurement error, whether they refer to an actual object or whether they represent an observation of the 2m limit. These errors are small (on the order of a couple of centimeters), but in principle some Ranges could exceed 2m, and in some cases a measurement of a Range smaller than 2m will nonetheless be associated with the robot seeing no obstacle or target. We will want to familiarize ourselves with the distribution of times, locations, ranges, and the errors in the measurements.

4.2.1 Reading an Entire Log File

Typically, we have to explore the actual data files in order to empirically discover and understand their structure. We have to identify the patterns and anomalies. In this case, the documentation for these files is quite explicit. While the structure of different lines is not the same, the data are very structured. As a result, with our understanding of the format of the log files, we can set about reading them into R [2]. We will write a function to do this so that we can reuse it for all 100 log files (and potentially others). In creating this function, we should try to keep in mind that in Section 4.4, we will want to sequentially read individual lines and not an entire file. If possible, we should try to structure the code so that we do not have to have separate functions for the off-line and the on-line processing. However, since there is almost one gigabyte of text to process, we also want the function to be reasonably efficient.

We cannot use any of the common functions such as read.csv() or read.table() to read data from a log file. Instead, the basic strategy we use is to read all of the lines with readLines(). We then discard all of the comment lines, i.e., starting with #. We can do this with a combination of calls to readLines() and grep() (or grepl()), such as,

```
grep("^#", readLines(filename), value = TRUE, invert = TRUE)
```

This returns the vector of lines from the file that are not matched by the regular expression ^#.

Once we have the record lines containing data, we break each of these into a collection of values using strsplit() and separating by one or more spaces. We can specify this separator/delimiter with a regular expression in the call

```
tokens <- strsplit(lines, "[[:space:]]+")
```

The result of this is a list with an element for each line. Each element is a character vector. The first from our sample file will appear as

```
[1] "0000000000.100" "16777343"       "6665"           "laser"
[5] "00"             "004"            "001"            "+0.000"
[9] "+0.000"         "0.000"          "0.156"          "0.155"
```

The first laser line, however, will have 735 elements.

With the data in this form, we can access the 7 values common to each record. This allows us to discard any record that doesn't have 001 as the value for the type field or whose interface value is neither position2d or laser. We'll also ensure that the laser line immediately follows the position2d line. This ensures that we do not include position2d or laser lines that are not part of the same position–look pair. To do all of this, we use sapply() to extract the two fields, i.e,

```
iface <- sapply(tokens, `[`, 4)
type <- sapply(tokens, `[`, 6)
```

Now we can find the indices of the lines with position2d and 001 values for interface and type. We can then subset this vector of indices to include only those for which the following line is an appropriate laser line, i.e.,

```
i <- which(iface == "position2d" & type == "001")
i <- i[ iface[i+1] == "laser" & type[i+1] == "001"]
```

The result is that i contains the indices of the position2d lines we want, and we can obtain the corresponding laser lines with i + 1.

Once we have the set of all of the individual values of interest, we need to organize the values into a data structure. We can return the records a list with an element for each record. Each element would be a vector with all of the individual values for that record. Accordingly, the elements corresponding to laser and position2d records will have different lengths. This is the form of the raw data and may well be the most convenient form when performing streaming/on-line analysis. However, it is not very convenient to work with for exploratory data analysis. Instead, we could represent each record/line as a row in a data frame. The data frame would have 8 columns consisting of the 7 fields common to all records and an eighth that contains the "payload" of the record. This eighth column would contain all of the values for a record as a single string. This gives us access to all the data and we can work easily with the common fields. Working with the range or positions values is somewhat awkward. Instead of storing the payload as a single string, we could store the vector of values using R's ability to store arbitrary objects in a column of a data frame. So we would have a column of type *list* and each element would be a vector of the remaining values in the record. This is better than having the values in a single string, but still not entirely convenient.

An alternative approach to organizing the records and values is to combine each position2d-laser pair into a single record. This makes sense as we think of this pair as representing all the information about the robot at an instance in time. We have the time, location, and the 361 range values for the "look." We'll use this approach and so return a data frame with 364 columns – time, x, y, and the 361 range values. This is a convenient format for exploratory data analysis (EDA).

To create this data frame, we can first create a data frame with just the time, x and y vectors from the lines indexed by i above. We then create a second data frame with all of the 361 range values for each of the laser records. We'll have a row for each of these laser lines immediately following the corresponding position2d lines. The rows in these two data frames correspond to each each other and so we then use cbind() to combine these two "side by side."

As we create these two data frames, we have to extract the relevant elements from each character vector and then convert these values to numeric values and combine the results into a data frame. There are various ways to do this. One approach is to use, for example, sapply(els[i], `[`, c(1, 9, 10)) to get the time, x and y values from position2d. Here we are referring the the [() function directly. This is a simpler version of

```
sapply(els[i], function(x) x[c(1, 9, 10)])
```

i.e., we didn't have to define a new anonymous function for sapply(). Since each call to [() in the sapply() call yields a vector with 3 values, sapply() simplifies the result into a 3-row matrix, with as many columns as there are records (i.e., the length of i). We can then transpose this matrix, convert the values to numeric, and then use as.data.frame() to create the data frame.

We write the function to read a log file as

```
readLog <-
function(filename = "logs/JRSPdata_2010_03_10_12_12_31.log",
         lines = readLines(filename))
{
  lines = grep("^#", lines, invert = TRUE, value = TRUE)
  els = strsplit(lines, "[[:space:]]+")

    # Get the interface and type so we can subset.
  iface = sapply(els, `[`, 4)
  type = sapply(els, `[`, 6)

    # Find the indices corresponding to a position2d
    # with a laser immediately after.
  i = which(iface == "position2d" & type == "001")
  i = i[ iface[i+1] == "laser" & type[i+1] == "001"]

    # Get the time, x, y, and then the range values
    # from the laser below.
  locations = t(sapply(els[i], `[`, c(1, 8, 9)))
  ranges = t(sapply(els[i + 1], `[`,
```

```
                            seq(14, by = 2, length = 361) ))

    # now combine these into a data frame
    locations = as.data.frame(lapply(1:ncol(locations),
                                     function(i)
                                        as.numeric(locations[, i])))
    ranges = as.data.frame(lapply(1:ncol(ranges),
                                  function(i)
                                     as.numeric(ranges[, i])))

    ans = cbind(locations, ranges)
    names(ans) = c("time", "x", "y",
                   sprintf("range%d", 1:ncol(ranges)))

    invisible(ans)
}
```

We've allowed the caller to specify the actual content of the file as a vector of lines. This makes it easier for us to test and debug the function by specifying a small subset of lines. We might also be able to use this for the on-line/streaming approach.

Q.1 When creating the two data frames in the function, we extract a subset of values from each record, create a matrix and then a data frame. This essentially involves creating at least 3 copies of the relevant values in memory. What are they? Are there better ways to do this? Consider the time, x, y data frame, for example. We could use lapply(), rather than sapply(), to return a list of vectors (say, listOfVectors) of length 3. We could then use as.data.frame(do.call(rbind, listOfVectors)) to create the data frame. Does this reduce the amount of memory used? Are there other ways to arrange the computations to avoid having 3 copies of the data in memory at once? If so, are these faster or slower than our approach above in readLog()?

As with all functions, and code generally, that we write, we need to test the readLog() function. For reading data, one obvious test is to explicitly and manually check the results with the contents in the file. This is very time consuming, but very important. We can look at a random line in a log file and then verify its values are in the data frame and correct. If we just look at the first record, we may miss some structural problems that occur in our code that do not apply to the first record. We must also test both the position and the range values. It is also important to test more than one file.

Testing the results manually is tedious and also error-prone, but very important before proceeding with subsequent analysis. However, we'd also like a more programmatic approach. We must use a different approach to extract the data rather than just emulate the function. This is important in order to verify that the logic of our function is correct. If we merely re-implement the same logic, we are not testing whether the approach is correct, but merely the implementation. One alternative approach is to use the shell to extract the different lines. We can retrieve all of the position2d lines that do not contain the 004 type with

Shell `grep position2d JRSPdata_2010_03_10_12_12_31.log | grep -v ' 004 '`

The -v flag for the *grep* command negates the pattern matching. This is equivalent to the *invert* parameter for the grep() function in *R*.

We can extract the time, x, and y values from these records and read them into *R* with

```
cmd <- "grep position2d logs/JRSPdata_2010_03_10_12_12_31.log |
        grep -v ' 004 ' | cut -f 1,8,9 -d ' '"
txt <- system(cmd, intern = TRUE)
```

This yields lines of the form "time, x, y". We can then read these into a data frame with

```
sh <- read.table(textConnection(txt))
```

A `textConnection` represents a stream of characters as if it were the contents of a file. Connections are abstractions of streams of data, as we will see at the end of this chapter.

We can now compare the data frame from our shell command and the one returned with our call to readLog(). If these two approaches contain the same basic data for these 3 fields, this would illustrate that our implementation of readLog() is correct. In fact, they do not even have the same number of rows! This is probably because our readLog() function discarded `position2d` lines that were not followed immediately by a `laser` line. How many `position2d` lines are not followed by a `laser` line? This is tricky to do with the shell as those utilities are line oriented. However, we can do some quick calculations with the shell and R. We can count the number of `laser` records and `position2d` records in the log file to see the difference. We can do this with

```
cmd <-
 "egrep 'laser|position2d' logs/JRSPdata_2010_03_10_12_12_31.log |
       grep -v ' 004 '| cut -f 4 -d ' '"
table(system(cmd, intern = TRUE))

    laser position2d
     7725      7732
```

Note that we use *egrep* to use extended regular expressions and we searched for either `laser` or `position2d`. We extracted only the interface values and then computed the counts for each word. We see that there are 7 more `position2d` lines. We also see that the number of `laser` lines is the same as the number of rows returned by readLog().

We leave it as an exercise to continue this shell-based approach to verifying the results from readLog(). Compare the time, x, and y values and also the range values that are more complex. To obtain the range values from a `laser` record, we need to extract 14th, 16th, 18th, ..., 734 elements. We can use *cut* to do this, but we would have to carefully specify each of these field numbers. Instead, we can create the shell command conveniently in R with

```
sprintf("cut -d , -f %s",
        paste(seq(14, by = 2, length = 361), collapse = ","))
```

4.2.2 Exploring Log Files

We explored testing the results from the readLog() function by comparing the values in the data frame with those in the files. We also test the results by exploring the data. On occasion, we will uncover anomalies during the EDA that result from bugs in reading the data. As a result, the process of reading the data and exploring it is highly iterative. We'll continue to verify the results from readLog(), but our focus is now on exploring and becoming familiar with the data.

We start by reading all of the log files:

```
system.time(logs <- lapply(ff, readLog))
names(logs) <- ff
```

This gives us a list of the 100 log files and takes about 6 minutes (on a Macbook Pro with 16Gb of RAM and a 2.6 processor).

We start by looking at the distribution of the duration of each experiment, i.e., how many seconds each lasted. We compute the difference between the first time and the last time with

```
dur <- sapply(logs, function(x) x$time[nrow(x)] - x$time[1])
```

(Note that we assume the records are ordered by time and we should verify this. How?) We can verify that all of the experiments are completed within 30 minutes by computing the range

```
range(dur)
```

```
[1]   33.8 1799.7
```

At least one experiment lasted almost the 30 minute maximum and at least one lasted only 30 seconds.

Q.2 Did the robot find the target in that experiment?

We can examine the distribution of times with

```
plot(density(dur), xlim = c(0, 30*60))
```

We have used our knowledge that the longest time for an experiment is 30 minutes to limit the range of the axes to avoid showing the smoothed density for infeasible values. In Figure 4.3, we see 3 different modes corresponding to approximately 7 minutes, 20 minutes, and 30 minutes. There are 4 runs that last more than 25 minutes (`sum(dur > 25*60)`). We may want to examine those to see why they lasted longer than the others. Did they fail to reach the target? Or did they fail to recognize it?

Let's look at the range of the locations to ensure that they all make sense. We want to verify that they are all within the course. We can check this with

```
range(sapply(logs, function(ll) range(ll$x)))
```

```
[1] -14.91  14.55
```

```
range(sapply(logs, function(ll) range(ll$y)))
```

```
[1] -7.713  7.316
```

Here we are computing the minimum and maximum for each log and then extremes of these. All appear consistent with our understanding of the course.

We expect the records in each log file to be ordered in time. However, we should verify this. We can check that the difference in times are all positive, e.g.,

```
table(sapply(logs, function(ll) all( diff(ll$time) > 0 )))
```

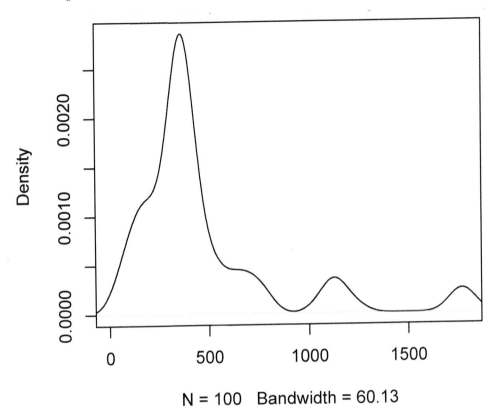

Figure 4.3: Elapsed Time of 100 Experiments in Seconds. *There appear to be 3 different groups in this distribution. Most of the experiments are completed between 1 and 16 minutes with a "center" of about 8 minutes. A smaller group of experiments is centered around 18 minutes. The final group includes those that do not find the target and use all of the 30 minutes allowed and end then.*

All of these are true, so the observations are ordered. Are there any lengthy gaps in time between successive records with a log? Let's compute all of the time differences between consecutive records and examine their distribution:

```
deltas <- unlist(lapply(logs, function(ll) diff(ll$time)))
summary(deltas)
```

```
   Min. 1st Qu.  Median    Mean 3rd Qu.    Max.
  0.100   0.100   0.200   0.175   0.200   8.800
```

This shows that we are getting many records per second from the robot. Also, there is an almost 9-second gap in one log file, which is curious. Are there other large values? We can look at the large quantiles in more detail:

```
quantile(deltas, seq(.99, 1, length = 11))
```

```
  99% 99.1% 99.2% 99.3% 99.4% 99.5% 99.6% 99.7% 99.8% 99.9%  100%
  0.5   0.5   0.5   0.5   0.5   0.5   0.5   0.5   0.5   0.5   8.8
```

This value of 8.8 is the only extreme one. Which log file is it in?

```
which.max(deltas)
```

```
logs/JRSPdata_2010_03_10_12_39_46.log2167
                                     49605
```

The name of the log file is in part of the names attribute. How many records are in this log file?

```
nrow(logs[["logs/JRSPdata_2010_03_10_12_39_46.log"]])
```

```
[1] 2184
```

How does this compare with other log files?

```
summary(sapply(logs, nrow))
```

```
  Min. 1st Qu.  Median    Mean 3rd Qu.    Max.
   208    1650    2180    2670    2840    9450
```

So this is very typical. We determine where in the log file the long delay occurs with

```
ll <- logs[["logs/JRSPdata_2010_03_10_12_39_46.log"]]
i <- which.max(diff(ll$time))
```

The pair is near the end of the log (record number 2167, as we could also have determined from the name of the value returned by `which.max(deltas)` above), but not the final records. We can then look at the two consecutive observations. They are at the same location and all of the range values are the same. So it appears that the robot stopped. By looking at the log file, we can see that the corresponding `position2d` record indicates that the robot's motor is stalled (see the file format documentation). This might explain the delay. However, other records in this time period also have this stalled flag set and they do not have the large time delay. So there doesn't seem to be anything unusual with these records.

Now that we know the records are ordered by time, let's look at how the robots move and see if there are any anomalies in this regard. We can compute the change in the x and y coordinates between each record with

```
delta.x <- unlist(lapply(logs, function(ll) diff(ll$x)))
delta.y <- unlist(lapply(logs, function(ll) diff(ll$y)))
```

and then plot them using a `lattice` plot:

```
library(lattice)
densityplot( ~ c(delta.x, delta.y),
            groups = rep(c("x", "y"), c(length(delta.x),
                                        length(delta.y))),
            plot.points = FALSE, xlab = "distance",
            auto.key = list(columns = 2,
                            text = c(expression(Delta[x]),
                                     expression(Delta[y]))))
```

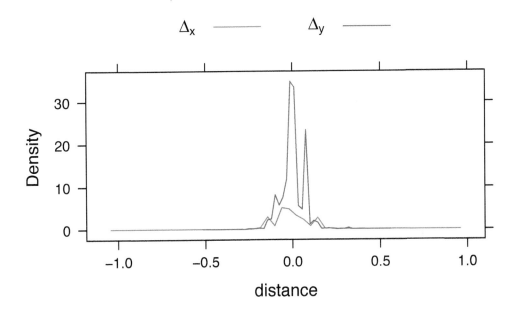

Figure 4.4: Distribution of the Changes in the Horizontal and Vertical Directions. *We compute the change in the horizontal and vertical directions separately for each pair of consecutive records in each log to explore how far the robot typically moves between records.*

The resulting plot in Figure 4.4 illustrates several features. a) Most changes are close to 0 and so the robot is moving small distances. There are no unreasonable values. b) The extreme changes in the vertical direction are smaller than those for the horizontal. We might expect this since the horizontal dimension is twice as long as the vertical dimension for the course. c) There are a several smaller modes in the densities. These suggest that several identical values occur frequently. We can find these with `tail(sort(table(¬ delta.x)))` and see the values ±0.16. These distances may be due to the search strategy or to the way the robot reports positions, or both.

Looking simply at the change in location ignores how long the robot took to travel this distance. We can compute velocity and again check that there are no unreasonable values that would indicate a robot moving an unusual distance. We compute distance divided by time with

```
velocity =
  lapply(logs, function(ll)
              sqrt(diff(ll$x)^2 + diff(ll$y)^2)/diff(ll$time))
```

Note that we are computing the distances between each consecutive pair of points in a vectorized manner. When we plot the density (Figure 4.5), we see two modes:

```
plot(density(unlist(velocity)), xlab = "meters/second", xaxs = "i",
     xlim = c(0, max(unlist(velocity))))
```

There are two clear modes corresponding to slow movement and faster movement. Again, the values seem entirely reasonable and we also now have a sense of how fast the robots are moving around the course.

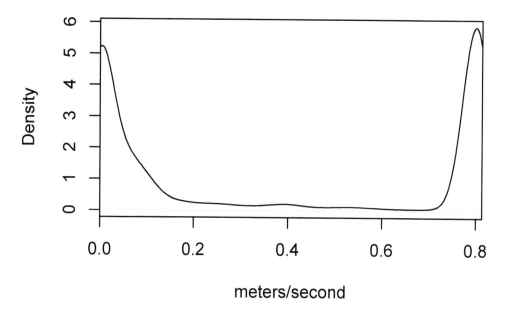

Figure 4.5: *Distribution of the Velocity of the Robots. This shows the bimodal distribution of the velocity of all of the robots across all log files. We compute the distance between consecutive points in each log and divide this by the time between these two records.*

We also want to examine the distribution of the range values of all of the looks. A simple summary yields

```
   Min.  1st Qu.   Median     Mean  3rd Qu.     Max.
  0.173    2.000    2.000    1.800    2.000    2.000
```

Three quarters or more of the values are 2, indicating nothing was detected. All are positive. These seem reasonable at this level of exploration. We need more context to understand and interpret the values together as segments as seen by the robot. We'll do this later.

We have looked at variables across all log files to validate the data and understand the aggregate distributions. We now turn our attention to individual logs. Specifically, we want to visualize the path of the robot, identify the obstacles in the course, and also explore the looks that seem to identify a target. By exploring looks that correspond to a target and also those that do not, we can hopefully develop an accurate classifier for identifying a target.

4.2.3 Visualizing the Path

By looking at the distributions of the number of records in each log, the total time for each experiment, the x and y coordinates, and the velocities, we have a coarse understanding of how the robots moved. However, we also want to understand the paths the robots took, see how much of the course they covered, and what general strategies they used. We also want to identify any potential problems or anomalies. To this end, we want to visualize the entire contents of each log file.

We can plot all of the x, y pairs for the robot with, for example,

```
plot(y ~ x, logs[[1]], xlim = c(-16, 16), ylim = c(-8, 8))
```

We specify the axes limits to show the entire course. This makes it possible to compare different experiments/logs that may operate in different parts of the course. We can see the result in Figure 4.6. The robot starts on the left side of the course and moves up and to the right and then gradually down as it moves all the way to the opposite side of the course. When it arrives at that side, we cannot tell if the robot moves up or down. We can see some overplotting so it may be that the robot moves up and takes large "steps" to the top right corner and then returns. However, it is hard to understand the details of the path. We cannot determine the start and end points or the direction of travel.

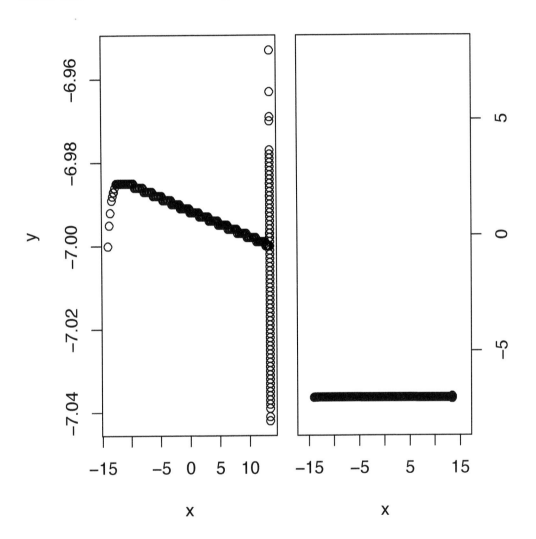

Figure 4.6: Robot Path for the First Experiment. *This displays the path from the first log file. The panel on the left shows the robot's movements from left to right across the course and then vertically along the side. The second panel shows this path relative to the entire course. This illustrates that the robot moved along the bottom side and only slightly vertically before the run terminated. This experiment lasted almost 19 minutes.*

We might improve the plot above by indicating the direction of travel. Firstly, we can mark the start and end points with, say, a circle and an x. We can do this with a call to points(). We can also change the plotting character from a circle to a disk or even a point (e.g., pch = '.'). This may help with the overplotting. However, it will not help identify the direction.

We might consider adding lines connecting the point to indicate direction. A simple line will not convey the direction of motion, however. We could use an arrow at one end. This, however, adds more content to the plot and may lead to confusion with the arrows obscuring some of the points. We might color each line segment connecting consecutive points with a gradient that changes from, say, green at the start to red at the end. When two of these lines intersect, the results may be confusing, but we can explore this to see. This leads us to another approach. Instead of drawing lines between consecutive points and coloring those, we could color each point with a color that shifts from, say, green at the start to red at the end. This avoids adding "ink" to connect the lines but still conveys direction.

We can compute a sequence of colors for the rows in our data frame using the rgb() function. This takes 3 equal-length vectors specifying the red, green, and blue components of each color. We set the blue component to 0 for each color and vary the red and green components from 0 to 1 and 1 to 0, respectively. We can define a function to create the vector of color values as

```
makeColorRamp =
function(n)
{
  s = (1:n)/n
  zero = rep(0, n)
  rgb(s, (1-s), zero)
}
```

We pass it the number of observations and it returns the vector of colors. We can then define a function plot.RobotLog() to display the path for an experiment as

```
plot.RobotLog =
function(x, y, col = makeColorRamp(nrow(x)), ...)
{
   plot(y ~ x,  x, type = "p", pch = 20, col = col, ...)
   points(x$x[c(1, nrow(x))],  x$y[c(1, nrow(x))],
          pch = c("O", "+"), col = c("green", "blue"))
}
```

Note that we use ... in the signature of the function definition to pass any additional arguments to the call to plot(). This allows the caller to specify additional arguments for a title, to control the axes' ranges, etc. Our function doesn't attempt to process these arbitrary arguments, but R merely collects them and passes them on to plot().

We can use our new plot.RobotLog() to visualize all of the experiments in Figure 4.7 with the code

```
par(mfrow = c(10, 10), mar = rep(0, 4), pty = 's')
invisible(lapply(logs, plot.RobotLog,
                 xlim = c(-16, 16), ylim = c(-8, 8),
                 axes = FALSE))
```

From this, we can see some common patterns in the movement of the robots.

Our plots of the paths do not show any of the obstacles or the circular target, if the robot encounters it. In order to show these, we need to work with the individual looks, i.e.,

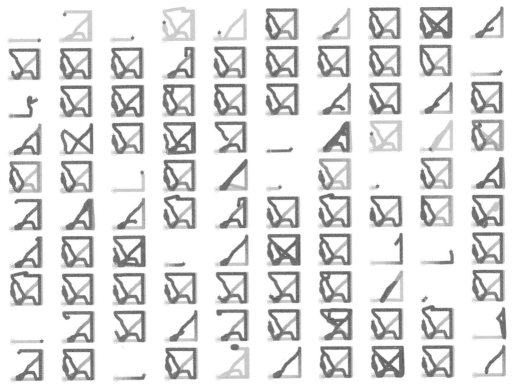

Figure 4.7: Display of All Experiments. *This displays the path of the robot in each of the 100 experiments. The starting point is displayed in green with a circle. The direction of the robot corresponds to the shift in colors from green to red. The final location is marked with a blue x.*

rows in the data frame, and determine where the robot saw an object. This will allow us to see if the objects are all in the same locations for each experiment and to get a sense of whether we can identify the circular target. In the next section, we examine the looks.

Q.3 Instead of changing the color uniformly across observations, we can change it based on the time difference between consecutive records. This will help to convey velocity. Implement this and visualize the experiments.

Q.4 We have colored the points in each path using a different, but related, sequence of colors. In colorRamp(), we compute a vector of colors based on the number of records in the log. This means that the i-th color is different across the logs. Consider using the same set of colors across all logs. Does this convey more information, e.g., allow us to compare the number of records in each log? Does this improve the overall display?

4.2.4 Exploring a "Look"

One of the primary tasks in this case study is to develop an approach to detecting a circular target, specifically one with a known radius. Before we can do this, it helps a great deal to become familiar with the "looks" and understand what the robot "sees." We want to consider the geometry of a "look" when we see nothing, the side of a rectangular or triangular obstacle, or the circular target. In the 100 logs, we have 266,000 looks. These are our data for developing our classifier.

First, we'll develop a function to plot a look. We have the location of the robot and then the range values for the 360 degrees, i.e., the distance from the robot to an object in that direction (or 2 meters). We use polar coordinates to render the path the robot sees. For each angle we multiply the range value by the cosine and sine of the angle and add these values to the robot's location. We'll plot the default "see nothing" path as a red circle and the actual path with a black curve. We can define a function to do this with

```
plotLook <-
function(row, ...)
{
  x = row[1, "x"]
  y = row[1, "y"]

  theta = seq(0, 2*pi, length = 360) - pi/2
  r = as.numeric(row[1, -c(1:3, 365)])
  x1 = x + r*cos(theta)
  y1 = y + r*sin(theta)
  par(pty = 's')
  plot(x + 2*cos(theta), y + 2*sin(theta),
       col = "red", type = "l",
       xlab = "x", ylab = "y", ...)     # 2 meter circle
  points(x1, y1, type = "l")            # what the robot sees
}
```

We subtracted 90 degrees ($pi/2$ radians) so that the points start behind the robot at $(0, 1)$, rather than at the coordinates $(1, 0)$.

Note that in the function we included the expression `par(pty = 's')` to ensure that the aspect ratio is square, i.e., the scales for the horizontal and vertical dimensions are the same. This ensures that the circle is not distorted to appear as an ellipse depending on the dimension of the graphics device.

When we set the *pty* graphics option in our function, it remains in effect after our function returns. This is generally bad practice. We can arrange to undo this setting by adding a call of the form `par(pty = oldValue)` to the end of our function. One problem with this is that it will not get called if the call to plot() or points() fail for some reason. For this reason, we use the on.exit() function. So we would add

```
oldValues = par(pty = 's')
on.exit(par(pty = oldValues))
```

Regardless of how the function returns – either correctly or due to an error – this code will be evaluated and restore the previous value. If we cannot specify all of the local settings for par() in a single call, we often use

```
old = par(no.readonly = TRUE)
on.exit(par(old))
```

and then one or more calls to par(). This arranges to set all of the values for par() back to their original values, as stored in `old`.

We call our plotLook() function with a row from one of our logs to show that look. For example,

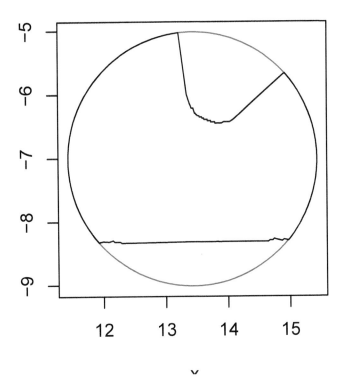

Figure 4.8: *Sample Final Look. This is the path/shape seen by the robot in the final look of the first log file, JRSPdata_2010_03_10_12_12_31. The robot is in the center of the circle. At the top right of the circle, we see a circular-like object that might be the target. A straight edge corresponding to a rectangular obstacle appears at the bottom of the circle.*

```
plotLook(logs[[1]][ nrow(logs[[1]]), ])
```

produces the display in Figure 4.8.

We can see the side of the rectangular obstacle on the bottom of the circle. These points lie on a line; however, the corresponding range values are not constant, since the distances from the center to these points vary. Note that at the ends of the line, the points vary from the linear path. This suggests some measurement error in the range values for the larger values closer to 2 meters.

Near the top right of the circle, we see the outline of part of a circular shape. The display is slightly misleading. We see straight lines connecting the edges of the circular obstacle to the outer 2 meter circle. These lines are an artifact of the way we drew the points. We connected them with lines, i.e., `points(, type = "l")`. This connects the end points of the arc to the points on the 2 meter circle giving the funnel shape. If we do not connect the points, we get a plot like the one shown in Figure 4.8. The correct plot is shown in Figure 4.9.

Q.5 Create a variation of the plotLook() code that produces the plot Figure 4.9. Divide the points into separate sub-groups that are all 2 meters or all less than 2 meters and use `points(, type = "l")` for each sub-group.

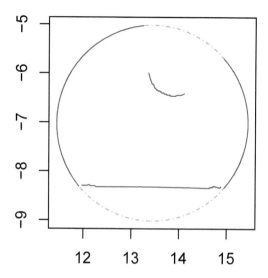

Figure 4.9: *Enhanced Display of a Look. This figure shows the improved display of a robot's "look." We remove the misleading lines connecting the edges of the circular target and the 2 meter arc. This uses* `points(, type = "l")` *but for each sub-group of points that form a contiguous sub-arc of points at 2 meters, and points that are less than 2 meters. It does not connect adjacent but disconnected sub-arcs. (See Q.5 (page 189).)*

Let's draw the final look for each of the 100 log files using our updated version of the function plotLook2() from the exercise above. To fit these onto a grid, we'll discard the axes labels, tick marks, margins, etc. We can do this with

```
par(mar = rep(0, 4), mfrow = c(10, 10), pty = 's')
invisible(lapply(logs, function(ll)
                         plotLook2(ll[ nrow(ll), ], axes = FALSE)))
```

Note that we are specifying a value for the *axes* parameter in the call to plotLook2() and this is passed on by our function to the call to plot(). This is why we added ... as a parameter for plotLook(). We can see the results in Figure 4.10.

4.2.5 The Error Distribution for Range Values

As we saw in the line and circular arc in Figure 4.8, the values for the range variable in a "look" may include a small random error. The points near the ends of the line show more "wiggle" than the others. The points on the circular segment also exhibit a similar "jaggedness." We would like to be able to quantify the magnitude of this error. We could use this when determining if we have found the target to quantify uncertainty. How can we estimate the standard deviation, say, of the error? Is the distribution of this error approximately symmetric? Is the standard deviation a good measure of the spread of the distribution? Is the distribution conditional on the value of the range, e.g., different for values of the range close to 2? The theory for estimating the distribution is straightforward – extract independent and identically distributed values measuring the quantity of interest. Unfortunately we do not have measurements that are identically distributed. The robot rotates and also

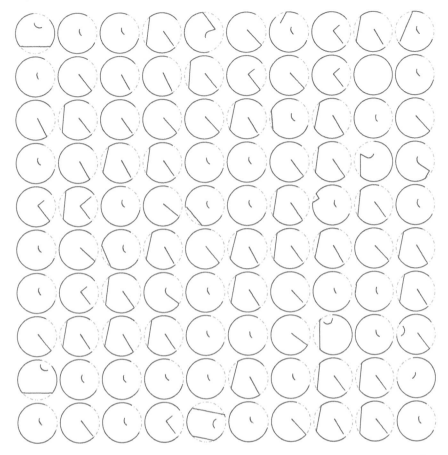

Figure 4.10: All Final Looks. *This shows the final look of each of the 100 log files. Which show a circle?*

moves around. Therefore, what it is attempting to measure is different at each recording. When the robot is at one location, we have measurements for different directions. For the same directions, the robot is at a different location and hence "seeing" different objects.

One situation in which we do have a repeated measurement of the same value is the the first and last range value in each `laser` record. By design, the robot measures 361 angles and so is measuring the same thing with the first and last value in the vector of ranges. We can compute the difference between these two values to estimate the error:

```
e = unlist(lapply(logs, function(ll) ll$range1 - ll$range361))
summary(e)
```

```
    Min. 1st Qu.  Median    Mean 3rd Qu.    Max.
       0       0       0       0       0       0
```

So these are all identically 0. Is there no error? These range values are reported to 3 decimal places, i.e., millimeters. So any errors based on these values appear to be below the millimeter level.

It is possible that the robot ensures that the start and end range values are the same. Can we obtain other repeated measurements of the same values? Perhaps a robot revisited precisely the same location in an experiment. If so, we can compare all of the range values at those positions. Alternatively, perhaps two robots visited the same location in different experiments. In this case, there may be a difference in the experiment (e.g., the location of the circular target), or a robot effect rather than a range effect. However, we may still be able to usefully compare the measurements.

Consider a single log file. How can we find locations that the robot visited more than once? We could compare each pair of x, y values to all other pairs. We have to compare the i-th location to all of the other locations in the log file. We could do this with nested loops, or a single loop and vectorize the comparisons. Alternatively, we could use a call to outer(). We have to compare both the x values and the y values. A useful general approach is to combine each pair of x and y values into a single value that uniquely identifies the combination. We can use a string such as "x, y", replacing x and y with the actual values for the location. We can quickly identify the duplicated locations with the function duplicated(). The location values are reported to 3 decimal places, so we should continue to use this format. For a given log file, we can find any duplicates with

```
ll$pos = sprintf("%.3f,%.3f", ll$x, ll$y)
w = duplicated(ll$pos)
```

If there are any duplicated points, we can gather them into groups and compute deviations from the average for each of the 361 range variables. We can use any of the by(), aggregate(), or tapply() functions to group the records based on their unique location identifier we created above (pos). Given a group of these observations at the same location, we want to subtract the mean for each column to find the "error" or deviation from the average value. We can do this directly or with the scale() function. Hence, we can compute the error values with

```
if(any(w)) {
  tmp = ll[ ll$pos %in% ll$pos[w], ]
  errs = unlist(by(tmp[, 4:364], tmp$pos, scale, scale = FALSE,
                   simplify = FALSE))
}
```

We put these computations into a function, say getRangeErrors() and then apply this to each log file:

```
rangeErrs = unlist(lapply(logs, getRangeErrors))
```

How many values do we have? Almost 39 million! We can compute the standard deviation and surprisingly it is 0.0675. This is quite a bit more than we expected. We can see the distribution in Figure 4.11.

We see that the distribution has very large tails and hence the standard deviation is not a good measure of the spread. Instead, we might use the inter-quartile range or perhaps compute the length of the interval containing the central 68% of the distribution. In either case, the resulting interval is effectively 0. After all of our work, our estimate of the error in the range values is effectively 0. However, we have found some extreme variability and we might want to explore those further. For example, does the distribution of this error depend on the value of the range variable with larger range values having more error as we saw suggested in Figure 4.8?

We do have the potential confounding factor that the yaw of the robot may be different across the different measurements at the same location. We discarded the yaw measurement when we read each log. We could refine our readLog() function and reread the log files. We could then group by x, y, and yaw value. We leave this as an exercise.

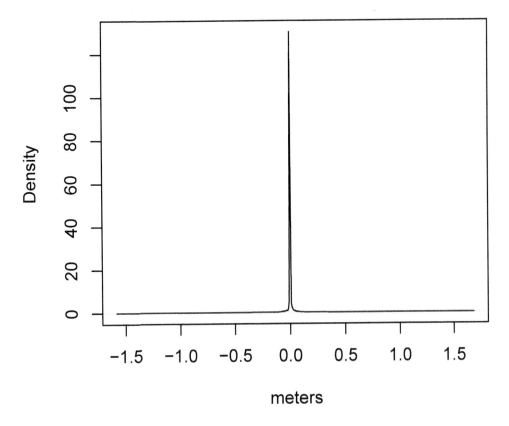

Figure 4.11: Density of Repeated Range Values. *This shows the distribution of the repeated measurements of the range values when a robot revisited the same location. These are deviations from the mean of nominally identical values. There are some very extreme values (-1.58 and 1.68). The distribution has very large tails. Most of the observations are exactly 0.*

Q.6 How many different robots were used for these experiments? Are there any differences in their operating characteristics? That is, do different robots have a different error distribution?

We have now familiarized ourselves with the data and verified that it seems correct, and we haven't introduced anomalies when reading it into R. We have done a reasonable amount of exploratory data analysis. We can now move on to our goal of developing a mechanism to identify the circular target.

4.3 Detecting a Circular Target

For a given "look," how can we determine if the robot sees the circular target with a 0.5 meter radius? Take some time to study each of the 100 final looks in Figure 4.10. Identify qualitatively different characteristics and groups of looks that we have to process. One advantage of the final looks is that they are more likely to contain the circular target of interest as each experiment set out to find this and then terminate. So those experiments that successfully found the target should have a final look that contains the target. We could and should look at many other looks, not just the final looks, so that we see different types of looks. But for now, we focus on these final looks and, for convenience, create a list containing each of them:

```
finalLooks = lapply(logs, function(ll) ll[nrow(ll), ])
```

Figure 4.12 shows several of the final looks from Figure 4.10. We have selected examples from 4 types exhibiting different characteristics. The first thing we recognize is that to find the target, we are looking for a reasonably short sub-arc/segment of the 360 degree view, not the entire 360 degrees. Secondly, we are only interested in segments all of whose range values are less than 2, i.e., the robot sees something. Each short contiguous segment corresponds to some object on the course, i.e., the target, an obstacle, or a side. We start with the 361 range values and we want to find these shorter segments of potential interest. We know from earlier explorations that the first and 361st elements for each look are identical. Accordingly, we can discard the 361st element for each set of ranges.

To identify the segments, we can explicitly loop over the range values in our vector and find the first element that is less than 2. This is the start of the first segment. Then we can find the next element that does have a value of 2. This identifies the end of the first segment (i.e., the previous element). Then we continue from this element and find the next value less than 2 and so on.

We have to find a way to implement this search. We can easily compute the logical vector `range < 2` and then find the index of the first TRUE or FALSE value and do this sequentially to find the segments. We can use `which(v)[1]` to find the next TRUE value and then repeatedly subset to remove all of the values up to that point. Here we will loop over the number of segments, not the number of elements in the vector. Accordingly, the number of iterations is smaller.

Fortunately, however, there exists a function in R that will do the computations for us and significantly quicker. This is the rle() function, whose name stands for "run-length encoding." rle() finds contiguous segments of a vector that have the same value and returns the lengths of these segments and the value of each separate segment. This allows us to identify the indices of the elements in each segment. Consider the simple vector `c(1, 1, 2, 3, 3, 3, 3, 5, 5, 5, 1)`. When we call rle() with this vector, we get

```
Run Length Encoding
  lengths: int [1:5] 2 1 4 3 1
  values : num [1:5] 1 2 3 5 1
```

There are 5 segments. The first contains 2 elements, the second just 1, and the third is the sequence of four 3s. Note that this includes the sequences with just one element.

When we call rle() with a logical vector, it identifies the contiguous homogeneous blocks of the TRUE and FALSE values. For example,

```
rle(c(TRUE, TRUE, FALSE, FALSE, FALSE, TRUE, TRUE, TRUE))
```

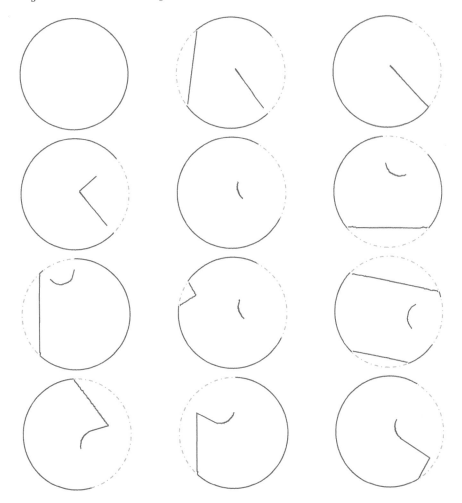

Figure 4.12: Characteristics of Looks. *This shows 4 different types of looks. In the first look, the robot sees nothing and so there is no circular target present. In the next 3 looks (moving row-wise), the robot sees a straight side or two straight sides that intersect. Again, there is no circular target. All of the remaining looks appear to contain a circular target. In the fifth look, the robot only sees the circle, while in the next 4 looks, it sees a circle and part of one or two obstacles. The last 3 looks are more complex. We see the circular target but, drawn in this manner, the circle appears to be connected to obstacles.*

yields

```
Run Length Encoding
  lengths: int [1:3] 2 3 3
  values : logi [1:3] TRUE FALSE TRUE
```

If we only want the segments corresponding to the TRUE values, we have to construct the indices for the elements in each of these sequences from the object returned by rle().

For our purposes, we call rle() with the logical vector range < 2 to find the regions where the robot sees something and does not see something. We can then use the lengths of these segments to compute the sequence of positions/indices within the vector that make up each object the robot sees. We can do this with the function getSegments() below. The key to the function is the call to rle(). We then process each segment and keep track of its starting position relative to the previous segments. We can do this with a simple `for` loop. For each block of TRUE values, we compute the vector of indices. We skip the FALSE blocks, but update the current location to its end position. The entire function is

```
getSegments =
    # return a list with elements being integer vectors
    # giving the indices of each contiguous segment
    # with values less than the threshold.
    #
    # We discard the 361st element.
function(range, threshold = 2)
{
  if(length(range) == 361)
      range = range[-361]

  rl = rle(range < threshold)

  cur = 1L
  ans = list()
  for(i in seq(along = rl$lengths)) {
     if(!rl$values[i]) {
        cur = cur + rl$lengths[i]
        next
      }
     ans[[length(ans) + 1L]] = seq(cur, length = rl$lengths[i])
     cur = cur + rl$lengths[i]
  }

  ans
}
```

Note that we allowed the caller to specify a different value for the threshold to identify when a robot saw something. Instead of using 2 meters, we could use, say, 1.9 meters since a value of 1.99 may actually correspond to a true value of 2, but is smaller due to measurement error. We can adjust the threshold to potentially account for measurement errors and somewhat control false positive sightings.

Q.7 Can we vectorize the computations in getSegments() to avoid the `for` loop? Given that we are looping over the number of objects we see, not the 360 range values, does vectorizing significantly improve the performance? Think about how often we call getSegments() – once for each look in each log file.

Q.8 The getSegments() function has similarities to the plotLook2() function (the function we developed in Q.5 (page 189)). Rewrite plotLook2() to call getSegments(). Is this a good idea? Does it simplify the code? avoid repeating the same computations in two places? avoid testing similar code?

As usual, we now need to test the getSegments() function. We can do this using the looks in our existing log files. For these, we need to manually determine the segments of interest.

This is a good thing to do. For this purpose, we temporarily modified our plotLook2() function to show each of the segments in a different color and to return the number of segments it encountered. This simplified matching the output of getSegments() with what we expected from the plots.

Alternatively, we can test the getSegments() function by creating look/range vectors with specific segments whose characteristics we define and then verify that getSegments() returns them correctly. We'll start with the degenerate look where we see nothing (i.e., all range values are 2):

```
length(getSegments(rep(2, 360)))
```

We get no segments back as we expect.

Let's create a look with 3 positive segments of length 10, 25, and 41, e.g.,

```
x = c(rep(2, 20), seq(1.7, 1.9, length = 10),
      rep(2, 50), seq(1.4, 1.6, length = 25),
      rep(2, 59), seq(.3, .5, length = 41))
x = c(x, rep(2, length = 361 - length(x)))
```

Note that we pad the remainder of x with values of 2 to have length 361. There is no need for us to manually calculate how many of these we need; instead, we compute this in R. Alternatively, we could have created this with

```
x = rep(2, 361)
x[21:30] = seq(1.7, 1.9, 10)
x[81:105] = seq(1.4, 1.6, 25)
x[165:205] = seq(.3, .5, 41)
```

The advantage of this approach is that we explicitly know the correct locations of the segments. This will make it easier to verify the results. The output from getSegments() is

```
getSegments(x)

[[1]]
 [1] 21 22 23 24 25 26 27 28 29 30

[[2]]
 [1]  81  82  83  84  85  86  87  88  89  90  91  92  93  94  95
[16]  96  97  98  99 100 101 102 103 104 105

[[3]]
 [1] 165 166 167 168 169 170 171 172 173 174 175 176 177 178 179
[16] 180 181 182 183 184 185 186 187 188 189 190 191 192 193 194
[31] 195 196 197 198 199 200 201 202 203 204 205
```

which correspond precisely to the locations we specified.

We can also test the function with a different value for *threshold*, e.g.,

```
sapply(getSegments(x, 1.6999), length)

[1] 25 41
```

We can see that this ignores the first segment from our previous call since all the values in that segment are greater than the threshold.

With the getSegments() function tested on synthetic data, we can now apply it to looks from our log files. For example, for the first look in the first log file, we can call getSegments() with

```
getSegments(as.numeric(logs[[1]][1, -(1:3)]))
```

We have to discard the timestamp, x, and y values from the record and then transform the one-row data frame to a numeric vector. This is cumbersome when compared with

```
getSegments(logs[[1]][1, ])
```

Furthermore, it is different from how we call other functions such as plotLook(). This makes it harder to remember how to call each of these functions. Accordingly, we should adapt getSegments() to accept a row from one of our data frames and perform the conversion to the numeric vector. We can implement this by adding the code

```
if(is.data.frame(range))
   range = as.numeric(range[1, -(1:3)])
```

to the start of the body of the function. We may want to add more checks to this if condition to ensure that the timestamp, x, and y fields are present. However, even this simple addition makes using the function a lot more convenient, especially interactively. Once again, we should retest the function and its new features.

4.3.1 Connecting Segments Behind the Robot

Consider the sixth look in Figure 4.12. At the bottom of the look, we see a horizontal line corresponding to the side of an obstacle. The first element of the range vector corresponds to the bottom of the circle and we then move counter-clockwise. As a result, this line is actually made up of two separate sequences of values in the vector – one at the beginning of the vector and the other at the end. The division is an artifact of the way the robot swivels and how the range values are represented/structured in our records. While we are not interested in the obstacle we see here, this "artificial" separation of the segment into two separate sequences at the beginning and end of the vector could potentially cause a problem. If the circular target appeared in this position, we would see two parts of the circle. We may not classify either sequence as a circle but we might classify the entire segment as the target.

To overcome this issue of a segment "wrapping" around from 360 degrees to 0 degrees, the pieces directly behind the robot need to be combined by being laid end-to-end. We can do this with `c(x, rev(x))`. We now have a vector of 720 elements. We can again find the segments where we see something using the rle() function. Our existing getSegments() function can work on the vector of 720 ranges. As we go through the vector, we stop when we find the first value of 2 after the 360th element. This avoids identifying segments a second time. We leave it as an exercise to implement this approach.

We will use a more direct approach. We note that we are only interested in a possible connection between the last segment and the first in our record. We should connect these two segments behind the robot if a) there are at least two segments, b) the first segment starts at position 1, and c) the final segment ends at position 360. This is equivalent to laying the vector end-to-end and computing the segments. However, it allows us to reuse the getSegments() function we defined earlier. We can call that to get all of the individual segments and then see if we should combine the first and the last. We can define this function as

```
getWrappedSegments =
function(range, threshold = 2,
         segments = getSegments(range, threshold))
{
   if(length(segments) > 1) {
      s1 = segments[[1]]
      s2 = segments[[length(segments)]]
      if(s1 == 1L && s2[length(s2)] == 360) {
         segments[[1]] = c(s2, s1)
         segments = segments[-length(segments)]
      }
   }

   segments
}
```

This merely implements the conditions a) – c) above. The call to getSegments() provides the default value for the *segments* parameter. This allows us to explicitly pass the segments if we have computed them in an earlier step, e.g., when we test the function or when we compute them in order to plot them. By using the existing getSegments() function, any improvements to that will automatically be propagated to this function. However, we can experiment with alternative implementations of computing the initial segments without changing getSegments() or this new getWrappedSegments() function.

Again, we need to thoroughly test our new function. Let's create a look that should be connected:

```
tmp = c(rep(1.5, 100), rep(2.0, 141), rep(1.5, 120))
```

This has values of 1.5 for the first 100 entries and for the last 120. It is good to make these lengths different so that we can identify any bug based on the common length. Evaluating getWrappedSegments(tmp) does indeed yield a single segment consisting of a combination of the first and last segment.

How does this function perform with the last look in our first log?

```
getWrappedSegments(as.numeric(finalLooks[[1]][1, -(1:3)]))

[[1]]
 [1] 312 313 314 315 316 317 318 319 320 321 322 323 324 325 326
[16] 327 328 329 330 331 332 333 334 335 336 337 338 339 340 341
[31] 342 343 344 345 346 347 348 349 350 351 352 353 354 355 356
[46] 357 358 359 360  50  49  48  47  46  45  44  43  42  41  40
[61]  39  38  37  36  35  34  33  32  31  30  29  28  27  26  25
[76]  24  23  22  21  20  19  18  17  16  15  14  13  12  11  10
[91]   9   8   7   6   5   4   3   2   1

[[2]]
 [1] 134 135 136 137 138 139 140 141 142 143 144 145 146 147 148
[16] 149 150 151 152 153 154 155 156 157 158 159 160 161 162 163
[31] 164 165 166 167 168 169 170 171 172 173 174 175 176 177 178
[46] 179 180 181 182 183 184 185 186
```

This does combine the two segments behind the robot and also identifies the other segment from 134 to 186 corresponding to a side/obstacle. Compare this with the sixth look in Figure 4.12, which shows a line behind the robot that consists of the first and last segments for that look.

Q.9 Test the function getWrappedSegments() for other looks, e.g., with no objects seen, more than 3 segments, different thresholds.

4.3.2 Determining If a Segment Corresponds to a Circle

Now that we have all of the segments that correspond to something visible in a look, we want to determine if they collectively correspond to seeing the circular object. We could consider all of them together. However, it is simplest and probably best to consider each individually. We can compare the shape defined by each segment to an arc/part of a circle. If we get a "good" match, we can conclude that we have found the target. If we get more than one good match within a look, we probably have a problem as there is only one target. We may want to take all of the matches into account, but we focus now on recognizing an individual segment as a circle or not.

We will focus on one contiguous segment with all of the range values less than 2 (or some threshold). We have the location of the robot, say pos_{robot}, and also the (x, y) coordinates of the points on the segment. We want to determine if the arc described by the segment is consistent with part of a circle with radius 0.5 meters. If we knew the center of the would-be target, say r_{target}, we could determine if all of the distances to each point on the arc from this center were the same and 0.5 meters. There are several issues. One is that we don't know the location of the center, r_{target}. Secondly, the range values, and hence the locations of the arc, contain slight errors, as we have explored earlier. If we determine r_{target} from the arc locations, r_{target} will contain random errors. Accordingly, we expect some variability in the distances from this center to each point and they won't all be the same, nor exactly 0.5 meters. Similarly, the circular target probably does not have a radius of exactly 0.5 meters and the x, y coordinates of the robot's location and even the angles probably contain some measurement error. Accordingly, there is noise in our measurements and we want to take this into account.

We can use simple geometry to estimate the center of the target, r_{target}. We can find the shortest distance from pos_{robot} to each of the points on the arc. The two points pos_{robot} and $(x, y)_{nearest}$ define a line and the center of the target should be 0.5 meters from $(x, y)_{nearest}$ in the direction of that line. From a statistical perspective, we are interested in understanding the characteristics of this estimate in terms of its bias and variance for the true location of the target. In reality, however, we are interested in how well it allows us to correctly classify a segment as the target, given the noise in the measurements.

Since the center of the target and its precise radius are unknown and there are random errors in the measurements of the arc locations, we may want to use a slightly different and more generic and commonly used statistical approach than our geometric approach above. We have n (x_i, y_i) locations for the arc. We know that, if they describe a circle, they should all be very close to $r = 0.5$ meters from the unknown center of the circular target. If we denote the center with (x_0, y_0), we then have

$$\sqrt{(x_i - x_0)^2 + (y_i - y_0)^2} - r \approx 0$$

Essentially, we want to allow x_0, y_0, and r to vary so that we find the best fit for our points as a circle. In other words, we want to find the values of x_0, y_0 and r that minimize

$$\sum_{i=1}^{n}(\sqrt{(x_i - x_0)^2 + (y_i - y_0)^2} - r)^2 \qquad (4.1)$$

This is similar to fitting a regression line. There we minimize

$$\sum_{i=1}^{n}(Y_i - \beta_0 + \beta_1 X_i)^2$$

for a simple bivariate regression. We vary β_0 and β_1 to find the values that minimize this sum of squared differences between what we observe (Y_i) and what we predict ($\hat{\beta}_0 + \hat{\beta}_1 X_i$).

In the case of linear regression, the solution for the values of β_0 and β_1 that minimize the sum of squares can be determined from a closed-form expression. In the case of finding the best fit for our circle, there is no simple closed-form solution. Instead, we have to vary x_0, y_0, and r and find the triple of values that minimize the sum of squares. There are various approaches to doing this. One is to search over a grid of feasible values for these 3 parameters. For example, we may consider that $6.3 \leq x_0 \leq 6.7$, $1.2 \leq y_0 \leq 1.45$, and $.475 \leq r \leq 0.51$. These constraints define a 3-dimensional region and we could divide this into, say, a $100 \times 100 \times 100$ grid. We would then evaluate the left-hand side of equation 4.1 at each point in this grid, i.e., (x', y', r') and find the value that gives the smallest sum of squares. We could then create a more localized and higher-resolution grid around this point to see if we can find a better solution.

The grid approach to finding a minimum works generally, but we often waste computations on regions of the grid that are unlikely to yield the optimum value. There are many approaches to numerical optimization that attempt to determine which direction to pursue next, based on evaluating the target function at one or more earlier candidate solutions. Some use the derivative to find the direction of descent. Some also use the second derivative to determine how far to move in the direction of the first derivative. Other approaches use stochastic search, which injects randomness into the directions we pursue, but in a way that speeds convergence to the result. Numerical optimization is a very rich topic and there are many available approaches. It can be important to explore different approaches and use one that best suits the problem at hand and the function being optimized. We will use a mechanism that uses approximations to the first and second derivatives (so that we do not have to calculate and implement the functional form of these). The R function nlm() does this for us.

To use nlm() to find the best fit for our potential arc of the circular target, we need to specify a function that computes the sum of squares of the points on the arc for a given triple of (x_0, y_0, r). nlm() will explore different values of this triple for us, but we need to provide the function that indicates how well it fits for a particular triple. This function needs to implement the left-hand side of Equation 4.1. nlm() will call this function with a vector specifying the current values of the triple (x_0, y_0, r). We will also arrange to have nlm() pass the vectors x and y giving the locations of the points on the arc. We can implement our sum-of-squares function as

```
circle.fit.nlm.funk <-
function (p, x, y)
{
    x0 <- p[1]
    y0 <- p[2]
    r <- p[3]
```

```
    actual.r <- sqrt((x - x0)^2 + (y - y0)^2)
    sum((r - actual.r)^2)
}
```

Again, we have to test this function and we leave this as an exercise for the reader.

In addition to the function that provides the measure of fit, nlm() also needs an initial guess for our triple (x_0, y_0, r). We can use our geometric reasoning from earlier to produce an estimate for the center of the would-be target, given the location of the robot and the points on the arc. We leave this as an exercise. Instead, we'll use the very simple approach of using the average of the x_i values and the y_i values in the arc as our initial guess. We do this with the expectation that nlm() will quickly converge to a reasonable estimate. We should explore and confirm this on actual data.

We now know the required inputs to our call to nlm(). Let's see how we put the computations together starting with a record from our log, i.e., a row in our 364-column data frame. We'll use the last look in the first log:

```
look <- logs[[1]][nrow(logs[[1]]), ]
```

We compute the segments from this with

```
segs <- getWrappedSegments(as.numeric(look[1, -(1:3)]))
```

This yields the two segments we discussed previously, i.e., the circular target to the northeast of the robot, and the straight line directly behind the robot that "wrapped" around the vector. segs is a list with two vectors, and each vector gives the indices of the elements of the range values for that segment/arc. Instead of the indices, we want the x_i and y_i values for the segment. We can compute these, as we did for the plotLook() function, with

```
i = segs[[2]]
range = as.numeric(look[, -(1:3)])[i]
theta = seq(0, 2*pi, length = 360) - pi/2
xi = look$x + range * cos(theta[i])
yi = look$y + range * sin(theta[i])
```

Again, we check our calculations by, for example, plotting the results:

```
plot(xi, yi, type = "l")
```

At this point, we have the essential inputs for nlm(). We can call it with

```
nlm(circle.fit.nlm.funk, c(x0 = mean(xi), y0 = mean(yi), r = .5),
    x = xi, y = yi)
```

The first argument is our sum-of-squares function to be minimized. The second argument is our initial guess for the 3 parameters. We use 0.5 as the value for r since we were told that was its (approximate) value. The results we get are

```
$minimum
[1] 0.003253

$estimate
[1] 13.8687 -5.9110  0.5387

$gradient
```

```
[1]    5.245e-05    5.115e-05   -7.121e-05

$code
[1] 3

$iterations
[1] 17
```

This tells us the sum of squares for our solution was 0.003. The `estimate` element gives us the values for the 3 parameters. The `gradient` element provides an estimate of the derivative of our function and `iterations` tells us how many steps in the numerical optimization nlm() took. `code` tells us about the reliability of the result. In this case, the value 3 indicates that either we have a local minimum as we want, or we should consider adjusting the *steptol* tuning parameter in the call to nlm(). *steptol* controls how small a step nlm() will bother taking. If we make this smaller, nlm() will consider taking more steps. If we try `1e-16`, nlm() uses 44 iterations and yields a value of 2 for `code`, i.e., "probably solution." (Consult *R*'s help page for the nlm() function [1].) However, the minimum value and estimates of the parameters are the same as with the default value for *steptol*.

These computations above show how to fit an arc to a segment. We should create a separate function that combines all of the computations that take a look, compute the separate segments of interest, and fit the arc for each segment. We first convert the look from robot location and range values into segments. Then we loop over the segments and fit the circle. To fit the circle, we only need the x and y vectors giving the coordinates of the segment. We can define the function to fit a circle to a segment as

```
circle.fit <-
function (x, y, initGuess = c(mean(x), mean(y), .5), ...)
    nlm(circle.fit.nlm.funk, initGuess, x = x, y = y, ...)
```

Again, this is a simple function that embodies important computations in the default values of the parameters, but allows the caller to override these. This directly returns the results from nlm(). We are leaving it to the caller to determine how appropriate the fit is.

We define the function robot.evaluation() to process an entire look. This function extracts the segments via a call to getWrappedSegments(). It then loops over these and determines which ones appear to be part of a circle. For each segment, it has to create the x and y vectors of the points on the segment/arc, as we did above. However, for each segment, we also evaluate 3 additional criteria that actually determine whether the segment seems to be part of the target.

Firstly, if there are too few points in the segment, we cannot reliably distinguish between part of a circle or any object. For instance, if we have 3 or fewer points, we cannot determine whether the segment is a line or part of a circle. We also cannot expect to get all of the points on the target. At most we can hope to see 180 points corresponding to half the target. The number of points we need is a tuning parameter that we would like to determine to obtain low type I and type II error rates for classifying a segment as the circular target. Accordingly, we'll allow the caller to specify the minimum number of points needed in a segment to even consider classifying it as part of the circular target. We will use 3 as the default value for this parameter *min.length*.

In addition to requiring a minimum number of points, we also want to evaluate how well the circle fits the segment and only accept those segments that yield a "close" fit. We need to specify what "close" means. There are various different metrics we could use. Some depend on how many points there are in the segment, and others depend on how close the robot is to the estimated center of the target. We will use the final value of the sum of squares function that we are trying to minimize to measure the goodness of fit. Generally, this will be larger with more terms in the sum as each term is non-negative. Accordingly, we divide the total sum of squares by the number of terms in the sum. This allows us to use a single threshold value for any segment. Since nlm() returns the minimum value of the function being minimized via the estimate element of the result, we can compare the goodness of fit to the threshold with

```
(out$estimate / length(segment)) > max.ss.ratio
```

The final criterion or constraint we will impose when evaluating how well the segment resembles an arc of a circle relates to the radius of the estimated circle. If the estimate of the radius is too different from the 0.5 meter value we expect, we won't consider the segment part of the circular target. We allow the caller to specify a lower and upper bound for the acceptable values of the radius. This allows for an asymmetric interval.

We could also impose a constraint on the estimate of the center of the target. If it is outside of the course, then we can eliminate it as being feasible.

We allow the caller to specify the values for testing each of these criteria. Additionally, we want the caller to be able to specify the maximum range value that constitutes the robot seeing something, e.g., 2 meters or some smaller value. So we will add each of these as parameters in our robot.evaluation() function. We'll also allow the caller to specify parameters for the calls to circle.fit(). Rather than adding each of these parameters from circle.fit() to robot.evaluation(), we can use the ... mechanism to pass any number of additional arguments in call to robot.evaluation() on to circle.fit(). We can implement this function as

```
robot.evaluation <-
function(look, min.length = 3, max.ss.ratio = 0.01,
         min.radius = .5, max.radius = 2,
         range.threshold = 2,
         segs = getWrappedSegments(range, range.threshold),
         ...)
{
  x = look$x
  y = look$y
  range = as.numeric(look[, -(1:3)])
  theta = seq(0, 2*pi, length = 360)

  for(s in segs) {
    if(length(s) < min.length)
       next

    xi = x + cos(theta[s]) * range[s]
    yi = y + sin(theta[s]) * range[s]
    out = circle.fit(xi, yi, ...)

    if(out$code > 3)
       next

    if((out$minimum/length(s)) > max.ss.ratio)
       next

    if(abs(out$estimate[3]) < min.radius
```

```
              || abs(out$estimate[3]) > max.radius)
    next

    return(list(x = xi, y = yi, range = range[s],
                robot = c(x, y), fit = out))
  }
}
```

Note that in addition to testing the 3 criteria we discussed, we also check the value of the code element returned by circle.fit() and nlm() to ensure the results are meaningful.

In our for loop within the function, we test each of the 3 conditions one after the other. If any of these fail, we move onto the next segment. We use the R keyword next to do this.

Note that we exit our loop and return the first segment that appears to be a circle, if there is one (and NULL otherwise). We may want to modify this function to return all segments that match. This depends on how we are going to deal with multiple matching segments. When exploring the data or validating our approach and functions, we may want to have all the matches that appear to be parts of a circle. However, when we analyze the data in real-time after we tune the approach, we may want just the first match.

Let's examine how our function robot.evaluation() performs on the last look of each of the 100 log files:

```
finalLooks = lapply(logs, function(x) x[nrow(x),])
circs = lapply(finalLooks, robot.evaluation)
```

circs is a list with each element either being a segment or NULL. We can examine those looks that were classified as the target by identifying the non-null entries:

```
unname(which(sapply(circs, length) > 0))
```

We can then compare these to the corresponding panels in Figure 4.10. Alternatively, we can plot these again to see them more clearly:

```
par(mfrow = c(7, 6), mar = rep(0, 4), pty = 's')
invisible(lapply(finalLooks[sapply(circs, length) > 0],
                 plotLook2, axes = FALSE))
```

The 40 looks appear in Figure 4.13.

In this, we see that many of the looks do indeed show the circular target. However, in the 5th, 8th, 14th, 22nd, and 37th looks, we seem to have mistaken a right angle corresponding to an obstacle as part of the circular target. These are type I errors – false positives. Also, the 18th look appears slightly strange, but there does appear to be a circular boundary near the center of the look.

We can look at the estimates and the minimum value of the sum of squares for these looks that are misclassified:

```
i = which(sapply(circs, length) > 0)[c(5, 8, 14, 22, 37)]
unname(sapply(circs[i], function(o) c(o$fit$minimum,
                                      o$fit$estimate[3],
                                      length(o$x))))
```

```
        [,1]    [,2]    [,3]    [,4]    [,5]
[1,]   0.589   0.589   0.674   0.511   0.512
[2,]   0.992   0.992   0.903   1.065   1.066
[3,]  83.000  83.000  84.000  83.000  83.000
```

Figure 4.13: Looks Containing a Segment Identified as a Circle. *These are the looks that were classified as containing the circular target. We see that most of the looks do indeed contain a shape that looks like the target. However, there are several that have confused a right angle corresponding to an obstacle in the course with the target and these seem to be false positives.*

The first row contains the minima of the sum of squares, the second the estimate of the radius, and the third row the number of points used to estimate the circular boundary. The minimum values for these are large relative to the other 35 looks that were classified correctly, the largest of these being 0.055. Furthermore, the estimates of the radii are large – close to 1. Perhaps we can further constrain the radius estimate to eliminate these, at the risk of generating false negatives.

Figure 4.14 shows the looks that were classified as not containing a circle.

Most of these contain obstacles or no objects. However, 8 of these looks seem to contain a circle. Again, we want to examine the nature of these fits. Unfortunately, we don't have

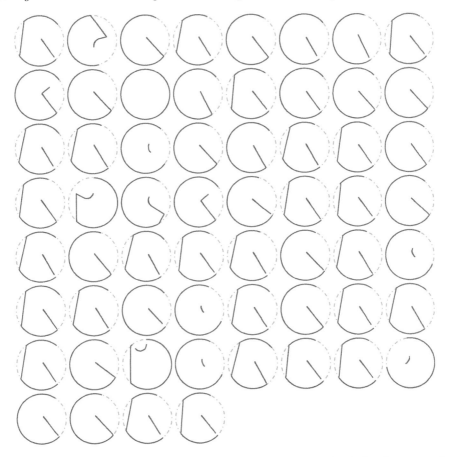

Figure 4.14: Looks Containing No Identified Circle. *These are the looks that were classified as not containing the circular target. We see that most of the looks do indeed contain no indication of the target. However, there are several that do and suggest false negatives.*

this information as we did not consider these a circle. We need to be able to compute the fits for the segments in these looks and see why they were rejected. We might decompose our robot.evaluation() function and move the body of the `for` loop into a separate function. Alternatively, we might try very lax thresholds in an effort to accept any segment.

Before we set about changing the code to fit circles to each of the segments in these looks to see why they were rejected as circles, let's examine the displays of these looks. At the risk of being redundant, we'll show just these in Figure 4.15 and we arrange them to show two patterns.

The first 3 of these looks have a line connecting the circular arc with a straight line corresponding to an obstacle. The fourth is similar but not quite the same. The remaining looks suggest a circular target but with very little curvature and almost two line segments that intersect. An important characteristic of these last 5 looks is that the circular target is very close to the center of the green circle, i.e., the location of the robot. The problem in identifying this arc as a circle is probably due here to the robot being too close to the target. When we are very close to a sphere or circle, we see a relatively flat curve. It would be interesting to look at the earlier looks in these log files to see if the robot could detect this circle from further away.

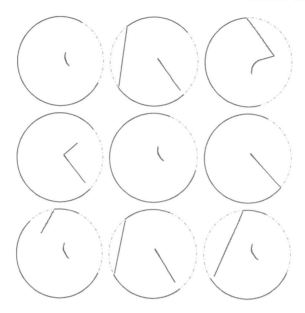

Figure 4.15: Patterns in the False Negatives. *This shows the 9 looks in which the circular target appears to be present but which were not detected by our robot.evaluation() function. They are arranged to show two characteristics. The first of these is a circular target "connected" to another obstacle. The second pattern is a circular target that is very close to the robot (i.e., the center of the look) and so does not appear circular.*

In the first 3 looks, we can see the circular arc and then a line from that which connects to another line. The problem here is our logic in computing the segments. Recall that we looked for contiguous values of the range vector that were less than our threshold, 2. The points in the circular arc correspond to a range less than 2. Likewise, the range values corresponding to the obstacle also have a range less than 2. However, in these cases, the end of the circular arc and the start of the obstacle are at neighboring angles and, hence, adjacent elements in the range vector. As a result, our getSegments() function assumes these two separate segments are just one segment characterized by all the points having a range value less than the threshold. What is clear to our human eye is that there is a large distance between the last point of the circular segment and the start of the obstacle's segment. This gives rise to the straight line connecting them in the plots.

Now that we have identified the problem for looks that have the circular target and an obstacle at adjacent elements of the range vector, we can modify our getWrappedSegments() function to correct for this. We can compute the distances between the pairs of consecutive points along a segment. If one distance is "excessively large," this identifies the divide between two sub-segments. We can implement this with the function

```
separateSegment =
function(idx, x, y, threshold = 0.15)
{
  xd = diff(x[idx])
  yd = diff(y[idx])
  d = sqrt(xd^2 + yd^2)
  if(any(d > threshold)) {
     i = which(d > threshold)[1]
```

```
      list(idx[1:(i-1)], idx[(i+1):length(idx)])
   } else
      list(idx)
}
```

This takes the x and y coordinates of the segment and also the indices in the range vector. It computes the distances between the consecutive points and compares these distances to a threshold. This is different from the 2 meter threshold for the robot "seeing" an object on the course. Instead, this threshold defines how far two points can be from each other to be considered part of the same arc/segment. If there are two significant segments, the function returns a list with two elements giving the indices of the two sub-segments. Otherwise it returns the original vector of indices for the segment.

We leave it as an exercise for the reader to adapt getWrappedSegments() to post-process the segments it originally produced to split segments by this distance approach. In our code, we allowed the caller to specify whether this was desired or not so that we could support the original and this enhanced approach.

How do we determine the appropriate value for the *threshold* parameter so that we can distinguish between two segments? We use trial and error on sample looks exhibiting the characteristic of having two different segments at consecutive angles.

Q.10 Experiment with values of the distance cut-off to find a reasonable value for separating segments that correspond to two separate obstacles in the same look.

Now that we have a new, more refined mechanism to compute the segments, we can reclassify the final looks for the 100 log files:

```
circs = lapply(finalLooks, robot.evaluation)
```

In order to simplify evaluating the results, we manually classified the 100 looks by deciding whether we thought there was a circle. Some of these were not obvious so we marked those as NA. We read these classifications into R with

```
hasCircle = as.logical(scan("logs/hasCircle100", 1L))
```

We can now compare the "truth" with the results of our classifier:

```
hasCircle.hat = sapply(circs, length) > 0
table(hasCircle, hasCircle.hat)
```

```
          hasCircle.hat
hasCircle FALSE TRUE
    FALSE    51    5
    TRUE      6   29
```

This is the "confusion matrix." The entries on the diagonal are correctly classified. The off-diagonal entries show 11 looks are misclassified. However, the total number of looks in this table is 91, not 100. We have to account for those looks we manually classified as NA. We'll consider those as containing the target, i.e., TRUE as we were conservative in classifying these. Accordingly, we'll change the values for hasCircle for these looks. With these changes, the confusion matrix is

```
              hasCircle.hat
hasCircle FALSE TRUE
    FALSE    51    5
    TRUE      9   35
```

Only 3 of these 9 looks are now misclassified.

We want to look at those looks that are incorrectly classified as not being a circle:

```
missed = hasCircle & hasCircle != hasCircle.hat
par(mfrow = c(3, 3), mar = rep(0, 4), pty = 's')
invisible(lapply(finalLooks[missed], plotLook3, axes = FALSE))
```

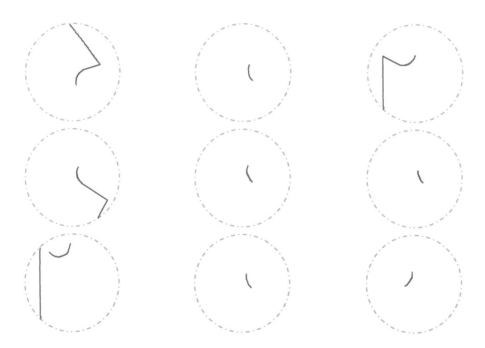

Figure 4.16: *Misclassified Looks with a Target. These 9 looks are those that were misclassified as not containing the target. In all but one of these, the target is very close to the center of the robot. The seventh look is more problematic.*

In 8 of the 9 looks shown in Figure 4.16, the circular target is very close to the center of the circle and hence very close to the robot. The robot cannot detect the curvature as it is too close and the arc looks more like a line. We want to examine the fit of the circle for each of these. To do this, we can modify robot.evaluation() to allow the caller to ask for the information for the fit for each segment rather than just the results for the segments that were considered circular arcs. Alternatively, we can move the code in the body of the loop in robot.evaluation() to a separate function, say evalSegment(), and adjust it to not simply return the first matching segment, but the list of all matching segments.

Q.11 Implement the evalSegment() function described above. Also, implement a function evalSegments() that takes a look and computes the segments and calls evalSegment() on each of these.

Using our evalSegments() from the exercise above, we can explore the segments for the looks that were not classified as containing a circle:

```
segs = lapply(finalLooks[missed], evalSegments, noCheck = TRUE)
```

The seventh of these corresponds to the "obvious" circle we missed. The details of the fit are

```
segs[[7]][[1]]$fit

$minimum
[1] 0.003707

$estimate
[1] -7.5589 -7.5258  0.4975

$gradient
[1] -9.958e-07 -1.123e-06 -1.552e-06

$code
[1] 2

$iterations
[1] 15
```

This was not classified as the target as the estimate of the radius is less than 0.5. If we relax the value for *min.radius* to 0.49, this will pass. Of course, we may need to accept lower type I rates also.

For the segments that were very close to the robot, we can look at the estimate of the radius of the would-be target and also the sum of squares for the fit (i.e, the minimum element):

```
unname(sapply(segs[-7],
       function(x) c(x[[1]]$fit$estimate[3],
                     length(x[[1]]$x),
                     x[[1]]$fit$minimum/length(x[[1]]$range))))

          [,1]     [,2]      [,3]     [,4]     [,5]     [,6]
[1,]    0.9340 1.42e+02    0.9907   1.7638 8.15e+01 5.16e+01
[2,]  167.0000 8.90e+01  139.0000 116.0000 8.70e+01 9.40e+01
[3,]    0.0181 4.08e-04    0.0263   0.0129 5.93e-04 1.59e-04
          [,7]     [,8]
[1,]    7.94e+01 5.43e+01
[2,]    8.90e+01 9.20e+01
[3,]    2.97e-04 3.47e-04
```

In several of these, the radii (first row) are very large. In the other cases, the sum of squares per point exceeds our 0.01 threshold for *max.ss.ratio*. If we increase this to 0.25, we would classify 3 of these as targets, but with a radius for one of 1.77! How would this affect our type I error rate?

Let's return to those 5 looks that were incorrectly classified as containing the target (type I errors) and examine their fit details:

```
missed = !hasCircle & hasCircle != hasCircle.hat
unname(lapply(circs[missed], `[[`, "fit"))
```

In all cases, the radii are close to 1:

```
summary(sapply(circs[missed], function(x) x$fit$estimate[3]))
   Min. 1st Qu.  Median    Mean 3rd Qu.    Max.
  0.903   0.992   0.992   1.000   1.070   1.070
```

We can reduce the value of *max.radius* to, say, 0.9 to eliminate these. Again, we have to consider how this would affect the type I and II error rates. We can check this with

```
circs = lapply(finalLooks, robot.evaluation,
               max.radius = 0.9, min.radius = .475)
hasCircle.hat = sapply(circs, length) > 0
table(hasCircle, hasCircle.hat)
```

```
         hasCircle.hat
hasCircle FALSE TRUE
    FALSE    56    0
    TRUE      9   35
```

We now have no type II errors. However, we have classified 9 looks as not having a circle. We accepted the seventh look we missed in our earlier classifier by reducing the value of *min.radius*. However, we added one more misclassified look, which again has the target close to the robot.

To handle these challenging cases where the robot is "too close" to the target, we might attempt to recognize the target in an earlier look. Alternatively, we could use a different criterion for classifying such a segment.

We have manually tuned the parameters of our classifier to fit our 100 looks. We are in real danger of overfitting our classifier to these looks. We should use a more robust mechanism such as cross-validation to determine these tuning values, and also examine other looks. Since we focused on the final look in each log file because we expected that there would be more targets found in these, we may have biased our classifier's ability to correctly identify looks with no targets. Accordingly, we need to consider using additional data.

Additional Data

We have looked at the final look in each of the 100 log files. We did this as these were the most likely to contain the target, as the experiment terminated if and when it located the target. We do not need to limit ourselves to just these final looks, however, when developing or understanding our classifier. We have the data for over 250,000 looks. We are only interested in those looks that actually contain an object the robot sees as we know we will not classify a look as containing the target if there is nothing seen in that look. We can identify these looks that see something with

```
allSegs = lapply(logs, function(ll)
                   lapply(1:nrow(ll),
                          function(i)
                            getWrappedSegments(ll[i,])))
```

This yields about 200,000 looks with visible objects of some sort. We can then sample from these and classify them ourselves. Once we know the "truth," we now have a regular data set with the response variable known. We can then use this to fit our statistical classifier. In order to determine the optimal values for our parameters for min.length, max.ss.ratio, min.radius and max.radius, we can use cross-validation. We can split this data set into training and test subsets and evaluate the classifier for different values of these parameters and find the best values for the type I and II errors.

A different approach to manually classifying actual looks to generate a complete data set with a known response is to simulate data. We can generate the location of a target and a radius close to 0.5 meters. From this, we can compute the coordinates of its circular boundary. We can then generate a location of the robot and simulate what it would see of the target's boundary. We have an understanding of the distribution of the errors in the measurement from our earlier exploration. We can sample directly from these observed errors and add these to the measurements. Recall that the overwhelming majority are 0. We then have a complete data set. We can also simulate looks in which the robot sees an obstacle and add these to the data set. We can then use this entire data set to tune our classifier.

Q.12 Our fit function used the sum of squares between the radii. We could use other criteria such as absolute value, i.e., the L_1 norm. Explore different criteria. How does this change the code we use to determine if the arc corresponds to part of a circle? How does it change how we classify different looks?

Q.13 Implement the geometric approach to estimating the center of the target and use that as the initial guess we pass to nlm().

Q.14 Modify robot.evaluation() to return all of the segments that could be considered part of a circle, and not just the first one.

Q.15 If the robot is very close to the target, it is hard identify the shape of the target as circular. The curvature of the part of the circle looks close to a line as the robot can see fewer points on the arc of the circle. How can we adjust the classifier or the criteria for accepting a segment as the target to handle this case? How do we guard against classifying an obstacle or side as the circular target in these cases?

Q.16 In the streaming data case where we process one look at a time and then determine where to move to next, we'd like to be able to identify a circle and move towards it if we are not certain whether it actually is a circle. How can we provide a level of confidence or uncertainty in our classification of a segment as the circular target?

Q.17 Use cross-validation on the 100 final looks to determine the best values of the tuning parameters *max.ss.ratio*, *min.radius*, *max.radius* and *range.threshold* for classifying a look. What are the corresponding error rates?

4.4 Detecting the Target with Streaming Data in Real Time

We have explored the different log files and their looks. We have used the data to develop a statistical classifier for identifying the circular target and gained an understanding of how well it works. This is the typical way we develop statistical methods, i.e., with off-line data that we use to fit and validate models. However, our original goal was to create a classifier that we could use in real-time, or on-line, while the robot is moving through the course to detect the target. This is to be used as the robot reports `position2d` and `laser` records one at a time. Not only do we want a good classifier to correctly identify the target, we also would like to be able to develop search strategies that combine information from a sequence of these records to tell the robot where to go next.

In this section, we process data from the robot line by line as if we were receiving the data from the robot in real time. There are many ways we might obtain the data from the robot as it reports its position and what it sees. Regardless of the specifics, the data would come to us not from a file, but from a stream of data from which we can read individual values or lines. R uses the concept of a *connection* to describe a general source of data that abstracts the source of the data, e.g., a file, a Web connection, the output from a shell command. We used readLines() to read all of the data from a log file. We can use the same approach with a connection. However, instead of reading all of the data from the fixed content (a file), we want to read one or more lines, process these, and then return to read the next lines as they become available from the robot. This is quite different from the fixed contents of files, but made possible with connections.

We first open() our connection. We'll use a regular log file to illustrate the approach, but this applies for any connection. The command

```
con = file("logs/JRSPdata_2010_03_10_12_12_31.log", "r")
```

creates and returns a connection object that is poised ready to read the first byte from the stream.

We will read one line at a time with `readLines(con, n = 1)`. Importantly, we specify the connection object from which to read. Each time we call readLines() or some similar function with the same connection, we read from where we left the connection in the previous call. In other words, we do not re-open the connection and start reading from the first line again. We merely keep reading from where the previous read operation finished.

To process the lines in the robot stream, we will mimic the computations from our readLog() function we developed earlier in the chapter. Instead of processing all of the lines with vectorized computations, we read one line at at time and interpret it, and then return to read the next line. If the current line is a comment line, we discard it. If the line is a regular record, we break its elements into separate values. We then examine the 7 common meta-fields to determine if the record is of interest to us, i.e., has a type value of `001` and an interface value of either `laser` or `position2d`. We can implement this via a `while` loop with

```
while(length(line <- readLines(con, 1))) {
  if(grepl("^#", line))
     next

  vals = strsplit(line, "[[:space:]]+")[[1]]
  if(vals[6] != "001")   # the type field
     next

  # update look with values
  # call robot.evaluation to determine if we are
  # looking at the target.
}
```

In the condition for the loop, we both read and assign one line from the connection. If we are at the end of the connection, readLines() will return an empty vector. Therefore, to detect the end, we test the length of the value of line. Note that we assigned the result of the call to readLines() in the call to length() and the condition for the while loop. This is shorter than explicitly assigning the value and testing it in the body of the loop.

To complete our loop, we need to add code to merge the information from the position2d and laser lines into a look data frame and then pass this to robot.evaluation(). If this returns information about a matching segment, then we exit the loop as we have found the circular target. We can implement this with

```
iface = vals[4]
if(iface == "position2d") {
  look[1, c("time", "x", "y")] = as.numeric(vals[c(1, 8:9)])
} else if(iface == "laser") {
  look[1, 4:360] = as.numeric(vals[seq(14, length = 360, by = 2)])
  ans = robot.evaluation(look)
  if(length(ans))
    break
}
```

The idea here is that we update a one-row data frame stored in look. We create this once outside of the loop but insert the values from the current records. When processing a position2d record, we update the location of the robot in look but do not call the robot.evaluation() function. We do that only when we get the range values from the next laser record and so have a complete record for the look.

Our loop provides the basic structure for processing streaming/on-line data. We can adapt it to use a sequential classifier that recognizes the target but does not announce it has found it until it has several confirmatory looks. Similarly, we can add functionality to provide suggestions to the robot for its next move so that we can get an improved look at the potential target based on the current view, e.g., move towards or away from the target. We can also use R's connections to write information to the robot and direct it as part of the actions of the loop.

Q.18 Study the code in this case study and modify it to allow a user to substitute his/her own functions for each of the different components we might want to change. For example, allow a different function for fitting the circle, or metric for evaluating the fit for a given triple of parameters. Similarly, allow for a different mechanism for computing an initial estimate of the center of the circle for the call to nlm().

Bibliography

[1] R Core Development Team. *R help page for the* nlm() *function*, 2000–2014.

[2] R Development Core Team. *R: A Language and Environment for Statistical Computing*. Vienna, Austria, 2012. http://www.r-project.org.

5

Strategies for Analyzing a 12-Gigabyte Data Set: Airline Flight Delays

Michael Kane
Yale University

CONTENTS

5.1	Introduction	217
	5.1.1 Computational Topics	218
5.2	Acquiring the Airline Data Set	219
5.3	Computing with Massive Data: Getting Flight Delay Counts	219
	5.3.1 The *R* Programming Environment	219
	5.3.2 The *UNIX* Shell	221
	5.3.3 An *SQL* Database with *R*	223
	5.3.4 The `bigmemory` Package with *R*	227
5.4	Explorations Using Parallel Computing: The Distribution of Flight Delays ..	229
	5.4.1 Writing a Parallelizable Loop with `foreach`	230
	5.4.2 Using the Split-Apply-Combine Approach for Better Performance ...	231
	5.4.3 Using Split-Apply-Combine to Find the Best Time to Fly	232
5.5	From Exploration to Model: Do Older Planes Suffer Greater Delays?	236
	Bibliography	238

5.1 Introduction

Anyone who has dealt with flight delays at the airport understands the associated inconvenience and aggravation. And while we might hope that delays are rare, they are probably more common than you think. Since October 1987, there have been over 50 million flights in the United States that failed to depart at their scheduled times. Around 200,000 of those flights were at least two hours late; some were much later. From these two simple facts we can surmise that delays are not isolated, rare events; they are routine. Since 1987 the number of flights per year has steadily increased and as this trend continues we expect to see more inconvenience, more aggravation, and more time lost.

But why do flight delays occur? Is it simply because there are more flights now than in previous years? Are delays caused by bad weather? Is enough time being scheduled for flights? Do single flight delays cause multiple flight delays later in the day? A better understanding of the cause of flight delays could allow the airline industry to intelligently react to issues such as bad weather, providing more flights with fewer delays.

This chapter presents a means for understanding flight delays by analyzing data. In 2009, the American Statistical Association (ASA) Section on Statistical Computing and Statistical Graphics released the "Airline on-time performance" data set [17] for their biannual data exposition. The data set was compiled and organized by Hadley Wickham [16] from the official releases from the US government's Bureau of Transportation Research and Innovative Technology Administration (RITA) Web site (http://www.transtats.bts.gov/DL_SelectFields.asp?Table_ID=236). The data include commercial flight information from October 1987 to April 2008 for those carriers with at least 1% of domestic U.S. flights in a given year. In total, there is information for over 120 million flights, each with 29 variables related to flight time, delay time, departure airport, arrival airport, and so on. In total, the uncompressed data set is about 12 gigabytes (GB) in size. The data set is so large that it is difficult to analyze using the standard tools and techniques we have come to rely upon. As a result, new approaches need to be utilized to understand the structure of these data. This chapter presents some of these new approaches along with an initial exploration of airline flight delays.

These new computational approaches are illustrated by taking the reader through the process of acquiring, exploring, and modeling the airline data set. The first section describes how to acquire the data. The second section demonstrates the use of 3 different computing environments to manage and access large data sets: *R* [7], the *UNIX* shell, and *SQL* databases. The third section offers easy-to-use parallel computing techniques for basic data exploration. No prior experience with parallel computing is required for this section. The fourth section provides an approach to answering the question, "Do older planes suffer greater delays?" This section synthesizes earlier material and demonstrates how to construct linear models with a potentially massive data set.

5.1.1 Computational Topics

- Big Data strategies.
- Shell commands and pipes.
- Relational databases.
- Parallel computing.
- External data representation.
- The split-apply-combine approach.

Question: What if the reader runs into software questions while trying to run the sample code?

Answer: Throughout the chapter an attempt has been made to anticipate some of the implementation issues that will arise as the reader works through the examples. These issues are presented as questions along with their solutions in boxes like this one.

5.2 Acquiring the Airline Data Set

The airline on-time performance data can be found at http://stat-computing.org/dataexpo/2009/the-data.html. The Web page contains the compressed airline delay data in comma-separated values (CSV) files, organized by year. The page also provides descriptions of each of the 29 variables associated with each flight. The following sections will assume that you have downloaded the data files and decompressed each of them, e.g., using the `bunzip2` shell command or potentially a point-and-click graphical interface.

5.3 Computing with Massive Data: Getting Flight Delay Counts

This section introduces methods for managing and accessing data sets that require more than a computer's available RAM (Random Access Memory). At the time this chapter was written, a 12Gb data set, such as the airline data, presents a computational challenge to most statisticians because of its sheer size. Admittedly, this may not always be the case. A future reader may scoff at the idea of having difficulty managing 12 gigabytes of data. However, it is assumed that there will still be a data set which, by virtue of its size, frustrates even this well-equipped statistician.

To begin our exploration, let's consider two simple questions:

- How many flights are in the data set for 1987?
- How many Saturday flights appear in the entire data set, i.e., all years?

The next four subsections show distinct approaches to answering these questions. The first uses the *R* programming environment with its native data structures and capabilities. The second uses the *UNIX* shell, independent of *R* or other programming environments. The third uses *R* to connect to a database where *SQL* queries are constructed to answer the posed questions. The fourth and final approach utilizes some of *R*'s more advanced functionality to show how the `bigmemory` package can be used to explore the entire airline data set from within the *R* environment.

5.3.1 The *R* Programming Environment

The downloaded files provide data for more than 20 years of airline traffic. Each file holds information for one year and each year contains information for approximately 5 million flights. The aggregate data set is larger than the amount of RAM on most single computers. However, to compute the flight count for 1987, we do not need to load the entire data set for all years into *R*. We can simply load the 1987 data and compute the number of observations with

```
x <- read.csv("1987.csv")
nrow(x)
```

```
[1] 1311826
```

To compute the number of Saturday flights across multiple files, we need to determine how many values in the column labeled `DayOfWeek` correspond to Saturday in each of the data files. The values in this column have values from 1 through 7, with 1 corresponding to Monday and 7 for Sunday. The value 6 indicates Saturday. At the same time, we need to be wary of the size of the data set. We can find the total number of Saturday flights by working on one file at a time, rather than trying to read all of the files at once. After a single file is read into a data frame, the number of Saturday flights will be calculated and saved as an intermediate result. This will be done for each file and, finally, these intermediate counts will be added together to get the total. This approach is sometimes called the *incremental* or *chunking* approach because a single, manageable "chunk" of data is processed and an incremental result is stored before aggregation, which yields the final result. Here, the individual files are natural chunks. In other cases, we divide a large collection of observations into smaller chunks by just reading/processing chunks of the observations sequentially. For this airline data, we can compute the total number of Saturday flights with

```
totalSat <- 0
for (year in 1987:1988) {
   x <- read.csv(paste(year, ".csv", sep=''))
   totalSat <- totalSat + sum(x$DayOfWeek == 6)
}
totalSat
```

[1] 15915382

We could improve the speed of this by specifying a vector of types for each column via the *colClasses* parameter. This helps read.csv() (and ultimately read.table()) so that it doesn't have to infer the type of each column and potentially have to reallocate memory if it guesses incorrectly from the first k observations. We could also provide the total number of observations via nrows if we already knew these, as we did for the 1987 data. Again, this helps R to allocate memory efficiently by doing it just once for each column/vector of values.

Instead of the `for` loop, we could also use

```
counts <- sapply(sprintf("%d.csv", 1987:1988),
                 function(f)
                    sum(read.csv(f)$DayOfWeek == 6))
sum(counts)
```

Each call to our function reads the appropriate CSV file and then discards it after computing the number of observations. This is the critical aspect of both approaches, i.e., to avoid having more than one year's data frame in memory at any time. The sapply() approach enables this; the `for` loop approach actually has two data frames in memory at times (when and why?). With the sapply() code, we also have the counts for each individual year and can examine these to see how they are distributed, e.g., does the number of flights in a year increase over the years? The `for` loop code is easily modified to create this vector also.

Although these questions were easy to answer, the approach of loading files as needed can be somewhat limiting. Also, the approach assumes that each data file can be stored as a data frame. When operations cannot be expressed incrementally/cumulatively, this approach becomes cumbersome. In the next sections, we will focus on approaches where the data are managed and accessed from a single source, rather than a set of files, allowing us to perform more sophisticated analyses.

> Question: Can R objects be destroyed explicitly when we are done with them for better performance?
> Answer: It is possible to force R to remove a variable and run the garbage collector, freeing all memory associated with that variable. For the vast majority of cases, little is gained by doing this. Even in cases where this helps, we will not see a dramatic increase in speed nor a dramatic decrease in memory usage.
>
> A call to the garbage collector may help in special cases where the variables held in an R session have memory requirements greater than the size of a computer's available RAM. When programs require more RAM than what is available, variables may be "swapped" from RAM to the disk and retrieved when needed. In some cases, even often-used variables can be swapped to disk. This causes calculations to run slower because accessing the data from the disk takes much longer than accessing them from RAM. In a few cases, calling the garbage collector may free unused memory in RAM, allowing other variables to take their place and reducing the amount of time spent swapping. It may also reduce memory fragmentation, which can be important.
>
> The function to remove a variable is rm() and the garbage collector is called with the gc() function. The following code shows how to use these two functions to ensure that only one data frame exists in RAM at any time when calculating the number of Saturday flights:
>
> ```
> totalSat <- 0
> for (year in 1987:1988) {
>     x <- read.csv(paste(year, '.csv', sep=''))
>     totalSat <- totalSat + sum(x$DayOfWeek == 6)
>     rm(x)
>     gc()
> }
> totalSat
> 
> [1] 15915382
> ```
>
> Here we explicitly remove the data frame object and force the garbage collector to run. R does this implicitly.

5.3.2 The UNIX Shell

This section uses the UNIX shell to answer the previously posed questions. The shell is a powerful interactive and scripting programming environment with a reasonably simple language and a rich set of computational resources. Many common operations are provided as built-in utilities, analogous to functions in R. For example, finding the number of flights in 1987 can easily be accomplished with the wc utility. wc stands for "word count" which is not quite what we want but the -l option can be used to count lines rather than words. To find out how many flights in 1987 were recorded we can simply type:

```
wc -l 1987.csv
1311827 1987.csv
```
Shell

Again we see that there are 1,311,826 flights that were recorded for 1987. The count from the shell command indicates one more, but this is because it includes the first (header) line of the file that lists the names of each column. The utility executes more quickly than the solution from the previous section and it was expressed with less code. In this subsection we will see that the shell is often a very good tool for file manipulation, simple subset selection, and simple summaries. Furthermore, we can invoke shell commands from R (and other languages) and read the resulting output back into R, giving us the best of both worlds.

For the rest of this section, we would like to work with a single CSV file, which contains all of the airline data. This file can be created by appending each of the airline files to a new file. However, when the files are being concatenated we need to make sure that the header information, which is given on the first line of each file, appears only once at the beginning of the new file.

The task of combining each of the CSV files into a single file can be accomplished with the following commands:

Shell
```
cp 1987.csv AirlineDataAll.csv
for year in {1988..2008}
   do
      tail -n+2 $year.csv >> AirlineDataAll.csv
   done
```

The resulting file begins with a header line containing the names of each column and then contains observations for each flight for every year.

This small script illustrates some of the features of the shell. We copied the first file (*1987.csv*) to our target file and then appended the contents of the other files to this new file. We can specify the years to iterate over using a list constructor `1988..2008`. This is analogous to `1988:2008` in R. Then, we can create an argument specifying a file name by appending **.csv** to the value of the shell variable *year*, i.e. `$year.csv`[1]. Next, we can extract the "tail end" of a file with the *UNIX tail* utility. The `-n+2` option specifies that *tail* will return all but the first line of the file. Finally, we can use the `>>` operator to redirect the output from the *tail* command so that it is appended to the *AirlineDataAll.csv* file.

The loop above is not our only option for aggregating information from a file. The *tail* command itself can take multiple files as input, outputting the specified lines for each of the files. For example, if were were interested in creating a single file, with no header information, for all files from the twentieth century we could simply use the command:

Shell
```
tail -n+2 19*.csv > All1900.csv
```

Like the last example, this one uses *tail* to return all but the first line of each of a set of files. Unlike the last example, we used **19*.csv** to specify all files that start with `19` and end with **.csv**. This example also made use of the `>` operator, which overwrites the existing contents of *All1900.csv* or creates the file if it does not already exist. This version does not include the header line at the top of the file. Depending on how we use this file, this may or may not be important, but it is something we need to know. Now that we have two different approaches for creating a single file, holding all of the airline data, let's use the *UNIX* shell to find the number of Saturday flights.

Calculating the total number of Saturday flights is trickier than simply counting the number of lines in a file. The day of the week column is the fourth one in *AirlineDataAll.csv*. (Examine the first line of any of the original data files to verify this.) So, we would like to extract only that column of data values from the file and see how many times 6 appears as the extracted value (corresponding to Saturday). To do this we'll need to introduce a few more *UNIX* shell utilities and concepts. First, we can use the *cut* utility, which extracts and outputs sections from each line of a file. We'll use it to extract the fourth column of each line and discard all other columns. This will create a large collection of values ranging from 0 to 6, i.e., the day of the week as a number, each on its own line of output. We can then use

---

[1]The shell knows that the variable is named year since a shell variable cannot have a "." in its name.

the *grep* command to match only those lines that contain the character 6. This produces a collection of lines that is a subset of that produced by the *cut* command. Finally, we can use *wc*, with which we are already familiar, to compute the total count for Saturday flights by counting the number of lines output by the *grep* command. We could implement this with 3 separate commands and output the results of each to intermediate files, e.g.,

```
cut -d , -f 4 AirlineDataAll.csv > tmp
grep 6 tmp > tmp1
wc -l tmp1
rm tmp tmp1
```
Shell

However, the shell explicitly supports directing the *output* from one command as *input* to another command without (explicitly) using intermediate files. This the purpose of the pipe operator |. We can write the entire computation to calculate the number of Saturday flights with the command:

```
cut -d , -f 4 AirlineDataAll.csv | grep 6 | wc -l
15915382
```
Shell

In this example the *cut* utility is passed 3 sets of parameters. The first set is *-d*, which specifies that the file consists of columns, each of which is delimited (*-d*) by a comma. The second set *-f 4* specifies that we want to extract the fourth column/field. The third specifies the name of the file whose contents we want to process. The output of the *cut* command is the fourth column of *AirlineDataAll.csv*. The lines from the *cut* command are then passed as lines of input for the *grep* command, which outputs all the lines that consist of the single value 6. All 15,915,382 of those rows are passed to `wc -l`, which counts the number and prints the result on the console.

We should note that we don't actually need to create this single file containing the data for all of the years in order to easily calculate the number of Saturday flights. We can have *cut* operate on all of the files with the command `cut -d , -f 4 *.csv` and then pipe this to *grep* and *cut* as before. However, in the next section we will need and use this single file containing all of the data.

The *UNIX* shell capabilities go far beyond the examples shown here. There are many other utilities included with the shell and users can even create their own utilities using almost any programming language. As a result, the *UNIX* shell is often a good choice for file manipulation and basic summaries. Moreover, we can use these shell commands from *R* via the system() and system2() functions.

5.3.3 An *SQL* Database with *R*

The *R* programming environment approach to answering the posed questions is simple for someone who is familiar with *R* and its functionality could easily be expanded to do more than simply tally the number of flights in 1987 or on Saturdays. However, the approach is somewhat limited in that it assumes each file is relatively small, i.e., the contents can be held in memory. The *UNIX* shell approach is also simple, if you are already familiar with its syntax and computational model. With only a single line we were able to answer each of the posed questions. Also, the results were returned more quickly than with *R*. However, the *UNIX* shell is not a familiar environment to many statisticians, it has a limited computational model, and it does not have built-in capabilities for performing statistical analyses. As a data exploration requires more sophisticated analyses, the corresponding *UNIX* scripts may become difficult to write. In practice, we often combine *R* and the shell to process data. However, there is an important alternative approach – databases.

A database provides a general solution for managing a large data set and extracting meaningful information. Where the previous approaches required that we manually open files and extract the data of interest each time we process the data, a database provides a means to ingest and structure the data once and reuse that structure each time we access the data. A database also provides the Structured Querying Language (*SQL*) that allows us to specify and efficiently extract the subset of data of interest with a powerful query. This approach creates a rich and general way of not only extracting potentially large and complex subsets of the data but also for computing basic summaries of these subsets. Just as we can call the shell from *R* or other languages, we can interact with a database from *R*, sending *SQL* commands to the database and accessing the output as *R* objects.

The examples in this section use SQLite [1], a lightweight (*SQL*) database engine. All interactions with the database are done in *R* with the `RSQLite` [2] package. However, these examples are not specific to SQLite or `RSQLite`. They will work with any *SQL* database engine and corresponding *R* database connector package.

> Question: How do I import the airline delay data into a database?
>
> Answer: We'll use SQLite as the database engine. You must have SQLite installed on the machine. The software can be downloaded from the SQLite Web page (http://www.sqlite.org/). You must also have the `RSQLite` package installed on your machine; it can be found on the CRAN Web site or installed directly with `install.packages("RSQLite")`. The following instructions are based on those given by Hadley Wickham on the 2009 Data Expo Web page.
>
> - Inside the directory where the data (CSV files) reside, type the command:
>
>   Shell
>   ```
>   sqlite3 AirlineDelay.sqlite3
>   ```
>
> - This will create the database and put you into a *SQL* console/read-eval-print-loop (REPL), with a prompt `sqlite<`. Next, create the table and its fields with the command
>
>   SQL
>   ```
>   CREATE TABLE AirlineDelay (
>     Year int,
>     Month int,
>     DayofMonth int,
>     DayOfWeek int,
>     DepTime  int,
>     CRSDepTime int,
>     ArrTime int,
>     CRSArrTime int,
>     UniqueCarrier varchar(5),
>     FlightNum int,
>     TailNum varchar(8),
>     ActualElapsedTime int,
>     CRSElapsedTime int,
>     AirTime int,
>     ArrDelay int,
>     DepDelay int,
>     Origin varchar(3),
>     Dest varchar(3),
>     Distance int,
>     TaxiIn int,
>   ```

```
            TaxiOut int,
            Cancelled int,
            CancellationCode varchar(1),
            Diverted varchar(1),
            CarrierDelay int,
            WeatherDelay int,
            NASDelay int,
            SecurityDelay int,
            LateAircraftDelay int );
```

- Now that the table has been created, import the data from the *AirlineDataAll.csv* file. This may take from 10 minutes to over an hour, depending on the speed of your machine.

```
    .separator ,
    .import AirlineDataAll.csv AirlineDelay
```

To extract all of the flights in 1987, start by opening an *R* session and connecting to the database:

```
library(RSQLite)
delay.con <- dbConnect("SQLite", dbname = "AirlineDelay.sqlite3")
```

The `delay.con` variable holds a connection to the database that we can use in subsequent commands.

Queries are expressed via *SQL* statements. These statements allow a user to describe the data of interest and perform operations (such as counting) on them. The SELECT statement is used to retrieve entries from a database and we can use it to retrieve all data of the flights from 1987:

```
delays87 <- dbGetQuery(delay.con,
                "SELECT * FROM AirlineDelay WHERE Year=1987")
```

dbGetQuery() is an *R* function. It sends an *SQL* query to the database to be processed there and dbGetQuery() waits for the result. It does not examine the query as that is written in a different language (*SQL*). The query above returns all of the variables in a data frame for those flights whose value for the year variable is equal to 1987. Now, we can find the number of 1987 flights in a familiar way:

```
nrow(delays87)
```

[1] 1311826

Equivalently, the *SQL* engine can do the counting for us by utilizing the *COUNT*() aggregator function:

```
dbGetQuery(delay.con, "SELECT COUNT(*), Year FROM AirlineDelay
                                        WHERE Year=1987")
```

[1] 1311826

It is reasonably clear from the examples above that *SQL* statements are useful as we perform more complex queries. What if, instead of getting the flight count for a single year, we want to get the flight count for each year in the database? We can do this using the GROUP BY clause in *SQL*. The following query groups all of the rows of the data set by year and counts the number of rows in each of these groups:

```
dbGetQuery(delay.con,
           "SELECT COUNT(*), Year FROM AirlineDelay GROUP BY Year")
   COUNT(*) Year
1   1311826 1987
2   5202096 1988
3   5041200 1989
4   5270893 1990
5   5076925 1991
6   5092157 1992
7   5070501 1993
8   5180048 1994
9   5327435 1995
10  5351983 1996
11  5411843 1997
12  5384721 1998
13  5527884 1999
14  5683047 2000
15  5967780 2001
16  5271359 2002
17  6488540 2003
18  7129270 2004
19  7140596 2005
20  7141922 2006
21  7453215 2007
22  7009728 2008
```

To find the number of Saturday flights, we can use a query similar to the previous one, but rather than grouping by the *Year* variable, we subset/filter by day of week using the *WHERE* clause and perform the calculations in *SQL* with

```
dbGetQuery(delay.con,
           "SELECT COUNT(*), DayOfWeek FROM AirlineDelay
                                      WHERE DayOfWeek = 6")
   COUNT(*) DayOfWeek
1  15915382         6
```

(Note the use of the single = operator in *SQL* for testing equality, different from == in *R*.)

A database provides a useful way of managing large data sets and it supports complex queries. However, it also comes with challenges. First, it requires the creation of the database and the importing of data. For the airline data this was not difficult, but it did involve an extra step. Second, the database approach also requires that the statistician is familiar with *SQL* to perform even simple operations. Finally, if an analysis is done in *R* and the data set is small, we would use a `data.frame` or `matrix` to manage it. When the data set gets big and we need a new mechanism to handle the larger volume, a database connection cannot simply be "swapped-in" in place of the familiar *R* data structures. Significant changes must be made in the code in order for the analysis to work with a database. Wouldn't it be nice if there was an *R* data structure that behaved like a `matrix`, but at the same time, managed large data sets for you?

5.3.4 The `bigmemory` Package with *R*

The *R* package `bigmemory` [4] provides matrix-like functionality for data sets that could be much larger than a computer's available RAM. This approach to computing with data provides several advantages when compared with other approaches. First, `bigmemory` provides data structures that hold entire, possibly massive, sets of data. As a result, there is no need to manually load and unload data from files. Second, the data structures provided by `bigmemory` are accessed and manipulated in the same way as *R*'s `matrix` objects. Our experience with *R* has prepared us to work with `bigmemory` and there is no need to learn a new language, such as *SQL*, to access and manipulate data. Third, `bigmemory` works with many of *R*'s standard functions with little or no modification, minimizing the amount of time required to perform standard manipulations and analyses. Finally, `bigmemory` was designed from the ground up for use in parallel and distributed computing environments. Later on we will see how `bigmemory` can be used as a basis for implementing scalable analyses for data that may be much larger than even the airline data set.

The essential data structure provided by `bigmemory` is the `bigmatrix`. A `bigmatrix` maintains a binary data file on the disk called a *backing file* that holds all of the values in a data set. When values from a `bigmatrix` object are needed by *R*, a check is performed to see if they are already in RAM (cached). If they are, then the cached values are returned. If they are not cached, then they are retrieved from the backing file, cached, and then returned. These caching operations reduce the amount of time needed to access and manipulate the data across separate calls, and they are transparent to the statistician. A `bigmatrix` object is designed to be a convenient and intuitive tool for computing with massive data. When using these data structures the emphasis is on the exploration, not the underlying technology.

As mentioned before, `bigmatrix` looks like a standard *R* `matrix`. It has rows and columns, and subsets of the elements of a `bigmatrix` can be read and set using the standard subsetting operator (`[]`). Like *R*'s `matrix`, a `big.matrix` object requires that all elements are of the same type. However, this leads to a challenge with the airline data set since it has columns that are character type as well as numeric. Before we can read the airline data into a `bigmatrix`, some preprocessing must be done so that all the columns and their values are numeric. For columns with character data, this preprocessing step creates a mapping between a unique numeric value and the character value for each row, much like *R*'s *factor* data type. Preprocessing is left as an exercise for the reader or a preprocessing script is available from the author of this chapter. For the rest of the chapter, we will assume that the preprocessing step has been performed and that the preprocessed file has been named `airline.csv`.

A user can create a `bigmatrix` from a CSV file with the function read.big.matrix() that is similar to *R*'s read.csv() function, e.g.,

```
x <- read.big.matrix("airline.csv", header = TRUE,
            backingfile = "airline.bin",
            descriptorfile = "airline.desc",
            type = "integer", extraCols = "age")
```

The *extraCols* and *descriptorfile* parameters used in the example will be explained later.

As we said, a `big.matrix` object x acts like a regular *R* `matrix` and commands such as dim() and head() give the appropriate results, e.g.,

```
dim(x)    # How big is x?
```

[1] 123534969   30

```
x[1:6,1:6] # Show the first 6 rows and columns.
```

```
     Year Month DayofMonth DayOfWeek DepTime CRSDepTime
[1,] 1987    10         14         3     741        730
[2,] 1987    10         15         4     729        730
[3,] 1987    10         17         6     741        730
[4,] 1987    10         18         7     729        730
[5,] 1987    10         19         1     749        730
[6,] 1987    10         21         3     728        730
```

At this point, we can compute the number of flights in 1987 in a familiar way:

```
sum(x[, "Year"] == 1987)
```

[1] 1311826

Similarly, the number of Saturday flights can be found using the command

```
sum(x[,"DayOfWeek"] == 6)
```

[1] 15915382

Depending on your hardware, the read.big.matrix() function could have taken over 30 minutes to complete. The prospect of waiting to load these data in future R sessions is very unappealing. Wouldn't it be nice if subsequent sessions could create a bigmatrix by simply attaching to the existing backing file and not have to wait to create the big matrix object? Fortunately, we can do this by using a descriptor file. This file contains all of the information needed to create a new bigmatrix from an available backing file. A new bigmatrix, named y, which uses the airline backing file, can be rapidly created with the attach.big.matrix() function:

```
y <- attach.big.matrix("airline.desc")
```

It is important to realize that the variables x and y now point to the same data set. This means that changes made in x will be reflected in y. To illustrate this point, let's create a new bigmatrix object which has 3 rows, 3 columns, and holds zero integer values.

```
foo <- big.matrix(nrow = 3, ncol = 3, type = "integer", init = 0)
```

We can look at the contents of foo by typing:

```
foo
```

```
     [,1] [,2] [,3]
[1,]    0    0    0
[2,]    0    0    0
[3,]    0    0    0
```

Now, let's create another variable bar:

```
bar <- foo
```

If `foo` and `bar` were *R* matrices, then `bar` would be assigned a copy of `foo`. However, since `foo` is a `bigmatrix` object, the assignment causes `bar` to point to the same data as `foo`. This is easily verified with

```
bar[1,1] <- 1
foo
```

```
     [,1] [,2] [,3]
[1,]    1    0    0
[2,]    0    0    0
[3,]    0    0    0
```

The fact that `big.matrix` objects can reference the same data can be extremely useful. By preventing *R* from making copies of a potentially large data set, calculations can be made more efficient, both in terms of memory usage and computing time. However, this functionality comes at a price. Because they do not follow the same copy semantics as *R* matrices, a statistician may have to make small modifications so that `bigmatrix` objects work correctly with existing *R* code.

Q.1 Using the *UNIX* shell, create the *AirlineDataAll.csv* file without using a loop.

Q.2 Write a preprocessing script (using the shell, *R*, *Python* or any tools) to create a file that can be used with `bigmemory`, i.e., convert the non-numeric values to numeric values in some well-defined manner.

Q.3 How many flights were there for each day of the week?

Q.4 For each year, how many flights were there for each day of the week?

Q.5 For each year, how many of the tail codes are listed as NA?

Q.6 Which year had the greatest proportion of late flights? Is this result significant?

Q.7 Which flight day is best for minimizing departure delays? Which time of day?

5.4 Explorations Using Parallel Computing: The Distribution of Flight Delays

Many calculations executing quickly on small data sets take proportionally longer on larger ones. When execution time becomes an issue, parallel computing can be used to reduce the time required by computationally intensive calculations. *R* offers several different packages for executing code in parallel, each relying on different underlying technologies. Because these parallel mechanisms, or *backends*, are different, their configuration and use are also slightly different. As a result, it has traditionally been cumbersome to migrate sequential code to a parallel platform, and even after this migration was successful, the resulting parallel code was usually specific to a single parallel backend. The `foreach` package [11] addresses this issue by providing one approach to standardizing the syntax for describing parallel calculations. The `foreach` package decouples the function calls needed to run code in parallel from the underlying technology executing the code. This approach allows a statistician to write and debug sequential code and then run it in parallel by registering an appropriate package such as multicore [14], snow [13] or nws [12]. Using the `foreach` package, R code can run sequentially on a single machine, in parallel on a single machine, or in parallel on a cluster of machines with no (or very minimal) code changes.

Although parallel computing can dramatically decrease execution time for many calculations, you should be aware that there are limitations. There are even cases where parallelization can increase execution time, not decrease it. When deciding whether or not to parallelize code there are a few things to keep in mind. First, there is some additional overhead associated with executing any parallel code (e.g., launching the worker processes, copying data to them, collecting the results). As a result, code should only be parallelized when each of the tasks run in parallel takes a sufficient amount of time to compute so that this overhead becomes a negligible part of the overall computational time. Second, you can expect a speed-up that is at most linear in the number of processor cores. If a snippet of code takes t seconds to execute on a single core and it is run in parallel on two cores, it will take more than $t/2$ seconds to execute. Speed gains often diminish as the number of cores being utilized increases. Finally, each R object in the main R session that is used in the parallel computations is typically copied to each of the parallel processes. If these copies cannot all be stored in RAM, then there will be significant overhead as the operating system uses the hard drive to manage these copied data structures.

Now that you are aware of the issues with parallel computing, we are going to explore the process of creating parallel code. The next subsection discusses the process writing code that can be executed in parallel. The following section presents an approach for optimizing potentially parallel code called *split-apply-combine*. The third and final subsection explores the use of the `foreach` and `bigmemory` packages to perform more sophisticated analyses with the airline data set in parallel.

5.4.1 Writing a Parallelizable Loop with `foreach`

Let's go back to the question from the last section's exercises, "For each day of the week, how many flights are recorded?" To compute the solution, we could use a `big.matrix` object to store the airline data and a `for` loop to iterate over each day, finding the number of flights, e.g.,

```
x <- attach.big.matrix("airline.desc")
dayCount = integer(7)
for (i in 1:7)
  dayCount[i] <- sum(x[,"DayOfWeek"] == i)

dayCount
[1] 18136111 18061938 18103222 18083800 18091338
[5] 15915382 17143178
```

You may notice that computing the number of Monday flights is completely independent of computing the number of Tuesday flights. Wouldn't it be nice if we could take advantage of the multiple cores in a machine to calculate day of the week counts at the same time? Well, we can; and loops like this, where each iteration is independent of other iterations, are so easy to execute in parallel that they are sometimes called "embarrassingly parallel."

It is important to understand that not all loops are embarrassingly parallel and some calculations must be run sequentially. A single Markov chain simulation is generally impossible to run in parallel. As an example, consider the following random walk on the integers with an initial state of zero, implemented with

```
state <- numeric(10)
for (i in 2:10)
  state[i] <- state[i - 1] + sample( c(-1, 1), 1 )
state

[1] 0 1 0 1 2 3 2 3 4 5 6
```

For the value of `state` at time `i` to be calculated, the value of `state` at time `i - 1` must be known. Information must be shared across loop iterations to perform the simulation, and as a result, iterations of the loop must be computed sequentially, not in parallel.

Getting back to the question at hand, let's start with a sequential solution to the problem of tallying flights in a given day. Unlike the previous implementation, let's use the `foreach` [11] package. This package allows us to define embarrassingly parallel loops either sequentially or in parallel. The new code uses the foreach() function and the previously created `big.matrix` object:

```
library(foreach)
dayCount <- foreach(i = 1:7, .combine=c) %do% {
                                    sum(x[,"DayOfWeek"] == i)
                                  }
```

Like the previous code example, loop iterations are indexed by an integer `i` going from 1 to 7. In each iteration of the loop, the number of times `DayOfWeek` is identical to the loop counter is calculated. Unlike the previous example, the calculated value is simply returned, not appended to the `dayCount` variable. The *.combine* parameter tells foreach() to combine the results from each iteration of the loop into a vector, which is returned and stored in the `dayCount` variable. You should also notice that after the foreach() statement, there is a `%do%` operator that tells the function to perform each loop iteration *sequentially*. (We'll use `%dopar%` later to perform the loop in parallel.) The result of this example and the previous one are the same; `dayCount` holds the number of flights recorded for each day of the week in a vector of numeric values.

Both the `for` and foreach() loop in this subsection process the entire `DayOfWeek` column 7 times to extract the number of delays for each day. For small data sets, each of these passes happen very quickly and the corresponding delay may go unnoticed. However, as the number of rows in the data set grows, each extraction requires more time, eventually delaying the exploration. This delay would be even more pronounced if we were finding the delay count for each day of the month. A day of the month count would require 31 separate passes through the data set and would take more than 4 times longer than finding the delay count for the day of the week. Wouldn't it be nice if we could perform these calculations while only passing through the data once to get the rows of interest and once to perform the calculation? In the next subsection, we will explore a different approach that only requires a single pass through the data offering significant performance gains.

5.4.2 Using the Split-Apply-Combine Approach for Better Performance

The task of counting the number of delays by the day of the week can be recast into separating all of the observations into 7 groups, one for each day of the week and then counting the number in each group. We can do this generally for records using the split() function, which passes through the data once. The split() function returns a named list. The names of the list, in our case, correspond to the day of the week. For each of day of the week, the list contains a vector of indices corresponding to the rows for that day:

```
        # Split the rows of x by days of the week.
dow <- split(1:nrow(x), x[,"DayOfWeek"])
        # Rename the names of dow
names(dow) <- c("Mon", "Tue", "Wed", "Thu", "Fri", "Sat", "Sun")
        # Get the first 6 rows corresponding to Monday flights.
dow$Mon[1:6]
```

```
[1]  5 11 19 26 32 38
```

Now that we have the row numbers for each day of the week, we can write a foreach() loop to get the counts for each day. Since all of the information to get the delay counts is contained in the dow variable, we don't need to include the data set in the calculation.

```
dayCount <- foreach(dayInds = dow, .combine = c) %do% {
                                          length(dayInds)
                                          }
dayCount
```

```
[1] 18136111 18061938 18103222 18083800 18091338
[6] 15915382 17143178
```

There are many calculations, like this, that can be accomplished by grouping data (called the *split*), performing a single calculation on each group (the *apply*), and returning the results in a specified format (the *combine*). The term "split-apply-combine" was coined by Hadley Wickham [18] but the approach has been available in a number of different computing environments for some time under different names.

There are several advantages to the split-apply-combine approach over a traditional for loop. First, as already mentioned, split-apply-combine is computationally efficient. It only requires two passes through the data: one to create the groups and one to perform the calculation. Admittedly, storing the groups from the split step requires extra memory. However, this overhead is usually manageable. In contrast, a for loop makes a costly pass through the data for each group, which can be significant if the number of groups is large. Second, calculations that can be expressed within the split-apply-combine framework are guaranteed to be embarrassingly parallel. Because the split defines groups on the rows of a data set, calculations for each group are guaranteed to be independent. When the calculation being applied to a group is intensive, as in the next section, parallel computing can dramatically reduce execution time.

5.4.3 Using Split-Apply-Combine to Find the Best Time to Fly

Now that we have gained some familiarity with foreach() and parallel computing, let's move on to the more difficult question, "Which is the best hour of the day to fly to minimize departure delays?" This was originally posed as one of the ASA Data Expo challenges and the solution lends itself to the split-apply-combine approach.

To answer this question, we need to start by determining what is meant by the "best" hour. About half of the flights in the data set do not have departure delays. Of the flights with departure delays, most are only a few minutes late. Is the best hour the one minimizing the chance of any delay? Is the best hour the one that tends to have fewer long delays? In this situation, we have all of the flights, not a sample. Accordingly, we do not need to perform hypothesis tests that take sampling variability of the statistics into account. Instead, we can examine the population distributions directly from the data.

Let's turn our attention to the worst delays. The question can be refined by asking, "How long were the longest 1% of flight delays for a given hour?" A frequent flyer might interpret this as, "How long could my longest delay be for 99% of my flights." Along with finding the longest 1% of departure delays, let's find the longest 0.1%, 0.01%, 0.001%. That is, let's find the quantile values for probabilities of 0.9, 0.99, 0.999, and 0.9999 in the departure delays.

To find the quantiles, we will start by splitting the data based on the hour of departure for each flight. Since there is no column that gives us the hour of departure for a given flight we will need to extract this information from the CRSDepTime column that encodes times, such as 8:30 AM as 830. The hour of departure can be calculated with

```
# Divide CRSDepTime by 100 and take the floor to
# get the departure hour.
depHours <- floor(x[,"CRSDepTime"]/100)
# Set the departure hours listed as 24 to 0.
depHours[depHours==24] <- 0
```

Now that we have a vector holding the departure hours for each flight, we can split on it and calculate the desired quantiles:

```
# Split on the hours.
hourInds <- split(1:length(depHours), depHours)

# Create a variable to hold the quantile probabilities.
myProbs <- c(0.9, 0.99, 0.999, 0.9999)

# Use foreach to find the quantiles for each hour.
delayQuantiles <- foreach( hour = hourInds, .combine=cbind) %do% {
                    require(bigmemory)
                    x <- attach.big.matrix("airline.desc")
                    quantile(x[hour, "DepDelay"], myProbs,
                                          na.rm = TRUE)
                  }

# Clean up the column names.
colnames(delayQuantiles) <- names(hourInds)
```

You may have noticed that for each iteration of the foreach() loop, we are ensuring that the bigmemory package is loaded and we are are calling attach.big.matrix(). These steps are not required when the loop is run sequentially but we will see later that they are required when the loop is run in parallel.

Now that we can calculate the delay quantiles sequentially, let's parallelize the code so that it runs faster. We'll start by registering a parallel backend. When this chapter was written, there were 5 different parallel packages that were compatible with foreach: doMC [8], doMPI [15], doRedis [5], doSMP [9], and doSNOW [10]. Each of these packages allow R users to exploit distinct parallel programming technologies. Packages like doMC and doSMP allow R users to take advantage of multiple cores on single machine. The other packages allow R users to create parallel programs for a single machine or even a cluster of machines. The doRedis package even supports programming in the "cloud." For this section, we are going to use the doSNOW package to perform parallel calculations on a single machine using multiple cores.

To run the foreach() loop in parallel, we need to determine the number of parallel R sessions that will be used to perform the calculation. In general, it is a good idea to use the total number of cores on the machine minus one. This allows the extra core to deal with some of the overhead associated with the parallel calculations. After the parallel workers are instantiated, they are registered with foreach. This step informs foreach how the iterations of the loop will be parallelized. Then, we change %do% clause in the foreach() loop to %dopar% in order to let the foreach() function know that code should be executed in parallel. We do all of this with

```
# Load the parallel package so we can find
# how many cores are on the machine.
library(parallel)

# Load our parallel backend.
library(doSNOW)

# Use the total number of cores on the
# machine minus one.
numParallelCores <- max(1, detectCores()-1)

# Create the parallel processes.
cl <- makeCluster(rep("localhost", numParallelCores),
                  type = "SOCK")

# Register the parallel processes with foreach.
registerDoSNOW(cl)

# Run the foreach loop again, this time
# with %dopar% so that it is executed in parallel.
delayQuantiles <- foreach(hour=hourInds, .combine=cbind) %dopar% {
            require(bigmemory)
            x <- attach.big.matrix("airline.desc")
            quantile(x[hour, "DepDelay"], myProbs, na.rm=TRUE)
        }
colnames(delayQuantiles) <- names(hourInds)
stopCluster(cl)
```

When you run this code you should notice that it runs significantly faster than the sequential code (depending on how many cores you have available). By using foreach(), we have reduced the effort needed to migrate between sequential and parallel code.

It is important to understand that when a foreach() loop is run in parallel each iteration of the loop is run in a separate R session in another process, sometimes referred to as a worker process. Variables used inside the loop are copied from the master R session to each of the worker sessions. This presents two challenges when computing with a big.matrix in parallel. First, the bigmemory package is not automatically loaded in each of the worker sessions when they are started. However, this issue is easily remedied by requiring that the bigmemory package is loaded in the worker, before a calculation begins. Second, a big.matrix object holds a pointer to a location in memory that is only valid in the process where the pointer is created. As a result, the mechanism foreach() uses to copy variables from master to worker sessions doesn't work for a big.matrix(). This issue is also easily remedied by using the attach.big.matrix() function in the worker process after the bigmemory package has been loaded and before the calculation.

Now that we have efficiently calculated the airline quantile delays we can visualize these delays using `ggplot2`:

```
library(ggplot2)
dq <- melt(delayQuantiles)
names(dq) <- c("percentile", "hour", "delay")
qplot(hour, delay, data = dq, color = percentile, geom = "line")
```

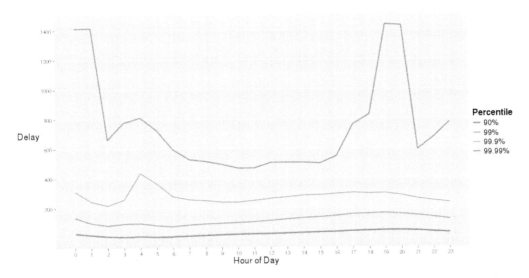

Figure 5.1: Hourly Delay Quantiles. *The airline delay quantiles (in minutes) for each hour of the day.*

Figure 5.1 shows the delays by hour. The graph indicates that delays are worse in the early hours of the morning, for these quantiles. These may correspond to red-eye flights that are delayed arriving at their destination and therefore cause delays for subsequent flights. The graph also indicates that flights leaving between 6:00 AM and 4:00 PM see the fewest lengthy delays.

For the examples in this section, a `big.matrix` object was used to hold the airline data because the data set is too large on many machines for a native R `matrix` or `data.frame` object. A `big.matrix` object manages a large data set by caching needed data and leaving the rest on disk. It has the advantage of using less RAM than what would be needed to hold the entire data set, but this isn't its only advantage. When performing parallel calculations, variables are copied from the master R session to worker R sessions. If a variable uses a large amount of memory, then a worker R session incurs the overhead of waiting for these variables to be copied. Also, each worker must hold a copy of the variable potentially using up all of the available memory on one machine. A `big.matrix` object does not suffer from either of these problems. Descriptors are small and are quickly copied from one process to another. Also, a `big.matrix` object in each worker session is not a copy of the original. Each is a reference to the same data used by the master. These two qualities mean that `bigmemory` provides an efficient solution to computing in parallel with massive data and that we are now able to explore and visualize these data sets more easily than before. The `bigmemory` approach can be used for more than exploration and in the next section we will use `bigmemory` to model data and provide techniques to help answer the question "Do older planes suffer greater delays?"

Q.8 Which is the best day of the week to fly?

Q.9 Which is the best day of the month to fly?

Q.10 Are flights being given more time to reach their destination for later years?

Q.11 Which departure and arrival airport combination is associated with the worst delays?

5.5 From Exploration to Model: Do Older Planes Suffer Greater Delays?

Another question posed by the 2009 Data Expo was "Do older planes suffer greater delays?" The fact that the airline data does not give a plane's age presents a difficulty in answering this. We might search for this information on the Web to see if we can find auxiliary sources of data to determine the age of each plane given its uniquely identifying tail number. However, in the absence of this auxiliary data, we can use the current data to approximate it. The year and month of each flight are available, and we have each plane's unique tail code. For each flight, we can get the number of months the plane has been used since the first time it appears in the data set. This approach does have an issue with censoring: if a plane appears in the first year and month of the data set, we don't know if the plane started service in that month (January, 1988) or sometime before then. Nonetheless, this approach is reasonable given the limited data we have.

How do we calculate the age of a plane? Using the `big.matrix` object from before, which holds the entire data set, we can quickly find that there are 13,536 unique tail codes that appear in the data set:

```
length(unique(x[,"TailNum"]))
```

```
[1] 13536
```

The task of finding the first time a tail code appears (in months A.D.) is independent across tail codes, so we'll split the data by the `TailNum` variable and use foreach() to find this value for each `TailNum` group:

```
planeStart <- foreach(tailInds = tailSplit, .combine=c) %dopar% {
    require(bigmemory)
    x <- attach.big.matrix("airline.desc")

        # Get the first year this tail code appears in the
        # data set.
    minYear <- min(x[tailInds, "Year"], na.rm = TRUE)

        # Get the rows that have the same year.
    minYearTailInds <-
        tailInds[which(x[tailInds, "Year"] == minYear)]

        # The first month this tail code appears is the
        # minimum month for rows indexed by minYearTailInds.
    minMonth <- min(x[minYearTailInds, "Month"], na.rm = TRUE)
```

```
          # Return the first time the tail code appears
          # in months A.D.
    12*minYear + minMonth
}
```

Remember the *extraCols* = *"age"* argument that was specified when the `bigmatrix` object was created in Section 5.3.4 (page 227)? This argument created an extra column named `age` in the `bigmatrix` object and this can be assigned a value with the single R command

```
x[,"age"] <- x[,"Year"] * 12 + x[,"Month"] -
                                  planeStart[x[,"TailNum"]]
```

Now that we have created a variable holding the age of a plane, what should we do with it? One approach to answering the posed question is to create a linear model with arrival delay modeled as a linear function of airplane age to see if there is an association between older planes and larger arrival delays. While the lm() function will not, in general, be able to handle this much data, there is a function, called biglm() in the `biglm` [6] package, designed to perform regressions in this setting. The biglm() function works on subsets of rows at a time so that a linear model can be updated incrementally with new rows of data. A wrapper for this function has been implemented in the `biganalytics` [3] package for creating linear models with `bigmatrix` objects. Regressing arrival delay as a function of age can be implemented as

```
library(biganalytics)
blm <- biglm.big.matrix( ArrDelay ~ age, data = x )
```

and calling `summary(blm)` gives a summary similar to that of an `lm` object:

```
Large data regression model: biglm(formula = formula,
    data = data, ...)
Sample size = 84216580
              Coef    (95%     CI)      SE    p
(Intercept) 6.8339  6.8229  6.8448  0.0055  0
age         0.0127  0.0126  0.0129  0.0001  0
```

The model indicates that older planes are associated with large delays. However, the effect is very small and there may also be effects that are not accounted for in the model. At this point though you have the tools and techniques to begin evaluating these issues, as well as pursuing your own exploration and analysis for the airline data as well as other massive data sets.

Q.12 One of the examples in this section creates a vector, named `planeStart`, which gives the first month in which a plane with a given tail code appears in the data set. Estimate the amount of time the loop to create this vector will take to run sequentially? and in parallel?

Q.13 How many of the planes ages are censored?

Q.14 How much do weather delays contribute to arrival delay?

Q.15 Along with age, which other variables in the airline delay data set contribute to arrival delays?

Bibliography

[1] The *SQ*Lite Web page. http://www.sqlite.org/, 12/9/2009.

[2] David James. RSQLite: SQLite interface for R. http://cran.r-project.org/package=RSQLite, 2011. *R* package version 0.10.0.

[3] Michael J. Kane and John W. Emerson. biganalytics: A library of utilities for big.matrix objects of package bigmemory. http://cran.r-project.org/package=biganalytics, 2010. *R* package version 1.0.14.

[4] Michael J. Kane and John W. Emerson. bigmemory: Manage massive matrices with shared memory and memory-mapped files. http://cran.r-project.org/package=bigmemory, 2011. *R* package version 4.2.11.

[5] B. Lewis. doRedis: Foreach parallel adapter for the rredis package. http://cran.r-project.org/package=doRedis, 2011. *R* package version 1.0.4.

[6] Thomas Lumley. biglm: bounded memory linear and generalized linear models. http://cran.r-project.org/package=biglm, 2011. *R* package version 0.8.

[7] R Development Core Team. *R: A Language and Environment for Statistical Computing.* Vienna, Austria, 2012. http://www.r-project.org.

[8] Revolution Analytics. doMC: Foreach parallel adaptor for the multicore package. http://cran.r-project.org/package=doMC, 2011. *R* package version 1.2.3.

[9] Revolution Analytics. doSMP: Foreach parallel adaptor for the revoIPC package. http://cran.r-project.org/package=doSMP, 2011. *R* package version 1.0-1.

[10] Revolution Analytics. doSNOW: Foreach parallel adaptor for the snow package. http://cran.r-project.org/package=doSNOW, 2011. *R* package version 1.0.5.

[11] Revolution Analytics. foreach: Foreach looping construct for R. http://cran.r-project.org/package=foreach, 2011. *R* package version 1.3.1.

[12] Revolution Computing with support and contributions from Pfizer Inc. nws: bounded memory linear and generalized linear models. http://cran.r-project.org/package=nws, 2010. *R* package version 1.7.0.1.

[13] Luke Tierney, A Rossini, Na Li, and H. Sevcikova. snow: Simple Network of Workstations. http://cran.r-project.org/package=snow, 2011. *R* package version 0.3-8.

[14] Simon Urbanek. multicore: Parallel processing of R code on machines with multiple cores or CPUs. http://cran.r-project.org/package=multicore, 2011. *R* package version 0.1-6.

[15] Steve Weston. doMPI: Foreach parallel adaptor for the Rmpi package. http://cran.r-project.org/package=doMPI, 2010. *R* package version 0.1-5.

[16] Hadley Wickham. Hadley Wickham's Homepage. http://had.co.nz, 19/9/2009.

[17] Hadley Wickham. Airline on-time performance Web page. http://stat-computing.org/dataexpo/2009/, 2009.

[18] Hadley Wickham. The Split-Apply-Combine Strategy for Data Analysis. *Journal of Statistical Software*, 40:1–29, 2011.

Part II

Simulation Studies

6
Pairs Trading

Cari Kaufman
University of California, Berkeley

Duncan Temple Lang
University of California, Davis

CONTENTS

6.1	The Problem	241
	6.1.1 Computational Topics	245
6.2	The Data Format	246
6.3	Reading the Financial Data	247
6.4	Visualizing the Time Series	250
6.5	Finding Opening and Closing Positions	251
	6.5.1 Identifying a Position	251
	6.5.2 Displaying Positions	254
	6.5.3 Finding All Positions	256
	6.5.4 Computing the Profit for a Position	257
	6.5.5 Finding the Optimal Value for k	260
6.6	Simulation Study	263
	6.6.1 Simulating the Stock Price Series	265
	6.6.2 Making stockSim() Faster	273
	Bibliography	276

6.1 The Problem

Finance and financial trading are important elements of all economies. Historically, people used various strategies and hunches to choose when to buy and sell stock and which stocks to trade. Increasingly, people try to use data, algorithms and, importantly, statistical methods to develop more automated strategies for determining what and when to trade in order to increase profits. Over the years, there have been many different approaches and algorithms in this direction, and recent developments illustrate computer trading is increasing rapidly. Some people are even developing statistical/machine learning algorithms that run on network routers to make sub-millisecond trades to exploit time delays other people experience.

In this chapter, we will explore one of the first trading strategies based on computer-intensive analysis of past stock performance. The approach is called pairs trading [3] and was developed at Morgan Stanley in the 1980s. The strategy involves two different stocks, hence the name pairs. The key behind pairs trading is to have two stocks whose prices are positively correlated over time. It is perhaps simplest to understand the idea via an example. The Dow Jones and S&P 500 indices are two "stocks" we can buy and sell. These are not regular stocks such as those offered by companies such as Google, YAHOO, and ATT. Instead, these are actually collections of stocks, but that detail doesn't concern us as the same principle applies for any pair of stock that can be bought and sold. As you can see for the time interval 1990–1995 in Figure 6.1, the prices of these two stocks are highly positively correlated. Figure 6.2 shows the daily ratio of the prices of the two indices over the same time period. The ratio seems to be fluctuating around a stable value, at least for most of the time period.

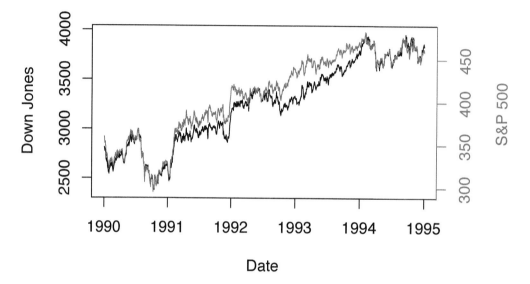

Figure 6.1: Historical Prices for the Dow Jones and S&P 500 Indices, 1990–1995. *This shows the time series of the two "stock" prices for the 5-year period. There is clearly a high correlation over time between the two financial indices. The values of the two series are quite different and plotted on different scales, with that for the S&P 500 on the right of the plot.*

The key idea underlying pairs trading is that the movement of the ratio away from its historical average represents an opportunity to make money. For example, if stock 1 is doing better than it typically does, relative to stock 2, then we should sell stock 1 and buy stock 2. This is called "opening a position." Then, when the ratio returns to its historical average, we should buy stock 1 and sell stock 2. This is called "closing the position." The reasoning is quite simple – when stock 1 is priced sufficiently higher than usual, it is likely to go down in value and the price of stock 2 is likely to go up, at least relative to the price of stock 1, since they are positively correlated. Of course, both could increase, but we are interested in relative change as we are looking at the ratio.

To implement this pairs strategy, we need rules for when we should open and close positions. (We'll assume we can always sell a stock, even if we don't actually own it. The mechanism for this is called "short selling," but the details aren't important here.) We can devise rules based on the current value of the price ratio and what it has been in the past. We'll use historical data to determine the "optimal" rule and then use this rule as a trading strategy for the future.

Pairs Trading

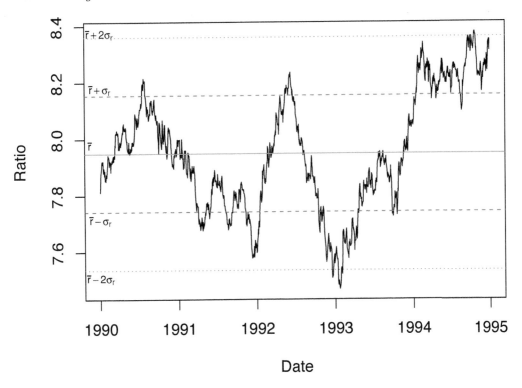

Figure 6.2: *Historical Ratio for the Dow Jones and S&P 500 Indices. This shows the ratio of the two "stock" prices from 1990 to 1995. The ratio appears to move around the mean until 1994 and then to rise above it. The horizontal lines show one and two standard deviations from the mean of the ratio.*

We'll refer to the mean and standard deviation of the ratio for this "training"/historical data as m and s, respectively. We'll consider the following class of rules:

- When the ratio of stock 1's price over stock 2's price moves above k standard deviations from the long term mean (i.e., $r > m + ks$), sell \$1 worth of stock 1 and buy \$1 worth of stock 2 (for simplicity, we are assuming we can buy fractions of a share). Then we wait until the ratio is less than or equal to m, at which point, we close the position by buying back however many shares of stock 1 we initially sold, and selling the shares of stock 2 we initially bought.

- Similarly, when the ratio is less than $m - ks$, do the same thing but reversing the roles of stock 1 and stock 2. In this case we wait until the ratio rises back up to be greater than or equal to m.

There are two things to note. First, k represents how extreme the ratio needs to be before we open a position. Second, \$1 is just an arbitrary amount to invest. Buying a fixed dollar amount, rather than a fixed number of shares, allows us to work with stocks that have very different prices, and also to deal with fractions of stocks.

Now let's see what happens if we implement this strategy with the Dow Jones and S&P 500 prices, going forward from 1995 with $k = 2$. Figure 6.3 shows the ratios for 1995 to 2010, with a horizontal line displaying the threshold of two standard deviations above and below the mean displayed on the plot. It is important to recognize that we are using the mean and standard deviation from the previous period 1990 to 1995. We are basing our trading strategy on historical values and so use these to determine the thresholds for our trading rules. Also, DJI and GSPC are the shorthand "ticker" names for Dow Jones and S&P 500.

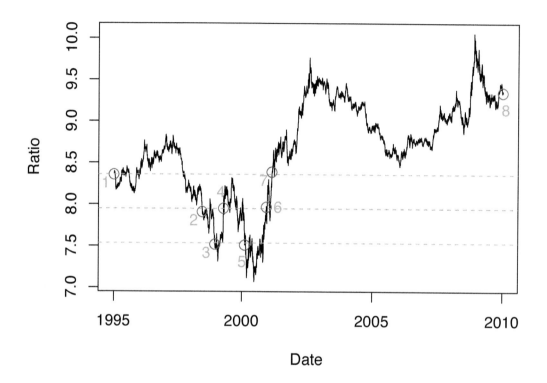

Figure 6.3: Ratio of Dow Jones to S&P 500 for 1995–2010. The circles show the starting and closing positions for trades. We open a position when the ratio is outside of the threshold lines. We close that position when the ratio returns to the mean. The green circles identify the opening of a trading position; the red circles the corresponding close of the position. Circle 7 opens a position but the ratio never returns to the mean. We close the position on the final day of the series. Note that the threshold lines use the mean and standard deviation from the period 1990 to 1995.

In this time series, we start with the value of the ratio already above our threshold, on the first trading day of 1995 (circle 1 on the plot). DJI is at $3838.48 and GSPC is at $459.11, so we sell $1's worth of the DJI corresponding to $1/3838.48 = 0.00026$ units, and we buy $1/459.11 = 0.0022$ units of GSPC. Now we wait until the ratio reverts to the mean, which happens on June 19, 1998 (circle 2 on the plot). At this point both indices have gone up: DJI is at $8712.87 and GSPC is at $1100.65. We close the position, meaning we buy back 0.00026 units of DJI for $2.27 and sell 0.0022 units of GSPC for $2.40. Our total profit is therefore ($1 − $2.27) + ($2.40 − $1) = $0.13. Note that we always subtract the buying price from the selling price, but we bought and sold in different orders for the two indices. We lost money on DJI but earned a bit more than that on GSPC. Pairs trading often works this way, with gains in one stock offsetting losses in the other. Also, note that to get the

Pairs Trading

profit for a different investment amount, you just multiply 0.13 by that amount. Therefore, using $1 as our baseline amount allows us to calculate the *percent* profit.

We have additional opportunities to trade: opening at point 3 and closing at point 4, then opening at point 5 and closing at point 6. Doing this, we end up with a cumulative/total profit at point 6 of $0.25. However, look at what happens beyond point 7. According to our rule, we open a position at 7, but the ratio does not revert back to the mean. We wait until the end of the entire time period and then close the position, even though the ratio hasn't reverted back to the mean by this point in time. We end up losing about $0.11 on that final trade, eradicating a significant portion of our profit.

In our example, we arbitrarily used the value $k = 2$. However, the "optimal" (i.e., to maximize earnings) value of k depends on many things. We want to "estimate" or determine the optimal value based on the historical data for our pair of stocks. For this, we use the data from previous years to determine what is the best value of k for the future. That is, we use the old data as "training" data.

One thing we haven't considered yet is that it costs money to make a trade. These costs can vary depending on whether you're an institutional trader or an individual. However, to keep things simple, let's assume that the costs are a fixed proportion p of the total money changing hands (in absolute value). For example, to buy $30 worth of stock and sell $100 worth of stock, we would pay $\$130 \times p$.

In this chapter, we'll develop functions to explore pairs trading. We start by retrieving and reading stock price data from the Web and exploring the time series of the prices and their ratios. We then develop functions to determine the start and end of a position and then all positions. In order to validate the code and understand the pairs trading approach, we define functions to visualize the positions. We also write a function to compute the profit for a position and then another for the entire period with multiple positions. We then use these functions to determine the optimal parameter value (for k) for pairs trading for two stocks. We divide the data into a training period, and the remainder as a test period. We use the former to determine the optimal value and use that value for the test data. Finally, we move from studying actual stock prices to simulating data from a mathematical model in an effort to understand the impact of within- and between-correlation for the two stock price time series on profit when using pairs trading. This is computationally intensive so we also attempt to improve the run-time speed of the functions we developed earlier and also for simulating the time series.

6.1.1 Computational Topics

This case study ranges from simple access to data directly from the Internet within R [2], to working with dates, writing fast code, simulating random processes, and "estimating" optimal parameter values for an objective function using a grid search. These include:

- Designing, writing and testing small, reusable functions.
- Developing efficient code.
- Using vectorized operations.
- Profiling code and finding the bottlenecks.
- Writing and interfacing to C code for speed.
- Numerical optimization by grid search.
- Dividing data into training and test subsets to determine optimal values.

- Simulation of a stochastic process.

6.2 The Data Format

Pairs trading works best with two stocks that are reasonably correlated. The common example is companies that are in the same business sector/category such as ATT and Verizon, or Google and YAHOO. When demand for more telecommunications services increases, both companies do well. However, they are also competitors. Companies whose products/services complement each other are also correlated, e.g., hard-drive manufacturers and computer/device vendors. As demand for devices grows, so too does the demand for storage.

We can select any two stock histories. We can use companies we know. Alternatively, we can explore many companies at http://finance.yahoo.com/stock-center/ or similar sites. For a given company name, we can retrieve its symbol name via the page http://finance.yahoo.com/lookup.

We can retrieve the historical data for a given company symbol via the Web page http://finance.yahoo.com/q/hp. In the form in the middle of that page, we enter the symbol name and this displays the data. That brings us to, e.g., http://finance.yahoo.com/q?s=GOOG, i.e., the same basic URL but with the symbol added as an argument to the request. On the left hand side of this page, there is a navigation bar with many links. One is entitled Historical Prices. When we click on this, we are brought to the page with rows of data corresponding to the different days. There we can enter the start and end dates of interest. However, the YAHOO site presents the resulting data in a sequence of *HTML* pages. We would have to extract the data from the first page, perhaps with readHTMLTable(), and then navigate to the next page and so on. Fortunately, there is a simpler, more direct mechanism.

When we enter the start and end dates, the resulting page provides a link to the raw data near the bottom of the initial page of historical prices for this symbol. Scrolling down to the bottom of the page, we see the link Download to Spreadsheet and we can explore that link. For Google, it is http://real-chart.finance.yahoo.com/table.csv?s=GOOG&a=02&b=27&c=2000&d=07&e=1&f=2014&g=d. This URL corresponds to an *HTTP* **GET** request that we send to the URL http://real-chart.finance.yahoo.com/table.csv along with named arguments. We can see the symbol name (GOOG) passed as the value for the argument named s. Similarly, we see there are 8 other named arguments. These are cryptically named a, b, c, d, e, f, and g. With minimal detective work, we see that a, b, and c specify the month, day, and year for the start date of interest, and d, e, and f specify the same for the end date. The argument g indicates we want daily prices, not monthly. We can download the data directly from within *R* and save the contents to a local file with, e.g.,

```
u = paste("http://real-chart.finance.yahoo.com/table.csv",
          "s=GOOG&a=02&b=27&c=2010&d=09&e=24&f=2014&g=d",
          sep = "?")
download.file(u, "GOOG.csv")
```

Then we can read this into *R* with read.csv().

Q.1 Instead of using download.file() and copying the data to a local file, we can use getForm() in the **RCurl** package to fetch the data directly into an R session. We can then pass the resulting content to read.csv() using a textConnection(). This approach is sometimes necessary when we need more control over how the request is sent to get the Web page. For example, some Web sites require a login or some form of authentication. Write the code to retrieve and read the data using this approach.

Once we manage to get the raw data, we can examine it to determine its structure. The first two rows (and header) of the Google data appear as

```
Date,Open,High,Low,Close,Volume,Adj Close
2013-11-07,1022.61,1023.93,1007.64,1007.95,1679600,1007.95
2013-11-06,1025.60,1027.00,1015.37,1022.75,912900,1022.75
```

Requests for different stock symbols will yield the same basic structure. For each record, we have the date and details about the price and the number of shares traded that day. There are various different prices reported (opening and closing, minimum and maximum during the day). We are interested in the adjusted closing price (Adj Close). This takes into account other aspects related to the stock that occur before the start of the next day's trading. These also include events such as stock splits and payment of dividends. See [1].

6.3 Reading the Financial Data

The first thing we do is write a function to read the data for a stock from a CSV file. Why bother? Well, we need to read data for at least 2 stocks and we don't want to repeat the code for each (i.e., the DRY principle). We want to handle reading the data into an R data frame and correctly representing the Date column as a Date object in R. This allows us to easily re-run the code from the very beginning of the computations, e.g., when we get new data for the same stocks or for new and different choices of stocks. Let's create a function readData() that takes the name of the CSV file and optionally a string or vector of string values specifying the format of the Date values, e.g., month/day/year. We'll make our function smart enough, by default, to try different common date formats, using each format successively until all of the values are valid dates, not NA (or we have tried all date formats). This may not be necessary if all of the data come in the same format, i.e., both the number of columns and the format of the Date values.

The function is quite simple. We read the data using read.csv(), ensuring that the Date values remain as strings and are not interpreted as a *factor*. Then we loop over the different common date format strings attempting to convert any date values that have not already been converted to Date values. Finally, we arrange for the observations to be ordered in increasing order of the date. We define the function with

```
readData =
  #
  # A function to read the data and convert the Date column
  # to an object of class Date.
  # The date values are expected to be in a column named Date.
  # We may want to relax this and allow the caller specify the
  # column - by name or index.
  function(fileName, dateFormat = c("%Y-%m-%d", "%Y/%m/%d"), ...)
```

```
{
   data = read.csv(fileName, header = TRUE,
                   stringsAsFactors = FALSE, ...)
   for(fmt in dateFormat) {
      tmp = as.Date(data$Date, fmt)
      if(all(!is.na(tmp))) {
         data$Date = tmp
         break
      }
   }

   data[ order(data$Date), ]
}
```

We use *stringsAsFactors* to keep the date values as strings rather than allowing R to convert them to a *factor* variable.

Q.2 What is the purpose of the ... in the function definition?

Now that we have a function to read our data and have manually downloaded the CSV files for the two companies, let's read the values for two stocks, ATT and Verizon:

```
att = readData("ATT.CSV")          # ATT symbol
verizon = readData("VERIZON.CSV")  # VZ symbol
```

As we mentioned in Q.1 (page 246), we can fetch the data directly from the Web site without having to manually download it. This reduces our labor but, more importantly, also avoids any mistakes by downloading the generically named *table.csv* file and changing its name to the corresponding stock symbol, e.g., VERIZON, GOOG. This will tightly couple retrieving the data by symbol name and assigning it to a variable in R, reducing the number of manual steps that might go awry.

Before we proceed, we have to restrict our attention to the common period of time for which we have prices for both stock. We need to compute the days in common between the two data sets. We can define a function that computes the subsets with common dates and returns a data frame with records/rows for each day with the adjusted closing prices for the two stock. We'll pass the two data frames from the calls to readData() as the inputs. Again, we want a function to do this rather than having to type the raw code for each data set. We would also have to adjust that code to use different variable names to avoid overwriting the variables. Using a function makes the computations local and allow us to easily assign the result of different calls to different variables.

Our function computes the set of dates that are common to both stocks and then the earliest and latest of these common dates. Then we subset the two stocks and create the data frame of common dates and pairs of stock values. We do this with the function definition

```
combine2Stocks =
function(a, b, stockNames = c(deparse(substitute(a)),
                              deparse(substitute(b))))
{
  rr = range(intersect(a$Date, b$Date))
  a.sub = a[ a$Date >= rr[1] & a$Date <= rr[2],]
  b.sub = b[ b$Date >= rr[1] & b$Date <= rr[2],]
```

Pairs Trading

```
    structure(data.frame(a.sub$Date,
                         a.sub$Adj.Close,
                         b.sub$Adj.Close),
              names = c("Date", stockNames))
}
```

Note the use of the default value for the *stockNames* parameter. It uses deparse() and substitute() to obtain the text versions of the argument values for the parameters a and b as they were specified by the caller of our function. For example, if we call combine2Stocks() with the expression `combine2Stocks(att, verizon)`, the default value for *stockNames* will be `c("att", "verizon")`. This is the same approach plot() uses to compute the labels for the horizontal and vertical axes. Since we made *stockNames* a parameter for the function, the caller can also specify the names explicitly to provide more meaningful names for the columns of the resulting data frame, if she desires.

Note also that our function assumes, or takes advantage of, the fact that the stocks will have a price for contiguous trading days in this period. If there were "holes" in either of the stocks, we would find the intersection of the dates in common and then `%in%` to subset the two data frames. Also, our function assumes the records in each data frame are sorted. We could sort them ourselves using order() if this was not the case. Recall that we did this in readData() for this very reason.

Let's test this function:

```
overlap = combine2Stocks(att, verizon)
names(overlap)
```

```
[1] "Date"      "att"       "verizon"
```

```
range(overlap$Date)
```

```
[1] "1984-07-19" "2013-11-07"
```

We can compare this with the dates from the two original separate data frames:

```
range(att$Date)
```

```
[1] "1984-07-19" "2013-11-07"
```

```
range(verizon$Date)
```

```
[1] "1983-11-21" "2013-11-07"
```

So we see that the earlier stock prices for Verizon are omitted and that we have data up to the same time point, November 2013.

We can now create the ratio of the adjusted closing price:

```
r = overlap$att/overlap$verizon
```

We might choose to also put this calculation into our combine2Stocks() function. This would make the function more specialized and potentially add a column we don't actually want. If we plan to use this function in contexts where the ratio is not of interest, we wouldn't want this. So we might add a parameter to combine2Stocks() that allows the caller to indicate whether to add this ratio column or not.

Q.3 Modify combine2Stocks() to accept an additional parameter that controls whether it adds the ratio to the data frame.

6.4 Visualizing the Time Series

We now have the inputs for our trading scheme, namely the prices and the ratio. Next, we want to see where the ratio goes outside of some range or limit that would make us open a position. This visualization doesn't define or implement the calculations for our rule. It merely displays the cutoff points. This will help us to understand the pairs trading strategy and also to debug the code when we develop it. So let's write a function to plot the time series for the ratio and also to draw the upper and lower horizontal lines for our pair trading rule, for a given k. This is very simple as we can just call plot() with the ratio and then abline() for the 3 lines at the mean, mean + k * sd and mean - k * sd. We'll optionally allow the caller to specify the dates to use on the horizontal axis, but also let them omit these if they don't care about the dates explicitly but only about the order. In that case, we just use 1, 2, ... as the values for the x variable in the call to plot(). We define the function as

```
plotRatio =
function(r, k = 1, date = seq(along = r), ...)
{
  plot(date, r, type = "l", ...)
  abline(h = c(mean(r),
               mean(r) + k * sd(r),
               mean(r) - k * sd(r)),
        col = c("darkgreen", rep("red", 2*length(k))),
        lty = "dashed")
}
```

Note again that we allow the caller to pass additional arguments to the plot() call via .... She can use this to specify a title, axis labels, etc.

We can now look at our ratio over time with

```
plotRatio(r, k = .85, overlap$Date, col = "lightgray",
          xlab = "Date", ylab = "Ratio")
```

Figure 6.4 shows the result.

Note that we wrote this function in such a way that allows the caller to pass a vector for k, not just a single value. The function then draws the bounds/lines for multiple trading rules. This is why we wrote the more general and complex expression rep("red", 2*length(k)) in the call to abline() rather than just c("red", "red") or rep("red", 2).

Q.4 In addition to drawing the lines, we might also want to identify them. Add code to display the value of k on each line.

Q.5 Instead of using rep() to repeat the color "red", can we use R's recycling rule and merely specify

```
col = c("darkgreen", "red", "red")
```

and have this work correctly with a vector of values for k?

Q.6 We have used the colors red and dark green to indicate the lines for the average ratio and the two extremes, respectively. Why might the choice of colors green and red be bad? How would we define our function to allow the caller to specify different colors?

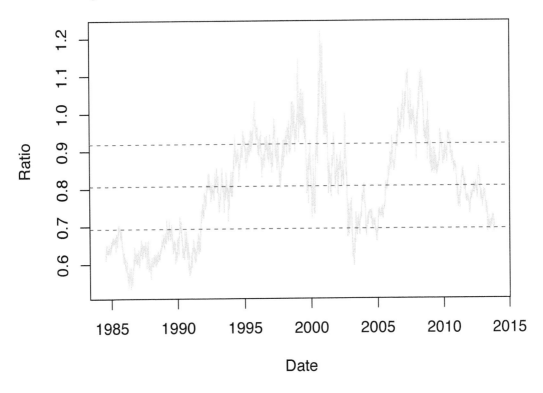

Figure 6.4: A Simple Plot of the Ratio of Stock Prices for ATT and Verizon. *The mean of the ratio (μ_{ratio}) is shown as a dashed line and thresholds lines indicating $\mu_{ratio} \pm k\sigma_{ratio}$, with $k = 0.85$.*

You should draw several plots or lines corresponding to different rules and go through the process of finding the opening and closing positions. It is important to do this manually so that we can write an algorithm to do it. We need to think how the calculations are done.

6.5 Finding Opening and Closing Positions

We are now ready to start writing code to calculate the opening and closing positions. We need to compute all the opening and closing positions for the entire period of interest, but we start by writing a function to identify just one. This is not the first position, but the "next" position given a starting position or day to start from. This will allow us to use the function to find all of the positions.

6.5.1 Identifying a Position

The function takes the entire ratio vector and an optional starting day that is an index into the vector at which we should start the search. To find the first position, we specify this position/index as 1. After we find the first position, we would specify the day of the end of that position as the starting point from which to search for the next position. In this way, we will move across the time series in blocks, starting from where we ended the previous position.

The function needs the value for k to define the extremes, and it can optionally accept the mean and standard deviation (m and s) of the ratio. This allows us to compute these statistics just once and avoid recomputing them for each call. However, this is cheap for even moderately sized vectors.

Our function starts by determining the upper and lower bound above and below which we would open a position. These are the variables up and down, respectively. It then discards all the ratio values up to (but not including) the startDay. The next step is the important one. We compute a logical vector as long as the remaining values in the ratio vector, which tells us if that ratio value is outside the bounds. These are the potential points at which we would open a position. We want the first of these as that is where we will open the next position. If there are no such points, we return an empty integer vector.

Assuming we have identified a starting position, we next need to find the end/close of that position. If we opened with a high ratio (i.e., ratio[start] > up), we need to find the index of the first ratio value after the start day that is at or below the mean m. Alternatively, if we started the position with a low value of the ratio, then we need to find the index of the first ratio that is at or above the mean. We compute this as which(¬ backToNormal)[1], having done the appropriate computation to obtain backToNormal. Of course, there is the possibility that there is no close to the position we opened within the period at which we are looking, as we saw in Figure 6.3. In this case, we use the index of the last element in the ratio. We might also have chosen to represent this explicitly as NA, as this would unambiguously indicate there was no closing position.

We need to do something to compute these start and end indices relative to the original vector. Since we subsetted the original ratio vector, we have to adjust the indices computed on this subset relative to the entire vector. Similarly, we have to adjust the index for the end of the position. In short, we have to add start to the end position index, and also add startDay to each term to keep all the which() calls relative to the same origin. With these additions, our function is defined as

```
findNextPosition =
  # e.g.,   findNextPosition(r)
  #         findNextPosition(r, 1174)
  # Check they are increasing and correctly offset
function(ratio, startDay = 1, k = 1,
         m = mean(ratio), s = sd(ratio))
{
  up = m + k *s
  down = m - k *s

  if(startDay > 1)
     ratio = ratio[ - (1:(startDay-1)) ]

  isExtreme = ratio >= up | ratio <= down

  if(!any(isExtreme))
      return(integer())

  start = which(isExtreme)[1]
  backToNormal = if(ratio[start] > up)
```

```
                    ratio[ - (1:start) ] <= m
            else
                    ratio[ - (1:start) ] >= m

    # return either the end of the position or the index
    # of the end of the vector.
    # Could return NA for not ended, i.e. which(backToNormal)[1]
    # for both cases. But then the caller has to interpret that.

    end = if(any(backToNormal))
                which(backToNormal)[1] + start
          else
                length(ratio)

    c(start, end) + startDay - 1
}
```

Now we have our findNextPosition() function and we can and need to test it. We specify k and then call it:

```
k = .85
a = findNextPosition(r, k = k)
```

This gives us the first position we open and close and we'll assign that to a. The contents of a are 10 and 276, indicating that we open the position on day 10 and close it on day 276.

For the next position, we call findNextPosition() again, but this time specify the starting point as the end of the previous position, i.e., `a[2]`

```
b = findNextPosition(r, a[2], k = k)
```

Similarly, we can find the third position with

```
c = findNextPosition(r, b[2], k = k)
```

Now we can check if these values are correct. How?

Q.7 Why don't we start at `a[2] + 1` rather than the same day we closed the position?

Note that our function findNextPosition() computes whether the ratio exceeds either threshold for *all* of the values in the vector. If we think about this, we are wasting computations. We could find the first index that exceeds the threshold and stop. However, to do that we would need to explicitly loop over the individual daily ratio values to find the first one that exceeds the threshold. Looping in R is significantly less efficient than vectorized operations. The expression `ratio >= up | ratio <= down` is vectorized. Even though it performs unnecessary computations, its speed means that these wasted cycles still result in a significant performance gain relative to looping over the individual values and stopping at the first one.

6.5.2 Displaying Positions

We can examine the start and end of each position at the R console to understand the process and the nature of the positions. This is a good idea and we can use the start and end positions to index into the r vector to examine how the positions work. We can then compare these ratio values to the mean and k * sd. However, another way is to plot the start and end positions of the positions on the plot of the ratio time series and examine them there as we did in Figure 6.3.

We could indicate these days on the plot as circles or lines using the symbols() function, e.g.,

```
symbols(overlap$Date[ a[1] ], r[a[1]], circles = 60,
        fg = "darkgreen", add = TRUE, inches = FALSE)
symbols(overlap$Date[ a[2] ], r[a[2]], circles = 60,
        fg = "red", add = TRUE, inches = FALSE)
```

We don't want to repeat this code twice and we certainly don't want to repeat it for each position we open and close. Instead, we can pass a vector of x and y coordinates in a single call to symbols() to draw two circles for the start and end dates. We can create a function for this as

```
showPosition =
function(days, ratios, radius = 100)
{
  symbols(days, ratios, circles = rep(radius, 2),
          fg = c("darkgreen", "red"), add = TRUE, inches = FALSE)
}
```

days is a vector of length 2 containing the dates of the starting and ending days of the position to be plotted. *ratios* is a vector with the values of the ratios for those two days. We call this function after a call to our plotRatio() function. That displays the time series of the ratio and showPosition() adds the circles to that plot. The *radius* parameter allows the caller to control the size of the circles.

If we call plotRatio() with the Date values for the horizontal axis, we must pass the actual Date values for the open and close dates for our trading position to showPosition(). Alternatively, if our call to plotRatio() uses the indices of the ratio vector (1, 2, ...), then we call showPosition() with the indices for the start and end of our position. In either case, we must pass the values of the ratio for the days.

A different approach to visualizing the start and end of a position is to put vertical lines at the opening and closing dates of that position. This avoids needing to know the value of the ratio. The basic idea of the function is just to add vertical lines at both the start and end positions and to color them green for open and red for close. The caller can change the colors. She can also specify additional arguments to abline(). We can implement this easily with

```
showPosition =
function(pos, col = c("darkgreen", "red"), ...)
{
  if(is.list(pos))
     return(invisible(lapply(pos, showPosition, col = col, ...)))

  abline(v = pos, col = col, ...)
}
```

Pairs Trading

The first two lines of the function handle the case where the caller specifies a list of individual position start and end vectors. This will be convenient when we compute all the positions and have them in a list. This part of the function arranges to loop over the individual start-end vectors and call the function with each vector to add just that line. Another, and possibly better alternative, is to collapse the list to a vector and draw a line for each of the elements in the resulting vector, i.e.

```
abline(unlist(pos), col = col, ...)
```

This is much simpler as we can remove the first two lines. The colors will be recycled across the resulting vector and so will work correctly. As a result, the function is more succinctly written as

```
showPosition =
function(pos, col = c("darkgreen", "red"), ...)
    abline(v = unlist(pos), col = col, ...)
```

Now let's use our original showPosition() to display our start and end points for our 3 positions.

```
plotRatio(r, k, overlap$Date, xlab = "Date", ylab = "Ratio")
showPosition(overlap$Date[a], r[a])
showPosition(overlap$Date[b], r[b])
showPosition(overlap$Date[c], r[c])
```

We see these first 3 positions in Figure 6.5. They appear to be correct, i.e., occurring at the points of intersection of the time series with the horizontal lines on the plot. We'll want to avoid repeating the calls to showPosition() for each position and instead, show all locations in a single call. Before we do this, we'll develop a function to compute all the positions.

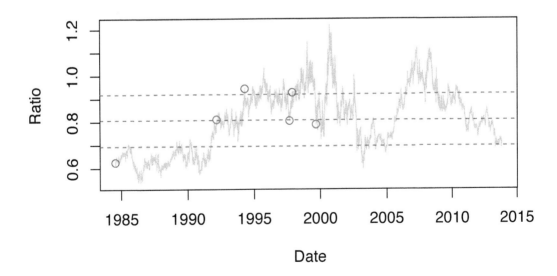

Figure 6.5: Visualizing the First Three Positions. *This shows the first 3 positions for the ATT/VERIZON stock price ratio. It allows us to verify that our findNextPosition() function is working correctly.*

6.5.3 Finding All Positions

To compute all the positions for a time series, we call findNextPosition() in much the same way as we did manually above to compute a, band c. Our function has to continue to call findNextPosition() while there are still more days to process. We can use a `while` loop to iterate over the blocks of the time series until we have processed all days and found all positions. We know there are no positions remaining when findNextPosition() either returns an empty vector to tell us there was no new position, or the position contains either an NA or the last day of the time series as the close of the next position. We collect the individual positions in a list in the variable when. We use cur to store the current day at which we start the search for the next position. We define the function as

```
getPositions =
function(ratio, k = 1, m = mean(ratio), s = sd(ratio))
{
   when = list()
   cur = 1

   while(cur < length(ratio)) {
      tmp = findNextPosition(ratio, cur, k, m, s)
      if(length(tmp) == 0)   # done
         break
      when[[length(when) + 1]] = tmp
      if(is.na(tmp[2]) || tmp[2] == length(ratio))
         break
      cur = tmp[2]
   }

   when
}
```

In each iteration of the loop (except the last), we append the new position to the list when. Ordinarily, appending objects is a bad approach. Instead, we should pre-allocate the vector/list with the appropriate number of elements and then fill these in as we iterate. In this case, we don't know ahead of time how many positions there will be, so we cannot create a list with the correct number of elements. Fortunately, the number of positions is reasonably small, so appending each to the end of the list does not incur significant overhead.

Again, let's check whether the new function behaves correctly. We'll do this by visualizing the positions using our existing functions:

```
pos = getPositions(r, k)
plotRatio(r, k, overlap$Date, xlab = "Date", ylab = "Ratio")
invisible(lapply(pos, function(p)
                      showPosition(overlap$Date[p], r[p])))
```

(The invisible() function just avoids printing the object without assigning it to a variable.)

We might want to verify the function works for a different value of k. Before we do this, however, let's change the definition of showPosition() so that we can pass the entire list of positions and plot them all in one step. We'll keep it so that we can call it with a single position, but also a list of the locations of multiple positions. There are various ways to implement this, e.g., looping over the positions. However, if we unlist the list of position indices, we can use the resulting integer vector to index the ratio. This allows us to keep the code almost the same as before. The primary difference is that we *unlist* the positions and

Pairs Trading

repeat the radius a suitable number of times. We can use R's recycling rule for the colors. Our enhanced function is

```
showPosition =
function(days, ratio, radius = 70)
{
  if(is.list(days))
     days = unlist(days)

  symbols(days, ratio[days],
          circles = rep(radius, length(days)),
          fg = c("darkgreen", "red"),
          add = TRUE, inches = FALSE)
}
```

Now we test the getPositions() function with a different value of k and show the positions with

```
k = .5
pos = getPositions(r, k)
plotRatio(r, k, col = "lightgray", ylab = "ratio")
showPosition(pos, r)
```

The results are shown in Figure 6.6. The smaller value of k leads to more positions being opened and closed.

6.5.4 Computing the Profit for a Position

The start and end day of each position is important. However, we also need to compute the profit for that position, and we need to write a function to do this. Our description of how to do this for the first position for the DJI and GSPC stock in Section 6.1 is quite explicit and now we need to map that into general R code. What we are buying and what we are selling can become somewhat confusing. Accordingly, let's take the time to write the code clearly to make the calculations explicit. Then let's test to make certain it works. The function's name is positionProfit(). It takes the position, which is the vector of the start and end day indices of that position. We can use these indices to get the stock prices for the two stocks. Therefore, we also need to pass these vectors of stock prices to the function as well. These are the parameters `stockPriceA` and `stockPriceB`. We also need the mean of the ratio (so that we can determine which stock we sell and which we buy) and the percentage we pay in commission of the total dollar amount.

Again, we start the function by handling the case that the position is actually a list of position vectors. We just loop over these using sapply() and then return the values. (We'll talk about the parameter `byStock` later.) After this `if` expression, we turn to the real computations for the start-end vector of a single position. Basically, we get the start and end prices for stock A and B by indexing using `pos`. We determine how many units of A and B we can buy for $1 at the start of the position. Then we determine how much that number of units of A and B cost at the end of the position. Given these values, we can now determine the profit for the position. We determine whether, at the opening of the position, we are selling A or selling B. This depends on whether the ratio is high or low, respectively. Then we know the terms for the profit calculation. By default, we return the sum of these profit terms. We define this function with

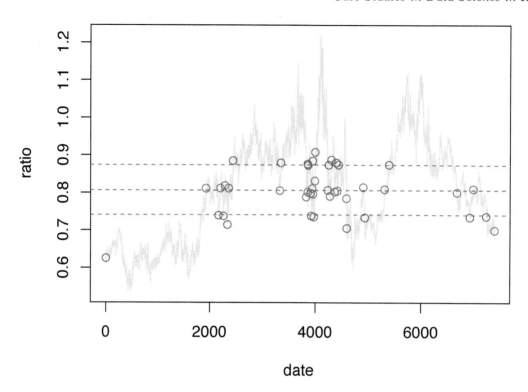

Figure 6.6: Positions for $k = 0.5$. With a smaller value of k, we get many more opening and closing positions.

```
positionProfit =
  #  r = overlap$att/overlap$verizon
  #  k = 1.7
  #  pos = getPositions(r, k)
  #  positionProfit(pos[[1]], overlap$att, overlap$verizon)
function(pos, stockPriceA, stockPriceB,
         ratioMean = mean(stockPriceA/stockPriceB),
         p = .001, byStock = FALSE)
{
  if(is.list(pos)) {
    ans = sapply(pos, positionProfit,
                 stockPriceA, stockPriceB, ratioMean, p, byStock)
    if(byStock)
       rownames(ans) = c("A", "B", "commission")
    return(ans)
  }
    # prices at the start and end of the positions
  priceA = stockPriceA[pos]
  priceB = stockPriceB[pos]

    # how many units can we by of A and B with $1
```

Pairs Trading

```
    unitsOfA = 1/priceA[1]
    unitsOfB = 1/priceB[1]

    # The dollar amount of how many units we would buy of A and B
    # at the cost at the end of the position of each.
    amt = c(unitsOfA * priceA[2], unitsOfB * priceB[2])

    # Which stock are we selling
    sellWhat = if(priceA[1]/priceB[1] > ratioMean) "A" else "B"

    profit = if(sellWhat == "A")
                c((1 - amt[1]),  (amt[2] - 1), - p * sum(amt))
             else
                c( (1 - amt[2]),  (amt[1] - 1), - p * sum(amt))

    if(byStock)
       profit
    else
       sum(profit)
}
```

The `byStock` parameter allows us to return not the total profit, but the individual components of the profit from stocks A and B separately. This might be of interest to us as then we can see the details of the gains and losses for each position.

The positionProfit() function is not very complex, but there are several steps. As a result, we really need to verify it is correct. Our example for the first position for DJI and GSPC works through a particular case and shows the calculations and the result. We'll try our function with that data.

```
pf = positionProfit(c(1, 2), c(3838.48, 8712.87),
                             c(459.11, 1100.65), p = 0)
```

This gives us a value of 0.12748, which we round to cents using `round(pf, 2)`, yielding $0.13 as we calculated previously.

Do we believe our function is working correctly? We can and did check the calculations manually for some of the positions we calculated for our ATT/VERIZON ratio with $k = 0.5$ above:

```
prof = positionProfit(pos, overlap$att, overlap$verizon, mean(r))

 [1]  1.067  0.097  0.108  0.122  0.155  0.174  0.087  0.078  0.088
[10]  0.101  0.119  0.113  0.091  0.091  0.090  0.069  0.179  0.092
[19]  0.137  0.101 -0.056

summary(prof)

    Min. 1st Qu.  Median    Mean 3rd Qu.    Max.
 -0.0559  0.0901  0.1010  0.1480  0.1220  1.0700
```

Look at the individual values for the profit. Do they seem reasonable? The only negative value occurs in the final position, which was closed by rule at the end of the series, not when the ratio returned to the mean. The overall profit is the sum over the individual profits (or losses) for the different positions, i.e., `sum(prof)`. Do we believe this result of 3.1?

6.5.5 Finding the Optimal Value for k

We are now ready to use these functions to find the best value of k from training data and the apply it to test data. We have the basic machinery in place with the functions we have defined. We just need to create the training and test data sets. We have almost 20 years of data for the two stocks. We can look at a particular period for the training data, or we can just split the data in two. We do the latter with

```
i = 1:floor(nrow(overlap)/2)
train = overlap[i, ]
test = overlap[ - i, ]
```

We can now compute the training and test ratio vectors.

```
r.train = train$att/train$verizon
r.test = test$att/test$verizon
```

Or we can subset r itself.

Instead of splitting the data in half, we may want to create the test and training data using specific dates. To do this, we use R's `Date` class. This represents a data as the number of days from some origin, e.g., January 1, 1970, or January 1, 1900. R displays this date in a more human-readable format such as `"1970-1-1"`. Furthermore, R's `Date` class handles all types of details such as leap years, number of days in each month, etc.

To create our training and test data, we can start at the first day we have both stock prices. Then we will create a 5-year period for the training data and the remainder of the data will serve as our test data. We can find the break point for the training and test data with

```
train.period = seq(min(overlap$Date), by = "5 years", length =2)
```

We now have a vector of length 2 of class `Date` containing the start and end dates of our training data. We can use this to subset the stock price vectors and compute the ratio time series for this period:

```
att.train = subset(att, Date >= train.period[1] &
                        Date < train.period[2])$Adj.Close
verizon.train = subset(verizon,
                       Date >= train.period[1] &
                       Date < train.period[2])$Adj.Close
r.train = att.train/verizon.train
```

We can compute the test data set similarly:

```
att.test = subset(att, !(Date >= train.period[1] &
                         Date < train.period[2]))$Adj.Close
verizon.test = subset(verizon,
                      !(Date >= train.period[1] &
                        Date < train.period[2]))$Adj.Close
r.test = att.test/verizon.test
```

We still have to combine the 4 vectors (the date, two stock prices and the ratio) into a data frame for the training and test data.

Regardless of how we split our data into training and test sets, we are ready to compute the overall profit for different values of k on our training data. Let's vary k over 1,000 values.

Pairs Trading

What values should we look between? The largest value we should bother with is the value of k, which gives us upper and lower bounds that lead to the ratio never crossing those lines. Any value greater than that will lead to no positions. We can calculate this maximum value with

```
k.max = max((r.train - mean(r.train))/sd(r.train))
```

Similarly, any value below

```
k.min = min((abs(r.train - mean(r.train))/sd(r.train)))
```

won't make any difference as all values of the ratio will be outside the bounds.

We compute the sequence of values for k that we want to explore with

```
ks = seq(k.min, k.max, length = 1000)
m  = mean(r.train)
```

We have also computed the mean of the ratio here as it doesn't vary and we can avoid recomputing it when we search for each position.

Q.8 Can we search over the different values of k more intelligently? For example, can we find just the set of unique values that will lead to different positions?

Q.9 Are we in danger of overfitting our training data?

Now let's loop over the different values of k and compute the profit for that strategy/rule defined by k:

```
profits =
 sapply(ks,
        function(k) {
            pos = getPositions(r.train, k)
            sum(positionProfit(pos, train$att, train$verizon,
                               mean(r.train)))
        })
```

We have 1000 values of k and 1000 corresponding profits. We can plot these and see the relationship between k and profit with

```
plot(ks, profits, type = "l", xlab = "k", ylab = "Profit")
```

and shown in Figure 6.7. Ideally, this is a simple concave function, or at least shows a single maximum for profit for a value of k.

We can see the maximum value in the plot and we can also calculate the corresponding value(s) of k that yield the maximum profit with

```
ks[ profits == max(profits) ]
[1] 0.610 0.613 0.616 0.619 0.622
```

Ideally, we would like a single value, but this gives us 5 contiguous values in our sequence. These correspond to the flat region of the curve in Figure 6.7 at $k = 0.61$. In some cases, we would explore this interval at greater resolution (in terms of k) to see if there were larger values for profit at other values for k. However, in this case, it is probable that all values of k within this interval lead to the same trading positions and hence the same profit. We can check this with

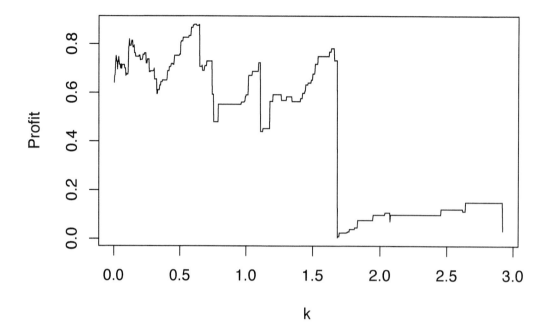

Figure 6.7: Profit for Different Values of k. *For this time training data, all values of k yielded a positive profit. For small values of k, there were many opportunities for positions and the profits were larger than for increasing values of k. The maximum profit of 0.88 occurs at values of k between 0.61 and 0.62.*

```
tmp.k = ks[ profits == max(profits) ]
pos = getPositions(r.train, tmp.k[1])
all(sapply(tmp.k[-1],
           function(k)
              identical(pos, getPositions(r.train, k))))
```

Indeed, all of the positions are the same as the first one.

Which value of k should we use from these 5 values? We could use any of them. Alternatively, we could take the mean or the median of these. We'll use the mean

```
k.star = mean(ks[ profits == max(profits) ] )
```

What is the rate of return, i.e, our profit on $1? We can calculate this directly with `max(profits)` and obtain a value of 88%, which is a very large return on investment (ROI).

Now that we have our "optimal" value of k for our training data (`k.star`), let's see how well we do on our test data using this value. We can compute the positions for our `k.star` using the mean and standard deviation from the training set. Then we compute the overall profits. We do this with

```
pos = getPositions(r.test, k.star, mean(r.train), sd(r.train))
testProfit = sum(positionProfit(pos, test$att, test$verizon))
```

This is our percent profit and yields 0.51 (51%), smaller than from our training data, but still a very good return.

Q.10 Download historical stock price information for many different stocks. Then explore the pairs trading strategy for each pair of stocks. Explore the distribution of profits. Are there characteristics of stocks that seem to lead to larger gains? Does this provide insight into when pairs trading might work or fail?

6.6 Simulation Study

We chose two stocks for which pairs trading yielded a high rate of profit, at least for the period of interest. For a different time period, the results may have been quite different. Similarly, for a different pair of stocks, the profits may also be significantly different, either higher or perhaps a lot lower or even negative, meaning a loss. A very reasonable question an investor should ask when considering whether to use a pairs trading approach is what are the characteristics of the two stocks that yield a high profit margin?

How do we set about addressing such a general question? It refers to all classes of stock time series. We need to make this more concrete and restrict ourselves to one or two specific classes of stock price series. How do we concretize these? It is hopefully natural to consider a mathematical model for a pair of time series. We can write an equation modeling the price of a stock, say $X^{(1)}$, on day t as a function of the price from previous days, i.e.,

$$X_t^{(1)} = f(X_{t-1}^{(1)}, X_{t-2}^{(1)}, \ldots, X_1^{(1)})$$

We could do the same for a second stock $X_t^{(2)} = g(X_{t-1}^{(2)}, X_{t-2}^{(2)}, \ldots, X_1^{(2)})$. However, if these are to be related in some ways, the value of each stock at time t must depend on the prices of both stocks from previous days, i.e.,

$$X_t^{(1)} = f(X_{t-1}^{(1)}, X_{t-1}^{(2)}, X_{t-2}^{(1)}, X_{t-2}^{(2)}, \ldots, X_1^{(1)}, X_1^{(2)})$$

and similarly for $X_t^{(2)}$.

Our equations above make matters somewhat more concrete, but what about the form of the mathematical functions f and g? These can be any meaningful functions. So again, we must be more specific to be able to address the investor's question. Two common simplifications are a) to use linear functions, and b) have the values $X_t^{(1)}$ and $X_t^{(2)}$ at time t depend only on the previous days values, i.e., $X_{t-1}^{(1)}$ and $X_{t-1}^{(2)}$. We might consider the following model, called a vector autoregression,

$$X_t^{(1)} = \rho X_{t-1}^{(1)} + \psi(1-\rho)X_{t-1}^{(2)}$$
$$X_t^{(2)} = \rho X_{t-1}^{(2)} + \psi(1-\rho)X_{t-1}^{(1)}$$

How do we interpret these two equations above and the mechanism for generating new values for time t? As we see, $X_t^{(i)}$ depends on the previous day's values of both stocks and we combine these two values by adding multiples of each together. ρ is a parameter that controls how correlated the price for, say, stock 1 is to the price for that same stock the previous day. If ρ has a value 1, the first equation indicates that $X_t^{(1)} = X_{t-1}^{(1)}$, i.e., each day will have the same value. This isn't a good model, but it does illustrate the role of ρ. If ρ is less than 1, the second term is non-zero and we add some proportion of the value from the second stock price for the previous day, specifically $\psi(1-\rho)X_{t-1}^{(2)}$. If ψ is positive, then larger values of $X_{t-1}^{(2)}$ will contribute more to the current day's price of stock 1. In short,

ρ controls the correlation between daily prices within a single stock, while ψ controls the between-stock correlation for the prices.

The equations define a simple but reasonably flexible model. We can generate various different types of time series by varying ρ and ψ. One important element missing from this model, however, is randomness. We saw that if $\rho = 1$, each day's price is the same as the last. Even for this specific case, we want some random fluctuation. Generally, we want our model to have a random component to have it be more realistic. This may complicate our mathematical analysis, but better address the actual question.

Again, in the interest of simplicity, we will introduce an *additive* random term to our equations:

$$X_t^{(1)} = \rho X_{t-1}^{(1)} + \psi(1-\rho)X_{t-1}^{(2)} + \epsilon_t^{(1)}$$
$$X_t^{(2)} = \rho X_{t-1}^{(2)} + \psi(1-\rho)X_{t-1}^{(1)} + \epsilon_t^{(2)}$$

We need to know the distributions of the $\epsilon_t^{(i)}$ if we are to analyze how pairs trading behaves under stock prices generated from this model. Again, for simplicity, we can use a Normal distribution for each of the error terms, say, $\epsilon_t^{(i)} \sim N(0, \sigma_i^2)$. Each can have a different standard deviation, so our model has 4 parameters: ρ, ψ, σ_1 and σ_2.

Each of these two time series fluctuate around a mean of 0. However, we want to allow the average price to change over time, so we add a trend. Again, we start with a simple model, specifically a linear trend in price. We'll define the actual stock price series as

$$Y_t^{(1)} = \beta_0^{(1)} + \beta_1^{(1)}t + X_t^{(1)} \qquad (6.1)$$
$$Y_t^{(2)} = \beta_0^{(2)} + \beta_1^{(2)}t + X_t^{(2)},$$

Note that we are using $Y_t^{(i)}$ to denote the two stock prices. We now have 4 additional parameters $\beta_0^{(i)}$ and $\beta_1^{(i)}$, for $i = 1, 2$, for a total of 8 separate parameters: ρ, ψ, σ_1, σ_2, $\beta_0^{(1)}$, $\beta_0^{(2)}$, $\beta_1^{(1)}$ and $\beta_1^{(2)}$. We'll denote this 8-dimensional vector as Θ. The linear term is a simple function of the day number, i.e., t. To this, we add the correlated components $X_t^{(i)}$ for each stock. This makes the $Y_t^{(1)}$ and $Y_t^{(2)}$ values correlated with each other and also with the values from the previous days $t-1, t-2, \ldots$.

We now have a concrete mathematical model for two time series. This defines a reasonable class of stock prices and we can use it to address the investor's question about how pairs trading would work for different time series. Ideally, we would be able to compute the optimal value of k and develop a closed-form solution for the profit as a function of the 8 parameters. Since there is randomness in the model, we will have to use our mathematical statistics skills to compute the distribution of the profit as a function of k and the 8 parameters. We may be able to compute the expected value (mean) and standard deviation of the distribution, but it may require approximations for the distribution. We may not be able to obtain a closed-form solution for these results. Even ostensibly simple mathematical models may prove to be inaccessible for mathematical analysis. In these cases, we can use simulation to explore the models and get answers to our questions.

Rather than working with the equations mathematically, we will simulate actual time series from the two equations. We want to understand how our pairs trading "algorithm" or rule performs as we vary the 8 different parameters. This will then help us understand the characteristics of the algorithm and be able to address the investor's question, i.e., how does profit vary with the values of ρ, ψ, σ_1, and σ_2, and the β vector? This will hopefully help us identify good pairs of stock to use together.

How will we use simulation? For a particular vector of the 8 parameters, say Θ, we will generate n observations for the two series $Y_t^{(i)}$ using Equations 6.1 above. We then divide

Pairs Trading

these into a training set and test set. As we did for the ATT and Verizon stock, we use the training set to find the best value of k for our pairs trading. Then we use that value with our test data to evaluate how well we would actually do. The value for profit we obtain is for a particular realization of the two time series. We need many independent replications to estimate the distribution of profit for this particular value of Θ. We may be interested in the average profit, some of the quantiles, the standard deviation, or some other statistics. We repeat this for many different values of Θ to understand the distribution of profit across the parameter space for Θ. This will allow us to understand how these factors change our expected profit and what characteristics of the stock work well for pairs trading, at least for our mathematical model.

6.6.1 Simulating the Stock Price Series

We now know the different steps involved in our simulation study. Our first task is to be able to generate the data for the two stock prices for a given value of Θ, i.e., the 8 parameters. This corresponds to implementing the two equations in Equations 6.1. Since the tth value of each series depends on the $(t-1)^{th}$ value of the time series, a loop is the most obvious way of creating the series. We may have concerns about speed. Instead, we could try using a call to cumsum() but it is difficult to see if this correctly implements the mathematics underlying the model. Let's write the function using a loop and see if it is fast enough.

We can clearly calculate the linear terms in Equation 6.1 using a vectorized approach in R. If we have a vector containing the two β_0 values and similarly a vector for the β_1 values, we can generate the values with

```
beta0[1] + beta1[1] * (1:n)
beta0[2] + beta1[2] * (1:n)
```

We may want to avoid repeating the expression `1:n`, but this is a minor detail.

We need to compute the vectors of length n for the $X_t^{(i)}$ terms that we will add to the $Y_t^{(i)}$ linear terms. We could write this as

```
x1 = x2 = numeric(n)
x1[1] = rnorm(1, 0, sigma[1])
x2[1] = rnorm(1, 0, sigma[2])
for(i in 2:n) {
    x1[i] = rho * x1[i - 1] + psi * (1 - rho) * x2[ i - 1]
                                      + rnorm(1, 0, sigma[1])
    x2[i] = rho * x2[i - 1] + psi * (1 - rho) * x1[ i - 1]
                                      + rnorm(1, 0, sigma[2])
}
```

This is a faithful mapping of the equations to code and is relatively easy to understand and verify. It is reasonably efficient in that it pre-allocates the vectors x1 and x2 and then inserts the values into those (rather than concatenating each value to the end and having to resize the vector in each iteration).

Our code is somewhat repetitive. We treat x1 and x2 separately even though the computations for each are very similar. We also have 4 calls to rnorm() in the code and this corresponds to 2*n calls when executed since two of these are within the loop. Each call to rnorm() generates a single value. We'd like to be able to vectorize these. We can exploit a powerful feature of many of R's random number generation functions. In the call rnorm(10, sd = c(1, 2)), R will generate 10 values. Note the fact that the argument for

sd is a vector of length 2. How does rnorm() interpret this? It will use R's recycling rule to make this a vector of length 10, i.e., c(1, 2, 1, 2, 1, 2, 1, 2, 1, 2). It then uses the i-th value from this vector as the standard deviation when generating the i-th random value. In other words, this call will generate 10 values alternating between a $N(0, 1)$ and a $N(0, 2)$ distribution. We could generate the first two $\epsilon_t^{(i)}$ values as a vector with rnorm(2, sd = sigma). However, we can generate all 2*n values as a matrix with

```
epsilon = matrix(rnorm(2 * n, sd = sigma), ncol = 2, byrow = TRUE)
```

This is significantly faster than the 2*n individual calls to rnorm().

Instead of having 2 separate variables x1 and x2, we could use a 2-column matrix to store the results. This also makes sense since we can return a single object containing the 2 series. We can pre-allocate this with

```
X = matrix(0, n, 2)
```

Then we can set the first day's values with

```
X[1,] = epsilon[1, ]
```

Q.11 In one implementation of our code, we used X = matrix(NA, n, 2) rather than X = matrix(0, n, 2). Why is the second slightly more efficient?

Since we are using a matrix to collect the results, we would write the body of our loop as

```
X[i, 1] = rho * X[i - 1, 1] + psi * (1 - rho) * X[ i - 1, 2]
                                                    + epsilon[i, 1]
X[i, 2] = rho * X[i - 1, 2] + psi * (1 - rho) * X[ i - 1, 1]
                                                    + epsilon[i, 2]
```

We have eliminated the calls to rnorm() within these expressions and we are extracting the relevant elements from epsilon matrix. However, we are still processing individual values (scalars) rather than vectors and the two expressions are very similar. With a little thought and framing the computations slightly differently, we can reduce the two expressions to one line of code and also improve performance. We can use matrix multiplication to compute the first 2 terms of each equation. Specifically,

$$\begin{bmatrix} \rho & \psi(1-\rho) \\ \psi(1-\rho) & \rho \end{bmatrix} \begin{bmatrix} X_{t-1}^{(1)} \\ X_{t-1}^{(2)} \end{bmatrix}$$

yields the vector of length 2 that we want. So we can implement this with

```
A = matrix(c(rho, psi*(1-rho), psi*(1-rho), rho), 2)
for(i in 2:n)
    X[i,] = A %*% X[i-1, ] + epsilon[i,]
```

When we put all of these changes together, we can define our function as

Pairs Trading

```
stockSim =
function(n = 4000, rho = 0.99, psi = 0, sigma = rep(1, 2),
         beta0 = rep(100, 2), beta1 = rep(0, 2),
         epsilon = matrix(rnorm(2*n, sd = sigma),
                          nrow = n, byrow = TRUE))
{
  X = matrix(0, nrow = n, ncol = 2)
  X[1,] = epsilon[1,]

  A = matrix(c(rho, psi*(1-rho), psi*(1-rho), rho), nrow = 2)
  for(i in 2:n)
      X[i,] = A %*% X[i-1,] + epsilon[i,]

         # Add in the trends, in place
  X[,1] = beta0[1] + beta1[1] * (1:n) + X[,1]
  X[,2] = beta0[2] + beta1[2] * (1:n) + X[,2]

  X
}
```

We have specified default values for the 8 different parameters in Θ. Interestingly, we have also allowed the caller to provide values for the epsilon variable. This allows us to specify the random values explicitly, which is useful for testing and reproducing results.

It is valuable to take the time to implement all 3 versions of our code. These are

1. the simple loop with 4 calls to rnorm(),

2. the vectorized call to rnorm() for all 2*n values, but using 2 expressions to compute the 2 time series, and

3. adapting the approach in b) to use matrix multiplication.

When we time these, we see that the last of these is approximately 3 times faster than the first. Of course, the important question is if each of the implementations is correct.

How do we verify that our function stockSim() is correct? It involves random values, so it is challenging. We can at least control the seed for the random number generator (see set.seed() and .Random.seed) to reproduce the exact same random values in the same calls to the random number generation functions. This may not allow us to guarantee the same sequence of random numbers. However, we allowed the caller of stockSim() to specify the matrix of random values via the *epsilon* parameter. This allows us to provide the same random values across different calls and so remove the randomness. We can also remove all of the randomness from the computations (e.g., with $\sigma_i = 0$) to see if the deterministic values are correct. Similarly, we can use fully correlated values, i.e., values of 1 for rho and psi. We can also specify 0 correlation and see if the results are uncorrelated.

Using the stockSim() function, we can generate stock prices for different values of rho, psi and the vector of betas with

```
a = stockSim(rho = .99, psi = 0)
```

We can display these with

```
matplot(1:nrow(a), a, type = "l", xlab = "Day", ylab = "Y",
        col = c("black", "grey"), lty = "solid")
```

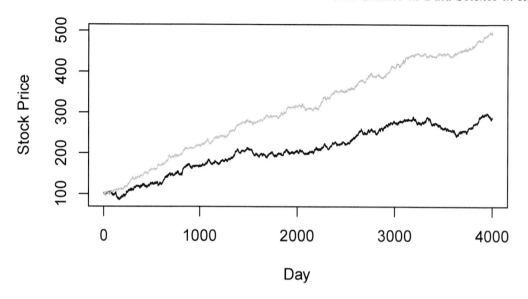

Figure 6.8: Two Simulated Time Series. *The two time series are generated from the model in 6.1 with $\rho = 0.99$ and $\psi = 0.9$ and $\beta_1 = c(.05, .1)$.*

as shown in Figure 6.8.

We can also generate values that have a different linear trend in the price. We'll use `beta1 = c(.05, .1)` for the slopes of our linear trends:

```
a = stockSim(beta1 = c(.05, .1))
```

The two series diverge and as we explore different slopes and intercepts, the series can be on quite different scales. As a result, when we plot them, one can appear to be almost constant as its variations are dominated by the scale of the other series. This merely makes it difficult to visualize them on the same plot with the same vertical scale. It does not affect our computations in any way, especially since we are focused on the ratio. However, it can be useful (and also potentially confusing) to plot the two series on the same plot using two different scales as we did in Figure 6.1.

Q.12 Write a function to plot two series on the same plot using two separate axes, one on the left and one on the right.

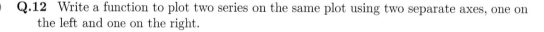

Now let's use our simulation function stockSim() to generate time series data and explore the effect of varying ρ, ψ, $\beta_0^{(i)}$, and $\beta_1^{(i)}$ on the profit rate for pairs trading. For a particular set of values for these 8 parameters, we want many independent and identically distributed values of the profit to provide an estimate for the distribution of the profit. We'll write a function simProfitDist() to generate these observations for a given set of values for the 8 parameters. We can then iterate over the different combinations of the parameter values and pass these to simProfitDist().

The idea for each sample replicate within the simProfitDist() function is as follows.

- Generate two time series using the 8 parameters.

- Split the time series into training and test datasets.
- Use the training data to determine the optimal value of k for the pairs trading strategy.
- Using that value of k, determine the profit for the test data.

We can use stockSim() to generate the sample data in the first step. We'll define two functions later – getBestK() and getProfit.K() – to compute the optimal value of k from the training data, and to calculate the profit for a list of trading positions. If we assume these exist, we can define a new function runSim() that implements steps 1 through 4 above as

```
runSim =
function(rho, psi, beta0 = c(100, 100), beta1 = c(0, 0),
         sigma = c(1, 1), n = 4000)
{
   X = stockSim(n, rho, psi, sigma, beta = beta0, beta1 = beta1)
   train = X[ 1:floor(n/2), ]
   test = X[ (floor(n/2)+1):n, ]
   m = mean(train[, 1]/train[, 2])
   s = sd(train[, 1]/train[, 2])
   k.star = getBestK(train[, 1], train[, 2], m = m, s = s)
   getProfit.K(k.star, test[, 1], test[, 2], m, s)
}
```

Note again that we use the mean and standard deviation from the training data to compute the positions for the test data. We can call runSim() with values for each of the 8 parameters in very much the same way as stockSim(). Indeed, we might consider using ... for the signature of runSim() and pass all of the parameters directly to stockSim().

For runSim() to work, we need to define getBestK() and getProfit.K(). These are closely related to the code we used for computing the profit for our ATT and Verizon stocks in Section 6.5.5. Indeed, we defined a function positionProfit() to compute the profit for one position, given the start and end of the position and the vectors of stock prices. This allows us to define our getProfit.K() function by computing all of the trading positions with getPositions() and then computing the total profit for each position with positionProfit(). Accordingly, we implement our function getProfit.K() with

```
getProfit.K =
function(k, x, y, m = mean(x/y), s = sd(x/y))
{
   pos = getPositions(x/y, k, m = m, s = s)
   if(length(pos) == 0)
      0
   else
      sum(positionProfit(pos, x, y, m))
}
```

Similar to getProfit.K(), our getBestK() function also needs to find the trading positions and compute the overall profit. However, it needs to do this for different possible values of k in order to determine the value of k yielding the highest profit for the training data. Again, we have all the building blocks and merely need to connect them:

```
getBestK =
function(x, y, ks = seq(0.1, max.k, length = N), N = 100,
         max.k = NA, m = mean(x/y), s = sd(x/y))
```

```
{
    if(is.na(max.k)) {
       r = x/y
       max.k = max(r/sd(r))
    }

    pr.k = sapply(ks, getProfit.K, x, y, m = m, s = s)
    median(ks[ pr.k == max(pr.k) ])
}
```

Note that we start the function by computing the default value for $max.k$ if it is not supplied, i.e., if it is NA. We compute this value from the vector of stock prices ratios. Importantly, we compute this before we refer to the argument ks. As a result, if the caller doesn't specify the value for ks, the default expression for ks will use the newly calculated value of $max.k$. This is lazy evaluation. The default value of ks is only computed when it is needed, i.e., first referenced, and at that stage $max.k$ has been calculated, if necessary.

With these two functions defined, we are now in a position to call runSim(). Before we do our simulations over the 8-dimensional grid for our model parameters, we need to verify that all of these functions work correctly. We can use our original ATT and Verizon data and results to test getBestK() and getProfit.K(). These should yield the same results for the training and test data that we obtained directly in Section 6.5.5.

Q.13 How can we test the runSim() function? Implement these tests.

We now return to define our top-level function simProfitDist(). Recall that we intended this to run multiple identically distributed simulations for a given vector of the 8 model parameters and return a collection of profit values. Given runSim(), this is quite easy to implement and may not even be worth having as a separate function. We can define it with

```
simProfitDist =
function(..., B = 999)
      sapply(1:B,  function(i, ...) runSim(...), ...)
```

B is the number of replicates we have in our simulation for the same parameter settings. This controls the variability of our estimate of the average.

... allows us to pass any arguments to runSim(). We use it to avoid having to explicitly copy the parameters and their default values from runSim(). Similarly, if we ever add a parameter to runSim(), we will not also have to add it here. Unfortunately, we cannot use replicate() and simply pass ... in a call to runSim(), e.g.,

```
replicate(B, runSim(...))
```

This does not work as replicate() uses non-standard evaluation and manipulates the expression to create a function. For this reason, we have to use sapply() or a `for` loop. With sapply(), we pass the ... parameter to sapply(), which in turn passes this to our anonymous function. This then calls runSim() and ignores the argument i. This is slightly strange gymnastics, but worth thinking about. It is equivalent to

```
function(..., B = 999)  {
   ans = numeric(B)
   for(i in 1:B)
```

Pairs Trading

```
        ans[i] = runSim(...)
    ans
}
```

which hopefully makes the idea clearer. In fact, we don't need the function simProfitDist() as we can use direct calls such as

```
replicate(B, runSim(rho = .9, psi = .95))
```

Q.14 In addition to the total profit for a simulation, we would like to examine the number of trading positions and the optimal value of k used for trading. Modify the functions to collect this information and explore the results.

With simProfitDist() defined, we can simulate the profit for a particular setting of our parameters:

```
system.time({ x = simProfitDist( .99, .9, c(0, 0)) })
```

On my machine, this took approximately 17 seconds to run for 999 samples or realizations of the simulation. Let's look at the distribution of the profit values:

```
summary(x)
   Min.  1st Qu.  Median    Mean  3rd Qu.   Max.
 -415.0    -0.1     0.0      0.3     0.1   579.0
```

Should we be suspicious of values such as 529 and -415? The former means that our profit was 529%! The density plot in Figure 6.9 shows very large tails and a strong concentration very near 0.

The percentage of times we actually lost money was

```
sum(x < 0)/length(x)
```

This is 48%. However, the median profit is 0. Should we be suspicious of getting values that are exactly 0? Do these come from having no positions? or do the amounts from different positions actually cancel each other exactly?

Now that we have our function simProfitDist(), we can explore the parameter space and the profit rate for different values of our 8 parameters. We'll vary psi from .8 to .99, and the slopes (beta1) from -.01 to .01. If we have 20 different values for each of these parameters, we have $20 \times 20 \times 20$ different combinations:

```
g = expand.grid(psi = seq(.8, .99, length = 20),
                beta1 = seq(-.01, .01, length = 20),
                beta2 = seq(-.01, .01, length = 20))
```

That is 8000 different settings. At 17 seconds each, that is about 2267 minutes. That's a little over $1\frac{1}{2}$ days to run. This does not include varying rho. If we had 20 separate values for it, our run time would be over 30 days. This is why we need to make our code run faster. We can use many computers and run the different simulations in parallel, or we can make our code faster, or both. We'll focus on making the code faster, and leave it as an exercise to use parallel computing for the different parameter value combinations.

To determine how to make the code faster, let's profile this code:

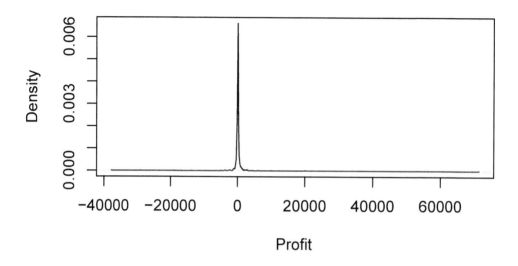

Figure 6.9: Density of Profit. *This shows the distribution for 999 values of profit simulated from our time series model in equation 6.1 with $\rho = 0.99$ and $\psi = 0.9$. We see a large concentration very close to 0. We also see some slightly more commonly occurring values corresponding to the small modes/"bumps." The extreme values are very extreme and also very variable across simulations, as we expect.*

```
Rprof("sim.prof")
system.time({x = simProfitDist( .99, .9, c(0, 0))})
Rprof(NULL)
head(summaryRprof("sim.prof")$by.self)
```

	self.time	self.pct	total.time	total.pct
"stockSim"	8.06	49.81	10.74	66.38
"%*%"	1.02	6.30	1.02	6.30
"FUN"	0.86	5.32	16.16	99.88
"+"	0.70	4.33	0.70	4.33
"findNextPosition"	0.64	3.96	2.20	13.60
".External"	0.58	3.58	0.58	3.58

As we might have expected, stockSim() is the expensive function, accounting for 50% of the total run time. Our findNextPosition() function appears as the fifth most expensive function, but it is only consuming 4% of the time and remember we are calling that a lot more often than stockSim(). The matrix multiplication is the second most expensive function, but its time is a lot less than stockSim() and again, we are calling it even more times than stockSim(). These data indicate that we should focus on improving the stockSim() function.

Before we make stockSim() faster, let's think about how we can count the number of times findNextPosition() and stockSim() are called. This will allow us to better understand the profiling data above. simProfitDist() calls stockSim() for each of the B samples it generates. So this count is simple to determine. How many times findNextPosition() is called

Pairs Trading

depends on the values of the ratio, i.e., the data themselves. If we have one very long position that lasts for 90% of the days, we'll only be looking for the remaining positions in the final 10% of the data. This will, on average, lead to only a few calls to findNextPosition(). Alternatively, if we have many short positions, we will call findNextPosition() many times. The number of positions and their length depends on the value of k. Furthermore, we search over different values of k for the optimal trading rule. The sequence of values of k we search also depends on the data. Accordingly, it is hard to know, a priori, how many times we call findNextPosition(). So we will count the number of calls.

We could modify findNextPosition() to add some code to update a variable each time it is called. However, there is a more general way than modifying the function directly. We can use the trace() function to perform arbitrary actions when a given function is called. In our case, our action will be to increment the value of a variable. We can do this with

```
counter = 0L
trace(findNextPosition, quote( counter <<- counter + 1L),
      print = FALSE)
```

We create the counter variable. The expression `counter <<- counter + 1L` increments this counter. We need the global assignment operator `<<-` to update the counter variable in our work space rather than creating a local value within findNextPosition(). We use quote() in the call to trace() because we don't want R to evaluate this expression as the value of the argument to trace(). Instead, we want this to be an expression that will be evaluated only when findNextPosition() is called in the future. If we omit the *print* argument, we will see thousands of messages of the form

```
Tracing findNextPosition(ratio, cur, k, m, s) on entry
```

Having called trace() with the incrementing expression, we can repeat our simulation

```
system.time({x = simProfitDist( .99, .9, c(0, 0))})
```

and then check the value of counter. This shows that we call findNextPosition() 100,899 times.

If we want to count the number of calls to findNextPosition() in a different simulation, we can reset counter to 0 and then run the simulation. When we want to stop collecting the counts, we use

```
untrace(findNextPosition)
```

Q.15 Rather than using the global variable counter in R's global environment, use a closure, in other words, lexical scoping, to manage a "private" variable that keeps the count of the number of times findNextPosition() is called. You do this by creating a function that returns a list with two functions. One of these functions increments the counter variable. The other function returns the current value of that counter. You can also add a third function that resets the counter. Use the incrementing function with trace() to count the number of calls to findNextPosition().

6.6.2 Making stockSim() Faster

How can we make stockSim() faster? We can think harder and find a faster equivalent algorithm to perform the computations. Alternatively, we can try to compile the code to make it faster. We might use the `compile` package to create byte-code or the very experimental

RLLVMCompile and Rllvm packages to create native machine code. Alternatively, we can write *C/C++* code to make this go quicker.

Let's try the `compiler` package and its cmpFun() function to compile the stockSim() function. We do this with

```
library(compiler)
stockSim.cmp = cmpfun(stockSim)
```

We can see how much this improves the speed of the function by calling the original and the new version several times and comparing the times:

```
tm.orig = system.time({replicate(80, stockSim())})
tm.compiled = system.time({replicate(80, stockSim.cmp())})
tm.orig/tm.compiled
```

```
   user  system elapsed
  1.455   2.000   1.454
```

This shows about a 45% improvement, which is significant. If we used longer time series by specifying a larger value for n in the call to stockSim(), we would see even greater improvements. This improved performance is good, but we may be able to get more with *C* code.

The *C* code below (which you don't have to understand in all its details) is basically the same as the original *R* version of the stockSim() function in which we did not use matrix multiplication, but implemented the two equations 6.1 directly.

C
```c
void
stockSim(double *ans, const int *len, const double *rho,
         const double *psi, const double *eps)
{
  int i, j;
  double psi_rho = (1 - *rho) * (*psi);
  for(i = 1, j = *len + 1; i < *len; i++, j++) {
      ans[i] = *rho * ans[i - 1] + psi_rho * ans[j - 1] + eps[i];
      ans[j] = *rho * ans[j - 1] + psi_rho * ans[i - 1] + eps[j];
  }
}
```

We can call this from *R* in a manner similar to calling an *R* function. We will actually pass an *R* vector (ans) into which the *C* code will insert the results. This is how the *C* code returns its results to *R*.

To use this, we have to compile, link and then load it into our *R* session. We use the shell command

Shell
```
R CMD SHLIB stockSim.c
```

to create the dynamically loadable/shared library (DLL/DSO). Then we dynamically load this into the *R* session with

```
dyn.load("stockSim.so")
```

(On Windows, we use the extension `dll` in place of `so`.) This makes the routine available so that we can invoke it from *R* using the .C() function.

We now write an *R* function that acts as a front-end, or wrapper, for calling this *C* routine from R. We define this much like the R stockSim() function. It has the same parameters as the R function and performs the same basic computations. However, it coerces the inputs to the appropriate type expected by the C code and allocates space for the two time series that the C routine will create. The R code is

```
stockSim.c =
function(n = 4000, rho = 0.99, psi = 0, sigma = rep(1, 2),
         beta0 = rep(100, 2), beta1 = rep(0, 2),
         epsilon = matrix(rnorm(2*n, sd = sigma), nrow = n))
{
  X = matrix(0, nrow = n, ncol = 2)
  X[1,] = epsilon[1,]
  X = .C("stockSim", X, as.integer(n), rho, psi, epsilon)[[1]]

       # Add in the trends
  X[,1] = beta0[1] + beta1[1] * (1:n) + X[,1]
  X[,2] = beta0[2] + beta1[2] * (1:n) + X[,2]

  X
}
```

The final steps in this function add the linear component, which we can do entirely in R very quickly.

We now have an alternative to the R version of stockSim(). We first need to ensure it is correct. One way to do this is check if the two implementations agree. We can test this by naming the two functions differently (as we did) and then passing the same matrix of epsilon values. This is why we made epsilon a parameter of the function. It allows us to control the randomness and hence make it the same in calls to the two functions. We call both versions with

```
e = matrix(rnorm(2*4000, sd = c(1, 1)), , 2)
tmp1 = stockSim.c(epsilon = e)
tmp2 = stockSim(epsilon = e)
identical(tmp1, tmp2)
```

This does indeed return TRUE, which confirms that the two functions yield the same results. We may want to test further.

Now that we believe stockSim() and stockSim.c() behave the same, we can replace the former with the latter with

```
stockSim = stockSim.c
```

This allows us to leave all of our code and functions that call stockSim() unchanged. We can now run our simulation again and profile this:

```
Rprof("sim.prof")
system.time({x = simProfitDist( .99, .9, c(0, 0))})
Rprof(NULL)
head(summaryRprof("sim.prof")$by.self)
```

This takes less than half the time of the original stockSim() function, down to 6.7 seconds. findNextPosition() is now consuming 10% of the time, and so is mean(). Accordingly, we may want to avoid recomputing the mean when we can avoid it. That is why I went back through the code and explicitly added a parameter to getBestK() and getProfit.K() so that we could pass the value and compute it in runSim() just once for a given training data set. This shaves 2 seconds off the run time.

We now have all of the functions to perform our simulations and understand pairs trading in the context of these simple time series models. For simplicity, we'll just look at 20 values for each of ρ and ψ and compute the median profit for 100 repetitions for each combination of these values. We use expand.grid() to compute all pairs of values:

```
p = seq(.5, 1, length = 20)
params = as.matrix(expand.grid(p, p))
```

We can then loop over each row of this matrix and separate the 2 values to call simProfitDist() with

```
profits = apply(params, 1,
                function(p)
                   median(simProfitDist(p[1], p[2], B = 100)))
```

We leave further explorations as an exercise.

Q.16 Our pairs trading approach uses the mean and standard deviation from the training data for the thresholds for the future/test data. We could, however, update the thresholds after we receive the new stock prices, e.g., each day. We can use the same mechanism to determine the optimal value of k with the updated training set, created by adding the new observation to the previous training set. This then changes when we open or close a position in the future. Implement this approach and explore how the profit rate changes.

Q.17 We have used a symmetric trading rule, i.e., $\bar{r} \pm k\sigma_r$. We could use an asymmetric rule with a different deviation from the mean above and below, i.e., $\bar{r} + k_a$ and $\bar{r} - k_b$. Implement a mechanism to find the optimal values for k_a and k_b, and then explore the properties of that trading approach and how it compares to the symmetric approach.

Bibliography

[1] Investopedia. Adjusted Closing Price. http://www.investopedia.com/terms/a/adjusted_closing_price.asp, 2005–2014.

[2] R Development Core Team. *R: A Language and Environment for Statistical Computing.* Vienna, Austria, 2012. http://www.r-project.org.

[3] Wikipedia. Pairs Trading. http://en.wikipedia.org/wiki/Pairs_trade, 2005–2014.

7

Simulation Study of a Branching Process

Deborah Nolan
University of California, Berkeley

Duncan Temple Lang
University of California, Davis

CONTENTS

7.1	Introduction	277
	7.1.1 The Monte Carlo Method	279
	7.1.2 Computational Topics	281
7.2	Exploring the Random Process	281
7.3	Generating Offspring	284
	7.3.1 Checking the Results	286
	7.3.2 Considering Alternative Implementations	287
7.4	Profiling and Improving Our Code	289
7.5	From One Job's Offspring to an Entire Generation	290
7.6	Unit Testing	292
7.7	A Structure for the Function's Return Value	293
7.8	The Family Tree: Simulating the Branching Process	294
7.9	Replicating the Simulation	299
	7.9.1 Analyzing the Simulation Results	301
7.10	Exercises	306
	Bibliography	308

7.1 Introduction

Parallel computing allows us to break up our programs into smaller pieces that can run simultaneously on different CPUs. Sometimes a program spawns another program/job that needs to wait before it can begin its work because it requires the results of another program that has not completed. In this situation even where there are 'infinitely' many CPUs available, a queue of interdependent tasks will form. Tsitsiklis, Papadimitriou, and Humblet [6] studied the behavior of systems of interdependent jobs. One of the questions that interested them was: what is the distribution of the length of time that a computational process, including all of its subtasks, takes to complete? The answer to this kind of question may help developers design code that can run efficiently in a parallel fashion or design queuing systems for managing jobs on a compute cluster. Similar problems arise with, e.g., Web servers with multiple queues and database requests where a request to update information must wait for earlier requests that are working with the same information.

In order to study this problem, Tsitsiklis et. al. proposed a probability model for the generation of jobs and their interdependencies. While it is relatively easy to state the assumptions of their model, answering questions about this random system is mathematically difficult.

Aldous and Krebs [1] considered a slight variant of the branching process proposed by Tsitsiklis et. al. (More recently, Bodenave [2] and Kordzakhia [3] have studied variants of the Aldous and Krebs process.) The Aldous and Krebs process can be viewed as one where:

- The initial job generates other jobs with the times between the start of jobs independent and identically distributed.

- These offspring jobs must all wait until their parent program completes before they can start running.

- Each job can spawn jobs of its own as soon as it is generated, i.e., it need not wait to begin running before it generates a job.

An illustration of this system appears in Figure 7.1. There, the initial job generated 3 jobs. The first of these 3 jobs did not generate any jobs of its own, the second spawned 3 jobs, and the third generated 2, the first of which spawned 2 jobs. In this figure, the system of jobs is still running. The initial job has completed, as have its first 2 child jobs so they are marked with an X. The third child, marked R, is still running so its children and grandchildren are waiting for it to complete before they can begin to run (they are marked W). As for the children of the second subtask, one has completed and the other two are running. Each program that is running or waiting to begin to run (i.e., the R and W nodes in the tree) may generate additional jobs at a time in the future that we have not yet observed. Aldous and Krebs were interested in studying the behavior of the time it takes for the entire system of jobs to finish running, particularly, which combination of job creation and job-completion rates leads to processes that definitely complete in a finite amount of time and which might never complete.

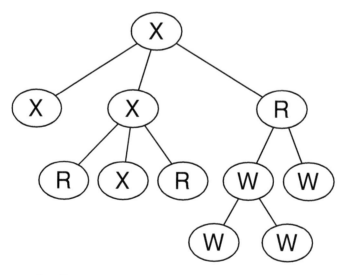

Figure 7.1: Diagram of an Example Branching Process. *This tree shows a possible realization of the stochastic process studied by Aldous and Krebs. Each node in the tree represents a program. Jobs marked with an "X" have completed running, those marked "R" are currently running, and "W" nodes are waiting for their parent to complete before starting to run.*

Both Tsitsiklis et. al. and Aldous and Krebs studied their proposed random processes via mathematics, where they found analytic solutions to features of the process. To derive these results, they considered ways to simplify the more complex problem to one where their analysis would still yield meaningful and potentially useful insights. Simulations can be used as an adjunct to, or substitution for, these closed-form analytic solutions when such solutions do not exist because, e.g., they may be very cumbersome to solve analytically or we might not yet have figured out how to solve them. Also, if we want to study how the behavior of the process changes when the assumed behavior is violated, then simulations can offer insights that might be difficult to obtain analytically. That is, simulation studies attempt to model a random process using the computer in order to provide insights about properties of the process. To do this, we use the Monte Carlo method.

7.1.1 The Monte Carlo Method

Monte Carlo simulation is very simple. To study a random variable, we repeatedly generate independent random outcomes from its probability distribution, and we use the properties of these empirical results as approximations to their expected properties. For example, if we are interested in studying the behavior of the sum of 3 Exponential random variables, then we generate, say, 6 thousand outcomes from this probability distribution, i.e., 6,000 sums of 3 exponential random variables. We do this with

```
empirical = replicate(6000, sum(rexp(n = 3)))
```

If we are interested in the expected value and standard deviation for the sum of 3 exponentials, then we estimate these quantities with the mean and SD of our sample, respectively. That is,

```
mean(empirical)
```

[1] 2.993

```
sd(empirical)
```

[1] 1.716

Of course, we can solve these questions easily analytically, and we know that the expected value is 3 and the SD is $\sqrt{3}$. Our simulated results are close to these values.

It is a bit harder to find analytically, say, the chance that the sum is at most 5. We can estimate this probability using our sample by finding the proportion of observed values that are at most 5, e.g.,

```
sum(empirical <= 5)/length(empirical)
```

[1] 0.874

If we run a larger simulation, then these sample statistics should be even closer to their expected value.

Additionally, we may want to understand how the distribution depends on the rate parameter of the Exponential distribution. In this case, we would select a set of values for the rate, e.g.,

```
rates = c(seq(0.1, 1, by = 0.1), seq(2, 7, by = 1))
```

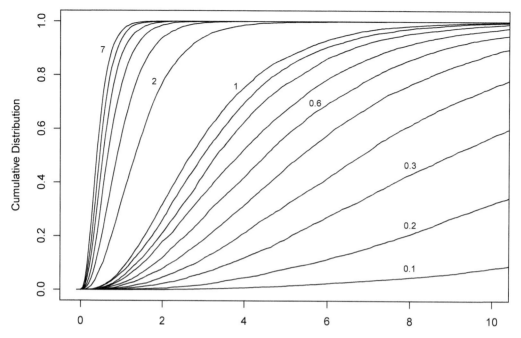

Figure 7.2: Empirical CDFs for the Sum of Three Exponentials. *In a study of the sum of 3 independent Exponential random variables with the same rate, 6,000 sample outcomes were generated and used to estimate the cumulative distribution function. The plot shows the empirical CDF for 16 values of the rate parameter, i.e., 0.1, 0.2, ..., 0.9, 1, 2, ..., 7. To help match each rate to its curve, a few of these rates are included in the figure next to their corresponding curves.*

Then we run the simulation for each value, i.e.,

```
samples = lapply(rates, function(r) {
          replicate(6000, sum(rexp(n = 3, rate = r))) })
```

And, we examine how various properties of this random quantity depend on the rate, e.g., Figure 7.2 shows the empirical cumulative distribution function of these distributions.

The branching process we study in this chapter is more complex than the sum of 3 independent random exponentials, so the Monte Carlo method is extremely useful for examining the behavior of this process. We use features of the samples of this random process to provide insights to the behavior of the process. We simulate the version of the stochastic process proposed by Aldous and Krebs. We wish to find conditions under which the system terminates in finite time. As we do this, we explore how to carry out a Monte Carlo study.

We can think of a simulation study as an experiment for which we design, conduct, and analyze the results. As with in-vivo experiments, we can determine the sample size needed to deliver acceptable precision, use statistical principles to summarize and analyze the results, and bring the principles of experimental design to determine how to study the process, e.g., what parameters to vary and how to vary them. With simulations we need to carefully establish the problem so that our results are reliable and reproducible. For example, testing and debugging code needs to be done carefully because the results of the simulation are not the same from one execution of the simulation to the next. In this chapter we:

Simulation Study of a Branching Process

- Precisely specify the model for the stochastic process. This includes understanding relationships between Poisson, Uniform, and Exponential distributions that underlay the Poisson process.

- Determine how to randomly generate the first generation of jobs, and then how to produce successive generations.

- Ascertain what information about each realization of the process we need to record and how to organize this information into a data structure so we can keep and easily summarize relevant information.

- Design and carry out a comprehensive study of this stochastic process.

7.1.2 Computational Topics

- The Monte Carlo method, simulation, and random number generation using R's [5] built-in probability distribution generators.

- Loops and recursion.

- Modular code and designing code in small testable steps.

- Debugging code via exploratory data analysis of the results.

- Data structures and deciding what representation to use for simulation results.

- Efficiency and profiling code.

- Three-dimensional visualizations and creating custom visualizations of simulation results.

7.2 Exploring the Random Process

As a first step in designing the simulation study, let's explore the process to get a better understanding of how it works. The process begins with a sole job that lasts for a random amount of time. The completion time (often referred to as lifetime) is determined by some probability distribution. Typically the exponential distribution is used to model lifetimes so we follow that approach here. The exponential distribution has a rate parameter, κ, where the density is

$$\kappa \exp^{-\kappa x}, \quad x > 0.$$

As mentioned in Section 7.1.1, the expected value for the exponential distribution can be easily found analytically to be $1/\kappa$.

Let's begin with a rate $\kappa = 0.3$ and generate the first job's lifetime as follows:

```
kappa = 0.3
d0 = rexp(1, rate = kappa)
d0
```

```
[1] 8.384
```

We see that our first job starts at time 0 and completes at time 8.384. Over its life span, the job spawns additional jobs. How do we generate these offspring?

According to Aldous and Krebs, births are to follow a Poisson process so we must figure out how to generate random values from a Poisson process. Note that the Poisson process is related to but different from the Poisson random variable. The process is a probability mechanism for generating random events in time, such as the starting times for jobs. The count of the number of events generated in a fixed time interval is a Poisson random variable, i.e.,

$$\mathbb{P}(k \text{ births in an interval of length } 1) = \frac{\lambda^k k!^{-\lambda}}{e}, \quad k = 0, 1, 2, \ldots$$

A property of the Poisson process that we can use to generate births is that the time between each pair of consecutive events, which is called the inter-arrival time, has an Exponential distribution and these inter-arrival times are independent of one another. This means that we can generate the inter-arrival times from the exponential distribution and piece them together to get the birth/start times of jobs. The initial job starts at time 0 so its first offspring's birth date is the first inter-arrival time. That is, we generate the first offspring's birth date as an exponential random outcome. Then, we generate the time between the birth of the first and second offspring with another independent realization from the exponential distribution, the time between the second and third offspring also has an exponential distribution independent of the others, and so on.

For example, we generate the time of the first offspring's birth using, say, a rate of `lambda = 0.5` as follows:

```
lambda = 0.5
birth1 = rexp(1, rate = lambda)
birth1
```

```
[1] 1.47
```

We generate the inter-arrival time between the first and second offspring with

```
itime = c(birth1, rexp(1, rate = lambda))
itime
```

```
[1] 1.470 4.084
```

This means the second job arrives 4.08 units of time after the first. Similarly, we generate the third inter-arrival time with

```
itime = c(itime, rexp(1, rate = lambda))
itime
```

```
[1] 1.470 4.084 2.052
```

We convert the inter-arrival times into birth times by computing the cumulative sum of these inter-arrival times. We do this with

```
cumsum(itime)
```

```
[1] 1.470 5.554 7.606
```

The birth of the third offspring is earlier than the completion time of the original job (recall the value of `d0` is 8.38) so we generate one more inter-arrival time and determine whether it occurs before `d0`:

```
itime = c(itime, rexp(1, rate = lambda))
btime = cumsum(itime)
btime
[1]  1.470  5.554  7.606 10.868
```

The fourth time occurs after the parent job has finished and, therefore, does not actually happen. This means the first job has only 3 offspring, i.e.,

```
btime = btime[ btime < d0 ]
```

In other words, once we get a time that occurs after the completion of our first job, i.e., after d0, then we stop generating offspring and keep only those times that occur before the parent job completes.

The process proposed by Aldous and Krebs says that once a program is spawned, it must wait until its parent completes before it begins running, and its run time follows the same distribution as its parent's. This means that we can think of each job's lifetime as consisting of two distinct parts.

- The time that the job's parent lives, and it waits to begin running.
- The time that the job lives beyond its parent, when it runs.

This two-part lifetime is why Aldous and Krebs dubbed the process a birth-and-assassination process. The idea is that the children in the process are protected by their parent. When the parent lives, its children cannot 'die,' but once the parent is 'assassinated,' then the children are no longer protected and they now can be assassinated. The time when they are unprotected and can be assassinated is the job's run time.

More formally, we can think of each child as living a certain random amount of time past the death of its parent. This time is treated as an independent increment past the parent's death, and we model it with the same Exponential distribution as the original job's lifetime, i.e., an exponential random variable with rate κ. This means that we can generate the completion times for all 3 of the offspring with one call to rexp(), e.g.,

```
dtime = d0 + rexp(n = length(btime), rate = kappa)
dtime
[1]  9.877  9.205 24.552
```

Notice that in this particular instance, the second-born is the first to die and the last-born lives much later than its siblings.

We could continue in this way to generate the offspring of each of these 3 children, i.e., the third generation of jobs. For example, the birth date of the first child of the original job's first born is:

```
btime[1] + rexp(1, rate = lambda)
[1] 6.234
```

and its time of completion is

```
dtime[1] + rexp(1, rate = kappa)
[1] 12.56
```

However, now that we have some experience generating a few of these jobs, we can outline a better approach to programming the stochastic process. For example, if we know the birth and completion time of a job, then we have the tools to generate the birth and completion times of that job's offspring. Let's make our first task to write a function to generate the birth and completion times of a job's offspring given the parent job's time of birth and completion. This is the topic of the next section.

7.3 Generating Offspring

We have identified a task that we would like to encapsulate in a function – generating random birth times and the corresponding completion times of a job's offspring. What should be the parameters for this function? As noted earlier, we need a) the parent's birth time and b) completion/death time in order to generate its offspring's lifetimes, so these two times should be input parameters to our function. Also, we can specify c) the rate for the inter-arrival of the offspring, and we can provide d) a rate for the run times of the offspring in order to parameterize the exponential distributions for the birth and completion times. In our exploratory code, we took these parameters to be $\lambda = 0.5$ and $\kappa = 0.3$, respectively. We want to parameterize these values so they can be easily modified when we study their impact on the longevity of the process. We now have identified 4 parameters for our function definition. Our function signature appears as

```
function(bTime, cTime, lambda = 0.5, kappa = 0.3)
```

We have provided default values for the two rates so that they need not be specified in a function call but can be easily overridden to get different distributions.

Next, we need to write the code in the body of our function. We saw in the previous section one approach to generating the birth of children. There, we generated an inter-arrival time and added this time to the latest birth time. If this new time occurs before the parent's completion time then it is the next offspring's birth time. If the time is after the parent's completion, then we stop generating birth times. Essentially, we manually performed a loop. We can do these computations with a `while` loop. For example, in pseudo-code this might appear as

```
while (most_recent_birth < parent_completion_time) {
  kidBirths = c(kidBirths, most_recent_birth)
  most_recent_birth = most_recent_birth + rexp(1, rate = lambda)
}
```

Notice that the while loop terminates once the most recent "birth" time occurs after the parent has died. Before checking this relationship, we must have generated that birth time, and if it occurs before the parent's death, then we can append it to the vector of birth times we have already observed. In this way, we incrementally build the birth-time vector. We might consider whether there are faster ways to generate these births and deaths. For example, if we expect a parent to have 4 offspring, then we could generate inter-arrival times 4 at a time, e.g., `rexp(4, rate = lambda)` and work with sets of 4 times. We leave this approach as an exercise. Later, we explore a different approach to generating offspring that utilizes a property of the Poisson process to generate all of a job's offspring without the need for looping. For now, we encapsulate this loop and the generation of the completion times of the offspring into the following genKids() function:

```
genKids =
function(bTime, cTime, lambda = 0.5, kappa = 0.3)
{
        # Parent job born at bTime and completes at cTime

        # Birth time of first child
    mostRecent = rexp(1, rate = lambda) + bTime
    kidBirths = numeric()
```

```
    while (mostRecent < cTime) {
      kidBirths = c(kidBirths, mostRecent)
      mostRecent = mostRecent  + rexp(1, rate = lambda)
    }

        # generate lifetimes for all offspring
    numKids = length(kidBirths)
    runtime = rexp(numKids, rate = kappa)
    kidCompletes = rep(cTime, numKids) + runtime

    data.frame(births = kidBirths,
               completes = kidCompletes)
}
```

Let's try out our function by calling it a few times.

```
genKids(1, 6)

[1] births    completes
<0 rows> (or 0-length row.names)


genKids(1, 6)

    births completes
1    2.47     9.42


genKids(1, 6)

   births completes
1   4.261   13.939
2   5.157    8.680
3   5.649    8.624
```

Our function calls do not give us any errors, but we should still examine the return values carefully to make sure that the function carries out the computations as expected. We consider this issue in the next section, but before proceeding we examine another way to generate the birth times.

In genKids() we calculate one birth time after another until a condition is met, i.e., until a birth time exceeds the parent's lifetime. This sequence of operations suggests an alternative way to successively generate the birth times via recursion. That is, we call a function, say genBirth(), to generate the next birth time, add it to the collection of previously generated birth times, and have the function call itself to generate the next birth time, and so on. The function continues calling itself until the next birth time exceeds the parent's completion time, at which point it returns the collection of birth times and the program ceases. Below is an implementation of genBirth().

```
genBirth = function(currentTime, cTime,
                    births = numeric(), lambda = 0.5) {

        # Generate birth time of next job after currentTime
    mostRecent = rexp(1, rate = lambda) + currentTime
```

```
    if (mostRecent > cTime)
      return(births)
    else {
      births = c(births, mostRecent)
      genBirth(currentTime = mostRecent, cTime, births, lambda)
    }
  }
}
```

Note that as with genKids(), we need the rate of the inter-arrival times and the completion time of the parent. Since we do not generate the completion times of the offspring, we do not need the parameter *kappa*. Also notice that we recursively add a birth time to a collection of birth times that have already occurred so we have the additional parameter, *births*, which we augment with the latest birth.

We can re-implement genKids() so that it uses genBirth() to obtain the birth times of the offspring as follows:

```
genKidsR =
function(bTime, cTime, lambda = 0.5, kappa = 0.3) {
       # Parent job born at bTime and completes at cTime

   kidBirths = genBirth(bTime, cTime, lambda = lambda)

       # generate lifetimes for all offspring
   numKids = length(kidBirths)
   runtime = rexp(numKids, rate = kappa)
   kidDeaths = rep(cTime, numKids) + runtime

   data.frame(births = kidBirths,
              completes = kidDeaths)
}
```

With the recursive function genBirth(), we no longer need the while loop in genKids(). We leave it as an exercise to confirm that this version of the function, i.e., genKidsR(), yields the same birth and death times as the original genKids(). The general concept of recursion can be difficult. It can also be inefficient, as, equivalently, can be `while` and `for` loops. In subsequent sections, we address the concept of efficiency, and consider how we might redesign our code.

7.3.1 Checking the Results

One simple check to determine that our function is behaving as expected is to confirm that the offspring are born after the parent's birth time and that they complete after their parent has completed. We can see that this is the case for our 3 sample sets of offspring, and we can imagine writing code to test whether or not this simple condition holds for all offspring of a job.

We additionally might take a statistical approach to check our function. For example, a random number of children are generated for the parent job. We can examine the distribution of the number of children born to this parent by generating many instances of its offspring and tallying the number of children in each instance. Since the inter-arrival rate is λ, probability theory tells us that we expect $\lambda \times (deathtime - birthtime)$ children. In our example, this is $0.5 \times (6 - 1)$, or 2.5, children. We can compare this expected value against our simulation results by repeating our simulation 1000 times and computing the average number of observed offspring with

```
numKids = replicate(1000, nrow(genKids(1, 6)))
mean(numKids)
```

```
[1] 2.503
```

The standard deviation of the number of children is `sqrt(2.5)`, or approximately 1.58, so the standard error for the sample average is `sqrt(2.5/1000)`, or 0.05. We see that our sample average is within 1 SE of what we would expect.

Furthermore, the distribution of the number of children should follow the Poisson(2.5) distribution. To confirm this, we can compare the empirical probability mass function for the number of children a job has to the Poisson distribution. We find the empirical proportions and the Poisson(2.5) probabilities with

```
eprobs = table(numKids)/length(numKids)
probs = dpois(x = 0:max(numKids), lambda = 2.5)
```

Then we plot these proportions and probabilities side by side with

```
plot(eprobs, type = "h",
     ylab = "Proportion", xlab = "Number of offspring")
segments(x0 = 0.1 + 0:max(numKids), y0 = rep(0, max(numKids)),
         y1 = probs, col="grey", lwd = 2)
```

In Figure 7.3, we see that the heights of the bars for the two distributions are very close. Additionally, we can carry out a chi-square goodness of fit test of the simulated counts of children to the expected counts for 1,000 observations from a Poisson(2.5) distribution, and we find a p-value of 0.994. The observed counts fit the expected counts very closely.

The property that we have been using to validate our code, i.e., that exponential inter-arrival times lead to a Poisson number of children, offers an alternative, possibly more efficient approach to generate offspring. We consider it next.

7.3.2 Considering Alternative Implementations

There are other ways than the approach taken earlier in this section to generate the birth and death times of a job's offspring. For example, since R is a vectorized language, we may ask ourselves if there is a way to modify our function genKids(), or design a new function, so that it generates the offspring in a vectorized manner, rather than one at a time in a while loop or via recursion. The problem is that we don't know how many children a job has so we cannot simply generate the correct number of inter-arrival times all in one go. As mentioned already, we can partially remedy this problem by generating multiple inter-arrival times, based on the expected number of children and its standard deviation. Then we would keep only those times that correspond to births before the parent's completion time. If we do not generate enough inter-arrival times, then we can generate another batch of inter-arrival times to add to our original set and check again. This way, we typically only need to generate a set of inter-arrival times once. We leave the implementation of this approach as an exercise.

Alternatively, we use the additional information about the Poisson process discussed in Section 7.3.1 and generate the offspring by a quite different method. When we observe a Poisson process over a fixed time interval, say, an interval of length t, then the number

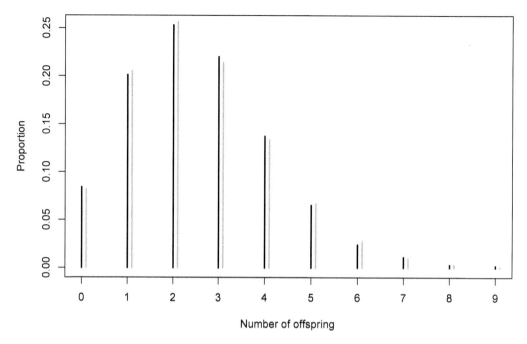

Figure 7.3: *Empirical Distribution of the Number of Offspring for a Job. This figure shows the observed proportion of the number of offspring randomly generated for a parent with a birth time of 1 and a completion time of 6. The inter-arrival times of the parent's children are independent and follow an exponential distribution with parameter $\lambda = 0.5$. The process was simulated 1,000 times, and the observed proportion of 0, 1, 2, ..., 9 offspring plotted (black line segments). In addition, the true probabilities from the Poisson(2.5) distribution are plotted next to the observed proportions (grey line segments).*

of events in that interval, e.g., the number of births, follows the Poisson(λt) distribution, where λ is the rate per unit of time. In addition, once we know the number of events in the time interval (e.g., the number of jobs generated by the parent), then the locations of these events are independent of each other and follow the Uniform distribution. We can use these two properties of the Poisson process to more simply generate the offspring. That is, rather than generating offspring one after another using the inter-arrival times, we first generate the number of offspring, and then we generate all of their birth times with one call to runif().

Suppose as before that the parent has a birth time of bTime and a completion time of cTime, then the number of offspring that this job has can be generated with

```
numKids = rpois(1, lambda = (cTime - bTime) * lambda)
```

Now that we have the parent's number of offspring, we can generate the offspring's birth times according to the Uniform distribution with

```
sort( runif(numKids, min = bTime, max = cTime) )
```

Note that we sorted these times so that they appear in birth order. We can update genKids() to use this probabilistic property and abandon the while loop. We leave this modification as an exercise, and call the modified function genKidsU().

7.4 Profiling and Improving Our Code

In this section, we compare two approaches for generating the offspring of a job to see if the approach that uses a `while` loop is significantly slower or faster than the approach that takes advantage of the Poisson and Uniform properties of the number and location of offspring (described in Section 7.3.2). We can use the system.time() function to make this comparison. Both versions of the genKids() function are quite fast for the values of the parameters in which we are interested, so how do we make an accurate comparison? We can repeat the call to these functions many times and determine the total time it takes to perform the multiple calls. We do this with:

```
time1 = system.time( replicate(4000, genKids(1, cTime = 9)) )
time2 = system.time( replicate(4000, genKidsU(1, cTime = 9)) )

time1/time2
```

```
   user  system elapsed
  0.823   0.375   0.815
```

The first approach with the `while` loop is nearly 20% faster!

This speedup seems counter-intuitive because it generates the birth dates of the jobs one at a time and incrementally expands the birthKids vector with each new child. However, since the arguments for the birth and completion time of the parent are 1 and 9, the parent typically has very few offspring so this incremental addition may not cost us much. What happens when we increase the completion time of the parent? Let's try a *cTime* of 100,

```
time1 = system.time( replicate(4000, genKids(1, cTime = 100)) )
time2 = system.time( replicate(4000, genKidsU(1, cTime = 100)) )
time1/time2
```

```
   user  system elapsed
  1.602  27.333   1.669
```

Now the genKids() function is slower than genKidsU() by about 60%. It makes sense that the gains from vectorizing the operations are realized when the parent has a large number of children.

We can investigate which operations in genKids() takes the most time by profiling the code. We use Rprof() to do this with

```
set.seed(seedx)
Rprof("profGenKids1.out")
invisible( replicate(1000, genKids(1, cTime = 100)) )
Rprof(NULL)
```

The summaryRprof() function tallies the timings for the various function calls and presents them in a table, e.g.,

```
summaryRprof("profGenKids1.out")$by.self
```

	self.time	self.pct	total.time	total.pct
".External"	0.18	34.62	0.18	34.62
"c"	0.08	15.38	0.08	15.38
"data.frame"	0.06	11.54	0.18	34.62
"genKids"	0.04	7.69	0.52	100.00
"rexp"	0.02	3.85	0.20	38.46
"make.names"	0.02	3.85	0.04	7.69
"paste"	0.02	3.85	0.04	7.69
"+"	0.02	3.85	0.02	3.85

...

We see from the profile that calling the rexp() function repeatedly is taking about 40% of the time. For a comparison, we see below that this operation is not among the top time-consuming computations for genKidsU(). It takes less than 6% of the time:

```
set.seed(seedx)
Rprof("profGenKidsU.out")
invisible( replicate(1000, genKidsU(1, cTime = 100)) )
Rprof(NULL)
summaryRprof("profGenKidsU.out")$by.self
```

	self.time	self.pct	total.time	total.pct
"genKidsU"	0.04	12.50	0.30	93.75
"match"	0.04	12.50	0.06	18.75
"as.data.frame"	0.02	6.25	0.12	37.50
"force"	0.02	6.25	0.10	31.25
"deparse"	0.02	6.25	0.08	25.00
"make.names"	0.02	6.25	0.06	18.75
"as.list"	0.02	6.25	0.04	12.50
"as.list.default"	0.02	6.25	0.02	6.25

...

On the other hand, the functions in which we are spending the most time are genKidsU() itself and match(). Also, calls to force() and deparse() take up a significant proportion of the time. Where are the calls to these functions? Additionally, it would be valuable to know how often these are actually called during the computations, and hence the time per call. Since the genKidsU() function takes the most amount of time and we wrote it, we may want to examine it more closely to see if improvements can be gained here. We leave it as an exercise to consider other modifications to speed up genKids() and/or genKidsU().

7.5 From One Job's Offspring to an Entire Generation

We have so far examined how to generate the offspring for one job. What are the next steps in our simulation? We can view the stochastic process as a family tree – the initial job has offspring, these offspring have their own offspring, and so on. The jobs in the process can be organized as a tree with a job being a node in the tree, branches from that job connect to its offspring, and a level/depth of the tree corresponds to a generation. This is why the process is called a branching process. (See Figure 7.1 for an example.) Our genKidsU() function generates the children for one job. A natural next step might be to generate the offspring for all of the jobs in one generation, and after that, to generate the next generation, and so on.

Let's first consider how we might generate all of the offspring for one generation. We might want to loop over each job in that generation and call genKidsU(), or we might modify genKidsU() so that it takes vector inputs, e.g., a vector of birth times for a set of jobs, rather than one birth time. To determine whether or not to vectorize genKidsU() requires us to decide how to implement the whole simulation. For now, let's explore the possibility of modifying genKidsU() to accept vector arguments for the birth and completion times.

We take bTimes and cTimes to be vectors of the same length where the ith element of each vector corresponds to the same job, $i = 1, \ldots, n$. That is, the ith element of bTimes and the ith element of cTimes contain the birth and completion times, respectively, of the ith of n jobs. Then, we can generate the number of children of each of the n jobs with

```
lifeTimes = cTimes - bTimes
numKids = rpois(n = length(lifeTimes), lambda = lambda * lifeTimes)
```

This code is nearly identical to that in genKidsU() because subtraction is a vectorized operation in R and rpois() accepts a vector of rates for $lambda$ (one for each of the n random outcomes).

Now that we have the number of offspring for each parent, we can generate the times for birth and completion of the offspring with runif() and rexp(). Unfortunately, runif() does not take vector values for its parameters so we instead use mapply(), i.e.,

```
kidBirths = mapply(function(n, min, max)
                   sort(runif(n, min, max)),
                 n = numKids, min = bTimes, max = cTimes)
```

Note that kidBirths is a list where each element contains the birth times of a job's children. We also must generate the completion time for these children. We can do this within the function provided in mapply(). Our vectorized version of genKidsU(), which we call genKidsV(), appears as

```
genKidsV = function(bTimes, cTimes, lambda = 0.5, kappa = 0.3) {
        # bTimes & cTimes - vector of birth and completion times

        # Determine how many children each job has
  parentAge = cTimes - bTimes
  numKids = rpois(n = length(parentAge),
              lambda = lambda*parentAge)

        # Determine the birth and completion times of the children
  mapply(function(n, min, max) {
            births = sort(runif(n, min, max))
            runtimes = rexp(n, rate = kappa)
            completes = rep(max, n) + runtimes
            data.frame(births, completes)
         },
         n = numKids , min = bTimes, max = cTimes,
         SIMPLIFY = FALSE)
}
```

At this point in the project, we may want to set up some test cases where we specify the inputs and know what to expect for the outputs of our function. This way we can run the test cases after each change to our function(s) and check that our code still runs as expected. This is the topic of the next section.

7.6 Unit Testing

When we design tests for code, we consider several sets of input values for our functions and examine the outputs. Since this is a simulation, we can't know exactly what the return values will be so we check that these make sense according to our understanding of the probability model. Additionally, we want to be sure to include scenarios that lead to special situations, e.g., a generation with no offspring, to make sure our code handles these cases correctly. As we debug our code and uncover new problems, we fix these problems and add more test cases to our suite of tests for these situations. We want to make sure that our code continues to work properly across a variety of scenarios.

With simulation studies, we often set the seed for the random number generator before we run our test cases. This way, the return values are the same each time we run the code so after we initially check that the return values are reasonable, then we only need to check that they match the return value from previous runs of the code.

Let's start with a simple small example with 3 jobs with the following birth and completion times:

```
bTimes1 = 1:3
cTimes1 = c(3, 10, 15)
```

We set the seed for the random number generator with

```
seed1 = 12062013
set.seed(seed1)
```

In our first call to genKidsV() we find:

```
kids = genKidsV(bTimes1, cTimes1)
kids
[[1]]
  births completes
1   2.94      8.92

[[2]]
  births completes
1   2.17     10.3
2   3.20     15.2
3   8.71     16.0

[[3]]
  births completes
1   3.17     16.8
2   5.50     15.5
3   9.78     20.5
4  10.33     24.8
```

We see that the first parent has 1 child, the second has 3, and the third has 4 children. The birth and completion times of these offspring are returned in a list of 3 data frames. We call the function a second time with the same input arguments, but since the seed has not been reset, a different collection of birth and completions times are generated:

```
kids2 = genKidsV(bTimes1, cTimes1)
sapply(kids2, nrow)
```

```
[1] 2 4 6
```

We find that this time the 3 parents have 2, 4, and 6 children, respectively. If we reset the seed, before calling genKidsV() again, we get the same results as the first call, i.e.,

```
set.seed(seed1)
kids3 = genKidsV(bTimes = bTimes1, cTimes = cTimes1)
identical(kids, kids3)
```

```
[1] TRUE
```

How might we improve the code in genKidsV()? One consideration is whether or not we want the return value as a list of data frames. The list may be a useful data structure for representing a tree, where each data frame holds the information for one generation. However, this return value can be awkward to work with when we are trying to examine the process across generations.

7.7 A Structure for the Function's Return Value

A single data frame provides a more compact format for the return value of genKidsV(). However, if we collapse the list of data frames into one data frame, then we lose the information as to which child belongs to which parent. We need to add an identifier for the parents as a column in the data frame. This identifier needs to be passed to the function. Additionally, it makes sense to assign an identifier to each offspring so that when we generate the children of these offspring, we have their identifiers to pass to genKidsV(). Our return value might be something like the following:

```
data.frame(parentID = rep(parentID, numKids),
           kidID = 1:sum(numKids),
           births = unlist(sapply(kidStats, "[[", "births")),
           completes = unlist(sapply(kidStats,"[[", "completes"))))
```

We leave it as an exercise to modify genKidsV() and confirm that we have the same results as earlier when the seed was set to seed1. That is, the births and completes should be the same as in our earlier call, but now formatted differently:

```
set.seed(seed1)
genKidsV(bTimes1, cTimes1, parentID = letters[1:3])
```

	parentID	kidID	births	completes
1	a	1	2.94	8.92
2	b	2	2.17	10.32

3	b	3	3.20	15.17
4	b	4	8.71	16.03
5	c	5	3.17	16.76
6	c	6	5.50	15.50
7	c	7	9.78	20.46
8	c	8	10.33	24.80

We have modified our function and confirmed that for this one test case the return value from genKidsV() remains the same, i.e., the first job produced 1 child, born at 2.94 and completed at 8.92, the second child had 3 children and their birth and complete times match those from the call to the earlier version of genKidsV() (Section 7.6). We want to develop additional test cases, and we may wish to programmatically check the return values rather than manually inspecting the output each time we run the test cases. We leave this as an exercise.

7.8 The Family Tree: Simulating the Branching Process

We are now ready to generate the whole process. We know from Aldous and Krebs that, depending on the parameter values for λ and κ, the process may continue on indefinitely or it may complete in a finite amount of time. How do we simulate this sort of process? Clearly we can't let the process run and run if it is never going to complete so how do we decide when to stop? We could run it for a fixed number of generations and if the process has not completed, then we terminate it; we could run it until we have generated a fixed number of offspring; or we could run the simulation for a fixed amount of time (i.e., we observe all births and completes before a particular time, T say). These are 3 reasonable possibilities and there are others. Additionally, we might check more than one of these approaches to decide whether or not to terminate the process. Our choice in part depends on the way we have implemented the process. With the current implementation, we can create an entire generation with one call to genKidsV so it makes sense to start by capping the number of generations. Of course, the number of offspring in a generation could become large, but let's try this approach and see how well it controls the process.

To simulate the process, we need to call the genKidsV() function for each successive generation and collect the return value from each call. We first get the process started with the single, lead job, and then we successively call genKidsV() for the specified number of generations. We also need to check to make sure that the process hasn't terminated before we ask genKidsV() to produce the next generation. Our input parameters to this function are the original λ and κ rates for the Poisson and Exponential random variables, and the limit on the number of generations to simulate. As for the return value, a list of generations seems simplest. One implementation might be as follows:

```
familyTree = function(lambda = 0.5, kappa = 0.3, maxGen = 10) {
      # maxGen - maximum number of generations to observe
      # Return value - a list with 1 data frame per generation.
   allGens = vector(mode = "list", length = maxGen)

      # Generate the root of the tree
   allGens[[1]] = data.frame(parentID = NA, kidID = 1, births = 0,
                             completes = rexp(1, rate = kappa))
```

```
    # Generate future generations, one at a time.
  for (i in 2:maxGen) {
    nextGen = genKidsV(bTimes = allGens[[ (i - 1) ]]$births,
                       cTimes = allGens[[ (i - 1) ]]$completes,
                       parentID = allGens[[ (i - 1) ]]$kidID,
                       lambda = lambda, kappa = kappa)
    if (is.null(nextGen)) return(allGens[ 1:(i - 1) ])
    allGens[[ i ]] = nextGen
  }

  return(allGens)
}
```

Let's try out our function, by setting the seed and calling familyTree() with

```
set.seed(seed1)
tree = familyTree(lambda = 0.4, kappa = 1, maxGen = 10)
```

Rather than poring over the numeric value of the birth and completion times, we can create a visualization to inspect the tree. For example, Figure 7.4 provides a plot of the contents of tree. It represents each job as a line segment and marks the birth of each job's children with an X on the segment. There we see that the process completed in 4 generations and that there are 9 jobs in this instance of the process. As an exercise, consider designing another custom visualization that explores the tree.

Let's try generating the process again, this time with a different seed and different values for λ and κ:

```
seed2 = 12212013
set.seed(seed2)
tree = familyTree(lambda = 0.3, kappa = 0.5, maxGen = 10)
```

The length of tree tells us how many generations are in this particular instance of the process, i.e.,

```
length(tree)
```

```
[1] 10
```

Since this tree has 10 generations, it may not have completed by the time it reached the maximum number of generations in the simulation. We can determine the number of children in each generation with

```
sapply(tree, nrow)
```

```
[1]   1   1   1   5   7   5  13  39  81 147
```

We see that the number of offspring in the seventh through tenth generations increases dramatically from 13 to 147. This implies that the process has not died out by the tenth generation, when we stopped generating offspring. Altogether, this particular random process includes 300 children, i.e.,

```
sum(sapply(tree, nrow))
```

```
[1] 300
```

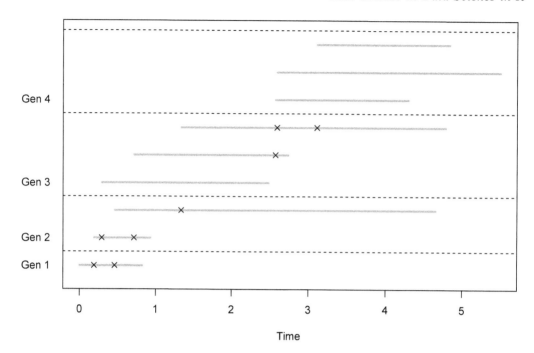

Figure 7.4: Visualization of a Randomly Generated Branching Process. *This plot shows the lifetimes of each member of a randomly generated birth and assassination process with a birth rate of $\lambda = 0.4$ and a completion rate of $\kappa = 1$. Each job's lifetime is represented by a grey line segment with endpoints at its birth and completion times. The Xs on the segment denote the birth times of the job's offspring. The dashed lines separate the generations. None of the jobs in the fourth generation of this instance of the process had offspring so the process terminated there. Notice that one job in the second generation ran for a very long time and had just one child.*

We can re-generate the process and this time allow for, e.g., 15 generations to see if the process runs for more than 10 generations, i.e.,

```
set.seed(seed2)
tree = familyTree(lambda = 0.3, kappa = 0.5, maxGen = 15)
sapply(tree[ - (1:9) ], nrow)
```

```
[1]   147   286   572  1130  2231  4666
```

Indeed, the tree has a large number of offspring in the 15th generation.

How do we summarize these trees for statistical analysis? The plot in Figure 7.4 is helpful for examining one or two trees, but it's not feasible to make thousands of these plots, especially when there may also be thousands of children in a tree. Before we decide on how to summarize a tree, let's experiment with other values for λ and κ and see what is generated:

```
set.seed(seed2)
tree = familyTree(lambda = 1, kappa = 0.5, maxGen = 10)
```

We can again determine the number of generations, children in each generation, and total number of offspring with the following computations, respectively:

```
length(tree)
```

```
[1] 10
```

```
sapply(tree, nrow)
```

```
[1]     1     3     7    19    99   464  2616
[8] 15155 90848 561791
```

```
sum(sapply(tree, nrow))
```

```
[1] 671003
```

This tree has over 670,000 jobs. Let's run the simulation again with the same values for λ and κ but without resetting the seed to see if this was an unusual occurrence or not:

```
tree = familyTree(lambda = 1, kappa = 0.5, maxGen = 10)
sum(sapply(tree, nrow))
```

```
[1] 989086
```

This time, there are nearly one million jobs in the first 10 generations.

We can imagine that other choices of λ and κ might lead to instances of the branching process that are so large as to cause problems with run time. We may want to implement a different approach to limiting the process, such as a time limit on observing the process. How would we go about modifying the code to implement this cap on the simulation? To answer this question, we further investigate the process. It would be helpful to know more about the birth and completion times across generations. We first examine the range of these with

```
sapply(tree, function(gen) range(gen$births))
```

```
     [,1]  [,2]  [,3]  [,4]  [,5]  [,6]  [,7]  [,8]  [,9] [,10]
[1,]    0 0.009 0.311 0.537  1.03  1.66  2.06  3.31  3.97   4.1
[2,]    0 0.009 3.388 6.689 12.62 20.30 27.31 37.17 49.84  58.4
```

```
sapply(tree, function(gen) range(gen$completes))
```

```
      [,1] [,2] [,3]  [,4]  [,5]  [,6]  [,7]  [,8]  [,9] [,10]
[1,] 0.214 4.48 5.26  5.93  6.46  6.85  7.28  7.94  8.34  9.49
[2,] 0.214 4.48 8.43 13.09 21.98 27.51 37.43 50.03 58.56 70.26
```

If we want to observe this process up until, say, a time of 12, we see that we would have to observe more than 10 generations because there are jobs in the tenth generation that are born as early as 4.1. On the other hand, there are jobs in earlier generations that die well after 12 so we would only know that they are still alive by 12 and we would only observe those jobs' offspring that occur before 12.

We need to develop an alternative version of familyTree() that accepts a cap on the time that the process is observed. If we wish to keep much of the approach the same as in the first implementation of familyTree(), then we need to drop the offspring in a generation if they are born after the specified time. And, we need to continue creating generations until all children born in a generation occur after the time limit. Figure 7.5 shows the first 5 generations of a process that is limited using this approach. With a cap of 8, i.e., we observe no births or deaths after 8, and with $\lambda = 1$ and $\kappa = 0.5$, we find that already in the third generation, we are not observing the entire lifespan of some jobs. We leave it as an exercise to implement this alternative stopping rule. Instead, we implement a simpler limitation, which terminates the process once a specified threshold for the number of offspring is exceeded.

This alternative stops producing generations once the total number of offspring in the generations simulated so far has reached a specified limit. We can incorporate this limit into our original familyTree() function so the simulation stops once either of these two limits is reached. Below is one implementation of familyTree():

```
familyTree = function(lambda = 0.5, kappa = 0.3,
                     maxGen = 10, maxOffspring = 1000) {

    # Return value - a list with 1 data frame per generation.
  allGens = vector(mode = "list", length = maxGen)

    # Generate root of the tree
  allGens[[1]] = data.frame(parentID = NA, kidID = 1,
                            births = 0,
                            completes = rexp(1, rate = kappa))

  currentNumOffspring = 0

    # Generate future generations, one at a time.
  for (i in 2:maxGen) {
    nextGen = genKidsV(bTimes = allGens[[ (i - 1) ]]$births,
                       cTimes = allGens[[ (i - 1) ]]$completes,
                       parentID = allGens[[ (i - 1) ]]$kidID,
                       lambda = lambda, kappa = kappa)
    if (is.null(nextGen)) return(allGens[ 1:(i - 1) ])
    allGens[[ i ]] = nextGen
    currentNumOffspring = currentNumOffspring + nrow(nextGen)
    if (currentNumOffspring > maxOffspring)
      return(allGens[1:i])
  }
  allGens
}
```

We can compare the output from the same settings of λ and κ from our earlier call to the version of the family tree function that caps only the number of generations. We set the seed to the same value as before and this time increase $maxGen$ to 100 but limit the total offspring to 1000. We call this new version of the function with

```
set.seed(seed2)
tree = familyTree(lambda = 1, kappa = 0.5,
                  maxGen = 100, maxOffspring = 1000)
```

The limitation on the process yields only 7 generations, i.e.,

```
length(tree)
```

[1] 7

Simulation Study of a Branching Process

We determine the number of offspring in each of these generations with

```
sapply(tree, nrow)
```

```
[1]    1    3    7   19   99  464 2616
```

Rather than generating nearly 700,000 jobs, we observe only 3209.

Now that we have developed and tested our functions to generate a random birth and assassination process, we turn to the question of how to design and carry out our simulation study.

7.9 Replicating the Simulation

We are now ready to study the branching process for various combinations of the parameter values λ and κ. To do this, we want to automate the call to familyTree(). Since it can easily become unwieldy to save the thousands of lists of data frames from each observed process, we want to store only relevant summaries. We have been examining the process through its number of generations and number of jobs. What other statistics about the tree might be useful to examine? The following simple function calls familyTree() and summarizes the return value by these two simple summaries:

```
exptOne = function(l, k, mG, mO){
       # Helper function to call familyTree
       # Returns - summary statistics for analysis,

   aTree = familyTree(lambda = l, kappa = k, maxGen = mG,
                      maxOffspring = mO)
   numGen = length(aTree)
   numJobs = sum(sapply(aTree, nrow))
   c(numGen, numJobs)
}
```

Let's try exptOne() with the previous settings of our seed, and the *lambda*, *kappa*, *maxGen*, and *maxOffspring* arguments:

```
set.seed(seed2)
exptOne(1, 0.5, 100, 1000)
```

```
[1]    7 3209
```

We obtain the same summary statistics from our earlier call to familyTree().

We want to call exptOne() many times for each of several values of *lambda* and *kappa*. We can specify the number of replications and the values for *lambda* and *kappa* via a function that calls exptOne(), i.e.,

```
MCBA = function(params, repeats = 5, mG = 10, mO = 1000){
       # params: matrix columns of lambda and kappa values
       # For each lambda and kappa pair, run "repeats" times

   n = nrow(params)
```

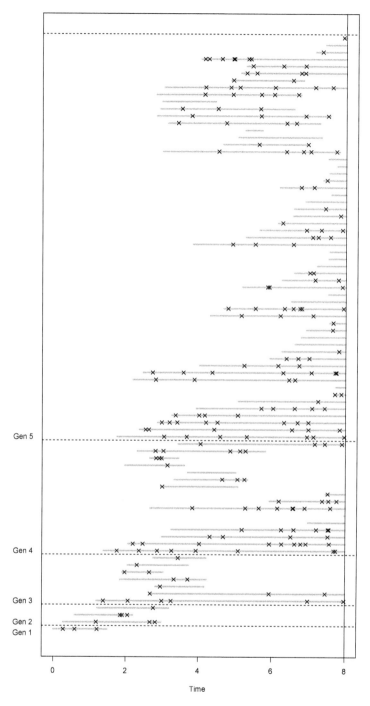

Figure 7.5: Visualization of a Randomly Generated Branching Process Over a Fixed Time Interval. *This plot shows the lifetimes of each member of the first 5 generations of a simulated birth and assassination process that has been observed up to time of 8. Each lifetime is represented by a grey line segment with endpoints at its birth and completion times. A job that has not completed by 8 time steps is censored and consequently, we see only its offspring born before 8.*

```
mcResults = vector("list", length = n)

for (i in 1:n) {
  cat("param set is ", i, "\n")
  mcResults[[i]] = replicate(repeats,
                             exptOne(l = params[i, 1],
                                     k = params[i, 2],
                                     mG = mG, mO = mO))
}
mcResults
}
```

Notice that as the code begins to run the simulation for a new row in the parameter set, it prints the row number to the console. This helps us follow the progress of our simulation. That is, we can determine if the simulation is 'stuck' on a particular set of parameter values or progressing through the simulation.

What values of λ and κ do we use for our simulation study? We can try a few values to determine a feasible region of the parameter space. For example,

```
trialKappas = c(0.1, 10, 0.1, 10)
trialLambdas = c(0.1, 0.1, 10, 10)
trialParams = matrix(c(trialLambdas, trialKappas), ncol = 2)
mcTrialOutput = MCBA(params = trialParams, repeats = 100,
                    mG = 200, mO = 100000)
```

We find that the second simulation is very fast and the others take several minutes to complete (running on a MacBook Air with 8GB of memory and a 2 GHz Intel Core i7).

We examine the output from the process by plotting the number of generations against the number of offspring.

```
oldPar = par(mfrow = c(2, 2), mar = c(3,3,1,1))

mapply(function(oneSet, lambda, kappa) {
  plot(x = oneSet[2,], y = jitter(oneSet[1, ], 1), log = "x",
       ylim = c(1,20), xlim = c(1, 10^7), pch = 19, cex = 0.6)
  text(x = 50, y = 15, bquote(paste(lambda == .(lambda))) )
  text(x = 300, y = 15, bquote(paste(kappa == .(kappa))) )
  },
  mcTrialOutput, lambda = trialLambdas, kappa = trialKappas)

par(oldPar)
```

We see in Figure 7.6 that these 4 sets of (λ, κ) pairs have very different behaviors. The pairs $(0.1, 0.1)$ and $(10, 10)$ behave similarly; some instances terminate before reaching 5 generations and others are still running when they reach the maximum of 100,000 offspring. On the other hand, none of the $(0.1, 10)$ simulations survive more than one generation. Conversely, it appears that all of the pairs $(10, 0.1)$ simulations run up against the limitation on the number of children before they reach the fifth or sixth generation.

7.9.1 Analyzing the Simulation Results

Based on our trial simulations for a couple of parameter values, we limit λ and κ as follows:

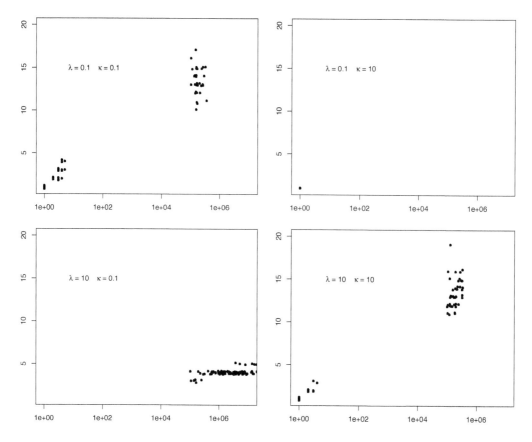

Figure 7.6: Scatterplots of the Number of Generations Against the Number of Offspring. These 4 scatter plots show the different behavior exhibited by the branching process as λ and κ vary. Each simulation terminates when the process dies out or one of the following limits is reached: 200 generations; 100,000 offspring. One hundred simulations are run for each (λ, κ) pair.

```
lambdas = c(seq(0.1, 0.6, by = 0.1), seq(0.8, 2, by = 0.2),
            seq(2.25, 3, by = 0.25))
kappas = c(lambdas, 3.25, 3.50, 3.75, 4.00, 4.50, 5.00)
```

We create all combinations of the values for these two parameters with expand.grid(), i.e.,

```
paramGrid = as.matrix(expand.grid(lambdas, kappas))
```

Additionally, we limit the number of offspring to 1,000 because the trial runs indicated that we should have a reasonable indication of the longevity of the process by the time 1,000 offspring have been generated.

We call MCBA() with the matrix of parameter values and the limitations on the size of the tree with

```
mcGrid = MCBA(params = paramGrid, repeats = 400, mG = 20,
              mO = 1000)
```

Note that for each (λ, κ) pair, 400 instances of the process are generated. We can summarize our simulation by, e.g., plotting the upper quartile of number of offspring in each of the 400 replications. Since we have 3 variables (λ, κ, and the number of offspring), we make a 3-dimensional scatter plot. We begin by computing the log upper quartile of the number of children for each parameter setting with

```
logUQkids = sapply(mcGrid, function(x)
            log(quantile(x[2, ], probs = 0.75), base = 10))
```

We also use color to distinguish between trees that die out and those that surpass the 1,000 offspring limit, i.e.

```
UQCut = cut(logUQkids, breaks = c(-0.1, 0.5, 2, max(logUQkids)) )
color3 = c("#b3cde3aa", "#8856a7aa", "#810f7caa")
colors = color3[UQCut]
```

The scatterplot3d package [4] offers functionality for making 3D plots. We use the package's function of the same name to plot the upper quartile values against their corresponding λ and κ pairs with

```
library(scatterplot3d)
sdp = scatterplot3d(x = paramGrid[ , 1], y = paramGrid[ , 2],
                z = logUQkids, pch = 15, color = colors,
                xlab = "Lambda", ylab = "Kappa",
                zlab = "Upper Quartile Offspring",
                angle = 120, type="h")

legend("left", inset = .08, bty = "n", cex = 0.8,
       legend = c("[0, 0.5)", "[0.5, 2)", "[2, 5)"),
       fill = color3)
```

In Figure 7.7 we see a pattern that is consistent with the observations made about Figure 7.6 where the process either dies out or generates a large number of offspring.

Alternatively, we can summarize these results by finding the proportion of families for each parameter set that hit the limit on the simulation, i.e., the families that contain 20 generations or that have more than 1000 offspring. We calculate these proportions as follows:

```
mcGridAlive = sapply(mcGrid, function(oneParamSet) {
   sum((oneParamSet[1,] == 20) | (oneParamSet[2,] > 1000)) /
   length(oneParamSet[2,]) })
```

Again, we must consider how to plot a triple, (λ, κ, p), where p is the proportion of the 400 simulations that reach the limits set in the simulation. One approach is to create a filled contour, or level plot, where the values for p are represented with color. We use filled.contour() to do this with

```
filled.contour(lambdas, kappas,
            matrix(mcGridAlive, nrow = length(lambdas),
                ncol = length(kappas)),
            xlab = "Lambda", ylab = "Kappa",
            xlim = c(0.1, 3), ylim = c(0.1, 3.1))
```

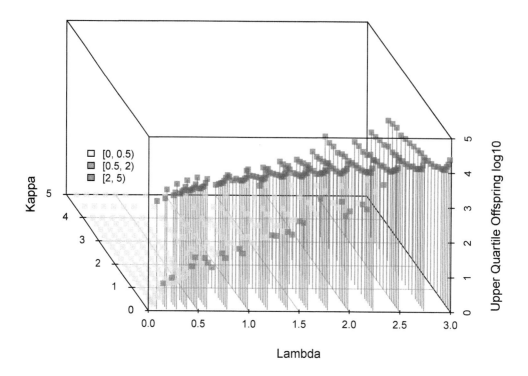

Figure 7.7: Three-Dimensional Scatterplot of the Number of Offspring by λ and κ. Each point in this scatter plot represents the upper quartile of the number of offspring in 400 random outcomes of the branching process for a particular (λ, κ) pair. The offspring are plotted on log base 10 scale, so the first category, i.e., [0, 0.5) corresponds to 1 to 3 offspring.

Figure 7.8 shows the results. There we see that nearly all of the simulated processes where κ is small are still 'alive' at the time we stopped the simulation and nearly all of the simulated processes for small λ values terminate before the limitations are reached. Also noticeable is the pale diagonal region in the plot where those pairs below this region are more likely than not to still be running and those above the diagonal are less likely than not to have completed.

Another measure of longevity that we can examine is the proportion of simulations that have 20 or more offspring. The simulation is run for at most 20 generations, so this seems like a reasonable threshold to examine. We compute this proportion as follows:

```
mcGridProp20kids = sapply(mcGrid, function(oneParamSet) {
   sum(oneParamSet[2,] > 19) / length(oneParamSet[2,]) })

mcGridProp20kidsMat = matrix(mcGridProp20kids,
                             nrow = length(lambdas),
                             ncol = length(kappas))
```

Simulation Study of a Branching Process

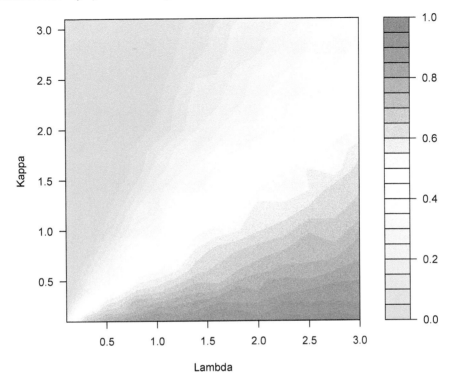

Figure 7.8: Image Map of the Proportion of Replicates That Reach the Simulation Limits. *This image map represents a smoothed contour of the proportion of the 400 simulations for each (λ, κ) pair that reached 20 generations or 1000 offspring and so were terminated.*

Rather than use filled.contour(), we make an image plot and again use color to denote p. This time, however, we choose 7 colors in the rainbow palette to represent the values of p. We make the high values correspond to red and the low values to blue in this palette, i.e.,

```
breaks = c(0, 0.10, 0.2, 0.3, 0.5, 0.7, 0.9, 1)
colors = rev(rainbow(10))[-(1:3)]

image(lambdas, kappas, mcGridProp20kidsMat, col = colors,
      breaks = breaks, xlab = "Lambda", ylab = "Kappa",
      xlim = c(0.05, 3.05), ylim = c(0.05, 3.05))

midBreaks = (breaks[ -8 ] + breaks[ -1 ]) / 2
legend(x = 0.1, y = 3.25, legend = midBreaks, fill = cols,
       bty = "n", ncol = 7, xpd = TRUE)
```

The image plot is shown in Figure 7.9. It has a similar appearance to Figure 7.8

With these visual summaries of the simulation study, we see that the relationship between λ and κ is key to the longevity of the process. For example, when $\kappa > \lambda$, the process dies out within a few generations. There appears to be a region where the process at times produces many thousands of children and seems to carry on indefinitely, and other regions (e.g., where λ is small) when the process never terminates, or nearly so. The plots we have made suggest that we may be able to parameterize the process in terms of the ratio λ/κ

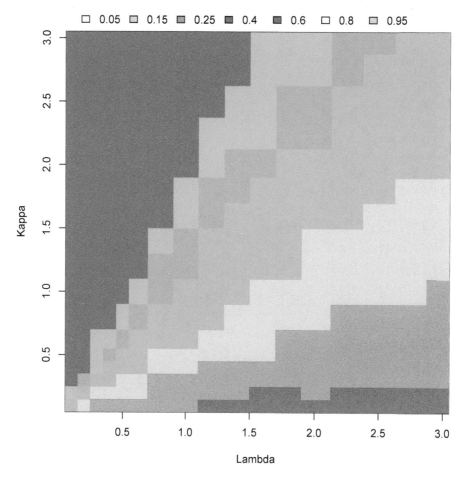

Figure 7.9: Proportion of Simulations with at Least 20 Offspring. *This image map uses color from the rainbow palette to represent the proportion of 400 random outcomes for each (λ, κ) pair that have at least 20 offspring.*

and work with only one parameter. This observation suggests alternative visualization for these simulation results may be useful to examine. If we want to carry out more simulations to further investigate this property, we can do this without modifying our code as we can simply set κ to 1. We leave it as an exercise to follow up on these insights.

7.10 Exercises

Q.1 Write an alternative function to genKids() (see Section 7.3), called genKidsB(), where the B stands for batch. In this function, generate the inter-arrival times in batches, where the size of the batch depends on the expected number of offspring for a job. The expected number depends on the rate λ and the birth and completion times of the parent. That is, for a job born at time α and complete at time β, we expect it to have

$(\beta - \alpha) \times \lambda$ offspring. The parameters to this function and their default values should be the same as genKids().

Q.2 Write the function to genKidsU() described in Section 7.3.2. This function generates the birth and completion times for a job that is born at α and completes at β. The number of children follow a Poisson($(\beta - \alpha) \times \lambda$). Once the number of children are known, their births can be generated according to the Uniform on the interval (α, β). As mentioned in Section 7.3.2, these times are not generated in order so they need to be sorted. The parameters to genKidsU() and their default values should be the same as genKids().

Q.3 Develop a test suite to confirm that the recursive function genKidsR() (see Section 7.3) is consistent with the while-loop version of genKids() (also in Section 7.3).

Q.4 Use Rprof() to profile the genKidsU() function in the previous exercise. As in Section 7.4, profile the code by calling genKidsU() 1000 times with the parameter settings: $bTime = 1$, $cTime = 100$, and the default values for λ and κ. Also profile the code with one simulation, where $cTime = 1000000$. Do the profiles look different? Try improving the efficiency of your code based on the profile information. Additionally, does the code include calls to force() and deparse()? Or some other unexpected functions? Can you determine why these functions are being called?

Q.5 Develop a set of test cases for genKidsV() (see Section 7.6). Write code to check the output from genKidsV() for these test cases.

Q.6 Figure 7.4 is a custom visualization of the birth and completion times for a tree. Design an alternative custom visualization of the return value from familyTree().

Q.7 Incorporate into familyTree() a limit on the time that the tree is observed. If we want to observe a process up until time t then those offspring with birth times after t are discarded. Also, the simulation stops once all of the observed births in a generation are after t. See Figure 7.5 for a visualization of the first 5 generations in a simulated process that is truncated at time 8.

Q.8 Update the genKidsV() function developed in Section 7.5 to return one data frame rather than a list of data frames. This data frame needs to include additional columns that supply the parent and offspring identifiers. See Section 7.7 for an example of the modified return value.

Q.9 The branching process was summarized by two statistics: the number of generations and the number of offspring (see Section 7.9). Consider other summary statistics for the process. Incorporate them into exptOne(). Carry out a simulation study and create a visualization of the simulation that uses these additional statistics. Do they confirm the earlier findings? Do they offer any new insights?

Q.10 Carry out a simulation study to see if the re-parameterization suggested in Section 7.9.1 is appropriate. For example, fix κ to be 1, and run the simulation for various values of λ. Compare the results to other simulations where κ is $c \neq 1$, but the ratio of λ/c matches one of the λ values from the earlier simulation when κ was 1.

Q.11 Consider other probability functions to describe the lifetime of a process. Revise familyTree() (see Section 7.8) and genKidsV() (see Section 7.5) to take as an argument the random number generator for any probability distribution. The functions familyTree() and genKidsV() are to use this probability distribution (with arguments that may be specific to the distribution) to generate the completion times of the jobs.

Q.12 Redesign the simulation study, where rather than generating one branching process at a time, the processes are generated in a vectorized fashion. This may require rewriting genKidsV() (Section 7.5) and familyTree() (Section 7.8).

Bibliography

[1] David Aldous and William Krebs. The 'Birth-and-Assassination' Process. *Statistics and Probability Letters*, 10:427–430, 1990.

[2] Charles Bordenave. On the birth-and-assassination process, with an application to scotching a rumor in a network. *Electronic Journal of Probability*, 13:2014–2030, 2008.

[3] George Kordzakhia. The Escape model on a homogeneous tree. *Electronic Communications in Probability*, 10:113–124, 2005.

[4] Uwe Ligges, Martin Maechler, and Sarah Schnackenberg. `scatterplot3d`: 3D Scatter Plot. http://cran.r-project.org/web/packages/scatterplot3d, 2014. R package version 0.3-35.

[5] R Development Core Team. *R: A Language and Environment for Statistical Computing.* Vienna, Austria, 2012. http://www.r-project.org.

[6] John Tsitsiklis, Christos Papadimitriou, and Pierre Humblet. The Performance of a Precedence-Based Queuing Discipline. *Journal of the Association for Computing Machinery*, 33:593–602, 1986.

8

A Self-Organizing Dynamic System with a Phase Transition

Deborah Nolan
University of California, Berkeley

Duncan Temple Lang
University of California, Davis

CONTENTS

8.1	Introduction and Motivation	309
	8.1.1 Computational Topics	310
8.2	The Model	310
	8.2.1 The Order Cars Move	312
8.3	Implementing the BML Model	314
	8.3.1 Creating the Initial Grid Configuration	314
	8.3.2 Testing the Grid Creation Function	318
	8.3.3 Displaying the Grid	321
	8.3.4 Visualizing the Grid	322
	8.3.5 Simple and Convenient Object-Oriented Programming	325
	8.3.6 Moving the Cars	327
8.4	Evaluating the Performance of the Code	334
8.5	Implementing the BML Model in C	346
	8.5.1 The Algorithm in C	348
	8.5.2 Compiling, Loading, and Calling the C Code	355
8.6	Running the Simulations	359
	8.6.1 Exploring Car Velocity	360
8.7	Experimental Compilation	362
	Bibliography	364

8.1 Introduction and Motivation

In this chapter, we will explore a very simple dynamic system. It exhibits two interesting characteristics – a phase transition and self-organization.

Most things in nature behave continuously. As inputs change a little, the outcomes change slightly. It is interesting when a slight change in an input causes a significantly different outcome. In some cases, as we vary an input over a range of values, we see continuous changes and then abruptly see qualitatively different outputs. This tipping point is called a phase transition. Basically, the system can jump between qualitatively different states as the inputs change very slightly in a continuous manner. One example is a liquid changing to a gas as we increase the temperature.

We are also familiar with systems which exhibit a global characteristic that is defined via set of equations or formulae controlling the global space. It is interesting when a system exhibits one or more global characteristics, but when the rules for the system are specified for local behavior. When elements change independently of the others, or only depend on a single neighbor, it can be surprising to see global patterns emerge. We will explore a system that uses very simple local rules and exhibits this *self-organizing* behavior.

Researchers are very interested in understanding systems that exhibit phase transitions and/or self-organization. They represent interesting opportunities and challenges in modeling systems. Complex systems are difficult to understand mathematically. Researchers try to find simple systems that exhibit a phase transition or self-organization and try to analyze the mathematics underlying these in order to potentially gain an understanding of more complex systems. In this case study, we will simulate one of the very simplest mathematical models that was thought to exhibit a phase transition. Via simulation and exploring the resulting data, researchers also found self-organizing behavior and other characteristics in this simple dynamic system.

8.1.1 Computational Topics

We will map a mathematical description of a dynamic process to code. We will focus on writing small functions in a flexible, reusable manner. We will validate the functions and combine them to create higher-level functions. We'll use R's [8] simple *S3* object-oriented programming model to define classes and methods for working with the data structures from our simulation of the dynamic model. We'll explore making the code faster in several different ways, including writing vectorized code and using compiled C code from within R. Finally, we'll use a simulation study to investigate the behavior of the dynamic model for different ranges of the inputs and different outputs.

- Developing small functions that we combine together to implement the entire dynamic process.
- Computational efficiency.
- Choice of data structures to simplify and improve computations.
- Comparing loops and vectorized operations and striving for vectorized code.
- Matrix subsetting operations in R.
- Classes and *S3* object-oriented programming in R.
- Profiling code to find where bottlenecks occur.
- The C programming language and interfacing to C code from R.
- Simulation.

8.2 The Model

The Biham-Middleton-Levine (BML) model [2] is a very simple 2-dimensional dynamic system that exhibits self-organizing behavior and a phase transition. We can describe the

A Self-Organizing Dynamic System with a Phase Transition

model as follows. We start with an m-by-n grid of cells. Each cell can contain at most 1 car. We have two types of cars — red and blue. Each car can move 1 cell in a given time interval. The red cars move horizontally to the right; blue cars move vertically upward. When a car reaches an edge of the grid, its next potential position is to wrap around to the other side/edge of the grid. In other words, when a red car reaches the right-most cell of the grid, it would next move to the cell of the first column of the grid, and in the same row. This wrap-around motion gives the grid a torus-like connection between the edges. A car cannot move if the cell to which it would move is currently occupied. There are useful interactive demonstrations of the model at http://www.jasondavies.com/bml/

We start with a (random) collection of cars located on the grid at time 0, for example, (a) in Figure 8.1. At time $t = 1$, all of the red cars move to the right or to the first column if they are already at the right edge of the grid, unless that cell is currently occupied. At time $t = 2$, the blue cars move. This continues with the red cars moving at the odd-numbered times and the blue cars moving at the even number times.

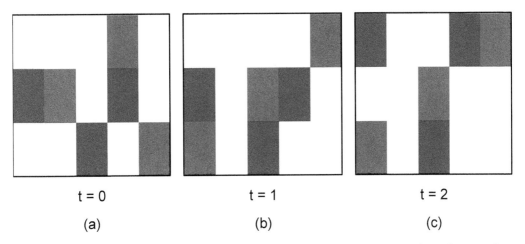

t = 0   t = 1   t = 2
(a)   (b)   (c)

Figure 8.1: Movement on a Sample Grid. *(a) shows the initial state of a 3-by-5 grid containing 3 red and 3 blue cars. At time $t = 1$, the red cars move horizontally. The red car on the bottom row "wraps around" to the first column on the same row. At time $t = 2$, the blue cars move up within the same column. The blue car in cell $(1, 3)$ is blocked by the red car above it that moved to that cell in at time $t = 1$. Accordingly, we obtain a sequence of grids indexed by time.*

How do the cars get on the grid? In other words, how do we create the initial state of the system? We randomly place n_R and n_B cars on the grid. We have to make certain that each car is in its own cell and that 2 cars don't occupy the same cell. This is the only random part of the dynamic system. After the cars are randomly located in the initial grid, how they move is entirely deterministic.

Suppose we only had 2 cars on the grid. Both are likely to be able to move at each time step as they probably will not be beside each other. Here, the cars move freely and their velocity is essentially 1 unit of distance for each unit of time. However, suppose we have a 10-by-10 grid and we have 100 cars. Clearly, no cell is vacant and no car can move. We have a complete traffic jam and the velocity is 0. Between these two extremes in the proportion of occupied cells, we will observe different behaviors and self-organizing schemes ranging from global free-flowing traffic to localized deadlock to global deadlock.

The behavior of this model depends critically on the number of cars on the grid. If we have very few cars, they are unlikely to be close to each other and so will probably move freely. Alternatively, if we have many cars relative to the number of cells in the grid, the cars are likely to form a traffic jam. So the density of cars — number of cars divided by the number of cells — is the the key parameter. Figure 8.2 shows a grid that has 25% of the cells occupied by cars. The locations of the cars are initially random. After 500 time periods of the cars moving, the cars appear to form regular diagonal lines and most of the cars are moving in each time cycle. The 3rd panel shows the grid after 1,000 iterations, confirming the free-flowing equilibrium.

Suppose we start with 40% of the cells in the grid randomly assigned a car (red and blue in equal numbers). The first panel in Figure 8.3 shows this initial configuration. After 500 iterations, we see emerging deadlock with some cars still in the middle of the grid and moving reasonably freely. After 1000 iterations, we see that the cars have formed deadlocked clusters.

What is the value of the density of the cars that changes the behavior of the dynamic system from free-flowing to jammed? How do the cars organize themselves ? What are the characteristics of these organizations? How do these depend on the dimensions of the grid? These are questions we want to be able to answer via simulation and data analysis.

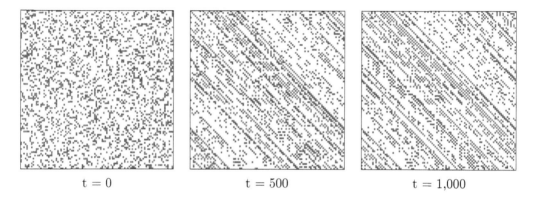

Figure 8.2: Sample Free-Flowing Traffic Grids. *The 3 panels show the initial grid at time t = 0, t = 500, and t = 1000. Cars occupy 25% of the cells. After 500 iterations, the cars start to organize along diagonal lines. After 1000 iterations, the lines are becoming clearer.*

8.2.1 The Order Cars Move

The details of how the cars move described above essentially characterize the BML model. However, there is one practical aspect that we need to discuss in order to program the model. Consider the very simple grid in Figure 8.4.

We have 3 adjacent red cars in a single row, which we refer to as A, B, C. At time $t = 1$, we attempt to move each of the red cars. If we move car C first, it will move horizontally to the empty fourth cell in the row. We can then move car B to the cell that C was previously occupying. Similarly, A can then move to B's previous spot.

If we had not moved C first, but processed the cars in the order A, B and then C, we would have a very different outcome. A would not be able to move since B is occupying the cell to which it would move. Similarly, B would be unable to move to C's current location. Finally, C would move to the fourth cell in the row, as before.

When we move the cars one at a time, the order in which we attempt to move the cars

A Self-Organizing Dynamic System with a Phase Transition 313

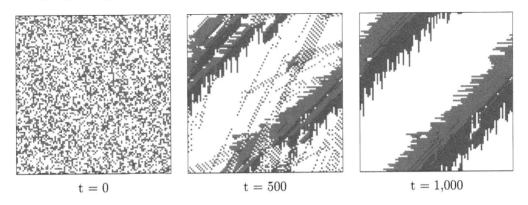

Figure 8.3: Sample Deadlocked Traffic Grids. *Three grids at different time steps showing the emergence of deadlocked traffic.*

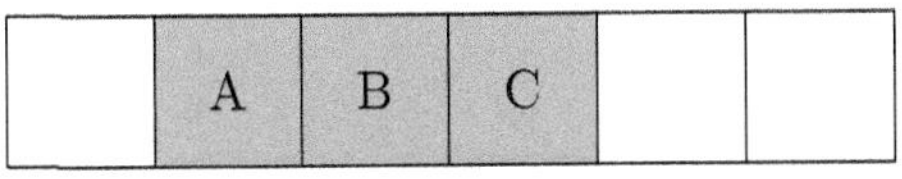

Figure 8.4: The order in which cars move. *If car A tries to move first, it cannot move as B is occupying the target cell. However, if C moves first, it can move to the right and then B can move and then A. The order in which the cars move leads to different outcomes.*

might change which ones actually move in a given time period. We have several choices for dealing with this when we implement our algorithm.

1. We can process the cars in the order in which they were placed on the grid and ignore any ordering effect.

2. We could randomize the order in which we process each car at each time period and again move them using this random order.

3. Within a time step, we could start by moving all those cars that can move and then move those cars that couldn't initially move. We would then continue to repeat this again within this time step until all the cars for this time step had either moved or were actually blocked, i.e., until none of the eligible cars moved in an iteration.

We would like to know if these different approaches lead to qualitatively different results. Initially, however, we will start with the simplest approach (1 above) but try to write our code so that we can easily substitute in any of the other different ordering schemes. Motivated by computational efficiency, we will ultimately use a vectorized approach that corresponds to a fourth approach. This moves all cars simultaneously. This effectively means that we determine which target cells are vacant in one step and then move only those cars that will move to one of these vacant cells. This does not allow a car to move into a cell if another car moves out of that cell in the same time period. This corresponds to moving car A, then B, and then C in Figure 8.4.

8.3 Implementing the BML Model

We now have a complete description of the BML model and can start to write R code to implement and simulate it. Let's think about the different steps we have and also what inputs we need at each step.

1. For the step at time t = 0, we need to create the grid.

2. We want to view the grid at the start and also at different time steps in the process as it evolves. Visualizing the state of the grid helps us to understand the process, but is also essential to help debugging the code.

3. At a time t, we need to move the cars. At odd numbered time steps, we move the red cars, and at even numbered time steps, we move the blue cars.

4. To run the process, we need to iterate over a sequence of time steps, not just a single time step.

We will need code for each of these steps. As we develop functions for these, we'll also try to make them flexible, extensible, and efficient. We'll also visualize the grids and compute summary statistics from them such as the proportion of cars that move at each time step, and hence average velocity.

8.3.1 Creating the Initial Grid Configuration

To create the grid, we need to know the dimensions of the grid and the number of cars to create and where to position them. We can think of the grid as being a square and use just a single dimension to describe it. However, this is unnecessarily restrictive. It costs us very little to allow the user of our function createGrid() to specify different lengths for the width and height of the the grid. We can allow the caller to specify 1 length and use that for both dimensions, but still have the option of specifying the 2 dimensions separately. Being able to vary the dimensions allows us to explore how the behavior of the process changes as we use, say, relatively prime dimensions. This does indeed turn out to be important. So we can use calls of the form

```
createGrid(100)
createGrid(c(100, 200))
```

Similar to the dimension, we can also specify the number of red and blue cars separately or specify a single value to be used for each, e.g., `createGrid(100, 50)` or `createGrid(100, c(50, 30))`. Rather than specifying the number of cars, it can be convenient to specify the proportion of the cells that should contain a car, e.g., `createGrid(100, .25)`. As we change the dimensions, this makes it easy to keep the same density of occupied cells. Given this proportion and the dimensions of the grid, we can compute the total number of cars. We can divide this in two to get the number of red and blue cars. Accordingly, we have 3 different ways to specify how the number of each type of car is computed: a single number used for both types, a vector of length 2 with a number for each car color, and a proportion between 0 and 1 used to compute the proportion of cells occupied by a car of either type.

So let's put these different forms of inputs for the caller together to define the skeleton of a function to create our initial (t = 0) grid:

A Self-Organizing Dynamic System with a Phase Transition

```
createGrid =
function(dims = c(100, 100), numCars = .3)
{
   if(length(dims) == 1)
      dims = rep(dims, 2)   # a square grid

   if(length(numCars) == 1 && numCars < 1)
      numCars = rep(prod(dims) * numCars/2, 2)

   ...
}
```

This function ensures that *dims* and *numCars* have the correct length and interpretation.

Q.1 Does the function above handle all input cases? If not, which ones could occur that haven't be handled?

We can now generate the actual locations of the cars. How should we do this? We know we need numCars[1] red cars and numCars[2] blue cars. For each car, we need its row and column numbers. We have to ensure that there is no existing car already at that location. We could generate the location one car at a time. We would generate a possible location (i, j) and then check to see if the cell is already occupied. If the cell is occupied, we generate another possible location and iterate until we place that car. As we place more and more cars, there will be fewer available locations for the next cars. As a result, we will spend more time/iterations finding an available location. Furthermore, if we place the red cars first, the blue ones have to fit into the remaining available cells. Will this introduce a bias?

Rather than using a loop to place each car, we want to use a vectorized approach. This is the R style of programming that makes for more succinct, flexible, and efficient code. So let's think of different approaches to placing the cars using a vectorized operation. Firstly, suppose we have a 2-column matrix of the locations (row and column indices) of all the cars without knowing the color of each car. We can randomly assign the red and blue labels to those locations/rows. This gives the same result probabilistically as if we allocate the colors before positioning them. We could generate the colors/labels for the $n_R + n_B$ cars with

```
carColors = sample( rep(c("red", "blue"), numCars) )
```

Note that we are using rep() to create a vector with the specified number of red and blue elements and then we are generating a permutation of this vector. This ensures we end up with the specified number of red and blue values, rather than a random sample from the 2 values, which allows the number of each to be different from what we specified for each. It is essential that numCars is a vector with 2 elements giving n_R and n_B. We can check carColors contains the correct number for each color with

```
table(carColors)
```

Q.2 What if we used

```
sample(c("red", "blue"), sum(numCars), prob = numCars,
       replace = TRUE)
```

to generate the colors for the sampled locations? Is this qualitatively the same as the previous code with `rep()`? If not, what is the difference? Will this matter for our simulations?

Once we know the colors of the cells with cars, we need to generate their locations. We might consider sampling the row indices and then the column indices and combine them together into a 2-column matrix, e.g.

```
N = sum(numCars)
rows = sample(1:dims[1], numCars, replace = TRUE)
cols = sample(1:dims[2], numCars, replace = TRUE)
```

Here `sum(numCars)` is the total number of cars to be placed. Unfortunately, we will probably end up with conflicts with 2 cars at the same location. We could determine which cars would be in an already occupied cell and resample the entire collection of cells again. We can repeat this until we eventually get no conflicts. This is similar to the approach and the problem we had above when we placed the cars separately and sequentially.

Let's consider a different approach. We could label each cell uniquely by its row and column numbers. We could then sample from these unique cell identifiers and decompose each identifier for a cell into its row and column. We would be guaranteed not to get any conflicts as we sample without replacement from all the available cells. This would allow us to do the sampling in one operation. We can create the unique identifiers with

```
ids = outer(1:dims[1], 1:dims[2], paste, sep = ",")
```

This yields character strings of the form `"1,1"`, `"1,2"`, ..., `"m,n"`. We can then sample these using

```
pos = sample(ids, sum(numCars))
```

and get the row and column values with

```
tmp = strsplit(pos, ",")
rows = as.integer(sapply(tmp, `[`, 1))
cols = as.integer(sapply(tmp, `[`, 2))
```

This works well, but we can make this much simpler.

Instead of explicitly creating all of the unique "row,column" pair identifiers, we could use the numbers 1, 2, 3, ..., `dims[1]*dims[2]` as unique identifiers. We then sample from these identifiers. Given these, we have to be able to map each identifier value back to a row and column. Fortunately, this is quite easy. To get the column number, we subtract 1 from the cell identifier value, divide this by the number of rows, and round this value down to the integer value, and add 1 to get the column number. As a computation, this is `floor( (value - 1)/dims[1]) + 1L`. To get the row number, we use the remainder from performing the division for the column number and adding 1, i.e., `(value - 1L) %%dims[1] + 1L`. For example, suppose we have a 10-by-9 grid and we sample an index 14. This would correspond to row 4 and column 5.

Again, this is relatively straightforward, but there is an even simpler mechanism. Let's take this idea of sampling the indices 1, 2, 3, ... further. Suppose we represent our grid of cells as an *R* matrix. We can use these positions to directly index into elements of the matrix. This is because a matrix is merely a vector of values with a dimension attribute that specifies the number of rows and columns. We can create our empty grid with

```
m = matrix(0, dims[1], dims[2])
```

We can then sample the indices at which the cars will be located with

```
pos = sample(1:(dims[1]*dims[2]), sum(numCars))
```

Then

```
m[pos] = 1
```

sets only the elements corresponding to our sampled cells to 1. This is much more succinct.

We have glossed over one detail in subsetting the matrix by the sampled indices. A matrix stores the values in column order and not row order. In other words, the matrix

```
matrix(1:10, 5, 2)
     [,1] [,2]
[1,]   1    6
[2,]   2    7
[3,]   3    8
[4,]   4    9
[5,]   5   10
```

is actually stored as an integer vector in the form

```
[1]  1  2  3  4  5  6  7  8  9 10
```

This is different from row-oriented storage, which would yield

```
[1]  1  3  5  7  9  2  4  6  8 10
```

In our case it doesn't matter whether we use rows or columns when indexing as the cells we sample are uniformly distributed throughout the entire collection of cells in the matrix. The order doesn't matter.

Let's create the final version of our function to create our sampled grid and color the cells red and blue using this final approach:

```
createGrid =
function(dims = c(100, 100), numCars = .3)
{
   if(length(dims) == 1)
     dims = rep(dims, 2)

   if(length(numCars) == 1 && numCars < 1)
      numCars = rep(prod(dims) * numCars/2, 2)

   grid = matrix("", dims[1], dims[2])

   pos = sample(1:prod(dims), sum(numCars))
   grid[pos] = sample(rep(c("red", "blue"), numCars))

   grid
```

We set the elements with a car to the corresponding color. Note that we explicitly returned `grid` at the end of the function. Sometimes people forget to do this and leave the last expression as the assignment `grid[pos] = sample(...)`. This would return just vector returned by the second call to sample().

8.3.2 Testing the Grid Creation Function

Before we move on to the next steps of moving the cars, let's verify that our function gives us sensible output. This is very important. If we get this wrong, the rest of our work will also be wrong. Any testing we do on subsequent code will be wasted and this can consume a lot of time and also even lead to erroneous code, results, and confusion in the programmer's mind. In short, we need a solid foundation for each next step.

We need to think of ways to test the createGrid()function. We can print the results, but for anything but very small grids, the output will be overwhelming. Therefore, we need some meaningful summaries. We also need to test it with different inputs, e.g., non-square grids, different numbers of red and blue cars, and different densities.

A simple test is whether the function returns what we expect

```
g = createGrid()
class(g)
```

[1] "matrix"

```
dim(g)
```

[1] 100 100

These are what we expect.

We should also check that the number of red and blue cars is the same and account for 30% of the available cells:

```
table(g)
```

```
     blue  red
7000 1500 1500
```

The first element (7000) is the count of the empty cells. The name appears blank as this corresponds to the value " " in the grid. Do we expect the numbers of red and blue cars to be identical?

Rather than simply evaluating these expressions and visually verifying the results are as we expect, we can raise an error if they are not. For example,

```
stopifnot(dim(g) != c(100, 100))
stopifnot(all(table(g) %in% c( 7000L, blue = 1500L, red = 1500L)))
```

The latter test doesn't check whether the individual elements in table(g) are the same as those in the vector we expect. Instead, it merely checks that all the values are in the vector we expect; the counts could be in a different order corresponding to different colors. To test they are exactly as we expect, we have to compare the corresponding elements. table() returns a slightly more complex object than a simple vector of counts. To test for equality of the object, we have to get the dimension, dimension names, and class to be the same, e.g.,

```
stopifnot(identical(table(g),
              structure( c(7000L, 1500L, 1500L),
                  dim = 3L, class = "table",
                  dimnames = list(g = c("",
                                       "blue",
                                       "red")))))
```

This is a much better test, but its added complexity runs the risk of introducing errors/bugs into the test itself.

Let's use our function to create a small grid with unequal dimensions:

```
createGrid(c(3, 5), .5)
```

This gives a warning

```
Warning in grid[pos] = sample(rep(c("red", "blue"), numCars)) :
 number of items to replace is not a multiple of replacement length
```

So good thing we checked! We use

```
options(error = recover, warn = 2)
```

to establish a debugging mechanism that allows us to explore the errors and warnings when and where they occur. If you are not familiar with the recover() function, take a moment to read its help page. Setting `warn = 2` causes warnings to be treated as errors. This allows us to stop at a warning and explore the current state of the computations where that warning occurs. We trigger the problem again by re-evaluating the same expression:[1]

```
createGrid(c(3, 5), .5)

Error in grid[pos] = sample(rep(c("red", "blue"), numCars)) :
  (converted from warning) number of items to replace is not
  a multiple of replacement length

Enter a frame number, or 0 to exit

1: createGrid(c(3, 5), 0.5)
2: #13: .signalSimpleWarning("number of items to replace is not
3: withRestarts({
   .Internal(.signalCondition(simpleWarning(msg, cal
4: withOneRestart(expr, restarts[[1]])
5: doWithOneRestart(return(expr), restart)

Selection:
```

When the warning occurs, R converts it to an error and calls stop(), which is intercepted by our error handler, the recover() function. We are presented with the stack of current function calls, i.e., the call stack. These are the items 1 through 5. The error is in the body of the function createGrid(). So we can enter 1 at the Selection: prompt. This will place us in the call frame for this function call. We can then examine (and even modify) the parameters and local variables that define the state of this call.

Within the debugging browser, when we look at the value of numCars, we see this is

```
[1] 3.75 3.75
```

and pos is something like

```
[1]  1 14  7  3 15  9 12
```

---

[1] Given the computations involve randomness, it is possible that the same call will not generate a warning each time. However, in this case it will.

What is the right hand side of the assignment? It is the value of the expression

```
sample(rep(c("red", "blue"), numCars))
```

and is something like

```
[1] "red"  "red"  "blue" "blue" "blue" "red"
```

The values are random, but the number of elements is not and is 6 rather than 7, which is the length of pos (since sum(numCars) is 7.5). That is the disparity. The problem is the fractional values in numCars, i.e., 3.75 and how this affects the call to rep(). Instead of using rep(), we can use rep_len() and ensure that we get the same number of elements as in pos:

```
rep_len(c("red", "blue"), length(pos))
```

But this doesn't allow us to specify a different number of cars for the red and blue cars, i.e., a vector for numCars. Instead, we might use

```
sample(rep(c("red", "blue"), ceiling(numCars)))[seq(along = pos)]
```

which rounds the number of cars for each type to the next largest integer and then subsets the result to have the same length as pos.

Q.3 Does this lead to an imbalance in the number of cars? Is this what we want?

We can make this change and redefine the function as

```
createGrid =
function(dims = c(100, 100), numCars = .3)
{
   if(length(dims) == 1)
     dims = rep(dims, 2)

   if(length(numCars) == 1 && numCars < 1)
      numCars = rep(prod(dims) * numCars/2, 2)

   grid = matrix("", dims[1], dims[2])

   pos = sample(1:prod(dims), sum(numCars))
   grid[pos] = sample(rep(c("red", "blue"),
                          ceiling(numCars)))[seq(along = pos)]

   grid
}
```

Now we need to rerun our tests that passed for the previous version and continue to add new tests. It is essential we do this. We have modified and added code. As a result, there is a good chance we have introduced a bug. Having the tests from earlier in separate files that we can source() into R and raise an error (via, e.g., stopifnot()) if there is an unanticipated result allows us to easily recheck our code and be somewhat confident about it before embarking on the next step.

Let's create a small grid. We may want to ensure we can get the exact values again for the cells and so set the random seed:

```
set.seed(1456)
createGrid(c(3, 5), .5)
```

```
     [,1]   [,2]   [,3]  [,4]   [,5]
[1,] "red"  ""     "red" ""     ""
[2,] ""     "red"  ""    "blue" ""
[3,] "blue" "blue" ""    "red"  ""
```

We can specify the number of red and blue cars with

```
set.seed(1234)
createGrid(c(3, 5), c(9, 3))
```

```
     [,1]   [,2]  [,3]   [,4]   [,5]
[1,] "red"  "red" "red"  "blue" "red"
[2,] "red"  "red" "red"  ""     ""
[3,] "blue" "red" "blue" ""     "red"
```

Again, we need to develop tests for corner cases, e.g., the caller asking for 0 cars of either or both types, or more cars than can fit on the grid, etc.

Q.4 Add code to the createGrid() function to raise an error or warning as appropriate when the caller specifies infeasible inputs for the number of cars or for the dimensions of the grid.

We can check the counts again

```
set.seed(1234)
g = createGrid(c(3, 5), c(9, 3))
table(g)
```

```
     blue  red
  3     3    9
```

Q.5 Should we allow the caller to use names on the vector of number of cars *numCars* to indicate the colors or values for the two types of cars in the matrix? Modify the createGrid() function to use these.

8.3.3 Displaying the Grid

When we display a BMLGrid object in the R console, it is shown as a matrix, as we saw in the 3-by-5 example above. There are several aspects of this that make it harder than we would like to quickly comprehend. Firstly, the row and column names are distracting, e.g., the [1,] and [,1] on the side and above the actual cell values. We could display these as 1, 2, 3, ..., without the surrounding [,]. More importantly, the rows appear in increasing order on the console. Specifically, row 1 appears above row 2, which appears above row 3, and so on. This is not how we think about the grid and can make it hard to reason about the movement of an individual car in the grid. When a blue car moves "up" from row 2 to row 3, it will actually move down as displayed on the console. Accordingly, we'd like the rows to appear in the opposite order. We can define our own function to display a BML grid object. We might implement it as

```
print.BMLGrid =
function(x, ...)
    print(structure(x[nrow(x):1,],
                    dimnames = list(nrow(x):1, 1:ncol(x))))
```

This creates a new `matrix` object in the appropriate manner so that the rows appear in the "correct" order and the row and column names are set appropriately. It then uses R's regular print() function to display this new representation of the original grid.

With this new print.BMLGrid() function, our grid `m` will appear as

```
   1       2      3       4
3 "blue"  "blue" ""      "red"
2 "blue"  ""     "red"   ""
1 ""      "red"  ""      "red"
```

Q.6 The quotes around the color names (e.g., `"blue"`) are distracting. Write a different version of print.BMLGrid() that produces the display without the quotes. Reuse existing functionality in R rather than writing code to format individual lines of output. Hint: when does R display character vectors/strings without quotes?

Q.7 Instead of using the names of the colors, use arrows such as ↑ and →.

8.3.4 Visualizing the Grid

In addition to printing the grid on the console, we can visualize it using R's graphics capabilities. This is useful so we can see that the locations of the cars and colors appear random and potentially identify any anomalies in our code. We should write a function to do this so that we don't have to remember the details. We could draw the cells as colored circles on a scatter plot. However, it seems to make more sense to display them as rectangles that occupy the grid's cell. We could use the rect() function to draw the rectangles. We need to create 4 vectors specifying the x and y locations for the two opposite corners of each cell, i.e., 4 n-length vectors where n is the total number of cars. We also have to create the initial coordinate system for drawing the rectangles by creating a new plot. This is quite simple but involves several steps. Instead, we should think about whether there is an existing high-level function in R or some R package that does what we want, or close to it, so that we could adapt it. Reusing functions is a good thing to do as they save us programming time but also are more likely to be correct and full-featured.

We might try to use the image() function to render the matrix for us. Before reading further in this chapter, consult the help page for image() and think about how we will use it to display the contents of the grid.

To use image(), we have to convert the string values in the matrix to numbers. We can do this by mapping the values `""`, `"red"`, and `"blue"` to a set of numbers, say, 1, 2, and 3, respectively. We can do this with the match() function via

```
z = matrix(match(g, c("", "red", "blue")), nrow(g), ncol(g))
```

The result corresponding to our grid `m` we created earlier is

```
     [,1] [,2] [,3] [,4]
[1,]   3    3    1    2
[2,]   3    1    2    1
[3,]   1    2    1    2
```

We can then call the image() function as `image(z)` This display in Figure 8.5 is somewhat difficult to interpret. We should control the colors explicitly. Based on our matching, the colors should be white for 1, red for 2, and blue for 3. Therefore, our plot command should be

```
image(z, col = c("white", "red", "blue"))
```

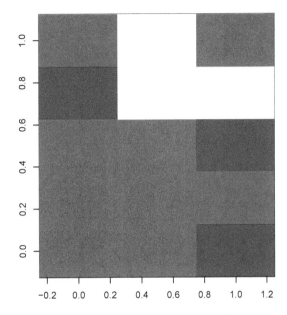

Figure 8.5: Simple Plot from image(). *Using a simple call to image() to display the grid uses the wrong colors and also does not display the cells in the order we expect or want.*

We have to carefully verify that the cells in the matrix correspond to those in the image. In fact, they do not. The problem is that image() essentially rotates the matrix by 90 degrees and displays that. Therefore, the rows and columns are exchanged in the plot. In order to obtain the same display as we have via print.BMLGrid() and as we would expect, we have to transpose the matrix before passing it to image(). We can use

```
image(t(z), col = c("white", "red", "blue"))
```

To view a grid, we don't want to have to remember and perform these calculations each time. We should write a function to plot it for us that uses and encapsulates these computations. Let's call our function plot.BMLGrid() and define it as

```
plot.BMLGrid =
function(x, ...)
{
   z = matrix(match(x, c("", "red", "blue")), nrow(x), ncol(x))
   image(t(z), col = c("white", "red", "blue"),
         axes = FALSE, xlab = "", ylab = "", ...)
   box()
}
```

We remove the horizontal and vertical axes as they have no meaning in this context. Alternatively, we could add the row and column numbers for the grid. Note also that we changed our code to refer to x, the function's parameter, rather than g, which we had in our interactive expression we used to experiment with the image() function. This style of interactively refining a command and then copying it to the body of a function can often lead to bugs due to referring to variables not defined in the new function. Our function would still work because g is available in the work space, but it ignores any grid we explicitly pass to it! It is good practice to use `codetools::findGlobals(plot.BMLGrid, FALSE)` to check for and identify any non-local variables.

Importantly, our function passes any arguments for image() that we don't use directly in our function to the call to image() via the ... mechanism. This allows the caller to customize the image() function with additional inputs. For example, she can specify a title for the plot with

```
plot.BMLGrid(g, main = "A sample title", sub = "A sub title")
```

This is good practice to make our function more flexible for the callers with little additional effort.

Q.8 How would we change our plot.BMLGrid() function to allow the caller to override the `axes = FALSE` argument in the call to image()? In other words, we want no axes to appear by default, but to allow the caller to optionally show them.

We can use our new function to display the grid with `plot.BMLGrid(g)`. Again, we have to verify that this code is correct and we leave it as an exercise.

Let's look at a larger grid:

```
g = createGrid(c(100, 100), .5)
plot.BMLGrid(g)
```

We show this in the first panel of Figure 8.6.

We can also use many more red than blue cars to see if this characteristic appears in the display. We'll create a 10,000-cell display with half of the cells being red and 500 being blue. The remaining 4,500 cells are white/empty:

```
g = createGrid(c(100, 100), c(5e3, 500))
plot.BMLGrid(g)
```

We can see this in the second panel of Figure 8.6.

Let's also create our own grid with no random values, but where we explicitly place the cars deterministically:

```
m = matrix("", 3, 4)
m[3, 1] = "blue"
m[1, 4] = "red"

m[row(m) == col(m) - 1] = "red"
m[col(m) == row(m) - 1] = "blue"
```

The final two expressions put red cars along the upper-off-diagonal and blue cars along the lower-off-diagonal. This is shown in the third panel of Figure 8.6.

Before we move on to the next topic, we note that we have now repeated our colors (white, red, and blue) in 3 different places: in createGrid() and in the calls to match() and image(). We are violating the **DRY** (Don't Repeat Yourself) principle. We have to ensure

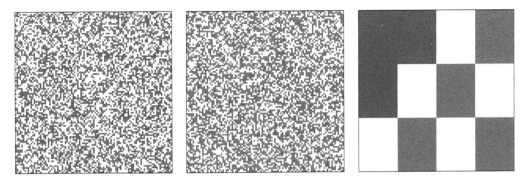

Figure 8.6: Sample Grid Displays. *The first panel shows a 100-by-100 grid with 50% of the cells occupied equally with red and blue cars. The second panel has the same dimensions but there are 5000 red cars and only 500 blue cars. The 3rd panel shows a small 3-by-4 grid where we placed the cars manually.*

they are all the same and in the same order. It also makes it harder for us to change if we want to use different colors, e.g., for issues with color-blindness, displaying in different media such as an overhead projector versus a monitor versus printing. Instead, we should define the color names as a vector and reuse this in each of these 3 locations. We can use a global variable and make this the default value for a parameter in each of these functions. (We leave this as an exercise.) In general, global variables are "bad;" however, here we are using this essentially as a global *constant*/immutable vector in a centralized location.

8.3.5 Simple and Convenient Object-Oriented Programming

We defined the plot.BMLGrid() function in the previous section. This made visualizing an arbitrary grid created with createGrid(), or even directly by specifying the values of the cells, very straightforward. Since we wrote the plot.BMLGrid() function, it is easy to remember its name. However, suppose that this function is in an *R* package. We'd have to find the function and remember its name. Instead, we'd like to be able to simply call the generic plot() function and have it display our grid/matrix using our plot.BMLGrid() function. In other words, we'd like to use the general verb plot() but have it understand to do something specific for our particular data structure. This is one important aspect of object-oriented programming and is quite simple using *R*'s *S3* classes and methods mechanism.

We don't want *R* to display all matrix objects using our plot.BMLGrid() function. Instead, we want it to be used for our grid objects that we generate with our createGrid() function or other similar specialized functions. In other words, we want to use the generic function plot() but have it be specialized for objects that represent BML grids. To do this, we need only make each of our grid objects have a class attribute named BMLGrid, the same suffix as we used in the name of our plot function, e.g., plot.BMLGrid(). We can associate the class name with a grid object easily with

```
class(g) = "BMLGrid"
```

Then, when we call plot(g), *R* will implicitly call our plot.BMLGrid() function and produce the plot we want. If you want to verify this, we can use trace() to show each call to a function, i.e., trace(plot.BMLGrid) and then plot(g).

How does *R* know to call plot.BMLGrid()? It is quite simple. The plot() function contains a call of the form UseMethod("plot"). When *R* evaluates this, it looks at the primary/first argument[2] in this call to plot() and obtains its `class` attribute. This is often a single string, but it can be a character vector with more than one element. R combines the name of the generic function being called — plot in this case — with each name in the class vector and also the word `default`, each separated by a period, i.e., `plot.BMLGrid` and `plot.default`. R then iterates over these names and searches for a function with that name. If it finds such a function, it passes control to that function. In our example, R will search for plot.BMLGrid() and plot.default(). Since we named our function plot.BMLGrid(), R will find and invoke that function. If we had named the function anything else, say `plot.CarGrid`, R would have ignored it and used plot.default().

The vector of class names for an R object also allows us to inherit methods. For example, suppose we evaluated the expression

```
summary(g)
```

summary() is also a generic function that calls UseMethod() to dispatch to class-specific methods. R then looks for summary.BMLGrid() but doesn't find one. Instead, it finds summary.default() and calls that since we inherited the default method.

Our grid object is stored as a matrix in R. However, when we set the `class` attribute, we overwrote the original `matrix` class value of the object with the single string `"BMLGrid"`. However, we could have set the class on our grid as

```
class(g) = c("BMLGrid", "matrix")
```

or more generally

```
class(g) = c("BMLGrid", class(g))
```

In either case, the class vector will include both `BMLGrid` and `matrix`. As a result, when R looks for a method for a generic function such as summary(), it will search for summary.BMLGrid(), summary.matrix(), and summary.default(), in that order. What this means is that if there is a summary.matrix() method, we will inherit that rather than using the default method. This is method inheritance by class. Our `BMLGrid` class is more specific than `matrix` and so we will use `BMLGrid` methods before a matrix method. However, if there is no `BMLGrid` method for a particular generic function, we will use a matrix method, if one is available. Otherwise, we'll use the default method or else fail.

The method dispatch mechanism also allows us to create specialized versions of our own `BMLGrid` class. For instance, we might define a `SquareBMLGrid` class that has all the characteristics of a `BMLGrid` object but which has the special property that the width and height of the grid are the same. Or we might define a class `CoPrimeBMLGrid`, which indicates the dimensions are relative primes of each other. We would set our class as

```
class(g) = c("SquareBMLGrid", "BMLGrid", "matrix")
```

and

```
class(g) = c("CoPrimeBMLGrid", "BMLGrid", "matrix")
```

respectively. This would allow us to define methods for each of these classes that required more specific computations, but also allow us to inherit more general methods, e.g., print.BMLGrid().

In order to make use of our `BMLGrid` methods, our grid objects need to have the class `BMLGrid`. We modify our function createGrid() to specify this class on our return object. We can replace the final expression returning the grid object with either

---

[2] Actually, R dispatches on the argument matching the first parameter, not necessarily the first argument as it may be named argument. Also, UseMethod() can identify the parameter on which to select the method.

```
class(grid) = c("BMLGrid", "matrix")
grid
```

or

```
structure(grid, class = c("BMLGrid", "matrix"))
```

Classes and methods help users by allowing them to use generic function names rather than having to remember names such as plot.BMLGrid(). They also allow programmers to simplify code. Rather than having numerous `if` statements in a function to handle different classes of inputs, these specializations can be handled in various separate functions that R automatically finds and invokes as we described. This makes it easy to extend an existing generic function without modifying it and also makes maintaining the functions easier and more reliable. It does make finding which code will actually be used a little more indirect. However, we can use R's methods() function to query the available methods for a particular generic function, e.g., `methods(plot)`.

The *S4* and reference class mechanisms are more powerful than the *S3* class mechanism. However, we won't explore these here. Instead, see Chapter 9. Most of the R modeling functions in R use *S3* and so these are powerful and effective, and also something useful to understand.

One of the powerful aspects of object-oriented programming is being able to define a new, more specialized class and provide a method for it. We did this with BMLGrid, print.BMLGrid(), and plot.BMLGrid(). We can introduce new classes that extend BMLGrid simply by prepending the new class name to the `class` vector, e.g.,

```
class(g) = c("ExtendedBMLGrid", "BMLGrid", "matrix")
```

We might also define a method for the ExtendedBMLGrid, e.g., plot.ExtendedBMLGrid(). However, suppose we wanted this method to first perform some computations and then call the regular method for the BMLGrid class. We might be tempted to explicitly call plot.BMLGrid(). Instead, a more flexible and general approach is to call NextMethod(), e.g., `NextMethod("plot")`. This arranges to call the next inherited method based on the class of the object (typically, the first argument in the current method). This avoids explicitly assuming the name of the next class, e.g., BMLGrid, and allows for other programmers to define further specialized classes and methods.

8.3.6 Moving the Cars

After all this work, we have now created our grid corresponding to the first of our 4 steps (see Section 8.3). The next task is to move the cars for a given time step. We have to move either the red cars east, or the blue cars north. We could write two separate functions, each handling the different colored cars. However, they are likely to share common operations and we would end up repeating code. This is bad. We'll probably cut-and-paste code from one function to the other. If there is an error in the original function, it will be present in the second. We often correct it in one but not the other. Additionally, if we improve the code in one place, we have to make the same changes in the other place. Furthermore, it makes reading, understanding, and maintaining the code harder. We have lost the explicit connection between the two that is clear when two functions call a shared function, but not when we have the code repeated. The general idea is the DRY principle — Don't Repeat Yourself. It is such an important concept in programming, we'll say it again — Don't Repeat Yourself!

Ideally, and ultimately, we would like to have a single function that we can call to move either set of cars, which color depending on the caller. Certainly, we would like to have a single function that identifies the common abstractions across the 2 colors, and perhaps specializes the actual motion for the different directions in other functions. It may be prudent not to be too ambitious at the very beginning. We may want to start by writing two separate functions to get things working. Then we could examine these and identify their common parts and combine them into one more general function. This is a good approach as we are striving for the general version ultimately, but making things simpler initially. Starting with the more general, abstract version may slow us down and be too difficult without having a working version for one type of car.

The most obvious way to move the cars is to process each car separately, determine whether the cell to which it would move is currently vacant, and if so update the location of that car. When a car moves, we have to clear the cell it currently occupies and set the color of the cell to which it moves. We can do all of this within a loop to process all of the cars of a given color. As usual, this is not ideal in R as it can be slow. Instead, we'd like a vectorized approach. However, let's implement this loop approach here as a) we want a version that we know is correct and which we can use to check a more ambitious, vectorized version, and b) we'll revisit this in another context later on in the chapter (when we implement a fast version in C). This approach treats the cars sequentially, rather than simultaneously. As a result, a car may not be able to move in this time step, but would if we changed the order in which we process the cars. This is fine for our implementation.

While the loop will be relatively easy to write, we immediately run into a problem with how we have represented our grid. We have the entire matrix but we do not know the location of the red or the blue cars. When we created the grid, we knew their locations. However, we then put this information into the matrix and discarded it. For each time step, we have to extract the information from the matrix about where the cars are currently located. We can do this with the row() and col() functions. We can get the locations of all the cars with

```
i = row(g)[g != ""]
j = col(g)[g != ""]
pos = cbind(i, j)
```

We can use pos to index our grid matrix to get the color associated with each of the cars:

```
colors = g[pos]
```

Note how we are using a (2-column) matrix to subset a matrix. This takes a little time to get used to. Take a moment to experiment at the R console with matrix subsetting using some small, simple matrices or these grids of cars.

We can combine the row and column information into a data frame representing all of the car locations and colors using

```
cars = data.frame(i = i, j = j, colors = colors)
```

We now have the locations and we can loop over the relevant subset (blue or red) to move those cars.

Q.9 Why should we use a data frame to represent this information? What are the alternatives? A *matrix*? A *list*? Three variables i, j, and pos?

A Self-Organizing Dynamic System with a Phase Transition

Our loop to move the cars of one color, say blue, can be implemented something like the following

```r
w = which(cars$colors == "blue")
for(idx in w) {
    curPos = c(cars$i[ idx ], cars$j[idx])
    nextPos = c(if(cars$j[idx] == nrow(grid))
                    1L
                else
                    cars$j[idx] + 1L,
                cars$j[ idx ])

    # check if nextPos is empty
    if(grid[ nextPos[1], nextPos[2] ] == "")  {
        grid[nextPos[1], nextPos[2]] = "blue"
        grid[curPos[1], curPos[2]] = ""
    }
}
```

The code is reasonably straightforward. We determine the indices for the blue cars and loop over these. For each blue car index, we extract its current position and compute its would-be next position. We adjust for reaching the top edge of the grid so that we wrap around to the lowest row on the grid, if this occurred. Then we test if the would-be position is available/empty and if so, update the current contents of the grid.

Q.10 Note that we cannot use a call to lapply()/sapply() in place of the `for` loop above. Why?

To move the red cars, we would have very similar code. We'd replace "blue" with "red" and also how we compute the `nextPos`. This allows us to see how to write a single function, moveCars(), which can move either set of cars — red or blue. Our moveCars() function takes the current grid and returns the updated grid. Note that these 2 grids will be separate copies, not a modification of a shared grid object. This allows us to build up a sequence of grids and visualize the progress and easily compare them to validate our code. Our function also needs to know which color of car we are moving. We define it by combining the code from the different steps above.

```r
moveCars =
function(grid, color = "red")
{
  i = row(grid)[grid != ""]
  j = col(grid)[grid != ""]
  pos = cbind(i, j)
  colors = grid[pos]
  cars = data.frame(i = i, j = j, colors = colors)

  w = which(cars$colors == color)
  for(idx in w) {

    curPos = c(i = cars$i[ idx ], j = cars$j[idx])
    nextPos = if(color == "red")
```

```
                           c(curPos[1],
       if(curPos[2] == ncol(grid))
                             1L
                           else
                             curPos[2] + 1L)
                 else
                    c(if(curPos[1] == nrow(grid))
                           1L
                      else
                           curPos[1] + 1L,
                      curPos[2])

       # check if nextPos is empty
    if(grid[ nextPos[1], nextPos[2] ] == "") {
       grid[nextPos[1], nextPos[2]] = color
       grid[curPos[1], curPos[2]] = ""
    }
  }

  grid
}
```

We could improve this function by separating the code to compute the data frame of car locations into its own function, e.g.,

```
getCarLocations =
function(g)
{
  i = row(g)[g != ""]
  j = col(g)[g != ""]
  pos = cbind(i, j)
  data.frame(i = i, j = j, colors = g[pos])
}
```

This allows us to test this code separately from moving the cars and also makes moveCars() easier to read.

We can also move the code to calculate nextPos to a separate function, getNextPosition(). That function would need the current position and whether we are moving horizontally or vertically. It also needs to know the dimension of the grid so that it can "wrap" cars around the edges, i.e., back to position 1 when they reach the edge of the grid. We represent the position of each car in the form c(row, column) where the row corresponds to the vertical position and the column corresponds to the horizontal position. We can define this function as

```
getNextPosition =
function(curPos, dim, horizontal = TRUE)
{
   if(horizontal)
       c(curPos[1],
          if(curPos[2] == dim[2])
             1L
          else
             curPos[2] + 1L)
```

```
    else
        c(if(curPos[1] == dim[1])
            1L
          else curPos[1] + 1L,
          curPos[2])
}
```

We'd call this in our moveCars() function as

```
getNextPosition(as.integer(cars[idx, 1:2]), dim(grid),
                color == "red")
```

or

```
getNextPosition(c(cars$i[idx], cars$j[idx]), dim(grid),
                color == "red")
```

Our refined version of moveCars() is now

```
moveCars =
function(grid, color = "red")
{
  cars = getCarLocations(grid)

  w = which(cars$colors == color)
  for(idx in w) {
    curPos = c(cars$i[ idx ], cars$j[idx])
    nextPos = getNextPosition(curPos, dim(grid), color == "red")

         # check if nextPos is empty
    if(grid[ nextPos[1], nextPos[2] ] == "")  {
       grid[nextPos[1], nextPos[2]] = color
       grid[curPos[1], curPos[2]] = ""
    }
  }

  grid
}
```

This is a lot more succinct and easier to read and follow. We can also test the getCarLocations() and getNextPosition() functions independently. This is an improvement in all regards with little mental effort or changes to the overall code.

We should verify that when we re-factored the code, we did not introduce references to parameters or variables from the old functions that are not defined in the new functions. Again, we use findGlobals() to verify this. Additionally, we need to write tests for getCarLocations() and getNextPosition() to verify they are working correctly for different inputs. Then we need to verify that the moveCars() function works correctly.

We'll use a small grid that we can inspect visually to test getCarLocations(). We'll use the grid shown below and assigned to the variable g:

	1	2	3	4	5
3					
2	blue	red		red	blue
1	red		blue	blue	

The output of getCarLocations() is

```
getCarLocations(g)
  i j colors
1 1 1    red
2 2 1   blue
3 2 2    red
4 1 3   blue
5 1 4   blue
6 2 4    red
7 2 5   blue
```

We leave it as an exercise to verify these are correct. It may be easier to verify this by displaying the grid as a regular matrix via `unclass(g)`.

We need to try other grids, e.g.,

```
a = createGrid(c(4, 5), .7)
pos = getCarLocations(a)
nrow(pos) == sum(a != "")
```

This doesn't test if the actual locations are correct, just that the total number of occupied cells is correct. We need to determine a good test to verify that the results from getCarLocations() are correct. We must use a different approach than implemented in getCarLocations() so we are not repeating the same logic. Again, we leave this as an exercise.

Q.11 Develop tests for the getCarLocations() function.

To test the getNextPositions() function, we can try different locations and directions (i.e., colors).

```
getNextPosition(c(2, 3), dim = c(4, 5), horizontal = TRUE)
```

should give (2, 4). However,

```
getNextPosition(c(2, 5), dim = c(4, 5), TRUE)
```

should wrap around and give (2, 1).

```
a = getNextPosition(c(2, 5), dim = c(4, 5), horizontal = FALSE)
```

returns (3, 5) as we are moving upwards. Moving that new position with

```
getNextPosition(a, dim = c(4, 5), horizontal = TRUE)
```

moves to (3, 1) due to the wrap around.

So all seems fine with our helper functions. Are we confident they are correct? As we have said, there is no point in moving to the next steps in the overall task unless we are sufficiently satisfied the functions that act as a foundation for those steps are correct. What other tests should we consider? When will we be satisfied/comfortable that our code is working correctly?

Now we need to test *moveCars*():

```
g1 = moveCars(g)
```

This moves the red cars in the grid. We can compare the 2 grids (the before and after) visually with

```
g
      1    2    3    4    5
3
2  blue  red        red  blue
1  red        blue blue
```

```
g1
      1    2    3    4    5
3
2  blue       red  red  blue
1        red  blue blue
```

We can see that the red cars moved right, except the one in position (2, 4). That is blocked by a blue car in (2, 5). If we move the blue cars in the *original* grid (out of order since we haven't yet moved the red cars) with

```
moveCars(g, "blue")
      1    2    3    4    5
3  blue              blue
2        red  blue  red
1  red              blue
```

the blue cars in columns 1, 3, and 5 move up. The other blue car in (1, 4) is blocked by the red car in (2, 4).

Instead of seeing how the blues move in the original grid, we'll move the red cars and then the blue cars. We do this by calling moveCars() twice — once for the red cars and once for the blue — making certain to pass the output of the first call as the updated grid in the second call, i.e.,

```
g2 = moveCars(moveCars(g), "blue")
g2
      1    2    3    4    5
3  blue              blue
2        red  red
1        red  blue blue
```

We can compare this to the output from g1 above. Again, things look correct. In this case, only the 2 blue cars in columns 1 and 5 were able to move. The other 2 were blocked by red cars in the second row. Moving the red cars in g2 with moveCars(g2) yields

```
      1    2    3    4    5
3  blue              blue
2        red        red
1        red  blue blue
```

So we see that the red car in (2, 4) has moved, but the red car at (2, 3) didn't. This is because the red car at (2, 4) was blocking it.

If we spend time verifying these functions, we will save ourselves time later on. Knowing these are correct allows us to use them to validate the results from other implementations, e.g., vectorized or compiled versions. So our time spent verifying these will be well spent and rewarded. We should note that in each call to moveCars(), we are recomputing the locations of the cars at each time step via getCarLocations(). However, in the previous time step we actually knew these locations so these computations are not really necessary. Instead, when we move a car on the grid, we could also update the location of that car in the associated data frame. This is duplicating information and, in a way, violating the DRY principle. However, here we are repeating data, not code. We would have to ensure that the two representations of the same information are synchronized at all times. We are also using more memory. However, it can remove the need for unnecessary, redundant computations. There is a trade-off.

8.4 Evaluating the Performance of the Code

Let's run our model through multiple time steps. We can write a function to do this as we repeat this regularly for different configurations:

```
runBML =
function(grid = createGrid(...), numSteps = 100, ...)
{
  for(i in 1:numSteps) {
    grid = moveCars(grid, "red")
    grid = moveCars(grid, "blue")
  }

  grid
}
```

Technically, we iterate over twice the number of time steps as in each iteration we move both sets of cars. We allow the caller to specify the initial grid. We also use ... to allow the caller to specify inputs to createGrid(), which we call on their behalf if they don't provide the grid.

We can use this function to run a simple simulation using the defaults for creating the grid:

```
g = createGrid()
g.out = runBML(g)
```

We can plot the initial and final grid side-by-side with

```
par(mfrow = c(1, 2), mar = rep(1, 4), pty = 's')
plot(g, main = "Initial Grid")
plot(g.out, main = "After 100 iterations")
```

Note that we changed the margins for each plot and also used a square plotting region so that the aspect ratio of each grid is square. The display is shown in Figure 8.7.

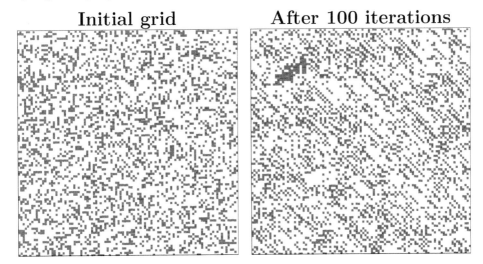

Figure 8.7: Sample Grid at Start and After 100 Iterations. *The left panel shows the initial grid with low density. The right panel shows the state of the grid after 100 iterations. The diagonal lines are already starting to emerge.*

Let's see how long this takes to run on a 100-by-100 grid, 50% filled with cars. We'll create this once and reuse it in all our timings so that we are comparing code on the same grid:

```
set.seed(1345)
g100 = createGrid(c(100, 100), .5)
```

We time the code with

```
tm1 = system.time(runBML(g100))
```

We create the grid separately from the timing so as to measure only the time to move the cars. The call to runBML() took 15.9 seconds on a Macbook Pro laptop running OS X Mavericks, with a 2.6Ghz Intel Core i7 processor and 16GB of memory. Note that this will get slower as the number of cars increases, i.e., the density gets larger. This is because our loop in moveCars() will loop over more items.

Let's find out where the computations spend most of their time. We can use profiling for this via the Rprof() and summaryRprof() functions, e.g.,

```
Rprof("/tmp/BML.prof")
g.out = runBML(g)
Rprof(NULL)
head(summaryRprof("/tmp/BML.prof")$by.self, 10)
```

	self.time	self.pct	total.time	total.pct
"moveCars"	5.58	37.55	14.86	100.00
"[[.data.frame"	1.16	7.81	3.54	23.82
"$"	0.90	6.06	5.88	39.57
"[["	0.88	5.92	4.42	29.74
"dim"	0.78	5.25	0.78	5.25

"match"	0.76	5.11	1.10	7.40
"getNextPosition"	0.74	4.98	1.64	11.04
"["	0.74	4.98	0.74	4.98
"<Anonymous>"	0.58	3.90	0.70	4.71
"$.data.frame"	0.56	3.77	4.98	33.51

As we expect, moveCars() is taking a large proportion of the overall time. So is accessing elements in the data frame (with [[and $). Where did we use match() other than in the plot.BMLGrid() function, which is not being called anywhere within the call to runBML()? Also, why is getNextPosition() taking almost 5% of the time?

Q.12 This information doesn't tell us how many times each function has been called. How could we calculate how often a function was called? Consider using trace() to gather this information. (See Chapter 6.)

We can make some simple improvements. We are calling dim() in each iteration of move-Cars() in the call to getNextPosition(). We could move this outside of the loop as the dimensions of the grid don't change. This is just a good thing to do in any computations — move invariants outside of the loop and compute them just once. Our new function definition is

```
moveCars =
function(grid, color = "red")
{
  cars = getCarLocations(grid)

  w = which(cars$colors == color)
  sz = dim(grid)
  horiz = (color == "red")
  for(idx in w) {
    curPos = c(cars$i[ idx ], cars$j[idx])
    nextPos = getNextPosition(curPos, sz, horiz)

       # check if nextPos is empty
    if(grid[nextPos[1], nextPos[2] ] == "")  {
       grid[nextPos[1], nextPos[2]] = color
       grid[curPos[1], curPos[2]] = ""
    }
  }

  grid
}
```

The small changes are highlighted.

We can compare the speed of our new function with our previous timing results in tm1 via

```
tm2 = system.time(runBML(g100))
tm1/tm2
```

```
   user system elapsed
  1.396  1.300   1.396
```

This does speed things up by approximately 40%, which is quite significant.

We can go further than taking the call to dim() outside of the loop in moveCars(). We can specify the actual dimensions for the grid as a parameter for moveCars() so that it can be computed once in runBML() and passed in each call to moveCars(). We can use a default value for this parameter in moveCars() so that callers don't have to specify it. Is passing the dimension to moveCars() likely to significantly reduce the overall computation time?

Unfortunately, these changes are not likely to speed up the computations tremendously. This is because the calls to dim() only account for 4% of the total time. To improve the performance significantly, we should focus on the first few functions in the output of summaryRProf(). Where do we subset the data frame? How can we speed up moveCars()?

The most obvious improvement we can make to our code is to remove the loops and attempt to vectorize the computations within in moveCars(). This also involves vectorizing getNextPosition(). We cannot remove the loop in the runBML() function as the grid that serves as the input for iteration t is the output from the t-1th iteration. So the iterations depend on each other.

Working with data frames can be expensive as they have a lot more structure and constraints than a matrix does. Let's see where [[.data.frame() is called. One way to do this is to trace calls to that function so that we can perform an operation each time it is called. We'll just print out the call stack.

```
trace("[[.data.frame", quote(print(sys.calls())))
gs = createGrid(c(3, 4))
moveCars(gs)
```

On the R console, we see output for each call to [[.data.frame() of the form

```
[[1]]
moveCars(gs)

[[2]]
which(cars$colors == color)

[[3]]
cars$colors

[[4]]
`$.data.frame`(cars, colors)

[[5]]
x[[name]]

[[6]]
`[[.data.frame`(x, name)

[[7]]
.doTrace(print(sys.calls()), "on entry")

[[8]]
eval.parent(exprObj)

[[9]]
```

```
eval(expr, p)

[[10]]
eval(expr, envir, enclos)
```

So [[.data.frame() is being called as a result of calls of the form `cars$varName`, e.g., `cars$colors`. This gives us a hint. What if we used a matrix to store the locations and we kept the colors in a separate but parallel vector, or as row names for the 2-column matrix of car locations? This involves modifying the code a bit. Accordingly, before we invest time in this, let's see if we can improve the performance by focusing on the most computationally expensive issues.

What we would like to do is work with all the locations for, say, the red cars in vector operations. Given their current positions, we would like to compute all of the "next positions" in a vectorized computation. Then we'd like to find out which of these are empty and then update just those in the grid. This would remove the loop over the individual cars. Let's try to do this within the current structure of the code.

Suppose we had 2 vectors giving us the row and column indices for all of the red cars. We can get these from `cars`:

```
rows = cars$i
cols = cars$j
```

These give us the current positions of the cars. The next positions can be computed for the red cars with

```
nextRows = rows
nextCols = ifelse(cols == ncol(grid), 1L, cols + 1L)
```

The row index doesn't change since the cars are moving horizontally. The ifelse() function is a vectorized version of an if-else statement. We could also compute the nextCols with

```
nextCols = cols + 1L
nextCols[ nextCols > ncol(grid) ] = 1L
```

Either approach allows us to vectorize the computations for calculating nextPos and we could still loop over the cars to determine if they could move. However, we can do this with the 2-column matrix subsetting we saw earlier, via

```
w = (grid[ cbind(nextRows, nextCols) ] == "")
```

to determine which of the target cells are empty. This is similar to how we obtained the colors of the cars in getCarLocations() (see page 328). The variable `w` is now a logical vector with as many elements as there are in both nextRows and nextCols, i.e., the number of red cars. So we can use this to assign new updated values to the matrix and also to set the old locations to "":

```
 grid [ cbind(nextRows[w], nextCols[w]) ] = "red"
 grid [ cbind(rows[w], cols[w]) ] = ""
```

To check if these computations are correct, let's create a simple grid and look at the results as we step through them. We'll create a new grid:

```
gs = createGrid(c(4, 7), .5)
gs
```

A Self-Organizing Dynamic System with a Phase Transition

	1	2	3	4	5	6	7
4	red	red	red		red	blue	
3							
2		blue	blue	red	blue	blue	red
1	red				blue		blue

We can see this in the left panel of Figure 8.8. Now we mimic the steps above for computing the next potential location of each red car but for this particular grid g. We compute the current rows and columns of the red cars with

```
pos = getCarLocations(gs)
red = pos$colors == "red"
rows = pos$i[red]
cols = pos$j[red]
```

Next we compute the target locations of each red car via

```
nextRows = rows
nextCols = ifelse(cols == ncol(gs), 1L, cols + 1L)
```

We check if the new columns are correct:

```
cbind(cols, nextCols)
```

```
     cols nextCols
[1,]    1        2
[2,]    1        2
[3,]    2        3
[4,]    3        4
[5,]    4        5
[6,]    5        6
[7,]    7        1
```

Note that the car in column 7 would move to column 1 as it wraps around and the other columns are simple updates.

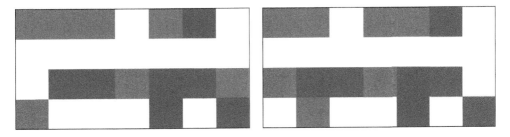

Figure 8.8: A Grid before and after Moving the Red Cars. *The left panel shows the initial 4-by-7 grid. The second panel shows the state of the grid after the red cars have moved.*

So how can we determine if the "next"/target cells are empty? We can look at their values

```
gs[ cbind(nextRows, nextCols) ]
```

```
[1] ""      "red"  "red"  ""    "blue" "blue" ""
```

Do these correspond to the values we expect?

We compute the logical vector of empty cells with

```
w = gs[ cbind(nextRows, nextCols) ] == ""
```

We then look at the positions of the red cars that will actually move:

```
cbind(rows, cols)[w,]
```

```
     rows cols
[1,]    1    1
[2,]    4    3
[3,]    2    7
```

These do indeed correspond to the only red cars that can move. They will move to

```
cbind(nextRows, nextCols)[w,]
```

```
     nextRows nextCols
[1,]        1        2
[2,]        4        4
[3,]        2        1
```

as we expect.

Updating the grid with

```
gs[ cbind(nextRows, nextCols)[w,] ] = "red"
gs[ cbind(rows, cols)[w,] ] = ""
```

yields the new grid

```
      1    2    3    4    5    6    7
4   red  red       red  red blue
3
2   red blue blue  red blue blue
1        red            blue      blue
```

This gives us the correct result as shown in the right panel of Figure 8.8.

So now let's rewrite our moveCars() function to use this vectorized approach:

```
moveCars =
function(grid, color = "red")
{
  cars = getCarLocations(grid)

  w = which(cars$colors == color)
  rows = cars$i[w]
  cols = cars$j[w]

  if(color == "red") {
    nextRows = rows
    nextCols = ifelse(cols == ncol(grid), 1L, cols + 1L)
```

```
    } else {
      nextRows = ifelse(rows == nrow(grid), 1L, rows + 1L)
      nextCols = cols
    }

    w = grid[ cbind(nextRows, nextCols) ]    == ""
    grid[ cbind(nextRows, nextCols)[w, , drop = FALSE] ] = color
    grid[ cbind(rows, cols)[w,, drop = FALSE] ] = ""

    grid
}
```

Note the use of `drop = FALSE` when subsetting the matrices created with `cbind()`.

Q.13 Why is the use of `drop = FALSE` important? Under what circumstances will it yield a different result than `cbind(nextRows, nextCols)[w, ]`?

Again, we need to test our computations and function moveCars() thoroughly, including degenerate cases, e.g., where no car can move, where only one car can move, or with 1-by-1 grids. We can compare the output from this function with our previous version. These should yield the same results for arbitrary inputs. If they do agree for several different inputs, we can be confident our new function is correct. However, even if they agree, both could be incorrect. We can rename the function we defined earlier before we overwrite it, and then call them both with the same inputs. We can display the 2 grids or compare them in the R console or use identical() or all.equal() to test for equality of results from the two versions of the function. We'll stop here, but you should not until you are satisfied the code is correct.

We can now time this new vectorized version of moveCars() and compare it to our previous timing:

```
tm_v = system.time(runBML(g100))
tm2/tm_v

   user  system elapsed
 21.281   0.794  20.148
```

This is a significant improvement. This is 20 times faster than the our previous implementation, which in turn was 30% faster than our initial version. Instead of running for an hour, this would be finished in 3 minutes!

Let's continue to profile our code and see if there are other improvements we can make:

```
Rprof("/tmp/BML.prof")
g.out = runBML(g)
Rprof(NULL)
head(summaryRprof("/tmp/BML.prof")$by.self, 10)
```

	self.time	self.pct	total.time	total.pct
"ifelse"	0.22	21.57	0.26	25.49
"getCarLocations"	0.12	11.76	0.52	50.98
"Ops.factor"	0.10	9.80	0.14	13.73
"data.frame"	0.06	5.88	0.32	31.37

"=="	0.06	5.88	0.20	19.61
"!="	0.06	5.88	0.06	5.88
"unique"	0.04	3.92	0.08	7.84
"deparse"	0.04	3.92	0.04	3.92
"NextMethod"	0.04	3.92	0.04	3.92
"unique.default"	0.04	3.92	0.04	3.92

What does this tell us? Calls to ifelse() takes a lot of time, over 20% of the overall time. Also, getCarLocations() is expensive.

We saw earlier (page 338) an alternative to using ifelse(), i.e., adding 1 to each column value and then changing those greater than ncol(grid) to 1:

```
nextCols = cols + 1L
nextCols[ nextCols > ncol(grid) ] = 1L
```

So let's try using this approach and see if this improves the overall time:

```
moveCars =
function(grid, color = "red")
{
  cars = getCarLocations(grid)

  w = which(cars$colors == color)
  rows = cars$i[w]
  cols = cars$j[w]

  if(color == "red") {
    nextRows = rows
    nextCols = cols + 1L
    nextCols[ nextCols > ncol(grid) ] = 1L
  } else {
    nextRows = rows + 1L
    nextRows[ nextRows > nrow(grid) ] = 1L
    nextCols = cols
  }

  w = grid[ cbind(nextRows, nextCols) ] == ""
  grid[ cbind(nextRows, nextCols)[w, , drop = FALSE] ] = color
  grid[ cbind(rows, cols)[w, , drop = FALSE] ] = ""

  grid
}
```

To time this, we use

```
tm_v2 = system.time(runBML(g100))
tm2/tm_v2
   user  system elapsed
  25.60    1.12   24.44
```

This does indeed improve the performance by a factor of approximately 24 on one machine, and 33 on another. In other words, this runs 24 times faster than the non-vectorized code. It is also 20% faster than our first attempt at vectorizing the code (see tm_v).

We can profile the code again and we end up with

	self.time	self.pct	total.time	total.pct
"getCarLocations"	0.16	20.51	0.50	64.10
"data.frame"	0.12	15.38	0.30	38.46
"Ops.factor"	0.10	12.82	0.12	15.38
"match"	0.10	12.82	0.10	12.82
"moveCars"	0.06	7.69	0.78	100.00
"=="	0.06	7.69	0.18	23.08
"as.data.frame"	0.02	2.56	0.16	20.51
"deparse"	0.02	2.56	0.04	5.13
"!="	0.02	2.56	0.02	2.56
"cbind"	0.02	2.56	0.02	2.56

Now getCarLocations() and data.frame() take most of the time. These two are related as getCarLocations() calls data.frame(). However, the times and percentages here are for each individual function and do not include the time they spent waiting for other functions they call.

As we discussed earlier (page 338), we might be better off to now switch to a matrix to store the positions of the cars and either return the colors separately or via the row names. The matrix would have 2 columns and as many rows as there are cars. This approach was not the highest priority earlier. However, given the changes we made to address these larger issues, this has now become more prominent.

We can implement this matrix approach with a new version of getCarLocations() defined as

```
getCarLocations =
function(g)
{
  i = row(g)[g != ""]
  j = col(g)[g != ""]
  pos = cbind(i, j)
  structure(pos, dimnames = list(g[pos], c("i", "j")))
}
```

This affects how moveCars() accesses the 2 columns and the colors. Accordingly, we have to redefine it also. This is quite simple. Instead of accessing the i and j elements, we access the first and second column. We could use names, but since we know they are the first and second elements this is even faster as we avoid matching the names, at the expense of clarity and ease of understanding of the code. So our updated function is

```
moveCars =
function(grid, color = "red")
{
  cars = getCarLocations(grid)

  w = which(rownames(cars) == color)
  rows = cars[w, 1]
  cols = cars[w, 2]

  if(color == "red") {
    nextRows = rows
    nextCols = cols + 1L
    nextCols[ nextCols > ncol(grid) ] = 1L
```

```
  } else {
    nextRows = rows + 1L
    nextRows[ nextRows > nrow(grid) ] = 1L
    nextCols = cols
  }

  w = grid[ cbind(nextRows, nextCols) ]  == ""
  grid[ cbind(nextRows, nextCols)[w, , drop = FALSE] ] = color
  grid[ cbind(rows, cols)[w, , drop = FALSE] ] = ""

  grid
}
```

If we time this and compare it to our previous implementation, we get

```
tm_v3 = system.time(runBML(g100))
tm_v2/tm_v3
```

```
   user  system elapsed
   1.35    1.33    1.35
```

So we see a 35% speedup by moving from a data frame to a matrix. This is just in the context of this particular problem. There are good reasons for using data frames for data analysis. They do incur some overhead, but avoid others. You might also explore the data.table [3] package if you need to use a data.frame for large data, but the computations are slow, e.g., with a lot of subsetting of rows.

We can again profile the new code and we get

	self.time	self.pct	total.time	total.pct
"moveCars"	0.12	21.43	0.56	100.00
"getCarLocations"	0.10	17.86	0.30	53.57
"=="	0.10	17.86	0.10	17.86
"!="	0.08	14.29	0.08	14.29
"structure"	0.06	10.71	0.06	10.71
"cbind"	0.04	7.14	0.04	7.14
"col"	0.04	7.14	0.04	7.14
"which"	0.02	3.57	0.10	17.86

moveCars() takes the most time but is reasonably efficient. We can improve getCarLocations(). It computes g != "" in two places. We can evaluate this once and assign it to a local variable. This is an invariant and we shouldn't recompute it. So we can define getCarLocations() as

```
getCarLocations =
function(g)
{
  w = (g != "")
  i = row(g)[w]
  j = col(g)[w]
  pos = cbind(i, j)
  structure(pos, dimnames = list(g[pos], c("i", "j")))
}
```

We already created pos once and used it twice. This is about as efficient as this function can be written for this approach.

In moveCars(), we compute cbind(nextRows, nextCols) twice. We could again use a local variable to store this and then reference that in both places. This amounts to changing

```
w = grid[ cbind(nextRows, nextCols) ]   == ""
grid[ cbind(nextRows, nextCols)[w, , drop = FALSE] ] = color
grid[ cbind(rows, cols)[w, , drop = FALSE] ] = ""
```

to

```
nextLocs = cbind(nextRows, nextCols)
w = grid[ nextLocs ]   == ""
grid[ nextLocs[w, , drop = FALSE] ] = color
grid[ cbind(rows, cols)[w, , drop = FALSE] ] = ""
```

in our most recent definition of moveCars().

Timing these enhancements, we get

```
tm_v4 = system.time(runBML(g100))
tm_v3/tm_v4
    user  system elapsed
    1.17    1.00    1.16
```

and a 16% improvement in the overall time.

Note that we have not had to change the runBML() function. We implemented the functions it calls in such a way that they continue to take the same inputs, but perform their computations differently. The caller does not need to know about these details.

So we have now gone from 41 seconds to 0.643 seconds for the initial loop-based version and the highly optimized, vectorized version of our functions, respectively. This is an improvement of a factor of 63. That's quite significant.

If we profile the code yet again, we get

	self.time	self.pct	total.time	total.pct
"moveCars"	0.14	29.17	0.48	100.00
"!="	0.08	16.67	0.08	16.67
"=="	0.08	16.67	0.08	16.67
"structure"	0.06	12.50	0.06	12.50
"which"	0.04	8.33	0.08	16.67
"cbind"	0.04	8.33	0.04	8.33
"getCarLocations"	0.02	4.17	0.18	37.50
"+"	0.02	4.17	0.02	4.17

We see that moveCars(), the primary function, is taking most of the time. The next important functions are low-level R functions and we can't improve them. We can avoid them, but how? getCarLocations() takes 4% of the time and it is responsible for the only call to the !=, which accounts for 17% of the time. So can we improve matters?

Let's think about what moveCars() does. It starts by calling getCarLocations() each time. This is fine if we call moveCars() just once. But in runBML(), we call moveCars() multiple times. At the end of each call to moveCars(), we know the updated locations of the cars. However, we compute them again each time. We could have moveCars() return these updated locations. Of course, we also want the grid and moveCars() needs that also. So we could have moveCars() return both the grid and the updated car locations, e.g.,

```
nextLocs[!w,] = cbind(rows, cols)[!w, ]
list(grid = grid, locations = nextLocs)
```

This could in fact define a BML grid. We could rethink how we represent our grid and in our function createGrid(), we could return the matrix and also the car locations. We already know where the cars are located and so we could return this information in different and redundant forms. The point is that we would avoid the time in recomputing the information in these different forms. This comes at the expense of extra memory. However, it is how I initially approached this problem!

Q.14 Modify the functions to use this approach of maintaining the locations of the cars across calls to moveCars(). Then determine how this changes the performance.

We have performed the timings, profiling, and performance improvements using grids with the same dimensions and densities. We should have explored other dimensions and densities to ensure that the changes to the code improved matters for all grids.

8.5 Implementing the BML Model in *C*

With our improved code, we can run a 100-by-100 grid with the (default) density 0.3 for 100 iterations in approximately 0.267 seconds. What about a 100-by-100 grid with density 0.6? This takes about 0.327 seconds. This is because there are 6000 cars as opposed to 3000 cars to move. If we increase the size of the grid to 1000-by-1000 and keep the density at 60%, the computations take about 37 seconds. This is because there are now 100 times the number of cells and cars. So we would expect the time to be about 100 times more than the smaller grid. These times will vary significantly across different computers, based on the capabilities of the individual machines. However, the key is that we are comparing them to each other on the same machine, not across machines.

While we can deduce that the computational time increases proportionally to the number of cells, we can explore this empirically. We can time the computations for different dimensions and then plot the times versus the dimension or number of cells. We can do this with

```
N = 2^(3:20)
timings =
  sapply(N,
         function(n) {
           print(n)
           g = createGrid(as.integer(rep(sqrt(n), 2)), .5)
           system.time(runBML(g))
         })
plot(N, timings[3,], type = "p",
     xlab = "Number of grid cells",
     ylab = "Elapsed Time (seconds)")
abline(lm(timings[3,] ~ N))
```

Note that the different values of N are used as the number of cells in the grid, not the value

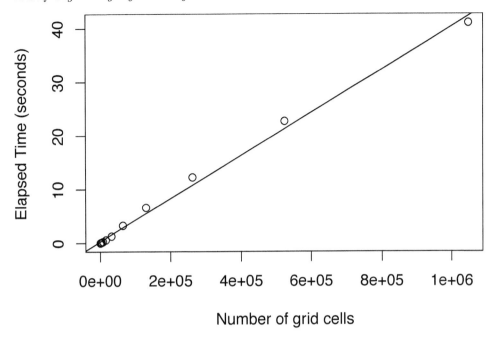

Figure 8.9: Run-Time for Vectorized Code as a Function of the Number of Grid Cells. *This shows that as the number of grid cells, and the total number of cars, doubles, the time taken approximately doubles. We have also added the least squares fit for the elapsed times as a function of grid size n. This plot was computed for square grids, but applies generally to the number of cells in a grid.*

of each dimension. This is why we take the square root of the parameter n in our function. The results are shown in Figure 8.9.

Q.15 When timing the computations, also vary the density of the cars and draw a 3-dimensional plot showing the relationship between number of cells, density, and time.

We want to explore the behavior for different densities of the cars and also for various grid dimensions. For each density-dimension configuration, we need to generate many different random starting grids and run the BML process for many time steps until the process is in a stable equilibrium. This involves a lot of computations. We can wait 30 seconds for one call to runBML() for 100 iterations on a small grid. However, if we want to run for 32,000 iterations, say, this will take approximately $2\frac{2}{3}$ hours. This is just for one instance of a grid with a specific dimension and density. We need to repeat this many times to average across the random initial configurations of the grids. With just 10 replications, the computations will take one day. However, we want more replications and also to repeat the entire procedure for many different dimensions and densities. This will take far too long and we need to make the computations a lot faster if we are to explore the different aspects of the process across the parameter space of inputs.

One approach to reducing the computation time is to run separate grids independently of each other on multiple processors. We can do this in a reasonably straightforward manner in

R with the `parallel` [7] package and several others. We leave this as an exercise described at the end of the chapter. Instead, we will focus on making the code significantly faster using compiled C code. This can also be used in the parallel approach and so we will achieve even greater speed improvements.

We have have already made the R code a lot faster by carefully profiling the code and using vectorized operations. If we want to make this significantly faster than this, we might try implementing the computations using a compiled language, specifically C or $C++$. We'll focus on implementing the important and bottleneck function moveCars() in C, but we'll also implement runBML() as a C routine since it is quite simple. We'll also compute additional information for each time cycle, namely how many cars of each type actually moved. This will allow us to explore different aspects of the BML process.

One benefit of implementing moveCars() in C is that we can write the code in the "obvious" manner that we started with, i.e., a loop to move each car. This means we can deal with individual cars rather than vectorized code that moved all the cars in one step. We still have to decide, however, whether to move the cars sequentially or simultaneously. Since our runBML() function uses the simultaneous approach and the original description and the existing research of the BML process uses this definition, we will also implement this. It is marginally more involved and very slightly slower. These are not important as we want to get the same and correct results.

In the next section, we will discuss how to implement moving the cars using C code. We'll also outline how to call C code from R. This involves transferring inputs from R to a C routine (function), and accessing the results returned from C. This is intentionally very high-level. We will not cover these in great depth, and we certainly are not attempting to teach programming in C. However, it provides concrete examples of the basic steps that one can often mimic relatively easily. There are many good references on different aspects of these topics, including books, on-line documents, and one of the the R manuals [6]. We'll discuss the basic mechanisms provided by R. There are others, however, such as `Rcpp` [5] and other foreign function interface packages, e.g., `rdyncall` [1] and `Rffi` [9].

8.5.1 The Algorithm in C

Before we discuss the details of calling C from R and "marshalling" data between the two languages or writing C code, let's focus on how to implement the simultaneous approach. There are two issues — one is how we find the cars we want to move, and the second is the computation to ensure they move simultaneously, not sequentially. These are orthogonal to each other, i.e., how we solve one doesn't impact how we solve the other. We'll start with the first issue and also consider the slow way of doing this.

We start with just our grid, and suppose we want to move the red cars. We can loop over all of the cells in the grid, visiting them one at a time. If that cell contains a red car, we determine whether that car can move to the target cell and update the grid if it can. If the cell we visit is empty or contains a blue car, we continue on to visit the next cell. This approach involves looping over all cells in the grid. If the grid is large and the density low, most of the cells we visit will be empty (or contain a blue car). We waste a lot of time.

A potentially better approach than looping over all the cells in the grid is to maintain the locations of the red cars in a 2-column matrix, as we discussed earlier. To maintain the current state of the BML grid, we will use a grid and two 2-column matrices containing the locations of the red and blue cars separately. Each row in these location matrices contains the row and column index for the corresponding car. When moving the red cars, we loop over each row in the corresponding location matrix, and retrieve the location of that car. Then we determine whether it can move and update the state of the grid if it can. This involves not only updating the 2 cells in the grid matrix (where the car currently is and its new position), but also updating the location of the car in the current row of the red car locations matrix. At the next iteration, we will be able to use this new car location without having to find that car.

How do we deal with moving the cars simultaneously rather than in order? Essentially this relates only to 2 red cars at the end points of a given row, or 2 blue cars at the edges of a given column. If we were to move 1 red car in the first column and later attempt to move a red car in the final column of the grid in the same row, we would need to identify that this second car could not move. Its target cell is occupied when it attempts to move at the same time as the first car. If we moved the cars sequentially rather than simultaneously, the target cell would be vacant. This is a not a concern for cars in adjacent cells not at the edges since we process these in the order they appear in the matrix, i.e., column-wise, and row-wise within each column. This means that, for example, a red car in column 2 cannot move to column 3 if there is a car there and we will not move that car in the third column before the car in the second column.

How do we guard against moving a red car on the right edge of the grid into a cell that has just become vacant in the same time step? One approach is to keep a vector for the original state of the first column. Then we check if the corresponding element is empty in this column when moving to the car on the right. In other words, we don't check the cell in the current state of the grid as that might have been vacated in the same time step; instead, we check the original state of that cell.

A second approach is to use 2 grids in our computations. We have an original grid and our new grid that we update. The original grid remains unchanged as we move the cars within an iteration. When determining if we can move a car, we check the original grid. When we do move a car, we update the new grid (and the row in the matrix of car locations). When we are finished moving all the cars within this iteration, the new grid becomes our current grid and we use it as the input to the next step.

To make the *C* code marginally simpler, we will use the second approach, i.e., 2 grids. We'll also create this second grid in *R* and pass it to the *C* code. We want to avoid managing (allocating and freeing) memory in *C* code when we can. That is why we do this in *R*. Hopefully the basic algorithm for moving either collection of cars is now reasonably clear. We maintain 2 grids and a location matrix for that color's collection of cars. We loop over each car and get its location (i, j). We then determine the location of where that would move, say (ni, nj). We then consult the original grid and check the value `originalGrid`¬ `[ni, nj] == 0`. If this is true, we move the car by updating the new grid and the car location in its matrix. After potentially moving all of the cars in this location matrix, we set `originalGrid = newGrid`.

It is easier for most people to use loops, rather than vectorized operations. The cost of the simplicity of using loops in *C* rather than matrix subsetting in *R* is that we have to learn the initially complex interface between *R* and *C* code. *R* provides two interfaces to invoke compiled routines — .C() and .Call(). The latter is more flexible and powerful, but more involved and specialized to *R*. In general, you should use .Call() whenever the *C* code needs to allocate results as *R* objects and also when the code needs richer access to *R* objects than simple vectors and matrices. When we are dealing with vectors or matrices of data and inserting values into existing vectors or matrices to perform our computations, we can use the .C() interface. This is useful when exactly the same code can be used in *R*, *Python*, *MATLAB*®, etc. Typically, we have one high-level worker routine in *C* that manipulates these basic types and performs the actual computations of interest, and an additional "proxy" *C* routine that we call from *R*. This proxy routine passes its arguments to the worker routine in the appropriate form. The worker routine can be shared across different environments such as *R*, *Python*, stand-alone applications, etc. The proxy routine can be written to use either the .Call() or the .C() interface.

We'll use the .C() interface to implement our BML routines. This allows us to pass data values from *R* as pointers to primitive *C* types (rather than as *R* objects, which the .Call() interface enables and expects). We can pass an *R* integer vector to *C* as a pointer to a collection of *int* elements, and similarly a numeric vector as a pointer to a collection of *double* elements. A matrix is passed as a pointer to its elements, which are stored as a vector, arranged in column order. In each of these cases, we also have to pass the number of elements in the vector as a separate argument. This is typically passed as an integer vector of length 1, i.e., a pointer to a sequence of integers declared as *int* * in *C*. Similarly, we can pass the dimensions of a matrix as an integer vector with 2 elements. In this way, the length and dimension are treated as regular data vectors, which simplifies the interface.

When reading this section, it will clearly help if you are familiar with the *C* programming language. However, hopefully you can understand the big picture and the different parts and steps by which we develop, compile and use the *C* code from *R*. The syntaxes of the *R* and *C* languages and their basic programming models are quite similar. Don't worry about the details. You can learn about those later. Knowing how to use *C* code from *R* can greatly simplify speeding up bottlenecks in your code. Hopefully, you can use other people's existing *C* code. In that case, you just need to know enough to be able to invoke that *C* code from *R*.

Our strategy is to write code in *C* and an *R* function to invoke this code. We will implement two *C* routines. The first is a high-level routine named R_BML that we call directly from *R* with our grid, the number of iterations and additional information we can implement in *R* for simplicity. This will actually perform the iterations by moving the red and then blue cars as many times as desired. The second *C* routine, named moveCars, will move the actual cars and update the grid. The *R* function we write will take our initial grid and the number of iterations and call the R_BML routine. We will create the auxiliary information, i.e., the matrices containing the locations of the red and the blue cars, and pass all of these to R_BML. R_BML is a little more than the proxy function we mentioned above since it actually performs the loop for the number of iterations and is more than a direct call to moveCars. The moveCars routine is quite similar in nature to our moveCars() function. It attempts to move all of the cars of a particular color and to update the grid.

For simplicity in the *C* code and for speed, we'll represent our grid as matrix with integer rather than string values ("red" and "blue"). A value of 0 corresponds to empty, 1 to red, and 2 to blue. We'll also pass the locations of the blue cars and the red cars separately as matrices, each with 2 columns and a row for each of the cars of that color. We want to know the number of cars that are moved at each time step and we pass a 2-column matrix for this via the **velocity** parameter. We also need to specify how many iterations to perform, where each iteration moves the red cars and then the blue cars, i.e., 2 time steps per iteration. Accordingly, we can declare our *C* routine to move the cars in 1 time step as

```
void
R_BML(int *grid, int *newGrid, int *dims,
      int *red, int *numRed, int *blue, int *numBlue,
      int *velocity, int *numIters)
```

We haven't mentioned the parameters **dims**, **numRed**, **numBlue**, or **newGrid**. Since the .C() interface passes a matrix such as our grid as a pointer to the contiguous collection of its data elements (*int* *), we no longer have access to its dimensions. We would if we used the .Call() interface, but then our code would be slightly more complicated. When using the .C() interface, we will pass the length or dimensions of vectors and matrices as separate arguments in the form of integer vectors. Therefore the **dims** parameter has type *int* * in *C* and we use this to pass the value of dim(grid). Similarly, we pass the number of red cars and the number of blue cars via the parameters **numRed** and **numBlue**. We compute these with `nrow(red)` and `nrow(blue)` from the car location matrices. While each of these dimensions is a single number, they are passed generally as *integer* vectors and so have type *int ** in *C*.

The final parameter to explain in the signature of our routine is **newGrid**. This is the second grid that we will use to update the current grid as we move the cars. We will allocate this in *R* as an empty grid (i.e. all 0 values) with the same dimension as our initial grid.

Because we create this second grid in *R* and also pass the 2-column matrix for storing the velocities of the time steps, our *C* code will not have to allocate any memory. This greatly simplifies matters as this is a source of common problems for *C* programmers. We allocate the memory in *R*, pass it to our routine, and when control is returned back to *R*, we collect the results and *R* releases the memory as it does for any *R* object no longer in use.

Rather than launching each time step from *R* within a loop as we did for runBML(), we'll implement this loop in *C* for additional speed. **numSteps** specifies the number of iterations, or time steps, we perform. The parameter **verbose** is a logical vector of length 1 (i.e. a scalar) and so is passed to *C* as an *int **, like an integer vector. This controls whether we emit a message to the console for every 100-th time step. This can be useful for lengthy runs to show that the computations are proceeding and not caught in an infinite loop or stopped in any other way.

The final parameter — **ans** — is used to return the number of cars that moved for each type in each time interval/cycle. *ans* is an integer matrix with 2 columns. With the .C() interface, the caller in *R* pre-allocates the *R* objects to store the results and passes these to the *C* routine. That routine can then insert values into this space and the .C() function takes care of passing the values back to the caller as regular *R* objects. The R_BML routine inserts the number of cars moved at each time into this memory.

Similarly to updating **ans**, the routine bml_move that actually moves the individual cars updates the state of the grid by modifying the contents of the **grid** parameter directly. This is returned to *R* containing the new state. This is also true of the locations of the red and blue cars in the parameters **redLocations** and **blueLocations**. The .C() function makes a copy of its arguments and so we are not modifying the objects in our *R* session, but copies of them. The ability to change the state of memory passed by the caller has both advantages and disadvantages. It reduces the amount of memory needed for computations and provides a different computational model than *R*. However, it also makes for very complex bugs and difficult debugging.

We now know how we will define the R_BML *C* routine that we will call from *R*. Often, this is just a simple routine that passes its argument to a another routine that is not directly accessible to *R*. This is because routines that we invoke via the .C() interface have a very specific signature. They return nothing explicitly (i.e. have a return type of *void*) and can only accept *int ** and *double ** types (along with a few other limited types). So let's write the body of this routine *R* invokes:

```
void
R_BML(int *grid, int *newGrid, int *dims,
      int *red, int *numRed, int *blue, int *numBlue,
      int *velocity, int *numIters)
{
    int dir = RED, i;

    for(i = 0; i < *numIters; i++) {
        // move the red cars
        dir = RED;
```

```
        moveCars(grid, newGrid, dims, &dir, numRed, red,
                velocity + i);

        // next the blue, but copy the contents of newGrid
        // to grid and reset newGrid.
        memcpy(grid, newGrid, sizeof(int) * dims[0] * dims[1]);

        dir = BLUE;
        moveCars(grid, newGrid, dims, &dir, numBlue, blue,
                velocity + i + numIters[0]);
        if(velocity[i] == 0 && velocity[i + numIters[0]] == 0)
           break;

        // again copy the newGrid back to
        // grid for the next iteration.
        memcpy(grid, newGrid, sizeof(int) * dims[0] * dims[1]);
    }
}
```

This is conceptually quite simple. We create a loop that iterates over $0, 1, 2, \ldots,$ numIters-1. Hence there are **numIters** iterations in total. We have to dereference the value in **numIters** since we are passed a pointer to this scalar (an R integer vector with 1 element). We do this with ***numIters** or equivalently **numSteps[0]**. Remember that C uses 0-based indexing, i.e., the first element is at position 0, while R uses 1-based indexing. This can be the cause of many hard-to-see bugs when writing C code to be called from R as we switch between the two programming models.

The real work in each iteration is to call moveCars to move the red cars and then again to move the blue cars. Before we move the blue cars, we copy the updated contents of the grid in **newGrid** back to **grid**. This means **grid** will start each call to moveCars containing the current state of the grid. We use memcpy to copy the elements from **newGrid** to **grid**. This is different from R where we use the simple assignment grid = newGrid. This is not how C works. We have to explicitly copy each element in **newGrid** to the corresponding element in **grid**.

Q.16 We can avoid these calls to memcpy, both simplifying the code and also making it slightly faster by not copying the contents of **newGrid** to **grid**. Make the appropriate changes and verify that the results are the same and correct.

This is the entire routine, analogous to runBML() in R. It merely loops over the time steps. The real work is done in the calls to moveCars.

The idea of moveCars is that we give it the grid of current positions of the car and the locations of the cars we want to move. We also specify which direction we are moving, corresponding to red or blue. The routine loops over each car we want to move, computes its target next position, as we did in R, and then checks to see if that cell is currently vacant in the current grid. If it is, we update the **newGrid** and also the position of the current car in the locations matrix; if not, we move on to the next car. This is quite straightforward and very similar to our initial implementation in R in which we used loops to iterate over each car. However, before we start moving the cars, we have to initialize our target grid to have the contents of the current grid. So our routine is defined as

```
void
moveCars(int *grid, int *newGrid, const int *dims, int *dir,
         int *numCars, int *carLocations, int *ans)
{
    int k, i, j, ni, nj;

    int colorToMove = *dir;
    int numMoved = 0;

    /* Copy the old grid to the new grid and then we update
       the positions on the new grid only for the cars that
       can move, relative to the old grid. */
    memcpy(newGrid, grid, sizeof(int) * dims[0] * dims[1]);

    for(k = 0; k < numCars[0]; k++) {
        i = carLocations[k] - 1;
        j = carLocations[k + numCars[0]] - 1;

        if(colorToMove == RED) {
            ni = i;
            nj = j + 1;
            if(nj == dims[1])
                nj = 0;
        } else {
            nj = j;
            ni = i + 1;
            if(ni == dims[0])
                ni = 0;
        }

        if( grid[ ni + nj * dims[0] ] == 0 ) {
            // moving the car
            numMoved++;
            newGrid[ ni + nj * dims[0] ] = colorToMove;
            newGrid[ i + j * dims[0] ] = 0;

            if(colorToMove == RED)
                // update column
                carLocations[k + numCars[0]] = nj + 1;
            else
                carLocations[k] = ni + 1;   // update row
        }
    } // end k loop

    ans[0] = numMoved;
}
```

Within our loop over the cars being moved, we fetch the row and column of the current car. Next, we compute the target row and column based on whether the car is red or blue. Then we use these new coordinates to query if that cell in the current grid is vacant. If it is, we update the corresponding cell in **newGrid** and mark the current location as empty

(0). We also update the corresponding column in the `carLocations` to reflect the new location. Unlike R, the changes to `newGrid` and `carLocations` will be seen in the caller and so persist to the next call to `moveCars`. This is a very different computational model than in R. There are some points to note that may not be entirely obvious for those less familiar with C. `carLocations` is a pointer to a sequence of *int* values. This corresponds to a 2-column integer matrix in R. Matrices are stored in column order, as shown in Figure 8.10. However, to get the current location of a car, we want the 2 elements in the i-th row. We use `locations[i]` for the i-th element in the first column, but `locations[ i + *num ]` to get the i-th element in the second element. This is because we have to jump to the i-th element of the second column. Since the elements of the first column are stored first, we have to skip all the elements in the first column and start from the beginning of the second column. These start at offset `num[0]`, which is the number of rows in the matrix. This is a common idiom when accessing elements in a matrix and one we use to access elements in the grid. Note that when we compute the current position of the i-th car (`r` and `c`), we also subtract 1 from the row and column positions. The reason for this is that the locations in R use 1-based indexing. However, in C we use 0-based indexing and so have to subtract 1. We use these offsets to index the grid array element to check the current state of the cell.

We compute the row and column of the next position of the car and store these in `ni` and `nj`. We do this differently for red and blue cars to reflect how they move. So in this expression, we check to see if adding 1 to the column would bring it beyond the right edge. If it would, we wrap the car to the other side of the grid. Otherwise, we just add 1 (`speed`) to the column value. We use a general increment value here, `speed`, so that the cars can move more than 1 cell in a cycle. However, this is always 1 in our simulations.

We have used the data type *Direction* in the signature to indicate whether we are moving red or blue cars, and we also use the value RED in our code. These are not defined by the C language. Instead, we defined them ourselves as an enumerated constant type, i.e. an *enum*. We do this with

C
```
typedef enum {RED = 1, BLUE = 2} Direction;
```

This is much better than using the ad hoc convention 1 and 2 to indicate RED and BLUE, respectively. These are hard to understand as plain numbers and also easy to confuse. Symbolic names are much more clear and robust. We can also define RED and BLUE as R variables for the same reasons. However, it is imperative that we ensure that they have the same values as in the C code if we are to pass these values from R to C.

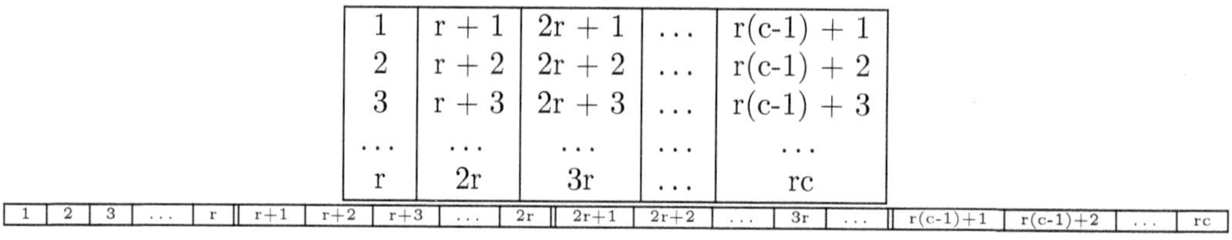

1	r + 1	2r + 1	...	r(c-1) + 1
2	r + 2	2r + 2	...	r(c-1) + 2
3	r + 3	2r + 3	...	r(c-1) + 3
...	...	...	...	...
r	2r	3r	...	rc

1	2	3	...	r	r+1	r+2	r+3	...	2r	2r+1	2r+2	...	3r	...	r(c-1)+1	r(c-1)+2	...	rc

Figure 8.10: Matrix Layout as a Vector. *A matrix in R is stored as a vector with elements ordered by column. The order of the elements from the matrix is shown in the vector. To compute the position in the vector of the (i, j) element in the matrix m, we use $i + (j - 1) \times nrow(m)$. In the first column, we just use the row offset i. In the second column, we have to add* `nrow(m)` *to skip over the elements in the first column. We use the same computation in C but subtract 1 since it is 0-based. Therefore, the index in C for the (i, j) element is $i - 1 + (j - 1) \times nrow(m)$.*

A Self-Organizing Dynamic System with a Phase Transition 355

We have arranged the code in two separate files — *RBML.c* and *BML.c*. The former is for the routine that can be called directly from *R*; the latter is generic *C* code that can be reused outside of *R*. This is a good way to structure the code as we may want to use the code in *BML.c* in another setting such as from *Python* or *MATLAB*®. We might also want to put the loop over the time steps in *BML.c* for maximal reuse. However, we have put it in our *R*-callable routine for simplicity.

We define the enumerated type *Direction* and declare our routine `bml_move` in a header file, *BML.h*. This makes these available to the two separate *C* source files and ensures they both have a consistent view of the same entities. This is important so that the compiler can identify if we define `bml_move` in one way and call it in another or if we have different definitions for **RED** or **BLUE**.

8.5.2 Compiling, Loading, and Calling the *C* Code

Unlike with *R* code, we have to process the *C* code before we can use it. We have to compile the code in each of the two source files *RBML_simultaneous.c* and *BML_simultaneous.c*, and then we have to "link" them together to create a dynamically shared object (DSO), also called a DLL (Dynamically Loadable Library). If the code is in an *R* package (specifically, in the `src/` directory), all this is done for us via the package installation mechanism. So creating a package is typically a good idea. However, it is also good to see how we can compile, link, and load the *C* code directly and understand what the package installation mechanism does for us. This helps us when we encounter problems.

We can compile and link the code in the *C* source files with the shell command

```
R CMD SHLIB -o BML.so BML_simultaneous.c RBML_simultaneous.c
```
Shell

Here we call the *SHLIB* script provided by *R* and specify the two source files to compile and link. We tell *SHLIB* to write the resulting DSO/DLL to the file named *BML.so*. If there are any errors in compiling or linking the code, the DSO will not be created and error messages will be displayed. We would have to modify the *C* code to remove these errors and then re-run the shell *SHLIB* command.

Once the DSO is created, we can load it into an *R* session with

```
dyn.load("BML.so")
```

(If the code were in a package, we would load the code via the directive `useDynLib(`¬ `packageName)` in the package's *NAMESPACE* file.) If all the symbols in the code are defined, this will succeed. If not, we have to go back to the compilation and linking step and try to determine what is wrong. In our case, we have quite simple *C* code that does not refer to routines or data types in other libraries. As a result, this is simpler as we don't have to find other headers files or libraries at compile, link, or load time.

We can verify that our *C* routine `R_BML_simultaneous` is loaded and available with

```
stopifnot(is.loaded("R_BML_simultaneous"))
```

We are now ready to invoke our *C* routine. We'll create a grid and then convert it to an integer representation by mapping the color names to integers:

```
g = createGrid(c(1000, 1000), .5)
gi = matrix(match(g, c("red", "blue"), 0L), nrow(g), ncol(g))
```

Note that we mapped `"red"` to 1, `"blue"` to 2, and `""` to 0 (rather than NA) to be consistent with the *C* code. It is vital to get this correct, i.e., to synchronize the representation of the values in the two languages — *R* and *C*.

We can get the locations of the red and blue cars with

```
pos = getCarLocations(g)
red = pos[ rownames(pos) == "red", ]
blue = pos[ rownames(pos) == "blue", ]
```

Recall that our most recent version of the getCarLocations() function now returns a 2-column integer matrix so we have what we need.

We can invoke our C routine with the unwieldy

```
val = .C("R_BML_simultaneous",
         grid = gi, matrix(0, nrow(gi), ncol(gi)), dim(gi),
         red, nrow(red), blue, nrow(blue),
         1000L, FALSE, velocity = integer(2*1000))
g1 = val$grid
class(g1) = c("BMLGrid", "matrix")
```

The .C() call passes the R objects to the C routine R_BML. Since the C code could have changed any of its inputs, the .C() function returns (copies of) all of its arguments. So val is a list with 9 elements corresponding to the 9 arguments to R_BML. We have named two of the arguments, i.e., *grid* and *velocity*. This allows us to easily retrieve the final grid and also the velocity matrix. Since gi was an ordinary matrix when we passed it .C(), it is still a matrix. We need to set its class so that it is a BMLGrid.

Unfortunately, when we call plot() for g1 to see the final state of the grid, our plot.BMLGrid() tries to match the values of the cells to color names in order to map the matrix colors to integer values. Of course, g1 is already in this form. So we could enhance plot.BMLGrid() to recognize this. The simple fix is to define plot.BMLGrid() as

```
plot.BMLGrid =
function(x, xlab = "", ylab = "", ...)
{
   if(typeof(x) == "character")
     z = matrix(match(x, c("", "red", "blue")), nrow(x), ncol(x))
   else
     z = x
   image(t(z), col = c("white", "red", "blue"),
         axes = FALSE, xlab = xlab, ylab = ylab, ...)
   box()
}
```

Alternatively, we could define a new specialized class to indicate that the matrix is in named-color form or integer-form and then define methods for these.

Q.17 Implement a class and methods for the integer version of the BMLGrid matrix. Try to inherit methods as much as possible and avoid modifying existing code where possible. You can consider different class hierarchies or ancestry orders.

We certainly don't want to call the C routine directly via the .C() interface. We have to ensure that the inputs have the correct type, e.g., integers rather than numeric vectors. If they are wrong, we will get the wrong answer or crash the R process. The latter is greatly preferred as it identifies a problem; we may not recognize incorrect answers. For most C routines we call from R, we create a wrapper function that ensures the inputs are correct, allocates the space for the results, and so on. We can define our wrapper function by combining the computations we did above into a single function:

```
crunBML =
function(grid, numIter = 100L, check = FALSE)
{
    k = class(grid)
    gi = gridToIntegerGrid(grid)

    velocity = matrix(0L, as.integer(numIter), 2L,
                      dimnames = list(NULL, c("red", "blue")))
    pos = getCarLocations(gi)
    red = pos[rownames(pos) == "1",]
    blue = pos[rownames(pos) == "2",]
    ans = .C("R_BML_simultaneous", gi, grid = gi, dim(gi),
                                    red = red, nrow(red),
                                    blue = blue, nrow(blue),
                                    as.integer(numIter), FALSE,
                                    velocity = velocity)

    ans = ans[c("grid", "velocity")]
    class(ans$grid) = k
    ans
}
```

This function returns a list with 2 elements, one the updated `grid` and the other the 2-column `velocity` matrix.

Note that pass gi as two separate values to the *C* routine. The .C() function ensures that these two instances are first duplicated/copied and passed as separate instances of the integer matrix. By passing gi for the new grid, it is appropriately initialized with the current contents of the grid. This is important since the *C* code does not populate the cells in the new grid if a car cannot move in the old grid.

The gridToIntegerGrid() function that we call in crunBML() uses match() to map the color names "", "red", and "blue" to 0, 1, and 2, as we did manually above. It makes sense to define this as a separate function, again so we can test it and reuse it independently of crunBML(). Indeed, we can call it from plot.BMLGrid().

Before we consider how fast our crunBML() function and the associated *C* code run, we must verify that they give the correct results. Since we are implementing what we expect is the same algorithm as implemented in runBML(), we should be able to pass the same grid to crunBML() and runBML() with the same number of iterations and get exactly the same result. We need to do this for various different initial grids and explore corner cases, e.g., different number of blue and red cars, non-square grids, degenerate inputs (dimensions or number of cars). Since we need to compare many pairs of resulting grids, we should develop a function to compare 2 grids and verify that they are identical, not just similar.

Q.18 Write the function to compare 2 grids and verify they are the same. Raise an error if they are not. The error should identify in what way the grids are not the same. You can do this with the error message and/or by specifying a class for the error. (See the help for condition in *R*.)

Q.19 Use the function to compare the outputs from crunBML() and runBML() for a variety of different grids to verify they give the same results for the same inputs.

Now, let's call crunBML() and see how fast it is. We'll use the same sized grid as we did when timing our vectorized versions, i.e., a 100-by-100 grid with 50% density:

```
tm_c = system.time(o <- crunBML(g100))
tm_v4/tm_c
```

```
   user  system elapsed
   38.1     Inf    40.5
```

So we get a speedup of a factor 40 by using compiled code. The simple-minded compiled code significantly outperforms the carefully optimized vectorized version written in R. This doesn't mean R is a bad language. It merely means that its dynamic nature, which makes it easy to express computations, often leads to slower code than with fully type-specified, compiled languages. This is true of most interpreted languages such as *MATLAB*® and *Python*, i.e., they are slower than compiled languages such as *C*. It is often useful to rewrite small, relevant parts of an R computation in *C* code. We use profiling to identify the bottlenecks and only rewrite the time-critical parts, and only if run-time speed is a serious issue.

We should note that we recommend using the .Call() interface rather than .C() function. It is much more flexible. It requires learning the *C*-level API for R objects, but this is quite simple for the common tasks. The Rcpp package provides another approach that makes working with R objects in *C++* code more convenient.

Let's compare the timings of all of our approaches and implementations at this point. We can combine them into a matrix with

```
timings = rbind(loop1 = tm1, loop2 = tm2,
                vector1 = tm_v, vector2 = tm_v2,
                vector3 = tm_v3, vector4 = tm_v4,
                C = tm_c)[,1:3]
```

```
       user.self sys.self elapsed
tm1       16.337    1.218  17.557
tm2       12.364    0.027  12.391
tm_v       0.581    0.034   0.615
tm_v2      0.483    0.024   0.507
tm_v3      0.358    0.018   0.376
tm_v4      0.305    0.018   0.324
tm_c       0.008    0.000   0.008
```

It is far more effective to display these graphically. We can visualize these in many different ways. One possible approach is a dotplot of the times, e.g., `dotchart(timings[,3])`. However, it might be more meaningful to display the ratio of these times as a dotplot to show the relative times. We use the fastest implementation as the baseline to show how many times slower the others are relative to this. We do this with

```
dotchart(timings[,3]/timings[nrow(timings),3])
```

and the results are shown in the left panel of Figure 8.11. The plot illustrates the very large speedup by vectorizing over using loops. The different vectorization improvements and the transition to *C* don't seem very remarkable. However, this is due to the very large scale of the plot. If we look at just these (in the right panel of Figure 8.11), we can clearly see how performant the *C* code is.

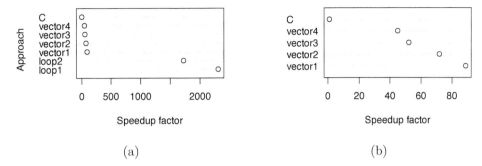

Figure 8.11: Comparison of Times for Different Computational Approaches. *This shows the relative speed of the different approaches, ranging from the naïve loop version, to the fastest vectorized approach, to using the C code. There is a very significant speedup from the loop-version of the code. However, this plot hides the significant speedup between the different vectorized approaches and especially the C code. Panel (b) shows these relative speedups when we omit the loop approaches and we can compare the remaining approaches on a more appropriate scale.*

8.6 Running the Simulations

Now that we have developed our functions to run the simulations rapidly, we can explore the BML model. We'll start by looking at how the model stabilizes for different densities of cars. For a given density and grid dimension, we create a grid and let it evolve for $2T$ time steps. Often we will plot the result, comparing the initial grid to its final state. We can encapsulate these steps in a function runGrid() so that we can then focus on varying the density and dimensions:

```
runGrid =
function(dims, numCars, numIter = 1000, plot = TRUE)
{
  grid = createGrid(dims, numCars)
  g.out = crunBML(grid, numIter)
  if(plot) {
    plot(grid)
    plot(g.out$grid)
  }

  invisible(list(initial = grid,
                 final = g.out$grid,
                 velocity = g.out$velocity))
}
```

We can call this function something like

```
par(mfrow = c(1, 3))
z = runGrid(c(1000, 1000), .35, plot = TRUE)
```

```
plot(rowSums(z$velocity), type = "l",
     main = "Number of cars moving in each pair of time steps")
```

to show the start and end grid states and also the number of red and blue cars that moved in each iteration.

```
set.seed(13123)
```

We now vary the density, while keeping the grid a 1024 square:

```
runs = lapply(c(.25, .33, .38, .38, .55, .65),
              function(density)
                 runGrid(1024, density, 30000, FALSE))
```

The final grids from each of these are shown in Figure 8.12. These illustrate the different equilibrium configurations of the system. We leave it to the reader to explore the behavior for densities between 0.25 and 0.33.

8.6.1 Exploring Car Velocity

Since we have added the ability to collect information about the number of cars that move in each time step, i.e., the velocity, we can explore how this changes across iterations within a run of the BML process. Let's create a small, low-density grid, e.g., a 20-by-20 grid with 10% of the nodes occupied. We create it and run it for 300 iterations with

```
run = runGrid(20, .1)
```

We can access the velocity matrix and compute the average velocity for each direction across all time steps:

```
colMeans(run$velocity)
```

```
[1] 19.86  19.84
```

We might want to normalize these values by the number of cars in each of these directions, which we can do with

```
tmp = colMeans(run$velocity)
tmp/table(run$final)[c("1", "2")]
```

```
  red  blue
0.993 0.992
```

[3] Since the density of cars on the grid is low, almost all of the cars can move at all times. Note how using names makes things simpler and more reliable.

Average velocity may not tell the entire story. We might want to see if it is relatively stationary across all iterations or if it changes as the system evolves. We'll combine the 2 directions (red and blue) and again look at percentages but this time displaying the proportion of cars that move in each iteration:

---

[3]Note that we used "1" and "2" to correspond to red and blue, respectively, in the output of table(). This is because the final object has been mapped to 0, 1, 2 rather than "", "1" and "2" in a call to gridToIntegerGrid().

A Self-Organizing Dynamic System with a Phase Transition

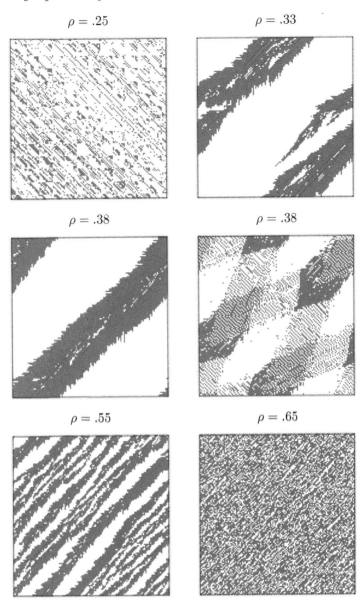

Figure 8.12: Sample 1024-by-1024 Grids with Different Densities. *Each grid has the same dimension and is run for 128,000 time steps. In the top-left, we have a free-flowing grid with a density of 0.25 in which the cars arrange themselves into diagonal lines. At a slightly higher density of 0.33 (top-right), we see one group of cars that are partially deadlocked but which continue to move. We also see bands of red and blue cars moving, but not parallel to each other. Some cars will move freely, at least until they meet the partially deadlocked group. The cars in that group move more slowly until they escape. This gives rise to different velocities. At a slightly higher density, we see additional groups but the same pattern of moving cars. For a different configuration at the same density (0.38) we see a very different structure with mostly deadlocked cars in two parallel groups running diagonally. Note the alternating colors of the diagonal bands. For higher densities, we see more localized groupings but the same deadlock.*

```
plot( rowSums(run$velocity)/sum(run$final != 0), type = "l",
      xlab = "iteration", ylab = "% moving",
      main = "Normalized velocity for low-density grid")
```

Clearly, the process takes some time to become stable but then reaches almost 100% of the cars moving at each cycle.

We would like to understand how velocity varies as we change the density of cars and the dimension of the grid. Figure 8.13 shows the average proportion of cars moving across all iterations for 4 different grid sizes and 20 different densities.

Q.20 Look at the velocity time series for a larger grid with the same density.

Q.21 Is there a difference in velocity for the vertical and horizontal directions? Does the fact that the horizontal moves first make a difference?

Q.22 Recreate Figure 8.13.

Q.23 Are there better ways to display the data from Figure 8.13?

Q.24 How long did it take to compute these results? How can we reduce this time?

Q.25 How do the dimensions of the grid affect velocity? What if the 2 dimensions are quite different? What if one is a multiple of the other? What if the 2 dimensions are relatively prime?

8.7 Experimental Compilation

Writing C code to speed up computations can be difficult, frustrating, error-prone, and distracting. It is important when we need to improve the run-time. However, it would be better if we could write code in R and have it transformed into faster-running code. The cmpFun() function in the compiler package is one approach. The experimental **RLLVM-Compile** package is another. RLLVMCompile can can be used to compile versions of the initial code that we developed using R loops. The package programmatically translates that R code into machine instructions similar to the C code. Importantly, we don't have to write this low-level code. Instead, an extensive collection of R functions do this for us by analyzing and mapping our existing R code and creating the native, fast code. *Currently,* this compilation approach results in code that can run BML simulations about 50% slower than the C implementation. It outperforms the fastest vectorized version (see tm_v4) by a factor of 27, at least on some machines.

The LLVM compilation approach is not yet very robust, or broad in its scope. It is not designed to compile all of the R language and is not likely to make all R code significantly faster. It is currently well suited to scalar operations and loops. It does mean we can potentially write non-vectorized R code and not pay an enormous penalty. It also means we don't necessarily have to move to C code to overcome significant bottlenecks. However, vectorization and using C code will always be important skills to master in R. Vectorization allows us to write efficient, but more importantly, highly succinct and expressive code to do things we do often in data analysis.

Q.26 Create an R package with all of the code to simulate a BML process.

A Self-Organizing Dynamic System with a Phase Transition

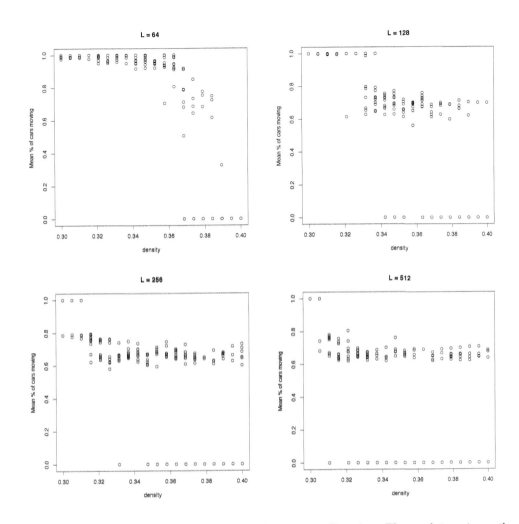

Figure 8.13: Average Velocity of Cars by Occupancy Density. *These plots mirror those in [4] (page 3), with apparently more replications. They show the average proportion of cars moving in each time step on the vertical axis. The horizontal axis shows a range of car densities for the different grids. The 4 panels correspond to the different sizes of the square grid: 64, 128, 256, and 512. Each point in a plot corresponds to running the initial grid through 64,000 total iterations of both red and blue cars, i.e. 128,000 time steps. We see variation for a given density corresponding to different random grid configurations. We see that the density at which deadlock occurs decreases as the grid size increases. Importantly, we see that there are different points of equilibrium, not simply free-flowing or deadlocked. These are the intermediate states that were previously unexpected until reported in [4].*

Q.27 Use *R*'s byte-compiler function cmpFun() to compile our different implementations of the functions moveCars(), getCarLocations(), etc. Compare the performance of these compiled functions to our different approaches.

Q.28 Run multiple BML processes in parallel to explore its behavior for different densities and dimensions. Explore the `parallel` package to do this.

Q.29 Make an animation of the BML process so that we could show this outside of *R*. By this, we mean a display that shows the sequence of grid states over time. This might be a video or an *SVG* file. You can show the changes to the grid at different time intervals.

Q.30 We might want to look at the time-series of the movement of individual cars across iterations. We could use this to see if particular cars are persistently stuck and others are not, or do they all go through periods of being stuck and then free-flowing. Change the code to allow capturing this information.

Q.31 If we look at displays of the BML model on, for example, Wikipedia, the patterns in the traffic flow appear in the opposite direction, i.e. from left to right. Why is this?

Q.32 Write code to find (contiguous) regions in a grid that have very low mobility, i.e., traffic jams. Can we characterize the shapes of these regions?

Q.33 For grid configurations that end in deadlock, how many iterations does it take to reach complete deadlock? How does this depend on the density of cars and grid size?

Q.34 Now that we have the tools for performing experiments, we can explore different behaviors of different configurations. There are many different interesting features uncovered in [4]. Explore the intermediate phases. Use co-prime grid dimensions and see if the behavior changes in any way.

Q.35 We mentioned that the choice of red and blue as colors may not be a good one, as some people have a rare form of color blindness that makes these 2 colors appear similar. Also, displaying colors on overhead projectors and other devices can give very different results than on computer screens. As a result, we may want to change the colors for the 2 different types of cars. What are better colors to use? Change the code in this case study so that the *R* user can specify different colors to represent the horizontal and vertical moving cars. Make certain that all of the code still yields the same results! In how many places does the caller have to specify the new colors? Ideally, they should only have to specify it in one place and the rest of the code should know which colors correspond to horizontal and vertical. How do we implement this?

Bibliography

[1] Daniel Adler. rdyncall: Improved foreign function interface (FFI) and dynamic bindings to *C* libraries (e.g., *OpenGL*). http://cran.r-project.org/package=rdyncall, 2012. *R* package version 0.7.5.

[2] O Biham, A Middleton, and D Levine. Self-organization and a dynamical transition in traffic-flow models. *Physics Review A*, 46:R6124–R6127, 1992.

[3] M Dowle, T Short, S Lianoglou, and A Srinivasan. `data.table` Package. http://cran.r-project.org/web/packages/data.table, 2014. R package version 1.9-4.

[4] Raissa D'Souza. Coexisting phases and lattice dependence of a cellular automaton model for traffic flow. *Physics Review E*, 71, 2005.

[5] Dirk Eddelbuettel and Romain Francois. `Rcpp`: Seamless R and C++ integration. http://cran.r-project.org/package=Rcpp, 2011. R package version 0.9.15.

[6] R Core Team. *Writing R Extensions*. Vienna, Austria, 2012. http://cran.r-project.org/doc/manuals/r-release/R-exts.html.

[7] R Core Team. `parallel`: Support for Parallel computation in R. http://www.r-project.org, 2014. R package version 3.2.0.

[8] R Development Core Team. *R: A Language and Environment for Statistical Computing*. Vienna, Austria, 2012. http://www.r-project.org.

[9] Duncan Temple Lang. **Rffi**: Interface to libffi to dynamically invoke arbitrary compiled routines at run-time without compiled bindings. http://www.omegahat.org/Rffi, 2011. R package version 0.3-0.

9

Simulating Blackjack

Hadley Wickham
RStudio

CONTENTS

9.1	Introduction	367
	9.1.1 Computational Topics	368
9.2	Blackjack Basics	368
	9.2.1 Testing Functions	370
9.3	Playing a Hand of Blackjack	372
	9.3.1 Creating Functions for the Player's Actions	373
9.4	Strategies for Playing	376
	9.4.1 Developing the Optimal Strategy	379
9.5	Playing Many Games	382
9.6	A More Accurate Card Dealer Shoe	384
9.7	Counting Cards	390
9.8	Putting It All Together	393
9.9	Exercises	394
	Bibliography	396

9.1 Introduction

Blackjack is an extremely popular casino game, in part because if the correct strategy is used, the house advantage is under 1% (meaning that in the long term, the gambler loses, on average, 1% of the amount he or she bets), and card counting techniques can give the motivated player the ability to swing the advantage even further so that it's actually possible for the odds to be in your favor. In this project, we develop a set of functions to simulate a game of blackjack, starting simply and working your way up to card counting.

Our overall strategy is to work in small steps and create code that we can test and explore at each step. Once we are satisfied with the code, we move on to the next step. This tutorial is a bit easier than situations where you need to design a simulation study on your own from scratch because we have laid out the steps in a logical sequence and we proceed in a (fairly) straight path from problem definition to solution. In real life, we often discover that we have missed something in a previous step, that we have gone down a dead end, or accidentally circled back to somewhere we have already been, and consequently we need to modify our previously written code or sometimes start again from scratch, taking into account what we have discovered. The exercises at the end of the chapter give some practice in a more realistic setting, where all the steps have not been laid out in advance.

We tackle the challenge of developing a realistic simulation of blackjack and the influence of card counting in 7 steps:

- **Blackjack basics**: We start by modelling a deck of cards, their value in blackjack, and figuring out who wins a hand.

- **Playing a hand**: We play a hand of blackjack, dealing the cards to a player and exploring what options they can take. Here we develop a blackjack game that we can play manually: our winnings are computed automatically, but we have to tell the computer what action to take.

- **Game strategies**: We implement some strategies so that we can automatically play a game without human intervention. We start with a simple strategy and work our way toward an optimal strategy.

- **Many games**: To see how well our strategy pays off, we play a few thousand hands of blackjack. This will involve some statistical thinking about variation and some visualisation to explore the results.

- **A more accurate shoe**: Before we can implement card counting, we need a model for the dealer's shoe (the device used to store and automatically shuffle and deal cards) that makes counting possible. Here, we learn about mutable objects and programming with reference classes in R [2].

- **Card counting**: We develop some functions to count cards and incorporate these into our strategy. That is, we adjust a bet based on the count.

- **Pulling it all together**: Finally we combine all of these elements to see how much of a difference counting cards actually makes in the long run.

9.1.1 Computational Topics

In this chapter, we cover the following:

- How to break down a complex problem into simple tasks.
- Ways to use data to make your functions simpler.
- Tips for developing robust functions in R.
- Strategies to make sure that our code actually works the way we expect it to work.
- Reference classes, for creating objects that can be modified in place.

9.2 Blackjack Basics

If you are not familiar with blackjack, there are many online resources, including the wizardofodds site at http://wizardofodds.com/games/blackjack/. You can play a few practice rounds online at http://www.onlineblackjackguru.com/ or http://www.hitorstand.net/.

We start our computer model of blackjack with a model for a deck of cards that is relevant to blackjack. A deck of cards has 52 cards with 13 cards from each suit — hearts, diamonds, spades, and clubs. Each suit has a card labeled 2, 3, ..., 10 and a Jack, Queen,

Simulating Blackjack

King, and Ace. The Jack, Queen, and King all have the same value (10). In blackjack, the Ace can count as either 1 or 11. For the moment, we assign an Ace a value of 1 since we do not interpret the card's value until later and we can determine if it is 11 from the value 1. Since the numeric worth of the card is all that matters in blackjack, i.e., the suit does not matter, we can represent a deck of cards in R with

```
deck = rep(c(1:10, 10, 10, 10), 4)
```

This in turn leads to a simple model for shuffling n decks of cards:

```
shuffle_decks = function(n) sample(rep(deck, n))
```

In blackjack, the player's hand has 2 or more cards, and if the total value of the cards in the hand exceeds that of the dealer's hand, but does not exceed 21, then the player wins. This means that we need to be able to compute the value of a hand of cards. This function should allow us to compare 2 hands of cards to see which is the winner. There are 3 things we need to account for:

- Busting (going over 21) loses to every non-busted hand.

- If a hand has an Ace and the total value of the cards in the hand (with the Ace counted as 1) is 11 or less, then we count the Ace as an 11 and increase the value of the hand without busting. If a hand has 2 Aces, only one can possibly switch to 11 because two 11s is a bust.

- The term "blackjack" refers to getting a hand worth 21 with only 2 cards. Blackjack is better than reaching 21 with 3 or more cards, so we need to distinguish between the different ways to achieve a value of 21.

To compute the value of a hand of cards, we can sum up the value of the cards, and then take into consideration the special cases described above. If our goal is to simply compare 2 hands to see which is higher, we can set the value of a hand to zero if it is a bust; add 10 to the value of a hand if it contains an Ace and the hand's value is less than 12; and add a small amount to the value, if it is 21 and contains only 2 cards, so that it beats other 21s. The following function takes in a vector of cards in a hand and returns its value:

```
handValue = function(cards) {
  value = sum(cards)

     # Check for an Ace and change value if it doesn't bust
  if (any(cards == 1) && value <= 11)
    value = value + 10

     # Check bust (set to 0); check black jack (set to 21.5)
  if(value > 21)
    0
  else if (value == 21 && length(cards) == 2)
    21.5 # Blackjack
  else
    value
}
```

The player's winnings depend on how much was bet and whether or not the hand is blackjack. To compute the winnings, we need to know the value of the player's hand and the dealer's hand. In blackjack, if the player goes bust then the player loses, even if the dealer also goes bust. If the player gets blackjack then the player is paid $1.5 to every $1 bet. All other winning bets are paid 1 to 1, meaning for every dollar the player bets, he or she keeps that dollar and the dealer pays another dollar. When the dealer and player tie, meaning their hands have the same value (and neither have busted), then the player keeps the bet and is not paid any winnings.

```
winnings = function(dealer, players) {
  if (dealer > 21) {
        # Dealer has blackjack, ties players with blackjack
    -1 * (players <= 21)
  } else if (dealer == 0) {
        # Dealer busts - all non-busted players win
    1.5 * (players > 21) +
      1 * (players <= 21 & players > 0) +
     -1 * (players == 0)
  } else {
        # Dealer 21 or below, all player values > dealer win
    1.5 * (players > 21) +
      1 * (players <= 21 & players > dealer) +
     -1 * (players <= 21 & players < dealer)
  }
}
```

Notice that this function makes use of the fact that in R, TRUE and FALSE are treated as 1 and 0 in arithmetic operations, e.g., FALSE * 2.5 is 0. Note the comments in the code: we use them to help remember the reasoning behind each condition in the if statement.

In the winnings() function, we have chosen to break the function down by the dealer's cards first, and then the player's cards. What would happen if we switched this order and started with the player's values? Is the function easier or harder to understand? We leave this as an exercise.

It's possible to write a very terse implementation of this algorithm by rearranging the conditions and taking advantage of logical to numeric coercion:

```
winnings = function(dealer, players){
  (players > dealer & players > 21)  * 1.5 + # blackjack
  (players > dealer & players <= 21) * 1   + # win
  (players < dealer | players == 0)  * -1    # lose
}
```

How does the function work? Does it do the same thing as the original? Is it easier to follow the logic in the first or second function? Why? When do you think using code like this is appropriate? We leave the exploration of this version of the winnings() function as an exercise.

Does our first version of the winnings() function work properly? That is, does it correctly calculate the player's winnings? What about the second version of winnings()? This is the topic of the next section.

9.2.1 Testing Functions

Rather than wait until we have finished all of our programming, we check the functions that we have written along the way. We want to make sure that these functions produce the

Simulating Blackjack

correct results. One way to check is to run the function for a subset of all possible inputs and make sure that the results are correct.

For the handValue() function, we need to create test cases that include blackjack, hands where an Ace counts as 11 and where it counts as 1, multiple Aces, and busts with and without an Ace in the hand. The following test cases cover these possibilities:

```
test_cards = list( c(10, 1), c(10, 5, 6), c(10, 1, 1),
                  c(7, 6, 1, 5), c(3, 6, 1, 1),
                  c(2, 3, 4, 10), c(5, 1, 9, 1, 1),
                  c(5, 10, 7), c(10, 9, 1, 1, 1))
```

We know that the value for each hand is

```
test_cards_val = c(21.5, 21, 12, 19, 21, 19, 17, 0, 0)
```

When tested, our function matches these values,

```
identical(test_cards_val, sapply(test_cards, handValue))
```

[1] TRUE

The handValue() function appears to be correctly determining the value of a hand of cards.

Next we set up test cases for the winnings() function. We want to cover the combinations of the dealer and the player busting, getting blackjack, or some other value, and ties. We generate these as a matrix of correct results for combinations of the following values: 0, 16, 19, 20, 21, and 21.5, for the player and the dealer, i.e.,

```
test_vals = c(0, 16, 19, 20, 21, 21.5)

testWinnings =
  matrix(c( -1,  1,  1,  1,  1, 1.5,
            -1,  0,  1,  1,  1, 1.5,
            -1, -1,  0,  1,  1, 1.5,
            -1, -1, -1,  0,  1, 1.5,
            -1, -1, -1, -1,  0, 1.5,
            -1, -1, -1, -1, -1,  0),
         nrow = length(test_vals), byrow = TRUE)
dimnames(testWinnings) = list(dealer = test_vals,
                              player = test_vals)

testWinnings

       player
dealer  0  16  19  20  21 21.5
   0   -1   1   1   1   1  1.5
   16  -1   0   1   1   1  1.5
   19  -1  -1   0   1   1  1.5
   20  -1  -1  -1   0   1  1.5
   21  -1  -1  -1  -1   0  1.5
   21.5 -1  -1  -1  -1  -1  0.0
```

We fill the check matrix with the results from calls to winnings() for the test cases with

```
check = testWinnings
check[] = NA

for(i in seq_along(test_vals)) {
  for(j in seq_along(test_vals)) {
    check[i, j] = winnings(test_vals[i], test_vals[j])
  }
}

identical(check, testWinnings)
```

```
[1] TRUE
```

Again, our function produces the correct answer for these test cases.

9.3 Playing a Hand of Blackjack

We now have the basic pieces set up to shuffle a deck of cards, score a hand of blackjack, and determine a player's winnings. Next we need to figure out how to model a game, i.e., the decisions that the player makes and also the dealing and process of delivering the next card. We start with a simple model that's easy to program, and gradually make it more sophisticated, getting closer and closer to the real situation.

There are two basic actions a player can take after the first 2 cards are dealt to the player.

Hit
The player can request "a hit." Then 1 card is removed from the deck(s), which is (are) stored in the shoe, and added to the player's hand.

Stand
The player can continue to request hits, receiving 1 card at a time from the shoe, until the player decides to end the turn, i.e., until the player stands.

Before requesting the first card, the player is allowed two other actions.

Double Down
The player can double his or her bet. If the player doubles the bet, then exactly 1 hit is made and the player must then stand.

Split
If the first 2 cards dealt to the player are identical in value, the player may choose to split them into 2 hands. The player must put up an additional bet of equal size to the original bet, each of these 2 hands is dealt the second card, and the player then makes decisions independently about hits, stands, etc. with these hands. There are many casino-specific rules for the actions a player can take after splitting. Typically, if dealt a blackjack from a split, the casino treats it as 21, not blackjack. Sometimes when splitting Aces, the player can only receive 1 additional card for a total of 2 cards per hand.

In a single round of blackjack, almost everything about the hand the player is dealt can change: cards can be added to the hand, the size of the bet can double, and the number of hands that a player has can increase. We need to keep this in mind as we proceed.

We start by modelling the shoe, which contains the shuffled cards to be dealt. We begin with a very simple model, a function that draws m cards with replacement from a deck.

Simulating Blackjack

Shoes typically contain more than 1 deck and the decks are reshuffled often so we start with the following function:

```
shoe = function(m = 1) sample(deck, m, replace = TRUE)
```

In practice, dealing is without replacement, but this shoe() is effectively equivalent to a shoe with a very large number of decks that is reshuffled after every hand. This is not a useful model for card counting, but is similar to the continuously reshuffled shoes used by some casinos.

Next we create a model for the hand. A hand has a bet, a shoe (our function for getting more cards), and some cards. Blackjack always starts with 2 cards so we default to drawing 2 cards from the shoe.

```
new_hand = function(shoe, cards = shoe(2), bet = 1) {
  list(bet = bet, shoe = shoe, cards = cards)
}
```

Even though the player does not get to choose cards in a real blackjack game, we keep cards as a parameter so that we can specify them when we start testing our functions.

We can make our new_hand() function a little more user friendly by using some *S3* object-oriented programming. The following code modifies new_hand() to return an object of class hand:

```
new_hand = function(shoe, cards = shoe(2), bet = 1, ...) {
  structure(list(bet = bet, shoe = shoe, cards = cards),
            class = "hand")
}
```

Then we can create a specialized print() method to give a nicely formatted display with

```
print.hand = function(x, ...) {
  cat("Blackjack hand: ", paste(x$cards, collapse = "-"),
      " (", handValue(x$cards), ").  Bet: ", x$bet,
      "\n", sep = "")
}
```

For example, we create a hand that has a bet of $7 with

```
myCards = new_hand(shoe, bet = 7)
```

Then we print myCards implicitly using the print.hand() method as follows:

```
myCards
```

```
Blackjack hand: 6-10 (16).  Bet: 7
```

9.3.1 Creating Functions for the Player's Actions

Now we can write functions for each of the 4 actions, which we call hit(), stand(), dd() for double down, and splitPair(). Each function takes a hand as input and returns a hand (or list of hands in the case of a split) as output. If you have programmed in other languages, your first thought might be to modify the existing hand rather than returning a new hand. This is not the most natural style of programming in R. It is generally better to stick to so-called "functional" programming, where the function doesn't have any side-effects, i.e., mutates its inputs, but just returns a new object. This makes it much easier to reason about a program, because you can understand each function in isolation.

Recall that the hit action draws 1 card from the shoe and adds it to those cards already in the hand. We can do this with

```
hit = function(hand) {
  hand$cards = c(hand$cards, hand$shoe(1))
  hand
}
```

We test hit with the hand that we created and stored in myCards

```
hit(myCards)
```

```
Blackjack hand: 6-10-7 (0).  Bet: 7
```

Our 2 cards are 6 and 10 and we are dealt a 7 so we go bust and the value of the cards is now 0.

The stand() action is very simple. It returns the unaltered hand. We write this function as follows:

```
stand = function(hand) hand
```

The double down function doubles the bet, requests 1 additional card, and then stands.

```
dd =  function(hand) {
  hand$bet = hand$bet * 2
  hand = hit(hand)
  stand(hand)
}
```

When we double down on our original hand, we find

```
dd(myCards)
```

```
Blackjack hand: 6-10-4 (20).  Bet: 14
```

Lastly, we write our splitPair() function, which splits a hand with 2 cards into 2 hands. We need to draw 1 card from the shoe for each new hand. The bet for each hand remains the same as the original bet. Our function is:

```
splitPair = function(hand) {
  list(
    new_hand(hand$shoe,
             cards = c(hand$cards[1], hand$shoe(1)),
             bet = hand$bet),
    new_hand(hand$shoe,
             cards = c(hand$cards[2], hand$shoe(1)),
             bet = hand$bet)
  )
}
```

We test our splitPair() function with myCards even though in a real game we cannot split cards that are not of equal value. We do this with

```
splitHand = splitPair(myCards)
splitHand
```

```
[[1]]
Blackjack hand: 6-10 (16).  Bet: 7

[[2]]
Blackjack hand: 10-10 (20).  Bet: 7
```

The basics of these functions (hit(), stand(), dd() and splitPair()) are working, but our testing is not very extensive.

Currently these functions do not check that they are given the correct input. How could we make the functions safer so, e.g., a hit could not be made if the hand has gone bust? Or as we just did, split when the 2 cards are not of equal value? What other "safety" features can we add to these functions? We leave these as exercises.

Now let's play a round of blackjack against a dealer. We set the seed of the random number generator with set.seed() to ensure the results are the same when the seed is set to the same value before running the code. This is a useful mechanism for making random results reproducible. We set the seed to, say, 1014 with

```
set.seed(1014)
```

Then deal a hand of cards to the dealer and a player:

```
dealer = new_hand(shoe)
player = new_hand(shoe)
```

When we play blackjack, we can see 1 of the 2 cards in the dealer's hand. Our dealer's top card is

```
dealer$cards[1]
```

```
[1] 5
```

And our hand is

```
player
```

```
Blackjack hand: 6-9 (15).  Bet: 1
```

We decide to take a card:

```
player = hit(player)
player
```

```
Blackjack hand: 6-9-1 (16).  Bet: 1
```

We were dealt an Ace, which must be used as a 1 in computing the total value of our cards, or otherwise we would bust. At this point, we stand. Now it is the dealer's turn so the dealer reveals the hidden card:

```
dealer
```

```
Blackjack hand: 5-5 (10).  Bet: 1
```

According to casino rules, the dealer must hit because the total value of the dealer's cards is less than 17.

```
dealer = hit(dealer)
dealer
```

```
Blackjack hand: 5-5-10 (20).  Bet: 1
```

At 20, the dealer must stand because the value is 17 or over. We know that we have lost, but we call *winnings()* to confirm that it works as expected:

```
winnings(handValue(dealer$cards), handValue(player$cards))
```

```
[1] -1
```

We leave it to the reader to check that all the action functions work correctly. You can do this by placing specific cards in a hand using the *cards* parameter of *new_hand()*. What cases do you need to test?

Our next step is to develop some code that plays a game automatically, using a strategy that says what to play given the dealer's top card and the player's cards. This is the topic of the next section.

9.4 Strategies for Playing

We start by developing a very simple strategy: hit if the dealer's top card has a value more than 6 and our total is less than 17, otherwise stand. This is a pretty poor strategy, but it allows us to build the scaffolding for developing better strategies. We use abbreviations to communicate our different actions: S for stand, H for hit, D for double down, and SP for split. Then our *simple_strategy()* function is

```
strategy_simple = function(mine, dealerFaceUp) {
  if (handValue(dealerFaceUp) > 6 && handValue(mine) < 17)
     "H"
  else
     "S"
}
```

This function is very simple, but it still has a bug in it! Can you spot it? If the player busts, then the value of the cards is zero, and the strategy will keep hitting forever. We need to add a check to ensure that if the player has busted then the player must stand. Below is our revision to *simple_strategy()*:

```
strategy_simple = function(mine, dealerFaceUp) {
  if (handValue(mine) == 0) return("S")
  if (handValue(dealerFaceUp) > 6 && handValue(mine) < 17)
     "H"
  else
     "S"
}
```

Simulating Blackjack

We also need a strategy for generating the sequence of dealer's cards. A dealer must follow the simple rule: hit if the value is less than 17, otherwise stand. We can implement the dealer's simple strategy directly:

```
dealer_cards = function(shoe) {
  cards = shoe(2)
  while(handValue(cards) < 17 && handValue(cards) > 0) {
    cards = c(cards, shoe(1))
  }
  cards
}
```

Notice that in one round of blackjack there is no back and forth between the dealer's and the player's actions. The round begins with the player and dealer each being dealt 2 cards, and the player sees only 1 of the dealer's 2 cards. The player's turn is first. He or she takes 1 card at a time and after each card is dealt decides whether to take another or to stand. After the player stands, it is the dealer's turn, and the dealer is obligated to follow the rule just described: take a card if the dealer's card total is less than 17 and otherwise stand, no matter what is in the player's hand. We have dealer_cards() carry out the dealer's entire turn because the dealer's strategy is not affected by the player's cards.

Now, we have all the pieces in place to write a function that plays a game of blackjack automatically. This function has 2 main parameters: *shoe*, where we pass in our function for getting new cards; and *strategy*, where we supply a function that, based on the dealer's top card and the player's cards, tells us what action to take. We also have parameters *hand* and *dealer* to provide the player's and dealer's hands, respectively. These hands have a default value of a new hand, and we override the defaults to control the inputs to the game.

We ignore the issue of splitting at first, and create a function, play_hand(), as follows:

```
play_hand = function(shoe, strategy,
                    hand = new_hand(shoe),
                    dealer = dealer_cards(shoe)) {

  face_up_card = dealer[1]

  action = strategy(hand$cards, face_up_card)
  while(action != "S" && handValue(hand$cards) != 0) {
    if (action == "H") {
      hand = hit(hand)
      action = strategy(hand$cards, face_card)
    } else if (action == "D") {
      hand = dd(hand)
      action = "S"
    } else {
      stop("Unknown action: should be one of S, H, D, SP")
    }
  }

  winnings(handValue(dealer), handValue(hand$cards)) * hand$bet
}
```

Notice that the winnings are computed based on a $1 bet, so they need to be scaled by the actual bet.

Let's try our new function. We first set the seed and then call play_hand(), passing it our shoe() and simple_strategy() functions as follows:

```
set.seed(1014)
play_hand(shoe, strategy_simple)
```

```
[1] 1.5
```

No errors occurred, and the winnings indicate we were dealt blackjack!

We cannot easily check if our simple strategy is working correctly without knowing what cards we were dealt and what the dealer was dealt. To figure this out we can augment our code to provide play-by-play information about the game. We would not necessarily always want to provide this information so we add another parameter to play_hand() called *verbose* to indicate whether or not we want to see this additional information. We set the default value of *verbose* to FALSE so the caller of the function must specifically request the extra information. Our augmented function is now defined as

```
play_hand = function(shoe, strategy,
                     hand = new_hand(shoe),
                     dealer = dealer_cards(shoe),
                     verbose = FALSE) {

  if (verbose) {
    cat("New hand \n")
    cat("  Dealer: ", paste(dealer, collapse = "-"),
        " (", handValue(dealer), ")\n", sep = "")
    cat("  Player: ", paste(hand$cards, collapse = "-"),
        ": ", sep = "")

  }
  face_card = dealer[1]

  action = strategy(hand$cards, face_card)
  while(action != "S" && handValue(hand$cards) != 0) {
    if (verbose) cat(action)
    if (action == "H") {
      hand = hit(hand)
      action = strategy(hand$cards, face_card)
    } else if (action == "D") {
      hand = dd(hand)
      action = "S"
    } else {
      stop("Unknown action: should be one of S, H, D, SP")
    }
  }
  if (verbose) {
    cat(action, " -> ", paste(hand$cards, collapse = "-"),
        " (", handValue(hand$cards), ")", sep = "", "\n")
  }

  winnings(handValue(dealer), handValue(hand$cards)) * hand$bet
}
```

Simulating Blackjack

Let's rerun our test to see if we indeed won with blackjack:

```
set.seed(1014)
play_hand(shoe, strategy_simple, verbose = TRUE)

New hand
  Dealer: 5-5-6-9 (0)
  Player: 1-10: S -> 1-10 (21.5)
[1] 1.5
```

The verbose option provides messages that are useful when we are testing our implementation, but impacts performance when we start to play thousands of games. Using the *verbose* parameter is a common pattern for this situation.

The final challenge in writing our play_hand() function is to handle the split action. Making it easy to implement the results of splitting depends critically on the output of this function. Here we make the function call itself for each of the new hands and then return the sum of the winnings. That is, we add up the winnings from the multiple independent hands created when splitting. The code that we add to play_hand() for the split action looks as follows:

```
hands = splitPair(hand)
one = play_hand(shoe, strategy, hands[[1]], dealer,
                verbose = verbose)
two = play_hand(shoe, strategy, hands[[2]], dealer,
                verbose = verbose)
return(one + two)
```

We leave it as an exercise to incorporate the split action into play_hand().

After adding the split action to our function, we rerun the earlier test to see that it still works the same way:

```
set.seed(1014)
play_hand(shoe, strategy_simple, verbose = TRUE)

New hand
  Dealer: 5-5-6-9 (0)
  Player: 1-10: S -> 1-10 (21.5)
[1] 1.5
```

Although we get the expected results, we have not tested the new code that performs the split action. We can set up an artificial hand with cards that a strategy would split to test this code. Unfortunately, our strategy is so simple that it will never determine that the hand should be split so we hold off further testing until we have a more complex strategy.

9.4.1 Developing the Optimal Strategy

Now that we have the ability to play a game of blackjack, we turn back to our strategy: what's the best action to take for each situation? It turns out that if the shoe behaves as an "infinite" deck then the optimal strategy has been known for quite some time. The optimal strategy was first derived in [1]. It is quite complex, but we can take a hint from the many Web sites that describe the strategy, and instead of describing an algorithm, work with a lookup table. Our lookup table is adapted from Wikipedia (see http://en.wikipedia.org/wiki/Blackjack). The table has 3 inputs: the value of the dealer's face-up card,

TABLE 9.1: Optimal Strategy

type	value	2	3	4	5	6	7	8	9	10	11
hard	20	S	S	S	S	S	S	S	S	S	S
hard	19	S	S	S	S	S	S	S	S	S	S
hard	18	S	S	S	S	S	S	S	S	S	S
soft	20	S	S	S	S	S	S	S	S	S	S
soft	19	S	S	S	S	S	S	S	S	S	S
soft	18	S	Ds	Ds	Ds	Ds	S	S	H	H	H
pair	20	S	S	S	S	S	S	S	S	S	S
pair	18	SP	SP	SP	SP	SP	S	SP	SP	S	S

This table displays a few of the strategies for playing blackjack. These are based on an infinite deck, i.e., the chance of a particular card on the next draw is determined by the proportion of such cards in a full deck. S means stand, SP means split, and Ds means double down if allowed, otherwise split. Columns correspond to the value of the dealer's face-up card. Rows correspond to the value of the player's hand and whether it is soft, hard, or composed of a pair.

the value of the player's cards, and the type of hand. A hand can be one of 3 types: a pair, hard, or soft. A soft hand is one with an Ace that is treated as an 11, rather than a 1. A hard hand is the complement of a soft hand. That is, a hard hand either has no Aces or every Ace must be treated as a 1 to avoid busting. We treat pairs specially as they can be split. For example, a soft 12 must be a pair of Aces, and the optimal strategy is to always split Aces. A few rows of this "strategy lookup table" appear in Table 9.1. The dealer values are encoded in the columns of the table and player's hand type and value in the rows.

Moving the algorithmic complexity to a lookup table makes our strategy function very simple. We first read in the strategies from a *csv* file with

```
lookuptable = read.csv("Data/strategy.csv", header = TRUE,
                stringsAsFactors = FALSE, check.names = FALSE)
head(lookuptable)

  type value 2 3 4 5 6 7 8 9 10 11
1 hard    20 S S S S S S S S  S  S
2 hard    19 S S S S S S S S  S  S
3 hard    18 S S S S S S S S  S  S
4 hard    17 S S S S S S S S  S  S
5 hard    16 S S S S S H H H  H  H
6 hard    15 S S S S S H H H  H  H
```

Our strategy function performs a table lookup based on the value of the cards in the player's hand, whether that value is hard or soft, and the dealer's face-up card. Our function appears below:

```
strategy_optimal = function(player_hand, dealerFaceUp,
                            optimal = lookuptable) {
    # Stand if 21 or already busted
    player_value = handValue(player_hand)
    if (player_value == 0) return("S")
    if (player_value >= 21) return("S")
```

Simulating Blackjack

```
    dealer_value = handValue(dealerFaceUp)
    loc_ace = player_hand == 1

    if (length(player_hand) == 2 &&
        player_hand[1] == player_hand[2]) {
      type = "pair"
      if (player_hand[1] == 1) player_value = 2
    } else if (sum(loc_ace) > 0 &&
                  (player_value - sum(loc_ace)) >
                              handValue(player_hand[!loc_ace])) {
      type = "soft"
    } else {
      type = "hard"
    }

    out = optimal[optimal$type == type &
                  optimal$value == player_value,
                  as.character(dealer_value)]
    if (length(out) == 0) browser()
    if (out == "Dh")
      if (length(player_hand) > 2) out = "H" else out = "D"
    if (out == "Ds" )
      if (length(player_hand) > 2) out = "S"  else out = "D"
    out
}
```

There's one important debugging feature of *R* that we use in this function:

```
if (length(out) == 0) browser()
```

This ensures that if the lookup is unsuccessful, we enter an interactive browser to diagnose the problem. It took a few iterations before we worked out all the kinks in using the table.

We examine the strategies selected for several cases to ensure that they have been correctly chosen. We do this by providing the player's cards and the dealer's face-up card in calls to strategy_optimal() and check the return value against the optimal strategy discerned from reading the lookup table. We use the stopifnot() function to check that all of the logical tests are true. We do this with

```
stopifnot(strategy_optimal(c(4, 1, 6), 3) == "S",
          strategy_optimal(c(3, 2, 6), 3) == "H",
          strategy_optimal(c(5, 6), 3) == "D",
          strategy_optimal(c(3, 2, 6), 1) == "H",
          strategy_optimal(c(5, 6), 1) == "H",
          strategy_optimal(c(6, 1, 5, 1), 7) == "H",
          strategy_optimal(c(6, 1, 5, 1), 6) == "H",
          strategy_optimal(c(1, 1), 7) == "SP",
          strategy_optimal(c(9, 9), 7) == "S",
          strategy_optimal(c(6, 6), 7) == "H",
          strategy_optimal(c(3, 3), 7) == "SP",
          strategy_optimal(c(7, 1), 5) == "D",
          strategy_optimal(c(7, 1), 7) == "S",
```

```
                 strategy_optimal(c(2, 1), 9) == "H",
                 strategy_optimal(c(2, 1), 5) == "D")
```
```
Error: strategy_optimal(c(6, 1, 5, 1), 6) == "H" is not TRUE
```

There is a problem with the strategy when the player holds cards with values 6, 1, 5, 1 and the dealer's top card is 6. The player's hand is worth 13 and despite having 2 Aces, it is hard. We re-check the lookup table and see that the correct action is not to hit, which is not what we have specified. After we correct this mistake, the tests all return TRUE.

Now that we have two strategies, we can compare them by playing many rounds of blackjack many times. This is the topic of the next section.

9.5 Playing Many Games

We have the pieces in place to easily play many rounds of blackjack. For example, we could use the base *R* function replicate() to play 10 hands with

```
set.seed(101451)
replicate(10, play_hand(shoe = shoe,
                        strategy = strategy_optimal))
```
```
[1]  1  1  1  0 -1  1  0 -1  1 -1
```

We see that in these rounds we lost 3 times, won 5 times, tied twice, and never beat the dealer with blackjack or doubled down. If we want to check that our code is working correctly, we can turn on the *verbose* option in play_hand(), e.g.,

```
replicate(3, play_hand(shoe = shoe, strategy = strategy_optimal,
                       verbose = TRUE))
```
```
New hand
  Dealer: 7-1 (18)
  Player: 6-10: HS -> 6-10-10 (0)
New hand
  Dealer: 4-5-1 (20)
  Player: 6-4: DS -> 6-4-2 (12)
New hand
  Dealer: 5-1-6-5 (17)
  Player: 1-3: DS -> 1-3-10 (14)
[1] -1 -2 -2
```

It appears that our code is working correctly.

We can play a thousand hands for each strategy and capture the results with:

```
set.seed(10114)
win_optimal = replicate(1000, play_hand(shoe = shoe,
                                 strategy = strategy_optimal))
set.seed(10114)
win_simple = replicate(1000, play_hand(shoe = shoe,
                                 strategy = strategy_simple))
```

Notice that we started each simulation at the same point by setting the seed between calls to the different strategies. We overlay density plots of the winnings from these results with

```
plot(density(win_optimal, bw = 0.25), col = "green", lwd = 2,
     xlab = "Winnings", xlim = c(-3, 3),
     ylim = c(0, 0.9), main = "")
lines(density(win_simple, bw = 0.25), col = "purple", lwd = 2)
legend("topright", col = c("green", "purple"),
       legend = c("Optimal", "Simple"), bty = "n", lty = 1)
```

The plot appears in Figure 9.1. It's difficult to tell whether the simple strategy loses more often than the optimal strategy. We can find the average net gain with

```
mean(win_optimal)
[1] -0.033
```

```
mean(win_simple)
[1] -0.056
```

The optimal strategy cannot make up for the casino's advantage, but it loses less on average than our simple strategy.

Now that we have verified that our gambling simulation works, we can turn this into a function that finds the average winnings and other summary statistics for a given number of rounds, e.g.,

```
payoff = function(n, strategy, shoe) {
  results = replicate(n, play_hand(shoe = shoe,
                                   strategy = strategy))
  c(avgGain = mean(results), sdGain = sd(results),
    medGain = median(results))
}
```

The real development cycle was not quite so smooth — we had a few problems that cropped up very rarely so we had to play a few thousand games to uncover them all. Systematic testing of the play_hand() function would have saved a lot of time! This is the topic of an exercise.

We can then use the payoff() function to compare our two strategies for 50 rounds of the game. For each strategy, we play 50 hands 1000 times with

```
win_simple50 = replicate(1000,
                         payoff(50, strategy_simple, shoe))
win_optimal50 = replicate(1000,
                          payoff(50, strategy_optimal, shoe))
```

We plot the distribution of these strategies' payoffs with a frequency polygon as follows:

```
df = data.frame(
  value = c(win_simple50[ "avgGain", ],
            win_optimal50[ "avgGain", ]),
  strategy = rep(c("simple", "optimal"), each = 1000))

library(ggplot2)
qplot(value, data = df, geom = "freqpoly", colour = strategy,
      binwidth = 0.05)
```

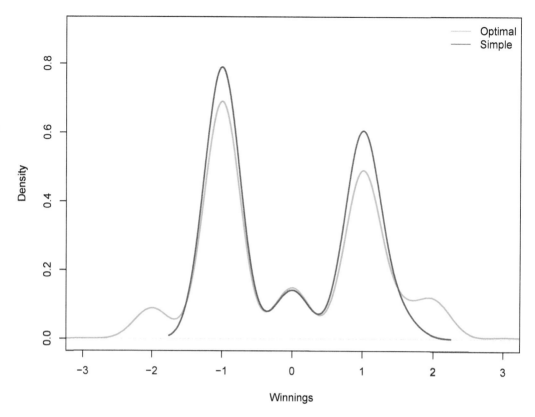

Figure 9.1: Density Plot Comparing the Winnings from Two Strategies. *This plot shows the density of the gain from 1,000 $1 bets for two different strategies: the optimal strategy for an infinite deck, and a simple strategy that hits when the dealer shows a 6 or higher and the player's cards are under 17, and otherwise stands. A gain of $2 or -$2 results from winning or losing a double down, respectively. The simple strategy never doubles the bet.*

If you haven't seen this type of plot before, it's a straightforward tweak to the histogram: instead of displaying the counts with a bar, we use connected line segments. (See [3]). This makes it easier to compare distributions. The plot appears in Figure 9.2.

Our optimal strategy is better than the simple strategy, but not by much. The optimal strategy is for an infinite deck. We might do better if we can use card counting techniques, but in order to do that we need a better model for the shoe. This is the topic of the next section.

9.6 A More Accurate Card Dealer Shoe

Our final task is to implement some card counting strategies. We have most of the pieces in place; we have implemented the optimal strategy and can play games automatically. However, our model of the shoe is too simple. The shoe currently just samples with replacement from a deck of cards, which does not match the behavior of most blackjack tables.

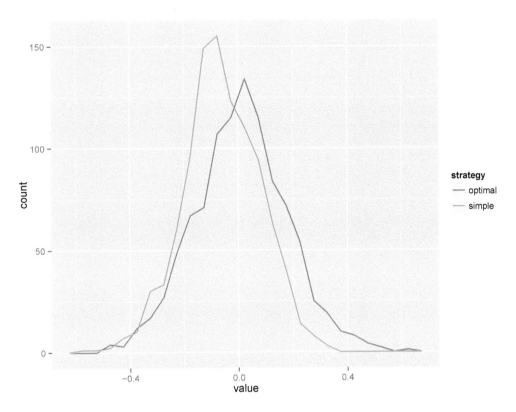

Figure 9.2: A Comparison of the Optimal and Simple Strategies. *The distribution of payoffs under the optimal strategy peaks at a higher value. Note that the variability is considerable: there is about a 40% chance of losing 5% or more of your money with the optimal strategy and about a 55% chance with the simple strategy.*

Instead, we want a shoe that contains a fixed number of decks of cards, deals cards without replacement, and reshuffles at known points.

Developing a better model for the shoe is challenging because now we must keep track of the state: what cards remain in the shoe. R is primarily a functional programming language, which means that it's not easy to modify an object within another function. For example, consider the simple function f() that modifies the element a in the input list x:

```
y = list()
y$a = 1
f = function(x) {
  x$a = 2
}
y$a

[1] 1

f(y)
y$a
```

```
[1] 1
```

The list y, which is passed into f(), remains unchanged because it is passed by value. The assignment within the call to f(), x$a = 2, occurs within the environment/call frame of the function call for f(). This sort of behavior makes it hard to modify our existing shoe() function to update the cards remaining in the shoe.

Generally, the only way a function can interact with the external environment is through the objects that it returns. Every time it looks as though we are modifying an object, R actually creates a modified copy. We can implement our new shoe() function by returning both the cards drawn from the shoe and the new state of the shoe every time we use it. This technique is called threading state, and would require extensive modifications to our existing code.

Instead, we're going to use a newer, less well known feature of R: reference classes (sometimes called R5 for short). Reference classes allow us to create objects that can be modified by other functions. Compare the behavior of the following reference class to the results from manipulating a list:

```
A = setRefClass("A", fields = c(a = "numeric"))
y = A$new()
y$a = 1
f(y)
y$a
```

```
[1] 2
```

Reference classes are a large topic so we give a brief introduction – enough to use them for the shoe. For our purposes, a reference class has 3 important properties: a name, a list of fields that store information about each instance/object of this class, and a list of methods that access and modify the object and the values in its fields. It is the ability of these methods to modify/update the values of the object's fields and for these changes to persist across calls that makes reference classes quite different than the regular computational model in R. We create our class by first figuring out what these should be and then filling in the details.

To make these ideas easier to discuss, let's create a simple reference class that will be useful to us later: a simple counter. The name of the class is Counter, it has one field, the count, a numeric vector, and it has two methods. One method, initialize(), sets count to 1. The second method, increment(), increases count by 1. We define Counter with

```
Counter = setRefClass("Counter",
  fields = list("count" = "numeric"),
  methods = list(
    initialize = function() {
      count <<- 1
    },
    increment = function() {
      count <<- count + 1
    }
))
```

Let's try out our new class by creating two counters, ctr1 and ctr2, with the following calls to Counter$new:

```
ctr1 = Counter$new()
ctr2 = Counter$new()
```

We also can simply call `Counter()`. Next we check the value of each of our counters with

```
ctr1$count
```

```
[1] 1
```

```
ctr2$count
```

```
[1] 1
```

The new() method of `Counter` sets up the counter for us and calls the initialize() method. We can increment the counters by calling increment(). Let's increment `ctr1` once and `ctr2` 3 times. We do this and confirm that they have been properly incremented with:

```
ctr1$increment()
ctr1$count
```

```
[1] 2
```

```
ctr2$count
```

```
[1] 1
```

```
ctr2$increment()
ctr2$increment()
ctr2$increment()
ctr2$count
```

```
[1] 4
```

There are a few important things to notice:

- The methods are associated with the object, so we call a method of a reference class object using the syntax `object$method()`. That is, we use the $ on the object like we would subset a list. The expression `ctr1$count()` calls the count() method of the `ctr1` object.

- The setRefClass() function creates reference classes by returning a special reference class that allows us to create new objects with this class's new() method. The example above calls new() without any arguments, but if supplied, the arguments override the default values of the fields.

- The object's fields are modified from inside the methods by using the special <<- operator. This operator makes the assignment in the parent environment of the function, which is the object's environment.

To design our `shoe` class, we need to work out what fields and methods we need. Figuring out precisely what you need to do to solve a problem is usually the main challenge for programming: once you've got that, you just need to turn it into code that the computer can understand.

Reshuffling rules vary from table to table and casino to casino, so we use a simple first-pass approximation: a multi-deck shoe that is replenished and shuffled whenever there are 52 cards left. This means that we need to know how many decks to use, which cards are left in the shoe, and when to reshuffle. We can use the following 3 fields to track this information:

- `decks`, a single integer, giving the number of decks in the shoe.

- `cards`, an integer vector containing the ordering of all the cards in the shoe.

- `pos`, a single integer giving our current position in the shoe.

Instead of modifying our vector of cards each time we draw them, we use the `pos` field to keep our place in the sequence of cards. This is a small performance optimization because we do not want to have to re-order or subset a (potentially large) vector of cards each time we draw 1 card.

In addition, we set up a field that we can use for debugging purposes:

- `debug`, a single logical value, which determines whether or not we should display useful debugging information on the R console from within methods.

As for methods, we need to shuffle the cards and to draw cards from the shoe. These are described below along with two additional methods that are useful to us.

- `shuffle()` generates a random ordering of the decks of cards, where the number of decks is provided by `decks` and the permutation is stored in `cards`. This method uses our existing `shuffle_decks()` function to update the cards field, and will reset `pos` to 0. If `debug` is `TRUE`, then it will also display an informative message.

- `draw_n()` returns a vector of n cards taken from `cards` beginning at `pos`. This method also increments `pos` by n so that we do not draw the same cards multiple times. It calls the helper method `decks_left()` to see if it is time to reshuffle. If there is less than 1 deck in the shoe then it calls `shuffle()` to shuffle the cards.

- `decks_left()` determines how many full decks of cards we have left in the shoe; this is just `floor(decks - pos / 52)`.

- `played()` displays what cards have already been played. This is a convenience function. It simply indexes into `cards` with the appropriate vector created from `pos`.

This description leads us to the following class definition. We have not provided the code for the `draw_n()` method and leave that as an exercise.

```
Shoe = setRefClass("Shoe",
  fields = list(
    decks = "numeric",   # number of decks
    cards = "numeric",   # vector of cards
    pos = "numeric",     # current position in shoe
    debug = "logical"    # display informative messages?
  ),
  methods = list(
    shuffle = function() {
      if (debug) message("Shuffling the shoe")
      cards <<- shuffle_decks(decks)
      pos <<- 0
    },

    draw_n = function(n) {
      # Return a subset from the field cards of the next n cards
      # begin at the field pos + 1. Increment pos appropriately
      # Exercise for you to provide code.
```

Simulating Blackjack

```
    },
    decks_left = function() floor(decks - pos / 52),
    played = function() cards[seq_len(pos)]
  )
)
```

We also create a new_shoe() function, which creates a new shoe and shuffles it. This function is defined as

```
new_shoe = function(decks = 6, debug = FALSE) {
  shoe = Shoe$new(decks = decks, debug = debug)
  shoe$shuffle()
  shoe
}
```

Let's examine a new shoe. We create a shoe with

```
my_shoe = new_shoe(decks = 3, debug = TRUE)
```

```
Shuffling the shoe
```

We see the message letting us know that our shoe was shuffled when it was created. Now let's deal a dozen cards, check the value of pos after, and see how many full decks we have left with

```
my_shoe$draw_n(12)
```

```
 [1]  4 10 10  1  8  4  5 10  2 10  3  1
```

```
my_shoe$pos
```

```
[1] 12
```

```
my_shoe$decks_left()
```

```
[1] 2
```

We can continue to draw more cards and trigger the shuffling with

```
my_shoe$draw_n(37)
```

```
 [1]  8  1  5 10 10 10 10  3  8  2  9 10 10  7
[15] 10  8  4  8  2 10  4  2  9  9  7 10  9  6
[29]  6 10  9 10  1 10  1 10  4
```

```
my_shoe$pos
```

```
[1] 49
```

```
my_shoe$draw_n(54)
```

```
Shuffling the shoe
[1] 3 1 ....
```

Or, we can simply shuffle the shoe with a call to `my_shoe$shuffle()`.

One of the important parts of this design is that we have a method, draw_n(), that works exactly the same way (i.e., it has the same interface) as our old shoe() function. This means we can easily plug the more accurate shoe() into our old code without changing any code that uses it. We do this with

```
my_shoe = new_shoe(decks = 6, debug = TRUE)
replicate(3, play_hand(shoe = my_shoe$draw_n,
                      strategy = strategy_optimal,
                      verbose = TRUE))
```

```
New hand
  Dealer: 3-3-10-2 (18)
  Player: 1-2: HS -> 1-2-6 (19)
New hand
  Dealer: 6-7-3-10 (0)
  Player: 10-6: S -> 10-6 (16)
New hand
  Dealer: 9-5-10 (0)
  Player: 8-10: S -> 8-10 (18)
[1] 1 1 1
```

We can also call payoff() with

```
my_shoe$pos = 0
payoff(50, strategy_optimal, my_shoe$draw_n)
```

```
Shuffling the shoe
avgGain   sdGain   medGain
 -0.04     1.16     0.00
```

```
my_shoe$pos
```

```
[1] 69
```

We reset pos to 0 and found that the shoe of 6 decks was shuffled once over the course of 50 rounds of blackjack. That seems about right because between the dealer and the player about 7–8 cards are used each round, so we expect to reshuffle after about 35 hands.

9.7 Counting Cards

We are ready to implement card counting. The idea of card counting is to reduce our risk by using our knowledge of what cards remain in the shoe. If there are many low cards left then we are safer when we hit; if there are many high-value cards left, then we are less safe. Card counting has two components. We generate a running count or tally of the cards that have been dealt since the last shuffle. Typically a counting system adds or subtracts 1 according

Simulating Blackjack

TABLE 9.2: Card Counting Strategies

strategy	1	2	3	4	5	6	7	8	9	10
Canfield Expert	0	0	1	1	1	1	1	0	-1	-1
Canfield Master	0	1	1	2	2	2	1	0	-1	-2
Hi-Lo	-1	1	1	1	1	1	0	0	0	-1
Hi-Opt I	0	0	1	1	1	1	0	0	0	-1
Hi-Opt II	0	1	1	2	2	1	1	0	0	-2

This table displays a few counting strategies. The columns of the table correspond to the value that we add to the running total for each card value. Rows correspond to the various counting regimes.

to whether the value of the card is low or high, respectively. Therefore, if the tally is high, then the remainder of the shoe is rich in 10s. The second component is a betting strategy that is guided by this running tally. The betting strategy can include both changing the size of the bet and the decision to hit or stand. We will examine only the effect of changing the size of the bet.

Many different counting systems have been developed. Table 9.2 shows a few, generated from http://www.qfit.com/card-counting.htm. To "count" we look at the cards that have been dealt, tally the values from the table that are assigned to each card, and then divide by the number of decks that have been dealt.

We leave it as an exercise to implement the hi_low() function that takes in a numeric vector, say *cards*, and returns the Hi-Lo count as described in Table 9.2.

This basic strategy extends well, if we want to make a function that takes the vector of cards and the name of a strategy. As with our playing strategy, we can use subsetting to index into the table of values. We leave it as an exercise to extend the hi_low() function to the any_count() function that takes a second argument, say *strategy*, which contains the card values for one of the strategies in count_table.

Next we need to connect the count to the shoe: when the shoe reshuffles, it also needs to reset all the counts. There are a few ways we could implement this:

- Each shoe keeps a list of all card-counting objects, and when the shoe is shuffled, it calls a function associated with each object to reset the count to zero.

- We create a new game function that plays multiple hands of blackjack and tracks when the shoe needs to be shuffled and the count to be reset.

- The card-counting function can look at the shoe to see if it has been reshuffled recently, and if so, reset its own count.

If we examine these possibilities, we see the three strategies correspond to the three main pieces of our code: the shoe, the game, and the count. When thinking about how to implement new behaviors, it is useful to systematically work through all our options.

A fourth option is to abandon the idea of a running count and compute the complete count whenever needed. Instead of providing a running count that is incremented or decremented each time a card is played, it would re-count all cards whenever asked. This would be less efficient, but much simpler because the count is insulated from the game, and doesn't need to be controlled by any of the elements in the game.

This is a sound design principle: if we can make something independent (even if it's less efficient), it's usually a good idea to do so. This approach makes our code easier to understand and to test, and once we have it working correctly we can use profiling tools to figure out where the slow parts are and then improve them. Generally our predictions about what parts of a program will be slow are poor, and we are better off using data to figure out what is slow. Otherwise, we easily can spend hours optimizing a part of the code that only takes seconds to run.

We make it easy to get the current count from the shoe by adding a count method to the Shoe class. This is also straightforward because we already have a way to get the count via the any_count() function for a set of cards played.

Lastly, we need a function that takes the count and figures out how much the player should bet. The basic principle is that the player should bet an amount proportional to the count, with the restrictions that the player must always bet some amount and can only bet integer amounts.

Here is a simple bet() function that implements this description:

```
bet = function(count) pmax(floor(count), 1)
```

We might want to check our bet() function with a plot. This is easy to do with a little-known feature of the base plot() function. We can give plot() a function and a range of values and it will create a plot of that function for us, e.g.,

```
plot(bet, from = -5, to = 10)
```

The results are shown in Figure 9.3. We see that bet() is a step function with increments of 1 at the integers.

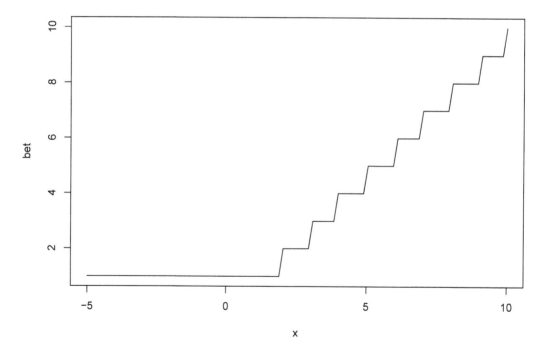

Figure 9.3: Plot of the bet() Function. *This plot shows the bet() function over the domain from -5 to 10. We can use this visualization to confirm that the function works as expected.*

9.8 Putting It All Together

Now we have a complete system to model the payoff from card counting. To make it easier to compare the bets from the counting strategy with the same bet every time, we take advantage of the fact that varying the bets changes our computation of the gain by a factor determined by bet(). We assume for simplicity that the initial bets are all $1, and we leave it as an exercise to modify the code to handle more complex situations. We define the count_payoff() function as a simple wrapper to play_hand() that also computes the count for each hand, which determines the size of the bet. The played() method is very useful here because we can pass that to our hi_low() function to ascertain the count. We do this as follows:

```
count_payoff = function(shoe, n = 100) {
  gain = numeric(n)
  count = numeric(n)

  gain[1] = play_hand(shoe$draw_n, strategy_optimal)
  count[1] = 0

  for (i in 2:n) {
    count[i] = hi_low(shoe$played())
    gain[i] = play_hand(shoe$draw_n, strategy_optimal)
  }
  c(sum(gain), sum(gain * bet(count)))
}
```

Now we can call count_payoff() many times and compare the gain of the optimal strategy with a fixed bet to one that varies the size of the bet according to the Hi-Low counting strategy. We run our simulation of 50 hands 1000 times with

```
set.seed(155100)
my_shoe = new_shoe()
payoffs = replicate(1000, count_payoff(my_shoe, 50))
```

In this simulation, the two strategies for bet size (Hi-Low versus a constant bet) differ by 0.8 on average:

```
apply(payoffs, 1, mean)
```

[1] 0.52 1.34

The standard deviation of gain is larger in the Hi-Low strategy:

```
apply(payoffs, 1, sd)
```

[1] 8.6 18.4

The difference in standard deviations makes sense since we are not changing the strategy, only increasing the size of the bet. The two sets of 1000 gains are not independent because they are based on the same set of cards and optimal strategy. The average and SD of the difference in gains across the 1000 bets are:

```
diffs = payoffs[2,] - payoffs[1,]
mean(diffs)
```
```
[1] 0.82
```

```
sd(diffs)
```
```
[1] 13
```

And the SE for the difference in these simulated means is about 0.41. We also make a polygon plot to compare the difference in the gains. The plot appears in Figure 9.4. It confirms our observations based on the mean and SD of the simulations.

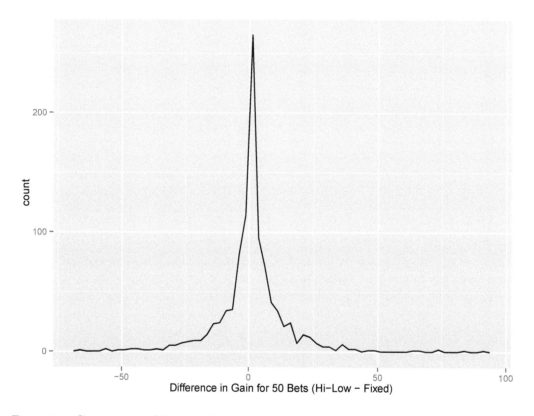

Figure 9.4: *Comparison of Average Gain for Card Counting. A comparison of the fixed and varying bet amount strategies. For each set of 50 hands, we take the difference in the gains between a fixed bet and a bet adjusted according to the Hi-Low counting strategy. Values above 0 are when the gain for the counting strategy exceeded the fixed bet.*

9.9 Exercises

Q.1 In the winnings() function in Section 9.2, we chose to break the function down by the value of the dealer's cards first, and then the player's cards. What happens when we switch this order and start with the player's values? Is the function easier or harder to understand? Write this alternative version of the function and test it using the test code in Section 9.2.1 to make sure that your code works as expected.

Q.2 How does the terse version of the winnings() function in Section 9.2 work? Is it consistent with the original version? Test this new version of the function with the test cases developed in Section 9.2.1. Is it easier to follow the logic in the first or second version of the winnings() function? Why? When do you think using code like this is appropriate?

Q.3 Consider how you might add checks to the hit(), dd(), and splitPair() functions (see Section 9.3.1) to ensure that they are being used properly. For example, add code to splitPair() so that an error message is returned if the player tries to split a hand when the 2 cards are not of equal value. Add other checks as well.

Q.4 Compile a set of test cases for the checks that you developed in the previous exercise. Write a function that calls these functions (hit(), dd(), and splitPair()) with the test cases and returns an informative message indicating whether the functions pass all of the tests and if not, where problems occurred. Create other test cases and test functions for the play_hand() function.

Q.5 Implement the draw_n() method for the Shoe class of Section 9.6. This function has one input: n, which indicates how many cards should be drawn from the shoe. It returns a vector of n cards taken from the cards field of the Shoe object. This vector of cards begins at the position indicated by the pos fields. This method also increments pos by n so that we do not draw the same cards multiple times. It also calls the helper method decks_left() to see if it is time to reshuffle. If there is less than 1 deck in the shoe then it calls the shuffle() method to shuffle the cards.

Q.6 Implement the hi_low() function from Section 9.7. This function takes one input, which is a numeric vector *cards*, and it returns the Hi-Lo count (see Table 9.2). If you want an additional challenge, write the function without using an if statement or any loops. For example, consider taking advantage of the fact that the vector of cards is a numeric vector, and use that to subset into a vector of count values.

Q.7 Implement the any_count() function for counting cards (see Section 9.7). This function has 2 arguments: a numeric vector *cards* that contains the cards played, and a numeric vector, *strategy*, which is of length 10 and contains the card values for one of the strategies in Table 9.2. Modify count_payoff() to use any_count(). This may require modifying the function definition to accept a *strategy* argument.

Q.8 Suppose a bet was larger than $1, say $4, how might the bet() function take this into consideration in deciding the size of the bet? Modify bet() (in Section 9.7) so that bets are integers, but utilize a standard bet size. What impact does this change have on other functions, e.g., winnings()? Is there a need to modify any of the earlier code?

Q.9 Implement other play options that are available in some games of blackjack, such as insurance and surrendering. Think about the play_hand() function. Can you simply augment this function with code for these new actions, or do you need to fundamentally re-structure how the function works? See Section 9.4 for the play_hand() function.

Q.10 A major simplification in our simulation is that we have effectively given the gambler an infinite amount of money. How do things change if we also account for ruin, i.e., the gambler running out of money?

Q.11 Can you make the code faster? Use profiling techniques to find the slow parts of the code. Also, consider how you might use vectorization to play multiple games at once.

Bibliography

[1] Roger Baldwin, Wilbert Cantey, Herbert Maisel, and James McDermott. The optimum strategy in blackjack. *Journal of the American Statistical Association*, 51:429–429, 1956.

[2] R Development Core Team. *R: A Language and Environment for Statistical Computing*. Vienna, Austria, 2012. http://www.r-project.org.

[3] Hadley Wickham and Winston Chang. ggplot2: An Implementation of the Grammar of Graphics. http://cran.r-project.org/package=ggplot2, 2011. R package version 0.9.3.1.

Part III

Data and Web Technologies

10

Baseball: Exploring Data in a Relational Database

Deborah Nolan
University of California, Berkeley

Duncan Temple Lang
University of California, Davis

CONTENTS

10.1	Introduction	399
	10.1.1 Computational Topics	400
10.2	Sean Lahman's Database	401
	10.2.1 Connecting to the Baseball Database from within R	401
10.3	Aggregating Salaries into Payroll	403
10.4	Merging Payroll Data with Information in Other Tables	408
	10.4.1 Adding Team Names to the Payroll Data	409
	10.4.2 Adding World Series Records to the Payroll Data	411
10.5	Exploring the Extreme Salaries	412
10.6	Exercises	415
	Bibliography	416

10.1 Introduction

Baseball fascinates many Americans and others, and many statisticians share this fascination with the sport. Nate Silver, who gained wide acclaim for the accuracy of his novel statistical methods to predict election outcomes, has also used statistics to study issues in baseball, such as the trade-offs in scheduling cold-weather games, home-field advantage, and effects of steroid use (see http://www.baseballprospectus.com/news/?author=59). This search for new baseball knowledge through statistical measures of in-game activity has been coined "sabermetrics" after the Society of American Baseball Research (SABR). In *Moneyball* [6], John Henry, former owner of the Florida Marlins, compared the baseball industry to the financial industry: "People in both fields [finance and baseball] operate with beliefs and biases. To the extent you can eliminate both and replace them with data, you gain a clear advantage."

In this chapter, we explore the change in player salaries and team payrolls over time via Sean Lahman's baseball data [5], a comprehensive baseball archive that is freely available online. This archive is organized as a relational database so a central focus of this case study is on the computational aspects of accessing data stored in a relational database, and through our explorations, we also gain practice with *SQL* (Structured Query Language), a language designed for accessing data stored in a relational database. As statisticians we often do not have a say in the format in which the data that we have been asked to work with are stored. For this reason, it can be useful to be familiar with various common formats so that we are not limited in the projects in which we can participate. On the flip side, we might be in a position to make recommendations about how to store data, so it is good for us to have some familiarity with the possibilities and their pros and cons.

Particular issues that arise when working with data stored in a relational database include figuring out where to perform the computations. For example, do we extract an entire data table from the database into *R* [9] and work with it there, or do we perform some subsetting, collapsing, or analysis in the database first and then retrieve the results back into *R* for further reduction and analysis? Additionally, do we retrieve the data in batches and carry out computations in *R* that update with each new batch or do we process all the results in one step? The answers to these and other questions, of course, depend on various circumstances and context, such as the size of the data, the complexity of the structure and organization of the data, and the functionality available in the database management system.

Section 10.2 provides a brief introduction to the baseball archive and its organization. There we determine where and how the information we need to analyze salaries is stored, and we see how to connect to the database from within *R*. In Section 10.3 we work with one of the tables in the database and extract and summarize salary information into team payrolls using simple *SQL* commands. In Section 10.4, we continue our investigation of salaries with tasks that require us to combine information that is stored in more than one table in the database. Additionally, at times we must perform parts of the analysis in *R* because the database does not offer the functionality required. In Section 10.5, we examine the players with the largest salaries, and we demonstrate how to retrieve intermediate results in batches and complete the analysis incrementally in *R*.

For those requiring an introduction to *SQL* we recommend [10]. We provide here only brief summaries of the basic features of the `SELECT` statement. In addition, we provide brief descriptions of *SQL* concepts, syntax, and statements as we encounter them.

Additional questions about baseball are provided in the exercises, and we encourage readers to pose their own questions to investigate.

10.1.1 Computational Topics

To explore the baseball data, we need to gain some experience working with a relational database. More specifically, we need to accomplish the following:

- Read schema for a database to understand how the database is organized.

- Use basic `SELECT` statements, including the *WHERE* clause, to access and retrieve subsets of data stored in a single table.

- Merge and combine information within and across tables with clauses such as `GROUP BY`, *HAVING*, and `LEFT OUTER JOIN`.

- Retrieve data from a query in blocks and further process it in *R*.

We use *R* functions to interface with the database, and we compare how we might perform some or all of the computations in *R*. This comparison also helps us understand the pros and cons of working in one environment over the other.

TABLE 10.1: Mapping Relational Database Terms to Statistics Terms

SQL	**Statistics**	**Description**
relation/table	data frame	rectangular arrangement of values
tuple	observation/record	row in a data frame
attribute	variable	column in a data frame
key	row name	variable or combination of variables that uniquely identifies a row in the table

This table provides a simple mapping between terms used in statistics and corresponding concepts in relational database theory and parlance.

10.2 Sean Lahman's Database

The baseball archive we work with in this chapter is available at `http://seanlahman.com/baseball-archive/statistics`. This archive was created and licensed by Sean Lahman so we are entitled to use it for research and teaching purposes, but cannot distribute it. We are very grateful for Lahman's efforts in compiling and managing this database. Other sites for baseball data include `http://asp.usatoday.com/sports/baseball/salaries/default.aspx` and `http://www.baseball-databank.org/`.

The terminology used to describe a relational database is different from the terminology statisticians typically use to describe data. We use common statistical terms throughout this chapter, but provide in Table 10.1 a mapping between database terminology and statistics terminology.

The variables in the different tables of Lahman's baseball archive are described online in `http://seanlahman.com/files/database/readme2012.txt`. As noted there, this database "can never take the place of a good reference book like The Baseball Encyclopedia [7]. But it will enable people to do the kind of queries and analysis that those traditional sources don't allow." We begin by accessing the basic descriptions of the tables and variables in the "readme" document. According to Section 2.0 of this document, called Data Tables, there are 4 main tables in the database: `Master`, which holds biographical information about the people in the database, and `Batting`, `Pitching`, and `Fielding`, which hold, respectively, batting, pitching, and fielding statistics for each player. Player salary and team payrolls are not included in the `Master` table so we must dig a little deeper. We can continue reading this documentation, or use *R* functions to explore the organization of the database.

10.2.1 Connecting to the Baseball Database from within *R*

Packages such as `RMySQL` [4], `ROracle` [8], and `RPostgreSQL` [2] offer the ability to connect with different implementations of relational database management systems (RDBMS), such as MySQL, Oracle, and PostgreSQL, respectively. All of these packages are compliant with the `DBI` package [3], which provides a common database interface for communicating between *R* and an RDBMS. For example, the `DBI` package provides functions to keep track of whether an *SQL* statement produces output, how many rows are affected by the operation, how many rows have been fetched (if the statement is a query), and whether there are more rows to fetch.

With each of these packages, we can initialize a driver to the relevant RDBMS and establish a connection to a particular database. For example, to access a MySQL database, we use the RMySQL package:

```
library(RMySQL)
```

Then, we initialize a driver for a MySQL database with

```
drv = dbDriver( "MySQL" )
```

and make a connection to the database with

```
con = dbConnect(drv, user = "login", dbname = "BaseballDataBank",
                host = "URL")
```

Here the *host* argument points to the location of the database server, which may be running locally or remotely. Also, *user* specifies the login name for the user.

At this point, we can submit commands for execution. For example, we find what tables are in this database with

```
dbListTables(con)
 [1] "AllstarFull"         "Appearances"
 [3] "AwardsManagers"      "AwardsPlayers"
 [5] "AwardsShareManagers" "AwardsSharePlayers"
 [7] "Batting"             "BattingPost"
 [9] "Fielding"            "FieldingOF"
[11] "FieldingPost"        "HallOfFame"
[13] "Managers"            "ManagersHalf"
[15] "Master"              "Pitching"
[17] "PitchingPost"        "Salaries"
[19] "Schools"             "SchoolsPlayers"
[21] "SeriesPost"          "Teams"
[23] "TeamsFranchises"     "TeamsHalf"
```

We see that there is a table called Salaries, and we can query the variable names in this table with

```
dbListFields(con, "Salaries")
[1] "yearID"   "teamID"   "lgID"     "playerID" "salary"
```

Although we do not have variable definitions, we see that this table is at the player–year level so we should be able to aggregate the salary values to create an annual team payroll. We can double check our understanding against the online documentation.

The documentation states: "Each player is assigned a unique number (**playerID**). All of the information relating to that player is tagged with his **playerID**. The **playerID**s are linked to names and birth dates in the Master table." Also, according to the documentation, the Salaries table contains the following information:

2.15 Salaries table

```
yearID          Year
teamID          Team
lgID            League
playerID        Player ID code
salary          Salary
```

This confirms our understanding of the variables in the `Salaries` table.

The process to access a relational database from within R is identical for all databases, i.e., the querying functions are the same across the R packages. For example, in the next section, we demonstrate how to use the `SELECT` statement to retrieve salary data from the `Salaries` table in our database. This query is independent of the RDBMS implementation.

10.3 Aggregating Salaries into Payroll

To calculate the team payrolls, we need to compute the sum of the salaries for the players on each team for each year. If this table is a data frame in R, then we can easily make these calculations with the tapply() or by() functions. That is, if the `Salaries` table is not very large, then we can retrieve it from the database as a data frame and perform these computations. To determine how large the table is, we can calculate the number of rows in the table with

```
query = "SELECT COUNT(*) FROM Salaries;"
dbGetQuery(con, query)
```

```
  COUNT(*)
1    23141
```

Here, con holds the connection to the database (see Section 10.2.1) and query is a character string consisting of the *SQL* statement.

The `Salaries` table has 23141 rows and, as we saw earlier, 5 variables. Since the table is not very large, we can retrieve the entire table with

```
salaryDF = dbGetQuery(con, "SELECT * FROM Salaries;")
```

Or we can retrieve only the variables necessary to compute the annual team payrolls with

```
query = "SELECT salary, teamID, yearID FROM Salaries;"
salaryDF = dbGetQuery(con, query)
```

The return value in R from the latter of these two queries is a data frame consisting of 3 columns, called salary, teamID, and yearID. These 3 elements in salaryDF correspond to the columns in the table in the database. We confirm this with

```
class(salaryDF)
```

```
[1] "data.frame"
```

```
dim(salaryDF)
```

```
[1] 23141     3
```

```
sapply(salaryDF, class)
```

```
    salary      teamID      yearID
 "integer" "character"   "integer"
```

We also see that the data types of the variables are as we expect so no conversion is necessary to calculate payroll from the salary values.

One advantage to retrieving the table in its entirety is that we can use familiar R commands to explore the data. For example, with this data frame, we can easily find the payrolls with

```
payrolls = tapply(salaryDF$salary, salaryDF[ -1], sum)
```

However, this approach to working with a database can be problematic or cumbersome, if the table is too large to retrieve from R or if the data reduction can be performed more efficiently in the database. For these reasons, we demonstrate how to compute the team payrolls in the database, rather than in R. For those familiar with R, the direct comparison of the two approaches can be helpful for learning *SQL*.

SELECT statements read a bit like an English sentence because *SQL* is a declarative language where the functions and clauses are regular English words and the delimiters are commas and white space. The *SQL* keywords are not case sensitive, but we use the convention of capitalizing them to make it easier to distinguish between the operations in the language and the variable and table names.

SELECTing Variables from Tables

The following simple SELECT statement demonstrates the basic features of the language:

```
SELECT  x, y    [1] [2]
  FROM table;   [3] [4]
```

[1] The names of the variables to be extracted/selected from the table appear in a comma-separated list. A * indicates all of the variables in the table are to be selected. An *SQL* function, such as *COUNT(*)* or *MIN(x)*, returns the number of rows or the minimum value in **x**, respectively.

[2] Blanks and new lines can appear anywhere in the statement so we format the statement for ease of reading.

[3] In *FROM* we specify the name of the table from which to select the variables. More than one table can be listed and these tables are joined together. (See Section 10.4.)

[4] Statements end with a semicolon.

The result is an *SQL* table with the specified columns.

Let's explore Salaries a little more before we compute the team payrolls. For which years do we have salary information? Most versions of *SQL* contain *MIN*() and *MAX*() functions, and we can use them to address this question with

```
dbGetQuery(con, "SELECT MIN(yearID) from Salaries;")
dbGetQuery(con, "SELECT MAX(yearID) from Salaries;")
```

or we can combine the two queries into one with

```
dbGetQuery(con, "SELECT MIN(yearID), MAX(yearID) from Salaries;")
    MIN(yearID) MAX(yearID)
1          1985        2012
```

The salary data covers 28 years.

To determine whether we have salary data for all years between 1985 and 2012, we can extract the unique values for year from Salaries with

```
query = "SELECT DISTINCT yearID from Salaries;"
years = dbGetQuery(con, query)
```

We can then examine years in R to see whether there are 28 values, one for each year from 1985 to 2012:

```
length(years[[1]])
```

```
[1] 28
```

Notice that we have performed some of the summarization in the database when we requested that the **yearID** values be *DISTINCT* and some in R when we used length().

We can also ask whether or not the other tables in the database cover the same time period. For example, we check the Teams table to see if it has earlier records:

```
dbGetQuery(con, "SELECT MIN(yearID), MAX(yearID) from Teams;")
```

```
  MIN(yearID) MAX(yearID)
1        1871        2012
```

Unlike the salary data, the team information dates back to 1871. If we examine **yearID** for the Pitching and Batting tables, we find that 1871 is the earliest and 2012 the most recent year for these tables as well.

We know there are over 23,000 records in the Salaries table covering the period from 1985 to 2012. Let's examine a few of these records with

```
query = "SELECT * FROM Salaries LIMIT 6;"
dbGetQuery(con, query)
```

```
  yearID teamID lgID  playerID  salary
1   1985    BAL   AL murraed02 1472819
2   1985    BAL   AL  lynnfr01 1090000
3   1985    BAL   AL  ripkeca01  800000
4   1985    BAL   AL  lacyle01  725000
5   1985    BAL   AL flanami01  641667
6   1985    BAL   AL boddimi01  625000
```

The *LIMIT* clause restricts the results to 6 rows. It is similar to the head() function in R and can be helpful for examining a few of the values in a table to confirm they are as expected. We see that we have each player's annual salary and we also have the team and league on which they played that year.

Let's first find the payrolls for the teams in a particular year, say 1999. We use a *WHERE* clause to select only those records from 1999, i.e.,

```
SELECT teamID, salary
  FROM Salaries
  WHERE yearID = 1999;
```
SQL

However, we want to aggregate all of the salaries for a team. We do this by including the GROUP BY clause as follows:

SQL
```
SELECT teamID, SUM(salary)
  FROM Salaries
  WHERE yearID = 1999
  GROUP BY teamID;
```

The GROUP BY clause in this statement collects together all records with the same value for **teamID**; it is similar to the tapply() and by() functions, e.g.,

```
tapply(salary, teamID, sum)
```

The results table has 2 variables: the **teamID** and the sum of salaries for all records with the same value of **teamID**.

If we prefer to use different names for the variables in the results table, then we can rename them using the AS clause as follows:

SQL
```
SELECT teamID AS team, SUM(salary) AS payroll
  FROM Salaries
  WHERE yearID = 1999
  GROUP BY teamID;
```

We do this here so that we can use the variable name `payroll` rather than `SUM(salary)`, which is cumbersome to use as a variable name in R.

The WHERE, GROUP BY, and HAVING Clauses

The SELECT statement has several optional clauses. Three commonly used ones are WHERE, GROUP BY, and HAVING.

SQL
```
SELECT * FROM table; WHERE x = 3 AND y > 10;
```

The WHERE clause subsets the rows in the table according to the specified logical expression. Logical expressions can be combined with AND and OR and negated with NOT. Note that because *SQL* does not have the notion of assignment, it uses = for the logical "equal to" operator in contrast to *R*'s ==.

```
SELECT   z, SUM(w)
  FROM table
  WHERE x = 3 AND y > 10
  GROUP BY z       [1]
  HAVING MAX(y) < 20;   [2]
```

[1] The GROUP BY clause collapses rows in the table that have the same value for z. The result is an *SQL* table with one row per unique value in z.

[2] The HAVING clause subsets the rows produced by the GROUP BY clause. The logical expression follows the same syntax as the expression in the WHERE clause. Since this clause selects rows from the collapsed table, it only appears when there is a GROUP BY clause.

Below is the sequence in which the clauses are applied.

1. FROM: The working table is constructed from the tables provided.

> 2. *WHERE*: The logical expression in the WHERE clause is applied to the working table and only those rows that test TRUE are retained.
>
> 3. *GROUP BY*: The results are split into groups of rows that have the same value for the variable(s) in the GROUP BY clause.
>
> 4. *HAVING*: This logical expression is applied to each group resulting from the GROUP BY and only those that test TRUE are retained.
>
> 5. *SELECT*: The variables not specified in the SELECT clause are dropped, aggregates are calculated, e.g., *MIN()*, and options such as *DISTINCT*, ORDER BY, and *LIMIT* are applied.

Now we retrieve the payrolls for all teams and years with the SELECT statement:

```
payrolls = dbGetQuery(con,
                "SELECT yearID AS year, teamID AS team,
                     lgID AS league, SUM(salary) AS payroll
                 FROM Salaries GROUP BY team, year;")
```

Note that we have used 2 variables **team** and **year** to aggregate (or group) the rows so that we have a payroll for each team for each year.

We want to adjust these payroll figures for inflation before comparing them. We use inflation rates relative to 1985, the first year of payroll data, found at http://www.bls.gov/data/inflation_calculator.htm. This information is available in inflation:

```
head(inflation)
1985 1986 1987 1988 1989 1990
1.00 1.02 1.06 1.10 1.15 1.21
```

We adjust the payrolls with

```
payrollStd = mapply(function(pay, year)
                         pay/inflation[as.character(year)],
                    payrolls$payroll, payrolls$year)
```

We prefer to perform these computations in R rather than in the database, where these simple algebraic calculations can be cumbersome.

Figure 10.1 displays boxplots of payroll versus year for each of the 2 leagues (American and National). The following code produced this plot:

```
payrolls$payrollStd = payrollStd/1000000

boxplot(payrollStd ~ league + year, data = payrolls, log = "y",
        ylim = c(5,125), col = gray.colors(2), axes = FALSE,
        ylab = "Inflation-adjusted Payroll (millions)" )

axis(1, at = seq(1.5, 55, by = 10),
     labels = seq(1985, 2010, by = 5))
axis(2)
legend("topleft",
       legend= c("American League", "National League"),
       fill = gray.colors(2), bty = "n")
```

The boxplots show a sharp rise in payroll in the early 1990s. In the 2000s the variability increases. (Note payroll is plotted on a log scale).

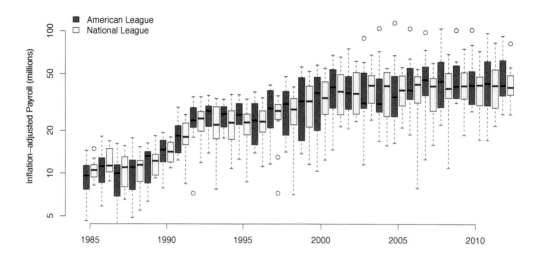

Figure 10.1: Boxplots of Team Payrolls by League. *These boxplots show the team payrolls for the American League (dark gray) and the National League (light gray) from 1985 to 2012. Payroll has been logged and is reported in millions of dollars. The American League appears to have greater spread in payroll than the National League. Also evident is the sharp increase in payrolls in the late 1980s and early 1990s.*

10.4 Merging Payroll Data with Information in Other Tables

We continue our investigation of team payroll with the question: Do teams with high payrolls more often win the World Series? The World Series results do not appear in the `Salaries` table. The World Series results are in the `SeriesPost`, which contains the post season wins and losses. It has the following variables,

```
dbListFields(con, "SeriesPost")
```

```
[1] "yearID"       "round"        "teamIDwinner" "lgIDwinner"
[5] "teamIDloser"  "lgIDloser"    "wins"         "losses"
[9] "ties"
```

We should be able to use the ID of the team winning the World Series in this table to augment the payrolls with this additional piece of information.

Relational databases are designed to reduce redundancy by storing information in multiple tables, so it's not uncommon to merge information contained in multiple tables to acquire the desired information. Before we do, we try a simpler task – adding the team name to the payroll information. For example, we might want to use the name of a team as an annotation or label in a graphic.

You might wonder why Lahman didn't include the team name along with/rather than the team identifier in the `Salaries` table. The team identifier is shorter than the name and having both is redundant. That is, the team is uniquely identified by its **teamID** so there's no need to also keep team name in this table, especially since the team appears many times in `Salaries` (once for every player-year combination) so the team name is repeated unnecessarily. The database saves space by having a smaller table that has only one row for each team's information, in exchange for the need to look in this table for any additional information about a team (other than **teamID**). Databases are designed for this type of look up/merger so they are fast at it. Design issues are an important part of building a database (see [10] and [1]).

10.4.1 Adding Team Names to the Payroll Data

We can include the team's name in our payroll data frame by "joining" the `Salaries` and `Teams` tables in *SQL*. First, we examine the variables in the `Teams` table with

```
dbListFields(con, "Teams")
```

```
[1] "yearID"      "lgID"       "teamID"       "franchID"
... "name"        "park"       "attendance"
```

Since teams can move, change names, close, etc., there is a team record for each year. This means that we need to match both the team identifier and the year across the `Salaries` and `Teams` tables. We do this as follows:

```
query =
  "SELECT Salaries.yearID AS year, Teams.name AS team,
      Salaries.teamID AS id, SUM(Salaries.salary) AS payroll
    FROM Salaries, Teams
    WHERE Salaries.teamID = Teams.teamID AND
          Salaries.yearID = Teams.yearID
    GROUP BY Salaries.teamID, Salaries.yearID;"

payrollWN = dbGetQuery(con, query)
```

This gives:

```
  year            team  id   payroll
1 1997 Anaheim Angels  ANA  31135472
2 1998 Anaheim Angels  ANA  41281000
3 1999 Anaheim Angels  ANA  55388166
4 2000 Anaheim Angels  ANA  51464167
5 2001 Anaheim Angels  ANA  47535167
6 2002 Anaheim Angels  ANA  61721667
```

The 2 tables to be merged are specified in the *FROM* clause as comma-separated table names. The resulting table consists of all possible combinations of 1 record from each table and is dubbed an "outer join" in relational database terminology. We specify which rows in this combined table to keep via the *WHERE* clause. Here we keep only those records where the team and year identifiers match across the 2 tables. Additionally, now that we are working with 2 tables, we need to keep straight which variable is from which table. To do this, we prepend the table name to the variable name, e.g., **Salaries.yearID** refers to the **yearID** in the `Salaries` table. This is similar to using the $ in *R*, e.g., `salaries$yearID`.

We can double check that the payroll values match our previously computed values with

```
all(payrollWN$payroll == payrolls$payroll)
```

```
[1] FALSE
Warning message:
In payrollWN$payroll == payrolls$payroll :
  longer object length is not a multiple of shorter object length
```

It appears that our joined tables have a different number of rows than the `payrolls` data frame we created earlier. (Note we also assumed that the rows of these 2 results tables are in the same order.) A call to dim() reveals that there is 1 fewer row in `payrollWN`:

`dim(payrollWN)`

`[1] 797    4`

`dim(payrolls)`

`[1] 798    5`

To figure out how we might be missing a record, we tally the number of records per team in each data frame with

`table(payrolls$team)`

```
ANA ARI ATL BAL BOS CAL CHA CHN CIN CLE COL DET FLO HOU KC  KCA...
  8  15  28  28  28  12  28  28  28  28  20  28  19  28   1  27
```

`table(payrollWN$id)`

```
ANA ARI ATL BAL BOS CAL CHA CHN CIN CLE COL DET FLO HOU KCA...
  8  15  28  28  28  12  28  28  28  28  20  28  19  28  27
```

The one payroll value for team KC looks suspicious. It is not in `payrollWN`. When we drop this record from `payrolls`, the payroll values all match:

`all(payrolls$payroll[payrolls$team != "KC"] == payrollWN$payroll)`

`[1] TRUE`

It looks as though that team identifier should be KCA. We leave it as an exercise to confirm whether or not this is the case.

In R we often use the merge() function to merge data frames. The outer join that we just completed is equivalent to

```
merge(Payrolls, Teams, by.x = c("yearID", "teamID"),
                      by.y = c("yearID", "teamID"),
                      all.x = FALSE, all.y = FALSE)
```

Here `Payrolls` is the aggregated salary data for annual payrolls. The *by.x* and *by.y* specify the keys for merging, and the *all.x* and *all.y* indicate that all records from both sources are to be kept. These 4 arguments yield an equivalent result to the WHERE clause in the SELECT statement.

Before we turn to the World Series question, we briefly consider the efficiency of performing a merge in R and SQL. Relational databases are designed to be efficient in merging so we typically merge tables in the database rather than in R. We compare the two approaches (R and SQL) by running 100 identical merges of the `Teams` and `Salaries` tables. In SQL, we find:

```
system.time(replicate(100, invisible(dbGetQuery(con, query))))
```

```
   user  system elapsed
  6.307   0.030   6.342
```

whereas, in R we have the following time:

```
system.time(replicate(100, invisible(queryFcn())))
```

```
   user  system elapsed
  8.606   0.462   9.063
```

Here, queryFcn() retrieves the Salaries and Teams tables into R and then calls merge() and tapply() to merge and aggregate the data. Performing the aggregation and merge in R is about 50% slower than in SQL. In this case, since the tables are not very large, the run times are not dramatically different.

Next we augment the payroll information with the World Series results. This involves a different type of table join.

10.4.2 Adding World Series Records to the Payroll Data

Let's begin by examining the data in SeriesPost for one year, say 2012,

```
dbGetQuery(con, "SELECT round, teamIDwinner, teamIDloser
                 FROM SeriesPost WHERE yearID = '2012';")
```

	round	teamIDwinner	teamIDloser
1	ALWC	BAL	TEX
2	ALCS	DET	NYA
3	ALDS1	NYA	BAL
4	ALDS2	DET	OAK
5	NLWC	SLN	ATL
6	NLCS	SFN	SLN
7	NLDS1	SLN	WAS
8	NLDS2	SFN	CIN
9	WS	SFN	DET

We see that there are records for the pennant races as well as the World Series in this table. We are interested only in the World Series winners, i.e., in the records where **round** is WS.

When we join the Salaries and SeriesPost tables, we want to keep all of the records in the Salaries and augment these with the information in the SeriesPost table. This type of join is called a left outer join, or simply a left join. Two other factors complicate matters. We still want to aggregate the salary records into team payrolls. And, we don't want to match all of the records in the SeriesPost table, only those for World Series results.

We build up the SELECT statement incrementally. First, we specify the variables that we retain in the results. These are the same as before with the addition of the **round**, e.g.,

```
SELECT                                                                    SQL
    Salaries.yearID AS year, Salaries.teamID AS team,
    SUM(Salaries.salary) AS payroll, SeriesPost.round as round
```

Next, for the *FROM* clause, we provide only the Salaries table and include SeriesPost in the LEFT OUTER JOIN clause as follows:

SQL   `FROM Salaries LEFT OUTER JOIN SeriesPost`

The values of the variables to use in the match are specified in the ON clause, rather than the WHERE clause as in the outer join. This is because the WHERE clause in this statement applies to the Salaries table. Also, we include the logical expression to limit the records in SeriesPost to those from the World Series in the ON clause. The ON clause is then:

SQL
```
ON SeriesPost.teamIDwinner = Salaries.teamID
   AND SeriesPost.yearID = Salaries.yearID
   AND SeriesPost.round = 'WS'
```

Notice that we match the identifier in **teamIDwinner** in SeriesPost to the value in **teamID** in Salaries. Finally, we include the GROUP BY clause to aggregate the records in Salaries as before. All together our SELECT statement appears as

```
query =
 "SELECT Salaries.yearID AS year, Salaries.teamID AS team,
    SUM(Salaries.salary) AS payroll, SeriesPost.round as round
  FROM Salaries LEFT OUTER JOIN SeriesPost
   ON SeriesPost.teamIDwinner = Salaries.teamID
      AND SeriesPost.yearID = Salaries.yearID
      AND SeriesPost.round = 'WS'
  GROUP BY team, year;"

payrolls = dbGetQuery(con, query)
```

Figure 10.2 shows a scatter plot of payroll by year. The points for the payrolls of the teams that won the World Series appear darker. Indeed, the high payroll teams tend to win the series.

10.5 Exploring the Extreme Salaries

Lastly, we investigate the largest salaries in the database. The salary table in the database is not very large, and we can easily pull the entire table into R to find, say, the 3 largest salaries in 2003 with

```
sals = dbReadTable(con, "Salaries")
sort( unique( sals$salary[sals$yearID == 2003] ),
      decreasing = TRUE)[1:3]

[1] 22000000 20000000 18700000
```

We can perform the same computation in *SQL* with the following:

```
query = "SELECT DISTINCT Salary FROM Salaries
         WHERE yearID = 2003 ORDER BY Salary DESC
         LIMIT 3;"
```

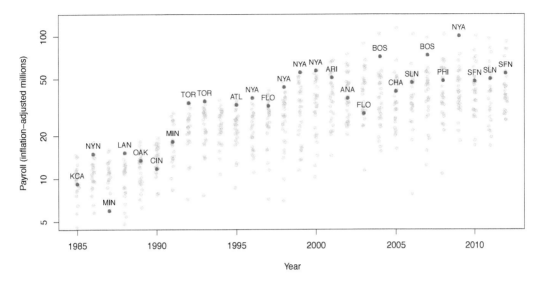

Figure 10.2: Scatter Plot of Team Payroll by Year. *In this scatter plot, the plotting symbol is a transparent gray and the year has been jittered slightly to avoid over plotting. The identifier of the teams that won the World Series are added to the corresponding point, which is colored red. In almost every case, these dots are at or above the upper quartile of team payroll. Payroll is reported in millions of dollars and plotted on a log scale.*

The result of this query yields the first 3 records from the ordered list of distinct values for salary.

If we want to get the next 4 salary values, we have to reissue the query and ask for the first 7 records. An alternative approach is to use the dbSendQuery() function, instead of dbGetQuery(). With dbSendQuery(), the query is performed in the database without waiting and bringing the results into R. Then we can transfer results in blocks with the fetch() function. We start by issuing a query for the sorted distinct salary values as follows:

```
query = "SELECT DISTINCT Salary FROM Salaries
        WHERE yearID = 2003 ORDER BY Salary DESC;"
res = dbSendQuery(con, query)
```

The results from this query have not been transferred into R, but they remain in the database for us to fetch. For example, we can retrieve the first 3 records with

```
topSalary = fetch(res, n = 3)
topSalary
```

```
    salary
1 22000000
2 20000000
3 18700000
```

These salary figures match those from the calls to sort() and unique() in R. We can then retrieve the next 4 records in the results table with

```
fetch(res, n = 4)
```

```
       salary
4   17166667
5   16000000
6   15714286
7   15666667
```

After we have completed our fetching, we clear the results object with

`dbClearResult(res)`

`[1] TRUE`

Alternatively, if we want to retrieve all the remaining results in one call to fetch(), we can do this with, e.g., `fetch(res, n = -1)`. Here the value `-1` indicates to retrieve all remaining rows in the table.

If the data are large, then sorting the salaries might be a very expensive operation in the database and it also would be difficult to bring all of the salaries into R for sorting. We consider a third approach to finding the 3 largest salary values that retrieves the unsorted salaries into R in blocks. For each block, we find the 3 largest salaries, and compare these with the 3 largest values seen so far, and so on. That is, we sort our data in blocks. To do this, we first determine the total number of records in the Salaries table so we can track our progress with

```
query1 = "SELECT COUNT(*) FROM Salaries WHERE yearID = 2003;"
totCount = dbGetQuery( con, query1)
```

Then we submit the simple query that obtains the salaries values for 2003 with

```
query2 = "SELECT Salary FROM Salaries WHERE yearID = 2003;"
res = dbSendQuery( con, query2)
```

The following `while` loop performs the fetching and batch sorting:

```
blockSize = 200
totRead = 0
top3Salary = NULL

while (totRead < totCount) {
  top3Salary = sort(unique( c(top3Salary,
                              fetch( res, n = blockSize)[[1]]) ),
                    decreasing = TRUE )[1:3]
  totRead = totRead + blockSize
}
```

`top3Salary`

`[1] 22000000 20000000 18700000`

`dbClearResult(res)`

The last batch of results may be shorter than the block size, but the fetch does not give us an error when we ask for more records than remain in the results table.

If the ultimate goal is to find the players that correspond to the 3 largest salaries, then we return to the database and query the `Salaries` table for the **playerID**s that correspond to the extreme salaries (there may be more than 3 because more than one player may have one of these salaries). One way to do this is to paste together a query that contains the 3 salary values:

```
charSalary = paste(top3Salary, collapse = ", ")
query3 = paste("SELECT playerID FROM Salaries WHERE yearID = 2003
                AND Salary IN (", charSalary, ") ;", sep = "")
dbGetQuery(con, query3)
```

```
  playerID
1 ramirma02
2 rodrial01
3 delgaca01
```

Notice that we have programmatically constructed an *SQL* query in *R* based on the results of an earlier *SQL* query.

Now that we have finished working with the database in this *R* session, we free up resources by disconnecting and unloading the driver with

```
dbDisconnect(con)
dbUnloadDriver(drv)
```

There are several additional functions that can help us manage the state of a query. For example, to determine whether or not there are more results to be fetched we call `dbHasCompleted()`. Also, if we are handling multiple queries in batch mode, we can keep track of them with `dbListResults(con)`, which gives a list of all currently active result set objects for the connection `con`. The call, `dbGetRowCount(results)`, provides a status of the number of rows that have been fetched so far in the query.

10.6 Exercises

The following exercises are designed to suggest additional exploratory analyses of the baseball archive. These questions have been loosely organized into 5 groups. In addition, we encourage you to make up your own questions that hopefully lead to fun explorations of the data. As you answer these questions, keep in mind the decisions as to where to perform the calculations, try addressing the questions using multiple approaches, and make comparisons of these approaches.

Instructions for how to set up a database can be found in Chapter 5.

World Series

Q.1 Which team lost the World Series each year? Do these teams also have high payrolls? Some argue that teams with lower payrolls make it into the post season playoffs, but typically don't win the World Series. Do you find any evidence of this?

Q.2 Do you see a relationship between the number of games won in a season and winning the World Series?

Team Payroll

Q.3 Augment the team payrolls to include each team's name, division, and league. Create a visualization that includes this additional information.

Q.4 One might expect a team with old players to be paying these veteran players high salaries near the end of their careers. Teams with a large number of mature players would therefore have a large payroll. Is there any evidence supporting this?

Q.5 Examine the distribution of salaries of individual players over time for different teams.

Players

Q.6 Not all of the people in the database are players, e.g., some are managers. How many are players? How many are managers? How many are both, or neither?

Q.7 What are the top 10 collegiate producers of major league baseball players? How many colleges are represented in the database? Be careful in handling those records for players who did not attend college.

Q.8 Has the distribution of home runs for players increased over the years?

Q.9 Look at the distribution of how well batters do. Does this vary over the years? Do the same players excel each year? Is there a clustering, a bimodal distribution?

Q.10 Are Hall-of-Fame players, in general, inducted because of rare, outstanding performances, or are they rewarded for consistency over years?

Q.11 Do pitchers get better with age? Is there an improvement and then a fall off in performance? Is this related to how old they are, or the number of years they have been pitching? What about the league they are in? Do we have information about each of these factors? If so, how can we combine them to present information about the general question?

Miscellaneous

Q.12 How complete are the records for the earliest seasons recorded in this database? For example, we know that there is no salary information prior to 1985, but are all of the other tables "complete"?

Q.13 Are certain baseball parks better for hitting home runs? Can we tell from this data? Can we make inferences about this question?

Q.14 What is the distribution of the number of shut-outs for a team in a season?

Bibliography

[1] Joe Celko. *SQL for Smarties: Advanced SQL Programming*. Elsevier, Burlington, MA, 2010.

[2] Jon Conway, Dirk Eddelbuettel, Tomoaki Nishiyama, Sameer Prayaga, and Neil Tiffin. RPostgreSQL: R interface to the PostgreSQL database system. http://cran.r-project.org/package=RPostgreSQL, 2013. R package version 0.4.

[3] David James. DBI: R Database Interface. http://cran.r-project.org/package=DBI, 2013. R package version 1.1-11.

[4] David James, Saikat DebRoy, and Jeffrey Horner. RMySQL: R interface to the MySQL database. http://cran.r-project.org/package=RMySQL, 2012. R package version 0.9-3.

[5] Sean Lahman. Lahman's Baseball Database. http://seanlahman.com/baseball-archive/statistics, 2014.

[6] Michael Lewis. *Moneyball*. W.W. Norton, New York, NY, 2003.

[7] MacMillan. *The Baseball Encyclopedia: The Complete and Definitive Record of Major League Baseball*. Simon and Schuster, New York, NY, 1996.

[8] Denis Mukhin, David James, and Jake Luciani. ROracle: OCI based Oracle database interface for R. http://cran.r-project.org/package=ROracle, 2014. R package version 1.1-11.

[9] R Development Core Team. *R: A Language and Environment for Statistical Computing*. Vienna, Austria, 2012. http://www.r-project.org.

[10] F.D. Rolland. *The Essence of Databases*. Prentice Hall, New York, 1998.

11
CIA Factbook Mashup

Deborah Nolan
University of California, Berkeley

Duncan Temple Lang
University of California, Davis

CONTENTS

11.1	Introduction	419
	11.1.1 Computational Topics	421
11.2	Acquiring the Data	421
	11.2.1 Extracting Latitude and Longitude from a CSV File	421
11.3	Integrating Data from Different Sources	423
11.4	Preparing the Data for Plotting	424
	11.4.1 Redoing the Merge of the Factbook and Location Data	428
11.5	Plotting with Google Earth™	430
11.6	Extracting Demographic Information from the CIA *XML* File	435
11.7	Generating *KML* Directly	442
11.8	Additional Computational Tasks	448
	11.8.1 Creating Plotting Symbols	448
	11.8.2 Efficiency in Generating *KML* from Strings	448
	11.8.3 Extracting Latitude and Longitude from an *HTML* File	450
11.9	Exercises	451
	Bibliography	454

11.1 Introduction

The tremendous increase in data that are freely available on the Web has created numerous possibilities for extracting data from different sources, putting them together, and creating exciting new types of visualizations. These visualizations, sometimes called "mashups," are typically interactive and displayed on the Web. According to Wikipedia [13],

> The term [mashup] implies easy, fast integration, frequently using open application programming interfaces (API) and data sources to produce enriched results that were not necessarily the original reason for producing the raw source data. ... The main characteristics of a mashup are combination, visualization, and aggregation.

This shift in the accessibility of data has changed the role of the statistician to one who actively seeks out relevant data and incorporates them into an analysis. Moreover, new technologies for presenting information have led to interesting ways to visualize data that are qualitatively different from the typical static 2-dimensional plot. As an example, Google

Earth™ [2] offers a canvas for presenting information that supports user interaction, easily incorporates contributions of others, and allows layering of information. In this chapter, we mash together data from a "Factbook" published by the Central Intelligence Agency (CIA) and a file of latitude and longitude pairs for countries around the world from MaxMind (http://www.maxmind.com) to create a visualization of statistics on infant mortality for display on Google Earth™ (see Figure 11.1).

More specifically, the data sources we work with to create this visualization are:

- Latitude and longitude of the geographic center of each country available in a comma-separated-value (CSV) format at http://dev.maxmind.com/static/csv/codes/country_latlon.csv;

- The CIA Factbook, which contains, among other things, the populations and infant mortality rates for hundreds of countries. The 2012 Factbook is available in an *XML* format from Michael Schierl's Web site on Source Forge at http://jmatchparser.sourceforge.net/factbook/data/factbook.xml.gz.

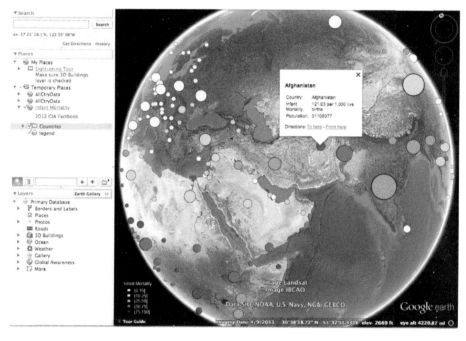

Figure 11.1: Display of Infant Mortality by Country on Google Earth™. *This screenshot of the Google Earth™ virtual earth browser displays circles scaled to the population size and colored according to the infant mortality rate for a country. The data are available from the CIA Factbook. The locations of the circles are determined from MaxMind's latitude and longitude of the country's geographic center. When the viewer clicks on a circle, a window pops up with more detailed information for that country.*

To create our visualization, we need to extract the relevant information from the data files. As mentioned, these data are in different formats, namely CSV and *XML*. We describe the CSV extraction in Section 11.2, and we postpone to later in the chapter in Section 11.6 the extraction of country statistics from the *XML* file. We also provide the data from the CIA Factbook in 3 data frames in R [10], in case the reader wishes to focus on the merging and plotting and not the extraction of the data from *XML*.

Once we have extracted these data, we integrate the data sources into a single structure where all of the information for each country (i.e., infant mortality, population, latitude, longitude, and country name) are merged together. We do this in Section 11.3. In addition, we transform the variables into formats that are amenable to plotting (Section 11.4). For example, we discretize a numeric variable in order to represent it via one of several colors for the plotting symbols.

Finally, we plot the data. We begin by making a static 2-dimensional map to which we add the information we have retrieved from the CIA Factbook. This task is addressed in Section 11.4. We then proceed to create an interactive visualization for display on Google Earth™. We take two approaches. The first uses the functionality in the **RKML** package [6]. For those who want more experience working with *XML*, we also create the Google Earth™ display more directly using the **XML** package [12]. These two approaches are the topics of Section 11.5 and Section 11.7, respectively.

11.1.1 Computational Topics

- Data manipulation: extract data from different formats, merge data from multiple sources, and transform variables.

- Programmatic access to data available on the Web: we write code that uses *HTTP* to acquire the data.

- Programmatic data retrieval to create a reproducible record of the data acquisition process.

- Visualization: maps, color, plotting symbols, labels, and legends.

- Interactive visualization on Google Earth™, including customized placemarks, pop-up windows, tool-tips, and legends.

- Modularity: design functions to create our visualization in pieces that correspond to different subtasks of the larger task of making a map, which facilitates testing and enables re-usability for different purposes.

- Tree structures in *XML* and *XPath* expressions to locate content.

11.2 Acquiring the Data

The data for our visualization are found in two different sources and formats. Both sources are available on the Web. The latitudes and longitudes for countries are available in a CSV file, and are easily read into *R* using the standard utility functions, e.g., read.csv() or read.table(). Extracting the country demographic information from the Factbook requires more work because these data are in an *XML* document. We postpone that task until Section 11.6.

11.2.1 Extracting Latitude and Longitude from a CSV File

The latitudes and longitudes for countries are available from many places on the Web. We found the site at http://dev.maxmind.com/static/csv/codes/country_latlon.csv to be easy to use. Below is a snippet of the CSV file found there:

```
"iso 3166 country","latitude","longitude"
AD,42.5000,1.5000
AE,24.0000,54.0000
AF,33.0000,65.0000
AG,17.0500,-61.8000
AI,18.2500,-63.1667
AL,41.0000,20.0000
```

We see that countries are identified by 2-letter codes, defined by the `iso 3166` country codes. The term "iso" stands for the International Organization for Standardization (ISO). This organization has created standard codes for country names, and these are available at http://www.nationsonline.org/oneworld/country_code_list.htm.

We can read the data into R with a call to read.csv() as follows:

```
latlonDF = read.csv(urlLatLon)
```

Here, `urlLatLon` is a character string containing the maxmind.com URL. When we examine the first few rows of the `latlonDF` data frame, we see that they match the above snippet of the CSV file:

```
head(latlonDF)
  iso.3166.country latitude longitude
1               AD    42.50    1.5000
2               AE    24.00   54.0000
3               AF    33.00   65.0000
4               AG    17.05  -61.8000
5               AI    18.25  -63.1667
6               AL    41.00   20.0000
```

In addition, we have two data frames with the CIA Factbook information, `infMortDF` containing the infant mortality values by country and, similarly, `popDF` for the population information. We have two data frames of demographic information, rather than one, because the countries' statistics are not presented in the same order in the Factbook. The population statistics are in order of the most populous country to the least and the infant mortality rates are in order of highest to lowest rate. We briefly examine these two data frames with

```
head(popDF)
         pop ctry
1 1349585838   ch
2 1220800359   in
3  316668567   us
4  251160124   id
5  201009622   br
6  193238868   pk
```

```
head(infMortDF)
  infMort ctry
1  121.63   af
2  108.70   ml
3  103.72   so
4   97.17   ct
5   94.40   pu
6   93.61   cd
```

It's not a surprise that China is the first country in the population data frame, India is second, and the USA third, and these 3 countries are clearly not among those with the highest infant mortality rates.

Notice that the CIA Factbook appears to also use the ISO codes to identify countries. We can match a country's latitude and longitude with its demographic information using this unique identifier. This is the topic of the next section.

11.3 Integrating Data from Different Sources

We have all of the pieces of information that we need to create our visualization, but they are not collected together in one data structure. We have no guarantee that `infMortDF`, `popDF`, and `latlonDF` contain information for the same countries. For example, we see that we have latitude and longitude for 240 countries, i.e.,

```
nrow(latlonDF)
```

```
[1] 240
```

However, we have infant mortality rates for 223 countries and population counts for 239 countries. Additionally, even though infant mortality and population figures come from the same source, the 239 countries for which we have population information need not be a superset of the 223 countries for which we have infant mortality data. Moreover as noted earlier, the population and mortality statistics are not provided in the same order. Additionally, the latitude and longitude are ordered alphabetically by ISO code.

Before we can create any plots, we need to match all of the relevant data values for each country. We can create a data frame for this purpose where we have latitude, longitude, infant mortality, population, and country code as variables and each country as a row in the data frame. Another issue that we must address is which countries should we include in this data frame, i.e., do we want to keep a country even if we do not have all of the information for it? If we are missing latitude and longitude, then we do not know where to place the plotting symbol for that country. And, if we are missing infant mortality or population size, then we do not have the information that we need to plot. How we handle the merging of these values depends on our purpose, and in this situation, we keep only those countries for which we have all of the relevant information.

Let's begin by combining the demographic data since these come from the same source and consequently should have fewer problems with matching. We merge these data using the country codes that appear in the variable `ctry` in each data frame. We can use the `merge()` function to do this with

```
IMPop = merge(infMortDF, popDF, by = "ctry", all = FALSE)
dim(IMPop)
```

```
[1] 222   3
```

We see that the resulting data frame is smaller than both the infant mortality and population data frames, indicating the intersection of these two sets of countries is a proper subset of both. We leave it as an exercise to determine which rows of `infMortDF` and `popDF` were excluded.

Matching longitude and latitude with demographic information requires that we match a record in `IMPop` with the correct record in `latlonDF`. The country code that identifies a country in `latlonDF` is in the column `iso.3166.country`. Unfortunately, when we read the CSV file into `latlonDF` we did not specify that this character data should be kept as strings so the country codes were converted into factors. We now convert them back to strings before using them to merge with `IMPop`. Also, notice that the codes in `IMPop` are lower case while those in `latlonDF` are upper case, so we also convert the codes in `latlonDF` to lower case. We do this with

```
latlonDF$code = tolower(as.character(latlonDF$iso.3166.country))
```

Then we merge the two data frames with

```
allCtryData = merge(IMPop, latlonDF, by.x = "ctry", by.y = "code",
                    all = FALSE)
```

We now have one data frame with all of the information for each country.

We soon see that we have made a big mistake in assuming that the two data sources use the same coding for country name. For example, `GB` stands for Great Britain in the Factbook and Gabon in the MaxMind file. When we discover this, we have more data extraction and merging to do. For now, let's continue with the data frame that we have created. We correct our mistakes later in Section 11.4.1.

11.4 Preparing the Data for Plotting

For our first visualization of infant mortality and population, we make a static map. After that, we tackle making a Google Earth™ visualization in Section 11.5. (We use much of the preparatory work for the static map in making the interactive visualization.) To include both infant mortality and population on the same map we can use discs where we scale each country's symbol according to population size, and color it according to infant mortality rate. To assign infant mortality to a color, we need to discretize the rates, i.e., we need to convert the numeric values into ordered categories because we can use only a finite number of colors. In this section, we discuss how we choose colors, create the categorical version of infant mortality, and translate population values into symbol size.

We choose colors from one of the color palettes offered by Cindy Brewer and made available in the `RColorBrewer` package [5]. Brewer's palettes have been developed especially for map making and to enable accurate comparisons. Several palettes are available in the package, and we prefer the palette called `YlOrRd` that ranges from pale yellow to orange to dark red because red typically connotes danger. This sequential palette is also useful for comparing ordered values where we want to focus on one end of the spectrum, e.g., those countries with high rates. Another reason for choosing the yellow-orange-red sequence is to make it easy for the viewer to distinguish the circles from the green and brown background colors used for rendering land areas in Google Earth™. Finally, we selected only 5 colors, and consequently 5 levels of mortality, because it's difficult for our eyes to distinguish between more than 5 to 7 colors.

```
library(RColorBrewer)
display.brewer.all()
cols = brewer.pal(9, "YlOrRd")[c(1, 2, 4, 6, 7)]
```

Now that we have our colors, we need to connect infant mortality to color. To do this, we can create an ordered factor using the cut() function where infant mortality values are categorized by the interval in which they fall. We can choose the end points of these intervals by either specifying the number of intervals or by providing the cut points. When we specify the number of intervals, the range is divided equally, e.g.,

```
newInfMort = cut(allCtryData$infMort, breaks = 5)
summary(newInfMort)
```

```
(1.68,25.7] (25.7,49.7] (49.7,73.7] (73.7,97.7]  (97.7,122]
         96          24          17           8            3
```

Alternatively, we can provide the cut points as

```
newInfMort2 = cut(allCtryData$infMort,
            breaks = c(0, 37, 50, 65, 80, 150))
summary(newInfMort2)
```

```
 (0,37]  (37,50]  (50,65]  (65,80] (80,150]
    106       15       12        9        6
```

Notice that there are only a few countries with very high rates.

In order to settle on the cut points, let's examine the distribution of infant mortality more closely. We make a histogram with

```
hist(allCtryData$infMort, breaks = 20, main = "",
     xlab = "Infant Mortality per 1000 Live Births")
```

We see in Figure 11.2 that infant mortality has a highly skewed distribution with a long right tail and that the vast majority of countries have infant mortality rates less than 15 per 1,000. When we use intervals of the same length, then the rightmost intervals have very few observations.

We could try using the empirical quantiles to determine the intervals,

```
quantile(allCtryData$infMort, probs = seq(0, 1, by = 0.2))
```

```
   0%   20%   40%   60%   80%  100%
  1.8   5.4  12.3  21.9  48.5 121.6
```

If we use these as cut-points, then the large values are lumped together in the top interval with other less alarming values. This makes it hard to focus on the highest countries. Let's try cut points made from a hybrid of these two approaches. We are interested in high and very high rates and less so in the lower values. We can choose these cut-points using quantiles that are finer at the right tail of the distribution and coarser on the left. We also want to round these cut-points to values that are easy for the reader to digest, e.g., 50 rather than 48.49. In addition, we might want to do a little research to see if there are commonly used rates set by, say, the United Nations or Centers for Disease Control that we might want to use for standard reference points. Taking these various constraints into consideration, we settle on breaks at 10, 25, 50, and 75:

```
InfMortDiscrete = cut(allCtryData$infMort,
                 breaks = c(0, 10, 25, 50, 75, 150))
```

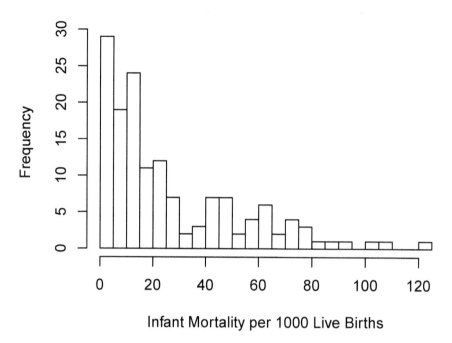

Figure 11.2: Distribution of Infant Mortality for Countries in the CIA Factbook. *This histogram of infant mortality rates shows a highly skewed distribution. Most countries have rates under 20 per 1000 live births and a few countries have rates between 80 and 125.*

Now that we have discretized the mortality rates, we are ready to make a world map. For this map, we want to place circles at the location of each country where the color filling the circle corresponds to infant mortality and the size of the circle corresponds to population. Specifically, we want the area of the circle to correspond to population, so the radius is proportional to the square root of population. After trial and error, we find that if we scale the square-root of population by 4000 then the circles are small enough that they don't overlap and occlude aspects of the map. We make our map with the map() function in the maps package [1] with

```
library(maps)
world = map(database = "world", fill = TRUE, col="light grey")
```

Then we add the discs of different size and color with

```
symbols(allCtryData$longitude, allCtryData$latitude, add = TRUE,
        circles= sqrt(allCtryData$pop)/4000, inches = FALSE,
        fg = cols[InfMortDiscrete], bg = cols[InfMortDiscrete])
```

And lastly, we place a legend on the map with

```
legend(x = -150, y = 0, title = "Infant Mortality",
       legend = levels(InfMortDiscrete), fill = cols, cex = 0.8)
```

Our choice for the radius has some difficulties because the range of population is several orders of magnitude:

```
range(allCtryData$pop)
```

```
[1]        5189 1349585838
```

The small countries have symbols that are so tiny that we can barely see them, if at all, and we certainly can't see their colors. On the other hand, if we make the circles larger so that we can see the small countries, then the circles for the larger countries such as India are so big that they cover too much of the map. We can confirm this problem via a histogram of the square-root of population (see Figure 11.3), where we create this histogram as follows:

```
hist(sqrt(allCtryData$pop), breaks = 20,
     xlab = "Square-root of Population", main = "")
```

One way to remedy this problem is to have a minimum radius so that any country with a population below a particular level is represented by a circle large enough to see it. Another way is to cap the size of the largest circles. We try the first approach and leave the second as an exercise. We set the minimum radius using the pmax() function as follows:

```
rads = pmax(sqrt(allCtryData$pop)/4000, 1)
```

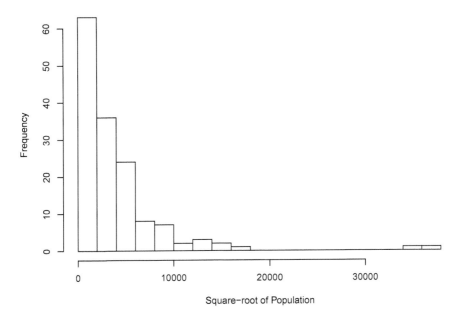

Figure 11.3: Population Distribution for Countries in the CIA Factbook. *This histogram of the square root of population size for countries shows a highly skewed distribution with a mode around 1000.*

Figure 11.4 uses these revised radii, and is an improvement. However, if we look a bit more closely at the map, we see that there appears to be something wrong. We would not expect the UK to have such a high infant mortality rate, the circle that represents China is very small, and there is a country in central Europe with an unexpectedly large population. What is going on? Something is amiss with the data.

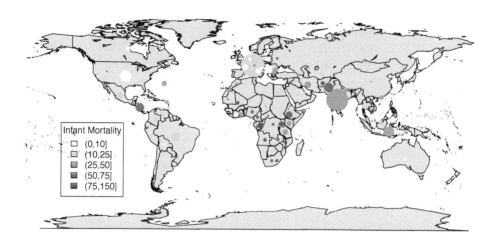

Figure 11.4: Incorrect Map of Infant Mortality and Population. *In this map each disk corresponds to a country's infant mortality and population. The size of the disk is proportional to the population and the color reflects the infant mortality rate. Notice the size of China's disk is too small – it has the highest population but one of the smallest disks. Other anomalies are apparent with closer inspection.*

11.4.1 Redoing the Merge of the Factbook and Location Data

Let's check the merging of the latitude and longitude data frame with the data frame of population statistics. We can query some of the countries from the merged data frame, allCtryData, and examine their values. We have assumed that, e.g., the code ch corresponds to China and af to Afghanistan and that the coding of the countries is the same in both sources, i.e., that they both use ISO 3166 country codes. The Web site https://www.iso.org/obp/ui/#search (see Figure 11.5) provides a lookup facility for country codes and names. We see that the United Kingdom maps to GB for Great Britain and that Switzerland, not China, maps to CH. Let's examine the entries in the data frame for these two country codes:

```
allCtryData[ allCtryData$ctry %in% c("ch","gb"), ]
```

```
    ctry infMort        pop iso.3166.country latitude longitude
29    ch   15.62 1349585838               CH       47         8
50    gb   49    1640286                  GB       54        -2
```

The latitude and longitude for these European countries look approximately correct, but the population and mortality rates look too high. Switzerland does not have 1.3 billion people and Great Britain does not have an infant mortality rate of 49 per 1000 births. Could the CIA Factbook be using a different set of country codes? When we examined the first few country populations, we saw the largest country had a country code of ch and a population that matches the one shown here for Switzerland. It must be that ch stands for China in the CIA Factbook and not Switzerland. The merge worked as instructed, but our instructions were wrong because the two data files use different codes for country name.

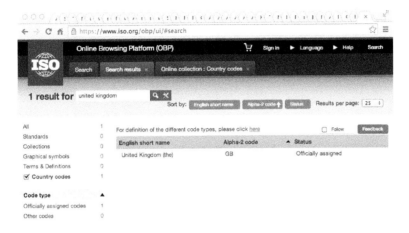

Figure 11.5: Screenshot of the ISO Country Code Mapping. *This screenshot of the ISO Web site shows the ISO code for the United Kingdom as* GB. *Codes are also available in XML and CSV formats at* http://www.iso.org/iso/home/standards/country_codes.htm.

When we re-examine the Factbook, we find a table near the end of the document that contains the mapping from the CIA code to the ISO code. We can extract this mapping and use it to merge the Factbook information with the latitude/longitude information. We can also retrieve the country name from this table to use later in the Google Earth™ mashup. These data are in the data frame codeMapDF:

```
head(codeMapDF)
```

```
    cia            name iso
1    af     Afghanistan  AF
2    ax        Akrotiri   -
3    al         Albania  AL
4    ag         Algeria  DZ
5    aq American Samoa  AS
6    an         Andorra  AD
```

The task of extracting this information from the Factbook follows the techniques of Section 11.3. We leave it as an exercise to redo the merge of these data sources. Below is the top of the revised data frame:

```
  iso ctry infMort      pop                name   lat    lon
1  AD   an    3.76    85293             Andorra 42.50   1.50
2  AE   ae   11.59  5473972 United Arab Emirates 24.00  54.00
3  AF   af  121.63 31108077         Afghanistan 33.00  65.00
4  AG   ac   14.17    90156 Antigua and Barbuda 17.05 -61.80
5  AI   av    3.44    15754            Anguilla 18.25 -63.17
6  AL   al   14.12  3011405             Albania 41.00  20.00
```

(Note the variable names were shortened for formatting purposes.) In addition, let's check that the confusion between China and Switzerland and between Gabon and Great Britain have been rectified:

```
allCtryData[allCtryData$ctry %in% c("ch", "sz", "gb", "uk"), ]
    iso ctry infMort        pop           name latitude longitude
37   CH   sz    3.90    7996026    Switzerland       47      8.00
42   CN   ch   15.62 1349585838          China       35    105.00
67   GA   gb   49.00    1640286          Gabon       -1     11.75
68   GB   uk    4.56   63395574 United Kingdom       54     -2.00
```

Our revised map appears in Figure 11.6

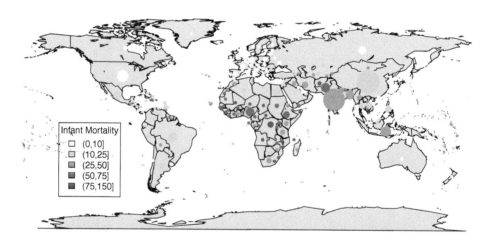

Figure 11.6: Map of Infant Mortality and Population. *This map correctly matches the country demographic information with latitude and longitude. Notice now the symbols for India and China are approximately the same size and the largest symbols on the map. Also note that the symbol for the United Kingdom is now pale yellow, the color we would expect it to be, because it is not being confused with Gabon.*

11.5 Plotting with Google Earth™

Google Earth™ [2] provides exciting new ways to display spatial and spatial-temporal data, whether they are locations of homes, stores, or hiking paths, or more scientific data such as earthquake locations, predictions from climate models, or as in our case, mashups of world health statistics. For example, we can load into Google Earth™ a *KML* document that contains the locations of the centers of the countries in the CIA Factbook. Google Earth™ renders these locations as pushpins on the surface of the earth. (See for example Figure 11.7.) Alternatively, we can mark each country with a circle, as in the map we created earlier, where the color corresponds to a level of infant mortality and the size relates to the country's population (see Figure 11.1). Moreover, we can augment each circle with additional information so that when we click on it, a small window pops up with this information, e.g., the window might show other health statistics or a plot of GDP per capita over time. If another organization has provided information and visualization elements for display on Google Earth™, we can layer this onto the earth browser along with our data. We can zoom in and rotate the earth to get a closer look at, e.g., the sub-Saharan countries. When zooming in, additional features, such as terrain, political boundaries, roads, and buildings, come into focus, and the circles appropriately scale to the new view. Finally, if we arrange other health statistics in folders by, say, level of infant mortality, then we can easily filter the view of the data by hiding/removing countries with rates under, say, 10, and focus on those countries with higher values.

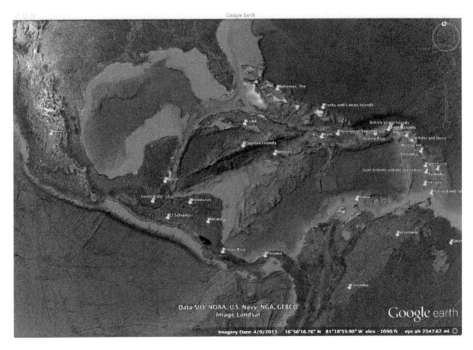

Figure 11.7: Default Google Earth™ Image. *This screenshot of Google Earth™ displays the location of each country with a pushpin. The locations are from our latitude and longitude file. A more informative Google Earth™ visualization appears in Figure 11.1.*

The earth browser is a technological development that allows us to move substantively beyond the static plot. Moreover, Google Earth™ is a well-established standard that makes it easy to incorporate information from other applications. It offers opportunities for interactive presentation-style spatial graphics that are not directly available in R. The **RKML** package [6] provides functionality to create displays for Google Earth™ and Maps. With **RKML**, we can produce plots on Google Earth™ for exploratory data analysis in an R session and for formal presentation of results. The plots can be simple displays of spatial data, richer displays that incorporate additional information via styles and balloon windows, and more complex interactive displays via links between Google Earth™, R plots, and *HTML* forms in Web pages (see for example [7]).

This package is called **RKML** because the Keyhole Markup Language (*KML*) is the language Google uses to render information on its earth browser. *KML* is an *XML* vocabulary. It is an open standard maintained by the Open Geospatial Consortium, Inc. (OGC) [9] (http://www.opengeospatial.org/standards/kml/). The functions in **RKML** produce *KML* documents that can be opened in Google Earth™ for display.

We can think of plotting on Google Earth™ as analogous to having a blank canvas made from a call to plot() with a *type* of 'n', and on this canvas, we add plotting symbols with a call to points(). With **RKML**, we call the kmlPoints() function to add what Google Earth™ calls placemarks to the surface of the earth. We can do this with

```
library(RKML)
doc = kmlPoints(allCtryData)
```

Since allCtryData contains variables called longitude and latitude, the kmlPoints() function uses these to position the placemarks. The next step is simply to save the document in the variable doc to a *KML* file for loading into Google Earth™. We do this with

```
saveXML(doc, "countryPlain.kml")
```

When we open this file in Google Earth™, we see the typical yellow pushpins arranged around the globe at the latitudes and longitudes of the countries in our data frame.

Of course, this visualization does not have any population and infant mortality information. We can substitute the yellow pushpins with our own circles that are filled with color representing the rate of infant mortality and sized according to population. This way, we make a visualization similar to the flat map that we constructed in Section 11.4.1. The color and size of each circle are determined as before by the country's infant mortality rate and population, respectively. See Figure 11.1 for a screenshot of the enhanced display. One difference from the static map, however, is that we also discretize population size for simplicity.

We can supply all of this style information via arguments to kmlPoints(), and as a result, customize a display to be more informative and aesthetic than the map with pushpins. However, we want to use the same style settings for many placemarks. *KML* supports style specifications by having a collection of descriptions in a separate part of the document that acts as a dictionary of styles. The approach is somewhat similar to styles in word processors or in Cascading Style Sheets. We create these styles and give them unique identifiers. Then, we use these identifiers to associate a style with each point in the call to kmlPoints(). The basic call in pseudocode is

```
kmlPoints(dataFrame, docStyles = styleList, style = pmStyleIds)
```

Here styleList holds a named list of styles that are stored at the top level of the document and available for use by all the graphical elements. The variable pmStyleIds is a vector of style names that associates each longitude–latitude pair with one of the styles in styleList. We use the *style* parameter here in a similar, but slightly more indirect, way as we use the *col* and *pch* parameters in plot(). The indirection comes from specifying a style name rather than the actual style information. The use of *docStyles* avoids redundancy in the *KML* document because we do not repeat the same information for multiple elements with the same appearance. This reduces the size of the file, avoids redundant information, and greatly simplifies changing the appearance of a set of points because we only need to change the information in one location in the file. Also, this way, we can combine styles.

CIA Factbook Mashup

We need 25 styles, one for each color-size combination, i.e., one for each mortality-population combination. More specifically, our goal is to create a *list*, `ballStyles`, to hold these 25 styles. We give each style a name, e.g., `ball_m-k`, where m ranges from 1 to 5 and corresponds to 1 of 5 colored circles (e.g., *yor3ball.png* when m is 3) and k also ranges from 1 to 5 and corresponds to a scaling of the circle, (e.g., 1.75 is the second smallest scale factor and is used when k is 2). In this example, the `"ball_3_2"` element of `ballStyles` is used to represent countries that fall into the third level of mortality and second population range; its contents are

```
ballStyles[["ball_3-2"]]

$IconStyle
$IconStyle$scale
[1] "1.75"

$IconStyle$Icon
         href
"yor3ball.png"
```

This is the format of the style information required by kmlPoints(), i.e., a list of lists with each element corresponding to a node in the *KML* style specifications. That is, `ballStyles` is in *R*, not *KML*. It will be converted to *KML* when the document is created.

We construct these 25 style settings with

```
popScales = as.character(1+ c(0.25, .75, 1.5, 3, 7))
icon = rep(sprintf("yor%dball.png",
                   seq(along = levels(InfMortDiscrete))),
           each = length(levels(popDiscrete)))
scale = rep(popScales, length(levels(InfMortDiscrete)))

ballStyles = mapply(function(scale, icon)
                      list(IconStyle =
                             list(scale = scale,
                                  Icon = c(href = icon))),
                    scale, icon, SIMPLIFY = FALSE)

g = expand.grid(seq(along = levels(InfMortDiscrete)),
                seq(along = levels(popDiscrete)))
names(ballStyles) = sprintf("ball_%d-%d", g[,2], g[,1])
```

With these style settings in hand, we match each observation to the name of its associated style. The discretized values of mortality and population for each observation are in `InfMortDiscrete` and `popDiscrete`, respectively. Together these values determine which style to use for each placemark. The code for generating the style id for each point is:

```
ctryStyle = sprintf("ball_%d-%d", InfMortDiscrete, popDiscrete)
```

We now have the list of 25 styles in `ballStyles` and the associated style for each longitude-latitude pair in ctryStyle. We call kmlPoints() with

```
kmlPoints(allCtryData, docStyles = ballStyles, style = ctryStyle)
```

However, before we create this *KML* document, we want to further augment the display with more informative descriptions that appear in the pop-up windows and Places panel, which is a hierarchical viewer that appears along the left-hand side of the viewer and is used to organize and show/hide individual placemarks and collections of them. Each placemark's description is displayed in a pop-up balloon-type window when the viewer clicks on its associated circle. For our descriptions, let's use the country's name and actual infant mortality rate and population, rather than the discretized versions. We format this information as an *HTML* table with

```
ptDescriptions =
 sprintf(paste(
   "<table><tr><td>Country:</td><td>%s</td></tr>",
   "<tr><td>Infant Mortality:</td>",
   "<td>%s per 1,000 live births</td></tr>",
   "<tr><td>Population:</td><td>%s</td></tr></table>"),
        allCtryData$name, allCtryData$infMort, allCtryData$pop)
```

We also could have used functions in the R2HTML package [3] to create an *HTML* table from our data frame. See Figure 11.1 for a screenshot that shows one of these pop-up windows.

Lastly, we set up some additional document-level information, e.g., the top-level document name and description and the folder name that appears in the Places panel. We specify these as strings with

```
docName = "Infant Mortality"
docDescription = "2012 CIA Factbook"
folderName = "Countries"
```

The docName variable is a label that appears alongside the blue and white icon for the document in the Places panel of the virtual browser, docDescription provides a brief description of the *KML* document that appears below the document label, and folderName is a label for a collection of placemarks within the document.

We now have prepared all of the information for our Google Earth™ mashup. The following call to kmlPoints() supplies this information to customize the display as well as the essential longitude-latitude pairs:

```
doc = kmlPoints(allCtryData, docName = docName,
                docDescription = docDescription,
                docStyles = ballStyles,
                folderName = folderName,
                style = ctryStyle,
                description = ptDescriptions,
                ids = allCtryData$ctry,
                .names = allCtryData$name)
```

Note that we override the default label that appears next to each placemark in the virtual-earth viewer via the argument *.names*. In this case, we supply the country name for the placemark's label.

Before saving our document, we add to it a legend detailing the meaning of the colors. We use the kmlLegend() function, which has arguments that are similar to those of the regular legend() function in *R*:

```
kmlLegend(x = 20, y = 20, title = "Infant Mortality",
         legend = levels(InfMortDiscrete), fill = cols,
         text.col = "white", dims = c(100, 108),
         parent = doc)
```

Google Earth™ adds the legend as an overlay to the viewing window. The *x* and *y* argument values in this function call provide the coordinates in pixels for the location of the legend. These coordinates are not supplied in longitude and latitude because the legend remains fixed on the 3D viewer so it can be seen in all views of the earth. The legend also remains the same size as the display zooms in and out. For this reason, it does not make sense to create a legend for circle size. The *parent* argument to kmlLegend() specifies the *KML* document to which the legend is to be added.

Finally, we save the document for viewing in Google Earth™ with

```
saveXML(doc, "ctryFancy.kml")
```

Although we specified the file name 'ctryFancy.kml', saveXML() creates the file named 'ctryFancy.kmz'. The **kmz** extension means the file is a zipped collection of files. This file contains the *KML* document and the 5 PNG files for the icons, e.g., *yor1.png*.

11.6 Extracting Demographic Information from the CIA *XML* File

The CIA Factbook is stored as a compressed (gzip'ed) file called *factbook.xml.gz* at http://jmatchparser.sourceforge.net/factbook/data/. Before we begin our search in this *XML* document for the desired information, we provide a little information about the structure of an *XML* document and how to work with it in *R* in the sidebar below. The functions presented there are in the XML package [12].

The Structure of an *XML* Document
  Consider the following simple small document:

```
<?xml version="1.0"?>
<a>
  <b><f/></b>
  <c>Some Text
    <d><f/></d>
    <e id="eId"/>
    <b id="bob">More Text</b>
  </c>
</a>
```

  This *XML* document can be represented as a tree, which represents the hierarchical structure of the document. There is a top node that contains all other nodes, and nodes are properly nested. Each node in the document (not the tree) begins with a start tag, e.g., `<c>`, and ends with a corresponding closing tag, e.g., `</c>`. When a node is nested within another, its start and closing tags must appear between the start and closing tags of its containing/parent node. If a node has no content (text or child nodes) then it can be contracted to one tag, e.g., `<b/>`.

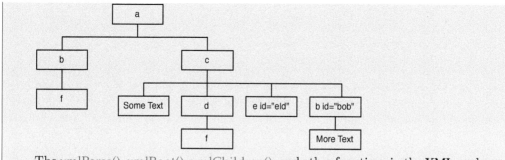

The xmlParse(), xmlRoot(), xmlChildren(), and other functions in the XML package can read and access the *XML* document as a tree structure within *R*. Additionally, since an *XML* document has a hierarchical list-like structure, we can operate on it as a list using the [and [[operators.

Before we embark on the task of reading the data from the Factbook into *R*, we download and unzip it and view this source document in a browser such as Chrome or Firefox to get a sense of its structure. The browser renders the document in its hierarchical form as shown in Figure 11.8.

The Factbook document is sufficiently large that it is hard to visually inspect it for the information that we want. By trial and error and searching for terms such as "infant" and "infant mortality," etc., we find the part of the document that contains the desired information. A relevant snippet of this document is shown here:

```
<factbook lastupdate="2013-03-20+01:00">
 <news date="2013-03-21+01:00">
  In 2012, fiscal and monetary policies shifted towards...
 </news>
 ...
 <category name="People and Society">
  <description>
   This category includes entries dealing with ...
   health and education indicators).
  </description>
 ...
 <field dollars="false" unit="(deaths/1,000 live births)"
   rankorder="1" name="Infant mortality rate" id="f2091">
  <description>
   This entry gives the number of deaths of infants
   under one year old in a given year per 1,000 live births...
  </description>
  <rank number="121.63" dateEstimated="true"
    dateLatest="2012-12-31" dateEarliest="2012-01-01"
    dateText="2012 est." country="af"/>
  <rank number="108.70" dateEstimated="true"
    dateLatest="2012-12-31" dateEarliest="2012-01-01"
    dateText="2012 est." country="ml"/>
 ...
```

We have discovered that the information about infant mortality is contained in a `<field>` node and this `<field>` node has an *id* attribute value of 'f2091'. Within this

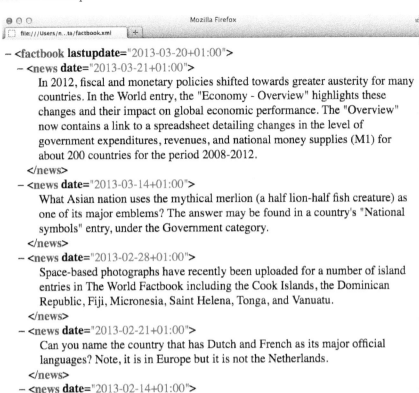

Figure 11.8: Screenshot of the CIA Factbook Rendered in Chrome. *The Chrome browser renders an XML file using indentation and color to highlight the structure of the document. The top or root node of the file is* <factbook> *and it has an attribute called lastupdate, which indicates how recently the information was updated. The* <news> *nodes are indented one space, corresponding to their depth in the hierarchy, i.e., they are children of* <factbook>.

<field> node there are several <rank> nodes, one for each country for which the CIA has infant mortality data.

The infant mortality value for a country appears as the value of the *number* attribute of a corresponding <rank> node, and the country code appears in the *country* attribute of this same node. Notice that the country identifier is 'af' for the first country, which we know is Afghanistan.

Exactly where in this document we can find this <field> node is not obvious by visual inspection due to the size of the document. Later we demonstrate how to use a powerful *XML* technology (*XPath*) to more easily locate this information, but for now, we explore the structure of the document with subsetting and other functions available in the XML package.

We first parse the downloaded document in *R* using the xmlParse() function as follows:

```
library(XML)
factbookDoc = xmlParse("Data/factbook.xml.gz")
```

We now have an *XML* tree structure in *R*, which we can explore to find the country statistics that interest us.

We begin our exploration at the root of the tree. We access the root, confirm that it is
<factbook>, and ascertain how many children it has with the following calls to xmlRoot(),
xmlName(), and xmlSize(), respectively,

```
factbookRoot = xmlRoot(factbookDoc)
xmlName(factbookRoot)
```

[1] "factbook"

```
xmlSize(factbookRoot)
```

[1] 228

We find the names of these 228 child nodes with

```
table(names(factbookRoot))
```

```
   appendix     category   definition  faqCategory
          7           10           42            7
       news       region
        149           13
```

We don't see any <field> nodes as children of <factbook> so they must be deeper in
the tree structure, e.g., grandchildren or great grandchildren of <factbook>.

We can return to the Chrome browser and more closely examine the document to see if
we can determine where the <field> nodes might be in the hierarchy, or we can take a
guess that the <field> nodes are within a <category> node because it seems unlikely
that they are in <appendix>, <definition>, or <news> nodes. If we do not find the
<field> node in a <category> node, we can expand our search. Let's examine the
children of these 10 <category> nodes:

```
sapply(factbookRoot["category"], function(node) table(names(node)))
```

```
             category category category category category ...
description         1        1        1        1        1
field               2       21       35       27       41
```

It appears that each <category> node has one <description> node and several
<field> children. Let's also examine the attributes on the <category> nodes to see
if they can help us locate the information we are seeking. We do this as follows:

```
sapply(factbookRoot["category"], xmlAttrs)
```

```
        category.name          category.name          category.name
       "Introduction"            "Geography"   "People and Society"
        category.name          category.name          category.name
         "Government"              "Economy"               "Energy"
        category.name          category.name          category.name
     "Communications"       "Transportation"             "Military"
        category.name
"Transnational Issues"
```

Of these 10 *name* attributes, the one called 'People and Society' seems most promising. We further explore this particular <*category*> node by examining the *id* attribute values on all of its <*field*> children. We are looking for one with the value 'f2091'. We access the index of the 'People and Society' category child with

```
categoryNodes = factbookRoot["category"]
w = sapply(categoryNodes, xmlGetAttr, "name")=="People and Society"
```

Then we extract the value for the *id* attributes of this node's children with

```
Ids = sapply(categoryNodes[[ which(w) ]] [ "field" ],
             xmlGetAttr, "id")
```

We search for the attribute value f2091 with

```
f2091Index = which(Ids == "f2091")
f2091Index
```

```
field
   17
```

We have found the desired <*field*> node. It is the 17th <*field*> node of the People and Society <*category*> node.

All that remains is to obtain the *country* and *number* attribute values on the <*rank*> children nodes of our f2091 <*field*> node. We obtain all of the <*rank*> nodes with

```
rankNodes =
  categoryNodes[[ which(w) ]][ "field" ][[ f2091Index ]]["rank"]
xmlSize(rankNodes)
```

```
[1] 223
```

Then we use xmlGetAttr() to retrieve the value of the *number* and *country* attributes with

```
infMortNum = sapply(rankNodes, xmlGetAttr, "number")
infMortCtry = sapply(rankNodes, xmlGetAttr, "country")
```

We examine the first few:

```
head(infMortNum)
```

```
    rank     rank     rank     rank     rank     rank
"121.63" "108.70" "103.72"  "97.17"  "94.40"  "93.61"
```

```
head(infMortCtry)
```

```
rank rank rank rank rank rank
"af" "ml" "so" "ct" "pu" "cd"
```

We have seen that it is possible to traverse the tree structure from within *R* using [and [[and xmlChildren(), but it is quite cumbersome. An alternative approach for extracting this information uses the *XPath* query language. *XPath* is a powerful language designed to locate sets of nodes and attributes in *XML* documents. We only briefly introduce *XPath* in the sidebar below to give a sense of the possibilities.

> **Simple *XPath* Expressions**
>
> *XPath* is a powerful technology for specifying queries to locate matching nodes and attributes in an *XML* document. We supply a set of examples here that give a few of the basics of the language. For a more comprehensive treatment see [11] and Chapter 3 of [8] for a description of the language and its use in R. The following sample tree hierarchy is used to demonstrate *XPath*.
>
>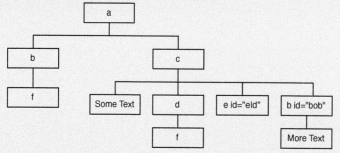
>
> /a/b All *<b>* children of the *<a>* root. This finds the first, leftmost *<b>* node in the document and not the *<b>* node with an *id* of 'bob' because it is not a child of *<a>*.
>
> /a/c/b All *<b>* nodes that are children of a *<c>* node, which is itself a child of the *<a>* root. This expression locates the rightmost *<b>* node, the one with the *id* of bob.
>
> //b All *<b>* nodes anywhere in the document. This expression locates both *<b>* nodes.
>
> //b[@id='bob'] All *<b>* nodes anywhere in document that have an *id* attribute with a value 'bob'. This locates the rightmost *<b>* node.
>
> //d/f All *<f>* children of any *<d>* node in the document. This expression locates one *<f>* node, the rightmost one.
>
> //b[@type='2']/f All *<f>* children of a *<b>* node anywhere in the document, where the *<b>* node has a *type* of 2. This expression does not locate any nodes in this document.
>
> //b/.. The parent of any *<b>* node in the document. This expression locates the *<c>* node and the *<a>* root node of the document.

The *XPath* expression that locates all *<field>* nodes anywhere in the document is simply, //field, and the following expression locates all *<rank>* nodes that are children of these *<field>* nodes: //field/rank. Of course, we want only the *<rank>* nodes that are children of the particular *<field>* node that has an *id* attribute value of f2091. We can add a conditional expression to restrict the *<field>* nodes to those with this *id* value: //field[@id='f2091']/rank. This expression essentially says, look through all levels of the document (//) for a node named *<field>* that has an attribute (@) named *id* with a value of f2091, and from within that node (or nodes if there is more than one that satisfies this constraint), locate all *<rank>* child nodes.

The getNodeSet() function accepts an *XML* node or document and an *XPath* expression and returns a list of the elements within the node that are located by the expression. We provide getNodeSet() the root node of the document and our *XPath* expression to locate the f2091 `<field>` node:

```
field2091 = getNodeSet(factbookDoc, "//field[@id='f2091']")
```

Let's examine the attributes of the returned node to confirm we have our desired node:

```
xmlAttrs(field2091[[1]])
```

```
        dollars                              unit
        "false"      "(deaths/1,000 live births)"
      rankorder                              name
            "1"              "Infant mortality rate"
             id
        "f2091"
```

It appears we have located the correct `<field>` node so let's continue with this approach to locate all of the `<rank>` nodes in the f2091 `<field>` node. We do this with

```
rankNodes = getNodeSet(factbookDoc, "//field[@id='f2091']/rank")
```

When we examine the attribute values of the first node in rankNodes we find:

```
xmlAttrs(rankNodes[[1]])
```

```
     number  dateEstimated     dateLatest   dateEarliest
    "121.63"         "true"   "2012-12-31"   "2012-01-01"
   dateText        country
  "2012 est."         "af"
```

This matches what we saw in the browser and in the alternative approach to extracting the information by traversing the tree via the `[[` operator.

Now we can use the xmlGetAttr() function as before to extract the desired attributes' values from these nodes. We do this with

```
infNum = as.numeric(sapply(rankNodes, xmlGetAttr, "number"))
infCtry = sapply(rankNodes, xmlGetAttr, "country")
```

Since the information in the *XML* file is plain text, we convert the mortality rate to a numeric vector. We leave the country codes as strings because it makes combining these data with population and location information easier.

Whether we use the powerful *XPath* expression or the subsetting approach, we now have the infant mortality data in infNum, and we have the corresponding country identifier in infCtry. We can combine these 2 vectors into a data frame with

```
infMortDF = data.frame(infMort = infNum, ctry = infCtry,
                       stringsAsFactors = FALSE)
```

We can use either of these two approaches (*XPath* or subsetting) to also extract country population values from the Factbook. This time we are searching for the `<rank>` nodes within the `<field>` node that has an *id* of 'f2119'. As before, we determined that 'f2119' is the value of *id* for the population values by investigating the tree. We use *XPath* with

```
rankNodes = getNodeSet(factbookRoot, "//field[@id='f2119']/rank")
popNum = as.numeric(sapply(rankNodes, xmlGetAttr, "number"))
popCtry = sapply(rankNodes, xmlGetAttr, "country")

popDF = data.frame(pop = popNum, ctry = popCtry,
                   stringsAsFactors = FALSE)
```

We have created the two data frames `infMortDF` and `popDF` that were used in Section 11.3.

We saw in Section 11.4.1 that we also need to find the mapping of the ISO country code to the country code used by the CIA in order to properly merge the latitude and longitude with this demographic information. This information is in another part of the Factbook, i.e., not in a `<field>` node. We leave it as an exercise to locate and extract this information. We have provided the results in the data frame `codeMapDF` for cross checking.

11.7 Generating *KML* Directly

We saw in Section 11.5 how the kmlPoints() function can create a *KML* document for display in Google Earth™. If we have knowledge of the *KML* node names and structure, we can generate the *KML* ourselves using the functionality in the *XML* package, specifically with the newXMLDoc() and newXMLNode() functions. A brief introduction to how to use these functions appears in the sidebar below. There are many more parameters and features to newXMLNode(), but these are all that we need to generate our *KML* document.

Generating a Simple *XML* Document

We demonstrate how to generate the following simple small document with the functions in the XML package:

```
<?xml version="1.0"?>
<a>
  <b/>
  <c>
Some Text
    <d/>
    <e id="eId"/>
  </c>
</a>
```

We use the tree representation of this document to organize the steps in creating this document.

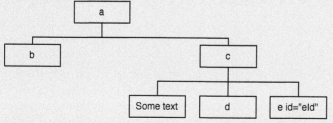

There are many ways to create this document. We provide a few possibilities as examples of the functionality in *XML*. A first step is to create the empty document with

```
doc = newXMLDoc()
```

Next we add the root node to the document with

```
aRoot = newXMLNode("a", doc = doc)
```

A child node can be added to a node with newXMLNode(). We need only provide the name of the node as a string and a reference to its parent. We add the <*b*> node to <*a*> with

```
newXMLNode("b", parent = aRoot)
```

Children can be added to the new node as it is created via the ... argument to newXMLNode(), e.g.,

```
cNode = newXMLNode("c", "Some Text", newXMLNode("d"),
                   parent = aRoot)
```

Here, a string in ... corresponds to a text node, not a node name. Also, attributes can be included in the node via the *attrs* parameter, e.g.,

```
newXMLNode("e", attrs = c(id = "eId"), parent = cNode)
```

We can also specify the parent of <*e*> as `aRoot[[2]]`, i.e., as the second child of the root node.

We save the *XML* document to a text file with

```
saveXML(doc, "sample.xml")
```

To generate a *KML* document, we need to know the basic structure of a document and the allowable tag names, attribute names, and values. We can ascertain the basic structure by examining an excerpt of our target document, e.g.,

```
<?xml version="1.0"?>
<kml xmlns="http://www.opengis.net/kml/2.2">
  <Document>
    <name>Infant Mortality</name>
    <description>2012 CIA Factbook</description>
    <LookAt>
      <longitude>-121</longitude>
      <latitude>43</latitude>
      <altitude>4100000</altitude>
      <tilt>0</tilt>
      <heading>0</heading>
      <altitudeMode>absolute</altitudeMode>
    </LookAt>
    <Style id="ball_1-1">
      <IconStyle>
        <scale>1.5</scale>
```

```
        <Icon>yor1ball.png</Icon>
      </IconStyle>
    </Style>
...
    <Folder>
     <name>CIA Fact Book</name>
...
      <Placemark id="uk">
        <name>United Kingdom</name>
        <description>
          <table>
            <tr><td>Country:</td><td>United Kingdom</td></tr>
            <tr><td>Infant Mortality:</td>
              <td>4.56 per 1,000 live births</td></tr>
            <tr><td>Population:</td><td>63395574</td></tr>
          </table>
        </description>
        <styleUrl>#ball_1-3</styleUrl>
        <Point>
          <coordinates>-2.000,54.000,0</coordinates>
        </Point>
      </Placemark>
...
    </Folder>
  </Document>
</kml>
```

To learn more about valid elements and hierarchy in a *KML* document, refer to the *KML* reference at http://code.google.com/apis/kml/documentation/kmlreference.html.

From the partial listing of the *KML* document above, we see that it begins with some document-level information, such as its name and description and where to orient the viewport (`<LookAt>`) when the document is first opened in Google Earth™. We also see that the general style information is placed at the top of the document. In this example, the style called "ball_1-1" provides a scale and an image/icon to be used by placemarks for countries in the lowest infant mortality and smallest population categories. These are in, respectively, `<scale>` and `<Icon>` in `<IconStyle>`.

We also see that one folder (the `<Folder>` element) contains the placemarks for all the countries. Each country's `<Placemark>` contains its name, the style to apply to the placemark, and the coordinates to position the placemark. In addition, the `<Placemark>` element for a country contains a brief *HTML* table with country-specific information that appears in a pop-up window when we click on its icon.

Before we start writing code to create our document, let's think about how we might organize our code into tasks and corresponding functions. We can easily identify one task to create the basic top-level document and a second task to create a placemark. We can design functions for each of these. Also, let's develop our functions incrementally, where our initial placemarks use the simple pushpins provided by Google Earth™ and have no information for the pop-up window.

Now that we have identified these first tasks, let's enumerate the inputs and output for each function. To create the template document, we provide the starting position of the Google Earth™ viewer and the document name and description. These appear in the side

CIA Factbook Mashup

panel called Places in the viewer. Let's make these the inputs to our function, but also provide default values so the user doesn't have to specify them. The return value should be the document tree to which we add our placemarks. The country placemarks are children of a `<Folder>` node so let's build that node in this initial function too. Our function is:

```
makeBaseDocument =
  function(docName = "Infant mortality",
           docDesc = "2012 CIA Factbook",
           lat = 43, lon = -121, alt = 4100000,
           tilt = 0, heading = 0)
{
  doc = newXMLDoc()
  rootNode = newXMLNode("kml", doc = doc)
  DocNode = newXMLNode("Document", parent = rootNode)
  newXMLNode("name", docName , parent = DocNode)
  newXMLNode("description", docDesc, parent = DocNode)
  LANode = newXMLNode("LookAt", parent = DocNode)
  newXMLNode("longitude", lon, parent = LANode)
  newXMLNode("latitude", lat, parent = LANode)
  newXMLNode("altitude", alt, parent = LANode)
  newXMLNode("tilt", tilt, parent = LANode)
  newXMLNode("heading", heading, parent = LANode)
  newXMLNode("altitudeMode", "absolute", parent = LANode)
  newXMLNode("Folder", parent = DocNode)
  return(doc)
}
```

We call our function makeBaseDocument(), using the default parameter values, to create the document template as follows:

```
baseDoc = makeBaseDocument()
baseDoc
<?xml version="1.0"?>
<kml>
  <Document>
    <name>Infant mortality</name>
    <description>2012 CIA Factbook</description>
    <LookAt>
      <longitude>-121</longitude>
      <latitude>43</latitude>
      <altitude>4100000</altitude>
      <tilt>0</tilt>
      <heading>0</heading>
      <altitudeMode>absolute</altitudeMode>
    </LookAt>
    <Folder/>
  </Document>
</kml>
```

Our stub of a *KML* document matches the target.

Our next task is to create the simple pushpin placemarks as shown in Figure 11.7. Let's have our function create one placemark. For inputs, we need to provide the latitude and longitude. We can also supply a unique identifier and a label for the placemark. The label appears next to the placemark on the earth browser. As in the visualization produced by **RKML** in Section 11.5 we use the country name for the label and country code for the identifier. We won't provide any default input values for these parameters. The `<PlaceMark>` node needs to be placed in the `<Folder>` node so we also want to supply the `<Folder>` node as an input to make it easy to add the `<PlaceMark>` node as a child of the `<Folder>`. Instead, we could have our functions return the `<Placemark>` node and assign its parent after creating it, but it seems simpler to pass in the parent node and add the new `<Placemark>` to its parent as we create it. When we do this, we don't need to return the updated `<Folder>` node because the original object is modified within our function. That is, the **XML** functions work with pointers to C-level structures and any change within a function to a node that has been passed in to the function is reflected in the original C-level structure. For more information about working with *XML* using the **XML** package in *R* see [8].

Our function, which we call addPlacemark(), is defined as

```
addPlacemark = function(lat, lon, id, label, parent){
  newXMLNode("Placemark",
             newXMLNode("name", label),
             newXMLNode("Point",
                        newXMLNode("coordinates",
                                   paste(lon, lat, 0, sep = ","))),
             attrs = c(id = id), parent = parent)
}
```

It might seem unnecessary to have this function because all it does is create a `<Placemark>` node by calling the newXMLNode() function with our input values. Let's wait to see if we still need it after we replace the pushpin with a circle and add a pop-up window.

We are now ready to add a placemark for each country to baseDoc, or more specifically its `<Folder>` element. We use our addPlacemark() function to do this. First we need to access the `<Folder>` node, which is to be the parent to all of our placemarks. We do this by accessing the root node of the document and then its child, the `<Folder>` node, as

```
root = xmlRoot(baseDoc)
folder = root[["Document"]][["Folder"]]
```

Then with mapply(), we add the placemarks:

```
mapply(addPlacemark,
       lat = allCtryData$latitude, lon = allCtryData$longitude,
       id = allCtryData$ctry, label = allCtryData$name,
       parent = folder)
```

If we save this document with saveXML() and open it in Google Earth™, we would see a yellow pushpin for each country as in Figure 11.7.

Now we consider how to change the appearance of the placemarks by providing style references on the placemarks. We can use the same references that we created in Section 11.5; these names are in ctryStyle. We also want to augment a placemark with additional information that appears in the pop-up window associated with the placemark. We can use the descriptions also created in Section 11.5, which are available in ptDescriptions. Since we have already constructed the descriptions, all that we need to do is add them as text content to a `<description>` node in each `<Placemark>`. We can either add another call to newXMLNode() within addPlacemark(), or add them to the <Placemark> nodes after they are created. It seems simplest to update our addPlacemark() function so that it also creates a <description> node. One consideration is whether or not the function has been in use for a while and other people's code depends on it. We wouldn't want to break their code unnecessarily. However, if we add an optional parameter to our existing function where the default value indicates that no description is needed, then that leaves the existing behavior unaltered. Similarly, we can add an optional parameter for style information.

The function definition for our updated function would be

```
addPlacemark =
 function(lat, lon, id, label, parent, style = NULL, desc = NULL)
```

We leave it as an exercise to modify the function. The augmented <Placemark> element would look like

```
<Placemark id="an">
  <name>Andorra</name>
  <description>
    <table>
      <tr><td>Country:</td> <td>Andorra</td></tr>
      <tr><td>Infant Mortality:</td>
          <td>3.76 per 1,000 live births</td></tr>
      <tr><td>Population:</td> <td>85293</td></tr>
    </table>
  </description>
  <styleUrl>#ball_1-1</styleUrl>
  <Point>
    <coordinates>1.500,42.500,0</coordinates>
  </Point>
</Placemark>
```

We apply the updated addPlacemark() with

```
mapply(addPlacemark,
       lat = allCtryData$latitude, lon = allCtryData$longitude,
       id = allCtryData$ctry, label = allCtryData$name,
       parent = folder, style = ctryStyle, desc = ptDescriptions)
```

We have one additional task – to add the <Style> nodes defining the actual styles to the document. Recall that this information is in a *list* called ballStyles. In this case, given the complexity of the <Style> element, we might choose to wrap these details into a separate function, say, makeStyleNode():

```
makeStyleNode = function(styleInfo, id){
  st = newXMLNode("Style", attrs = c("id" = id))
  newXMLNode("IconStyle",
             newXMLNode("scale", styleInfo$IconStyle$scale),
             newXMLNode("Icon", styleInfo$IconStyle$Icon),
             parent = st)
  return(st)
}
```

We can generate a list of these <Style> nodes with

```
styleNodes = mapply(makeStyleNode, ballStyles, names(ballStyles))
```

These elements are not yet part of the *KML* document because we did not specify a parent node in makeStyleNodes() when we created them. Our final task is to place these `<Style>` nodes in the document, between the `<LookAt>` and the `<Folder>` nodes. We can do this with the addChildren() function, which takes a list of children in its *kids* argument and a position as to where to add the children in its *at* argument. We call addChildren() with

```
addChildren(root[["Document"]], kids = styleNodes, at = 3)
```

We have built the *KML* document, and we now save it with

```
saveXML(baseDoc, file = "countryMashup.kml")
```

As mentioned in Section 11.5 the file name is changed to 'countryMashup.kmz' because saveXML() zips the *KML* file together with the 5 PNG files.

We should gather all of these calls to addPlacemark(), makeBaseDocument(), makeStyleNode(), saveXML(), etc., into a single function that creates the entire *KML* document for us. We leave this task as an exercise.

11.8 Additional Computational Tasks

There are several other computational tasks associated with the creation of the Google Earth™ display in Figure 11.1. For example, the circles representing each country were made using *R*'s plotting functions. Also, we sped up the process of creating the placemarks by a factor of about 200 by building them from strings. And, our first attempt at extracting the latitude and longitude for the center of each country was using an *HTML* file. We briefly describe each of these tasks here.

11.8.1 Creating Plotting Symbols

The colored disks used in the Google Earth™ display are created in *R* using its plot() function. That is, a disk is made by setting up a blank *R* canvas that has a transparent background and no axes and no labels. On this canvas we draw a circle with draw.circle() and fill it with the desired color. We leave it as an exercise to write a function that takes a vector of colors as input and creates a PNG file of a circle for each color.

11.8.2 Efficiency in Generating *KML* from Strings

There are occasions when using string manipulation to create *XML* content, rather than using newXMLNode(), can be much faster. These are typically when we need to create many nodes that have the same structure but with different values in the content or attributes. For this reason, the XML package includes the function parseXMLAndAdd(), which takes *XML* content as a string, parses it, and returns the parsed tree or adds the parsed nodes to a specified parent. When we develop a function to generate *XML*, the ideal approach is to use a string-based approach when it is significantly faster and to use a node approach otherwise because the tree structure can be more readily updated and modified than strings.

For our Google Earth™ visualization, it is convenient to generate the placemarks with strings. That is, we can create a series of `<Placemark>` nodes as a string in a vectorized manner as follows:

```
kmlTxt = sprintf("<Placemark><Point><coordinates>%.3f,%.3f,0
                 </coordinates></Point></Placemark>",
                 allCtryData$longitude, allCtryData$latitude)
```

This approach vectorizes the operation across all 200+ countries, which we cannot do with newXMLNode().

Let's compare these two approaches for creating the simple pushpin placemarks. We write two versions of a function that take the same inputs but differ in their implementation. The first, called addPlacemarks.fast(), creates placemarks using strings. This function is implemented as

```
addPlacemarks.fast =
function(lon, lat, parent)
{
  txt = sprintf("<Placemark><Point><coordinates>%.3f,%.3f,0
                 </coordinates></Point></Placemark>",
                 lon, lat)
  parseXMLAndAdd( paste(txt, collapse = ""), parent)
}
```

The second function, called addPlacemarks.slow(), creates placemarks using calls to newXMLNode(). This function is implemented as

```
makePM = function(x, y, parent) {
  newXMLNode("Placemark",
             newXMLNode("Point",
                        newXMLNode("coordinates",
                                   paste(x, y, 0, sep=","))),
             parent = parent)
}

addPlacemarks.slow =
function(lon, lat, parent)
{
  mapply(makePM, x = lon, y = lat, parent = parent)
}
```

For our test, we set up the document and the inputs to the functions as follows:

```
doc = newXMLDoc()
root = newXMLNode("kml", doc = doc)
folder = newXMLNode("Folder", parent = root)

lons = rep(allCtryData$longitude, 10)
lats = rep(allCtryData$latitude, 10)
```

Then we use system.time() to time the "slow" function with

```
system.time(invisible(
            addPlacemarks.slow(lons, lats, folder)))
   user  system elapsed
  4.580   0.016   4.645
```

Next, we time the "fast" function using the same latitudes and longitudes with

```
rm(doc)
doc = newXMLDoc()
root = newXMLNode("kml", doc = doc)
folder = newXMLNode("Folder", parent = root)

system.time(invisible(
            addPlacemarks.fast(lons, lats, folder)))
```

```
  user  system elapsed
 0.023   0.001   0.024
```

The string-based approach is 200 times faster!

11.8.3 Extracting Latitude and Longitude from an *HTML* File

As mentioned earlier, locations for countries are available from many places on the Web. For example, in addition to the CSV file, MaxMind also provides these data within an *HTML* page that is available at http://dev.maxmind.com/geoip/legacy/codes/country_latlon/. See the screenshot in Figure 11.9 for the layout. Since well-formed *HTML* is a special case of *XML*, we can extract latitude and longitude from this *HTML* file using the same tools that we used to find the demographic information in the Factbook. We leave it as an exercise to do this.

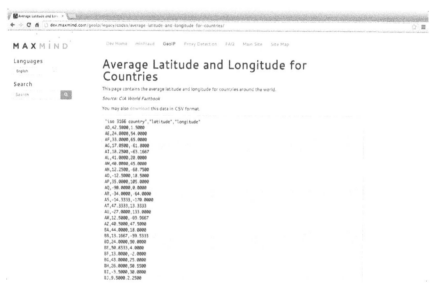

Figure 11.9: Screenshot of the MaxMind Web Page with Country Latitude and Longitude. *MaxMind makes available the average latitude and longitude for countries in several formats, including in a display on the Web as shown here in this screenshot.*

11.9 Exercises

Q.1 In the merge of `infMortDF` and `popDF` in Section 11.3 some rows in each data frame were excluded from the resulting data frame. Determine which rows in each data frame did not find a match in the other.

Q.2 The data frame `codeMapDF` contains a mapping of the ISO 3166 codes to the codes used in the CIA Factbook. It also contains country name. Below are the first few observations in this data frame:

```
head(codeMapDF)
  cia           name iso
1  af    Afghanistan  AF
2  ax       Akrotiri   -
3  al        Albania  AL
4  ag        Algeria  DZ
5  aq American Samoa  AS
6  an        Andorra  AD
```

Use this data frame to fix the problem with the merging of the location data, which uses ISO codes, with the demographic data, which uses the CIA Factbook coding. Be sure the final data frame contains both codes and the country name from `codeMapDF`, as well as the original variables.

Q.3 Remake the map in Figure 11.6 to include the countries for which there is latitude and longitude, even if there is no demographic information. To do this you need to modify the call to merge(). In particular, the *all.x* and *all.y* parameters are helpful here. In the map, use a special plotting symbol to denote those countries where demographic information is missing.

Q.4 To determine the color of a circle in the map in Figure 11.6, we discretized the infant mortality rates. Investigate alternative cut points for this discretization that derive from standards used by international organizations such as the United Nations, World Health Organization, and World Bank. Remake the map to incorporate these external values.

Q.5 To determine the size of a circle in the map in Figure 11.6, we used a scaling of the square-root of population. We found that the range in population was so large that when the largest countries had a reasonably sized disk, the disks for the smaller countries could not be seen. To address this problem we created a lower bound for the disk size so that no matter how small the country, a disk with this minimum size would be plotted. Find an alternative scaling for population that instead caps the size of the larger countries. That is, for very large countries, no matter how large the population, the disk plotted on the map is no bigger than this cap. Remake the map using this approach to scaling the circles.

Q.6 The `mapproj` [4] library contains various projections of the 3D earth into 2D maps. Install this package and investigate the projections. To do this you will want to read some resources and tutorials available online. Remake the map in Figure 11.6 using a projection.

Q.7 There are many other statistics available in the Factbook. Choose demographic information other than infant mortality to extract and visualize.

Q.8 Wikipedia tables offer another source of data. For example, Wikipedia has a list of infant mortality rates by country at http://en.wikipedia.org/wiki/List_of_countries_by_infant_mortality_rate. These data are from the United Nations World Population Prospects report (http://esa.un.org/unpd/wpp/Excel-Data/mortality.htm) and the CIA World Factbook. Another source of information is the World Bank, which provides infant mortality rates at http://data.worldbank.org/indicator/SP.DYN.IMRT.IN. Create a mashup with data from one of these alternative sources.

Q.9 *XPath* is a powerful tool for locating content in an *XML* document, and there are often many *XPath* expressions that we can use to access the information we want. For example, rather than use the *id* attribute of `<field>` to find the relevant data, try another approach, such as locating the desired nodes by filtering on the value of the *name* attribute.

Q.10 Use *XPath* to extract the mapping between the two types of country codes from the CIA Factbook. Once you have this mapping in a data frame, use it to re-merge the latitude/longitude data with the Factbook data. Then recreate the world map. It should look like Figure 11.6. To assist you in this task, we provide a snippet of the Factbook that shows a portion of the table that contains the various codes for countries:

```
<table title="" lettergrouped="1">
 <columnHeader colspan="1" title="Entity"/>
 <columnHeader colspan="1" title="FIPS 10"/>
 <columnHeader colspan="3" title="ISO 3166"/>
 <columnHeader colspan="1" title="Stanag"/>
 <columnHeader colspan="1" title="Internet"/>
 <columnHeader colspan="1" title="Comment"/>
 ...
 <row>
  <cell country="ch" content="China"/>
  <cell center="1" content="CH"/>
  <cell content="CN"/>
  <cell content="CHN"/>
  <cell content="156"/>
  <cell center="1" content="CHN"/>
  <cell center="1" content=".cn"/>
  <cell content="see also Taiwan"/>
 </row>
 ...
 <row>
  <cell country="gb" content="Gabon"/>
  <cell center="1" content="GB"/>
  <cell content="GA"/>
  <cell content="GAB"/>
  <cell content="266"/>
  <cell center="1" content="GAB"/>
  <cell center="1" content=".ga"/>
  <cell content=" "/>
 </row>

 <row>
```

```
    <cell country="sz" content="Switzerland"/>
    <cell center="1" content="SZ"/>
    <cell content="CH"/><cell content="CHE"/>
    <cell content="756"/>
    <cell center="1" content="CHE"/>
    <cell center="1" content=".ch"/>
    <cell content=" "/>
  </row>
```

Q.11 In Section 11.8.3 we saw that the latitudes and longitudes for the countries are available on the Web, embedded in the page at `http://dev.maxmind.com/geoip/legacy/codes/average-latitude-and-longitude-for-countries/`. A snippet of the *HTML* source for this page is shown in Figure 11.10. Extract these location coordinates from this *HTML* file.

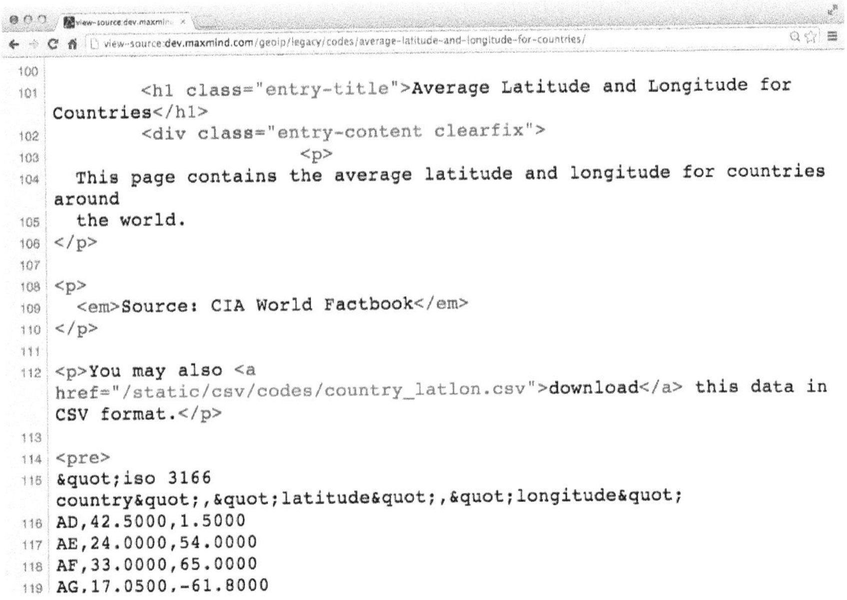

```
        <h1 class="entry-title">Average Latitude and Longitude for
Countries</h1>
        <div class="entry-content clearfix">
                  <p>
  This page contains the average latitude and longitude for countries
around
  the world.
</p>

<p>
  <em>Source: CIA World Factbook</em>
</p>

<p>You may also <a
href="/static/csv/codes/country_latlon.csv">download</a> this data in
CSV format.</p>

<pre>
&quot;iso 3166
country&quot;,&quot;latitude&quot;,&quot;longitude&quot;
AD,42.5000,1.5000
AE,24.0000,54.0000
AF,33.0000,65.0000
AG,17.0500,-61.8000
```

Figure 11.10: Screenshot of the *HTML* Source of the MaxMind Web Page. *This screenshot displays the HTML source of the Web page shown in Figure 11.9. Notice that the latitudes and longitudes for the countries appear within a <pre> node in the HTML document.*

Examine the *HTML* source, and notice that the data are simply placed as plain text within a *<pre>* node in the document. If we can extract the contents of this *<pre>* node, then we can place this information in a data frame. Begin by parsing the *HTML* document with htmlParse(). Don't download the document, simply pass the function the *URL*, `http://dev.maxmind.com/geoip/legacy/codes/average-latitude-and-longitude-for-countries/`.

Next, access the root of the document using xmlRoot(), and use an *XPath* expression to locate the *<pre>* nodes in the document. The getNodeSet() function should be useful here. The getNodeSet() function takes the *XML* tree (or subtree) and an *XPath* expression as input and returns a list of all nodes in the tree that are located with this expression.

Once you have located these `<pre>` nodes, check to see how many were found. It should be only one. The latitude and longitude values are in the text content of this node. Extract the text content with the xmlValue() function. It should look something like:

```
[1] "\n\"iso 3166 country\",\"latitude\",\"longitude\"\n
AD,42.5000,1.5000\nAE,24.0000,54.0000\nAF...
```

The content is one long character string containing all the data.

Complete the exercise by reading the plain text in your character vector into a data frame. Use read.table() to do this. The parameters, *text*, *skip*, *header*, and *sep* should be useful here.

Q.12 Collect all of the steps for generating the *KML* document (see Section 11.7) into one function, called makeCIAPlot(). Consider generalizing it so that it can be used with other Factbook variables or other data.

Q.13 The plotting symbols used in Figure 11.1 were created in *R*. Write a function as described in Section 11.8.1 to create a set of PNG files that can be used as placemarks on Google Earth™. Rather than simply making circles, consider making other shapes, including a miniature plot of the data. Also, use a different color scheme than the yellow-orange-red palette.

This placemark function creates the icons by setting up a blank canvas that has a transparent background and no axes or labels. On this canvas draw a circle or some other shape and fill it with the desired color. The function should take as input a vector of colors and create a set of circles (or some other shape) as PNG files, one for each color. The png(), plot(), and draw.circle() functions might be helpful. In addition, you might try making the symbols partially transparent so that when they overlap on the plot, they can still be seen.

Q.14 Make an *R* plot for the pop-up window of each country. That is, for each country, make a plot that is country-specific. Save this plot to a PNG file and add the file name to the `<description>` node, e.g., the content for the description might be something like

```
Afghanistan:
<img src = "af.png"/>
```

The resulting *KML* document should be a *.kmz* file, i.e., a **zip** archive, that contains all of these PNG files as well as the *KML* document.

Q.15 Rewrite the addPlacemark() and makeStyleNode() functions to use the technique described in Section 11.8.2. That is, construct the placemarks by pasting strings together, and then convert the strings into *XML* nodes with a call to xmlParseAndAdd().

Bibliography

[1] Richard Becker, Allan Wilks, Ray Brownrigg, and Thomas Minka. maps: Draw geographical maps. http://cran.r-project.org/web/packages/maps/, 2011. *R* package version 2.1.

[2] Google, Inc. : A 3D virtual earth browser, version 6. http://www.google.com/earth/, 2011.

[3] Eric Lecoutre. R2HTML: *HTML* exportation for *R* objects. http://cran.r-project.org/package=R2HTML, 2011. *R* package version 2.2.

[4] Doug McIlroy, Ray Brownrigg, Thomas Mink, and Roger Bivand. mapproj: Map Projections. http://cran.r-project.org/package=mapproj, 2014. *R* package version 1.2-2.

[5] Erich Neuwirth. RColorBrewer: ColorBrewer palettes. http://cran.r-project.org/web/packages/RColorBrewer, 2011. *R* package version 1.0-5.

[6] Deborah Nolan and Duncan Temple Lang. **RKML**: Simple tools for creating *KML* displays from *R*. http://www.omegahat.org/RKML/, 2011. *R* package version 0.7.

[7] Deborah Nolan and Duncan Temple Lang. Interactive and animated scalable vector graphics and *R* data displays. *Journal of Statistical Software*, 46:1–88, 2012.

[8] Deborah Nolan and Duncan Temple Lang. *XML and Web Technologies for Data Sciences with R*. Springer, New York, 2013.

[9] Open Geospatial Consortium, Inc. OGC *KML* standards. http://www.opengeospatial.org/standards/kml/, 2010.

[10] R Development Core Team. *R: A Language and Environment for Statistical Computing.* Vienna, Austria, 2012. http://www.r-project.org.

[11] John Simpson. *XPath and XPointer: Locating Content in XML Documents.* O'Reilly Media, Inc., Sebastopol, CA, 2002.

[12] Duncan Temple Lang. XML: Tools for parsing and generating *XML* within *R* and *S-PLUS*. http://www.omegahat.org/RSXML, 2011. *R* package version 3.4.

[13] Wikipedia. Mashup (web application hybrid). http://en.wikipedia.org/wiki/Mashup_(web_application_hybrid), 2012.

12

Exploring Data Science Jobs with Web Scraping and Text Mining

Deborah Nolan
University of California, Berkeley

Duncan Temple Lang
University of California, Davis

CONTENTS

12.1	Introduction and Motivation	457
	12.1.1 Computational Topics	459
12.2	Exploring Different Web Sites	459
12.3	Preliminary/Exploratory Scraping: The Kaggle Job List	465
	12.3.1 Processing the Text	469
	12.3.2 Generalizing to Other Posts	470
	12.3.3 Scraping the Kaggle Post List	473
12.4	Scraping CyberCoders.com	475
	12.4.1 Getting the Skill List from a Job Post	478
	12.4.2 Finding the Links to Job Postings in the Search Results	482
	12.4.3 Finding the Next Page of Job Post Search Results	487
	12.4.4 Putting It All Together	488
12.5	A Reusable Generic Framework for Arbitrary Sites	489
12.6	Scraping Career Builder	492
12.7	Scraping Monster.com	494
12.8	Analyzing the Results: The Important Skills	495
12.9	Note on Web Scraping	503
12.10	Exercises	503
	Bibliography	504

12.1 Introduction and Motivation

In this case study, we will explore on-line job postings for different professions or types of positions. We are interested in finding the set of skills that different types of positions expect and want, and which are valuable, but not required. We also want to find information about what educational level an applicant should have (i.e., BSc, MSc, or PhD) for different types of jobs, how many years of experience are needed, what the salary ranges are, and how these differ geographically. We will work with on-line postings so they are up-to-date and easily accessed programmatically. We expect the resulting information will be interesting for both students and instructors. Also, in the case study, you will learn some of the skills that are currently very much in vogue for data science and develop tools to continue to robustly gather these data in the future.

Ideally, there would be a single Web site with all job postings of interest. We could then query these with a rich query language and the results would be returned to us in a convenient form such as *JSON* or *XML*. We could then convert these results into data structures in any language such as *R* [6] or *Python* and start to explore the data. Each job posting would, ideally, have the same structure, with fields for salary range as a minimum and a maximum, location, list of required skills, educational background, and so on. We would be able to extract these easily, given the standard structure for each job. Furthermore, we could do text processing and even natural language processing (NLP) on the less structured aspects of each posting. We might even think of retrieving the job postings and housing them in a local database, or perhaps even better, a text search engine such as **Lucene** [1] or *ElasticSearch* [2].

Unfortunately, on-line job posting sites haven't yet moved to a programmatic interface that allows us to query the jobs easily. Indeed, they are almost exclusively sites for humans to visit and read individual job postings. The postings typically have little common structure. On a given site, each job posting is shown in approximately the same style, but there are significant differences in the structure and appearance of the postings. As we move from one site to another, the formatting and display of the postings are very different. We have to find the salary and the location in different places in the posts. The required skill set is typically in free-form text, sometimes in a bullet-list.

Basically, each job advertising site has a different format and there is potentially a lot of variability between the structure of each job posting. Instead of *JSON* or *XML*, the job postings are in *HTML* and the data of interest are intertwined with a lot of markup for formatting and rendering the postings, advertisements, etc. We have to figure out how to extract the actual data we want from this distracting formatting information. Ideally, we'd be able to write one function to handle any job posting. In reality, we will be fortunate if we can write one function for all postings on a particular site. We will have to explore a sample of postings from that site and try to identify commonalities across these, e.g., the class on an *HTML* element that identifies the salary, location, or set of skills. These commonalities may allow us to extract the primary information that is readily available, but potentially miss some that is too difficult to identify across different postings.

In addition to processing the content of each job posting, we have to locate and retrieve the individual job postings. Again, this is different for each Web site. We want to try to identify the commonalities to reduce the amount of code and testing we have to do.

To fetch the job posting documents from Web sites, we'll make *HTTP* requests from *R* (either implicitly or explicitly), just like your Web browser does. We'll process the *HTML* documents as hierarchical/tree-like structures, looking for elements and sub-trees that contain the information we want. To query the tree and retrieve these elements, we'll use the *XPath* [7, 5] domain-specific language (DSL). We'll use the **XML** [9] package to manipulate the *HTML* elements and evaluate *XPath* queries. When we have to explicitly make *HTTP* requests, we will use the **RCurl** package [10].

We will also briefly explore accessing job posts via a more structured Web service that provides an Application Programming Interface (API) to query jobs. This involves explicit *HTTP* requests to get the data. The results are returned as *JSON* content, which we process with either the **RJSONIO** [8] or `rjson` [3] packages.

This chapter does not provide an introduction to *HTML*, *XML*, *XPath*, *JSON*, *HTTP*, or Web services. Instead, we refer the reader to [5] for an overview of all of these topics, with an *R*-oriented perspective. There are also numerous tutorials and books on each of these different topics.

Before we start looking at any Web sites and details, we want to mention an important issue, namely reproducing the computations and results in this chapter. As we all know, Web sites and pages come and go, and their contents and appearances change, some very rapidly. In our case, job postings are often available for a short period of time and are then removed. We will be getting the data (i.e., the job postings) directly from the Web sites. So we cannot control whether the same job postings will be available and at the same *URLs* when we revisit the computations. Furthermore, owners of Web sites sometimes change the appearance and format of the pages. This means that when we develop code to extract data from pages based on a particular format, that code may not work in the future. This is an unfortunate problem with Web scraping, generally (and why we greatly prefer to use Web services and APIs). We have made some sample pages available on the book's Web site (`http://rdatasciencecases.org`) to allow readers to work with specific *HTML* formats, and to dynamically explore the computations we discuss in this chapter. We have also written the code with robustness in mind so as to be able to adapt to changes in the *HTML* representation. However, we cannot guarantee that the code will be able to read the same job postings in the future.

12.1.1 Computational Topics

In this case study, you will get some experience with the following.

- Scraping data from *HTML* pages.
- The basics of HTTP/Web requests.
- Using the *XPath* language and queries.
- Manipulating *XML/HTML* elements in *R*.
- Text processing, regular expressions, stop words, and word stemming.
- Reusable functions, modular software, and designing software for extensibility.
- Simple exploratory data analysis and visualization.

12.2 Exploring Different Web Sites

Let's start by looking at a collection of job postings. We will do this in our Web browser so that we can first get an understanding of the sites we are going to programmatically visit. There are several Web sites to explore. For example, linkedin.com, `monster.com`, `simplyhired.com`, `usajobs.gov`, `careerbuilder.com`, `indeed.com`, and more specific data mining and statistics oriented sites such as Kaggle's jobs forum and `statscareers.com` . We will look first at Kaggle, primarily because it is slightly smaller than some of the others and a little more focused on data science and statistical and machine learning.

> Before we explore Kaggle, we want to explicitly note that many Web sites actively prohibit users from scraping content. Several of the Web sites mentioned above do. We are discussing strategies to scrape Web sites generally by focusing on details for specific Web sites. If you want to scrape a Web site, you need to verify that you are permitted to do so by examining its `user agreement` or Terms of Service (ToS).

We visit Kaggle.com in our Web browser, and then click on the Jobs Board link. This brings us to http://www.kaggle.com/forums/f/145/data-science-jobs, a page similar to that shown in Figure 12.1.

Each row of the table on this page shows a job posting (and the first two are about the jobs forum itself). We can click on a job posting, say the one labeled "Singapore Data Analyst wanted." The page for this is shown in Figure 12.2. This is a slightly unusual posting. It is an informal posting with little description of the job; just a sentence or two by the poster. There are some responses by others commenting on the original posting.

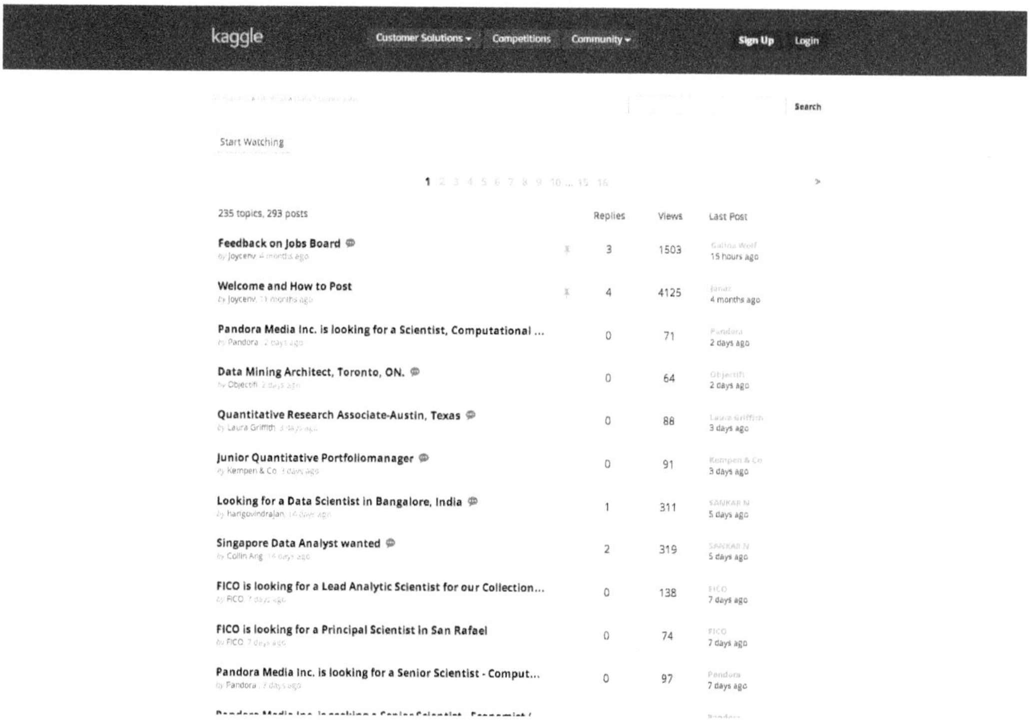

Figure 12.1: Kaggle Jobs Board Web Page. *This is the front page of the Kaggle jobs forum. All of the posts are available via this page and its successive pages via the links 2, 3, ... We can click on the link for a specific job to read that post.*

Let's look at another job posting, say, "Pandora Media Inc. is looking for a Senior Scientist – Growth Hacking." This is shown in Figure 12.3 and can be accessed directly at https://www.kaggle.com/forums/t/5140/pandora-media-inc-is-looking-for-a-senior-scientist-recommender-systems. This is a more typical posting. The post has several paragraphs describing aspects of the job. There is a separate bullet list of requirements.

Let's look at a second site – monster.com. Unlike Kaggle, which posts *all* jobs on its sole job list, the main page for monster.com provides a search form for querying jobs by a specific title, arbitrary keywords, or by location. We enter the search term "Data Science" as keywords, and we see a page like that shown in Figure 12.4.

We will look at the second post in this list, titled "Data Analyst for leading consumer science firm in Cincinnati..." This appears in Figure 12.5. As in the second Kaggle post we looked at, we see a few paragraphs describing the job, a few lists describing the position, the role, rewards, etc. The salary range is mentioned in the second item at the very top of the

Exploring Data Science Jobs with Web Scraping and Text Mining 461

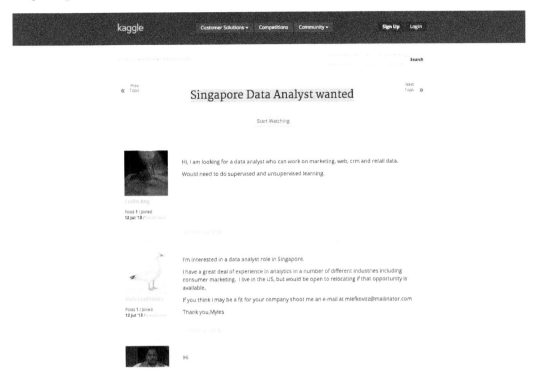

Figure 12.2: Kaggle Job Post and Comments. *This shows a very informal job posting and follow-up posts on the same page. There are very few details about the position being advertised.*

page, above the general text. We also see some metadata provided by `monster.com` about similar jobs (e.g., Clustering, Data Analysis, *SQL*) near the bottom of the post. This is separate from the content of the actual post and we can potentially use this to help cluster the jobs into groups and identify key skills. Indeed, can we think about how sites produce these associations, or are they specified manually?

We can explore other job posting sites and we will see different pages for different jobs. Before reading further, we suggest that you visit at least two other sites to see if we can identify commonalities in the job postings and how we find the postings. Kaggle is quite different from Monster.com in that we find jobs in the latter via a search form. This is the case for most of the other sites. Accordingly, we want to examine these search query forms and see if we can use the same approach, just with different details of the particular URL to which we submit the search query.

Generally, we will find the following pattern. We navigate to the first page of the job postings, either of the entire collection (on Kaggle) or for a particular search string or job title. This initial page typically contains multiple postings and also links to subsequent pages of postings. We want to collect information about each job post on that page, and then move to the next page. We use this approach to process the posts on each page and then visit the next page of results, continuing until there are no more pages of search results. Take a moment to identify these components on several of the different Web sites. Find the URL for submitting the search query, the link to one or more of the postings, and the link to the next page of postings. These are common aspects of Web scraping.

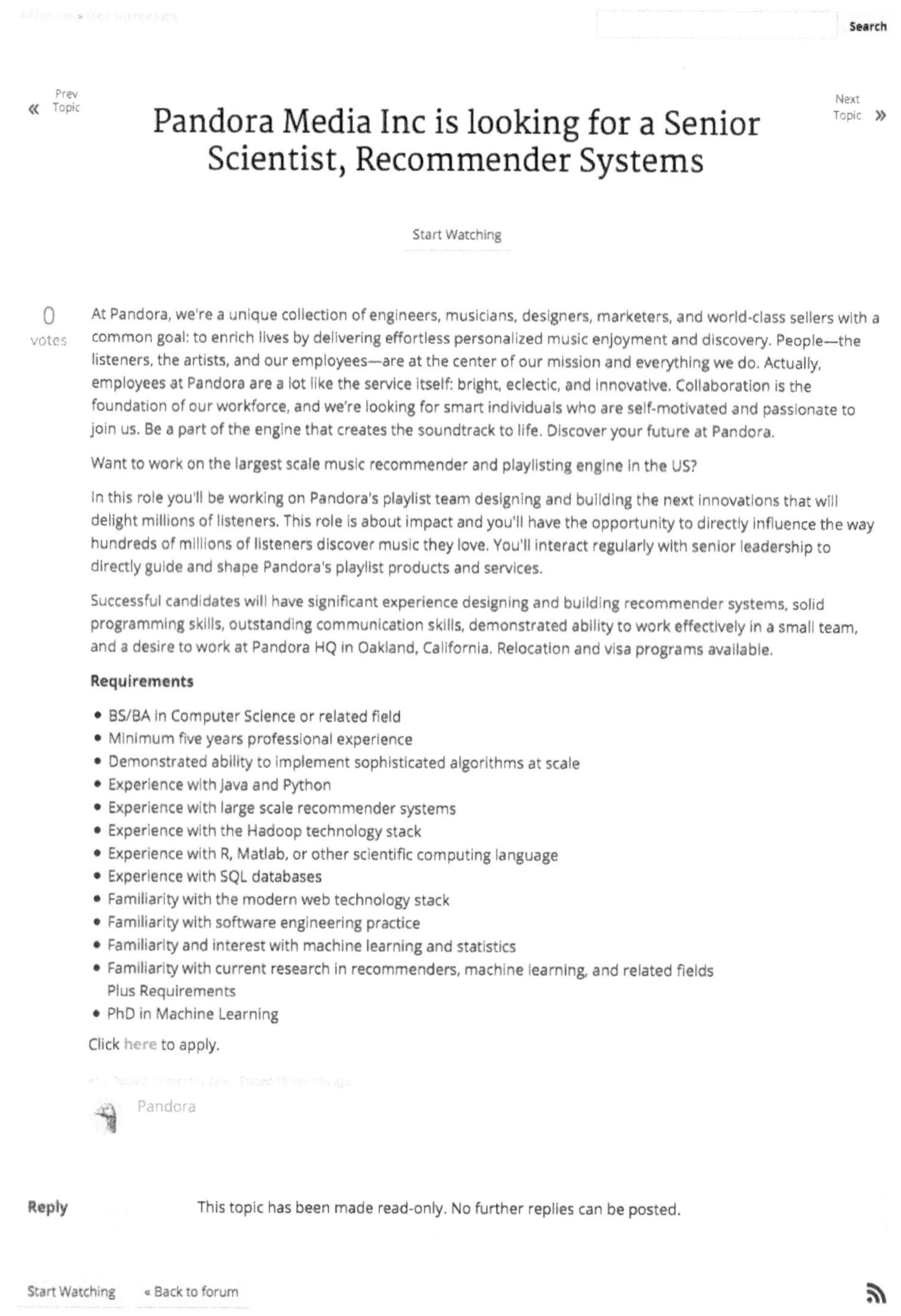

Pandora Media Inc is looking for a Senior Scientist, Recommender Systems

Start Watching

0 votes

At Pandora, we're a unique collection of engineers, musicians, designers, marketers, and world-class sellers with a common goal: to enrich lives by delivering effortless personalized music enjoyment and discovery. People—the listeners, the artists, and our employees—are at the center of our mission and everything we do. Actually, employees at Pandora are a lot like the service itself: bright, eclectic, and innovative. Collaboration is the foundation of our workforce, and we're looking for smart individuals who are self-motivated and passionate to join us. Be a part of the engine that creates the soundtrack to life. Discover your future at Pandora.

Want to work on the largest scale music recommender and playlisting engine in the US?

In this role you'll be working on Pandora's playlist team designing and building the next innovations that will delight millions of listeners. This role is about impact and you'll have the opportunity to directly influence the way hundreds of millions of listeners discover music they love. You'll interact regularly with senior leadership to directly guide and shape Pandora's playlist products and services.

Successful candidates will have significant experience designing and building recommender systems, solid programming skills, outstanding communication skills, demonstrated ability to work effectively in a small team, and a desire to work at Pandora HQ in Oakland, California. Relocation and visa programs available.

Requirements

- BS/BA in Computer Science or related field
- Minimum five years professional experience
- Demonstrated ability to implement sophisticated algorithms at scale
- Experience with Java and Python
- Experience with large scale recommender systems
- Experience with the Hadoop technology stack
- Experience with R, Matlab, or other scientific computing language
- Experience with SQL databases
- Familiarity with the modern web technology stack
- Familiarity with software engineering practice
- Familiarity and interest with machine learning and statistics
- Familiarity with current research in recommenders, machine learning, and related fields
  Plus Requirements
- PhD in Machine Learning

Click here to apply.

Pandora

Reply   This topic has been made read-only. No further replies can be posted.

Start Watching   « Back to forum

Figure 12.3: Screenshot of Kaggle Job Post by Pandora, the free, personalized radio Web site. *There are several paragraphs describing the company and the position. There is a bullet-list of required skills.*

Exploring Data Science Jobs with Web Scraping and Text Mining 463

Figure 12.4: List of Search Results for Data Science Jobs on Monster.com. *We can enter the search term in various fields to obtain a list of matching job posts. These are shown below the textfields. Each entry corresponds to a job posting. These is some metadata in these items, but none of much interest to us. There are also numerous advertisements on the page, distracting us from the data of interest.*

In short, we need to find two sets of information for each page of posts: the links to the individual posts, and the link to the next page. Instead of the link to the next page, we can also find the links for all of the pages as these are typically presented as pages 1, 2, 3, ..., 10 or some similar maximum number. However, when there are more than this maximum number of pages, we have to find these incrementally, i.e., find the next page of results after this one. Accordingly, it is often more robust to find the next successive page, until there are no more.

In addition to finding the links to each post and the link to the next page of posts, we also have to process the contents of each post. This will differ for each site. Hopefully we can identify some similarities. After looking at some representative pages, we can identify these potential common patterns, and also the anomalies. We can start to develop strategies to extract the information programmatically. As we do this, we also want to consider how we will represent the data we scrape from each page. Should we just store the entire page, unprocessed? Should we break the text into words? What about the links it contains? Are these of interest? Can we identify, and reliably extract, the different parts of the posting such as the skills needed, the salary, the educational background necessary, keywords? Do we want all the words, or perhaps the important words? How do we differentiate between words like programming and program? Are they different or do they indicate the same concept?

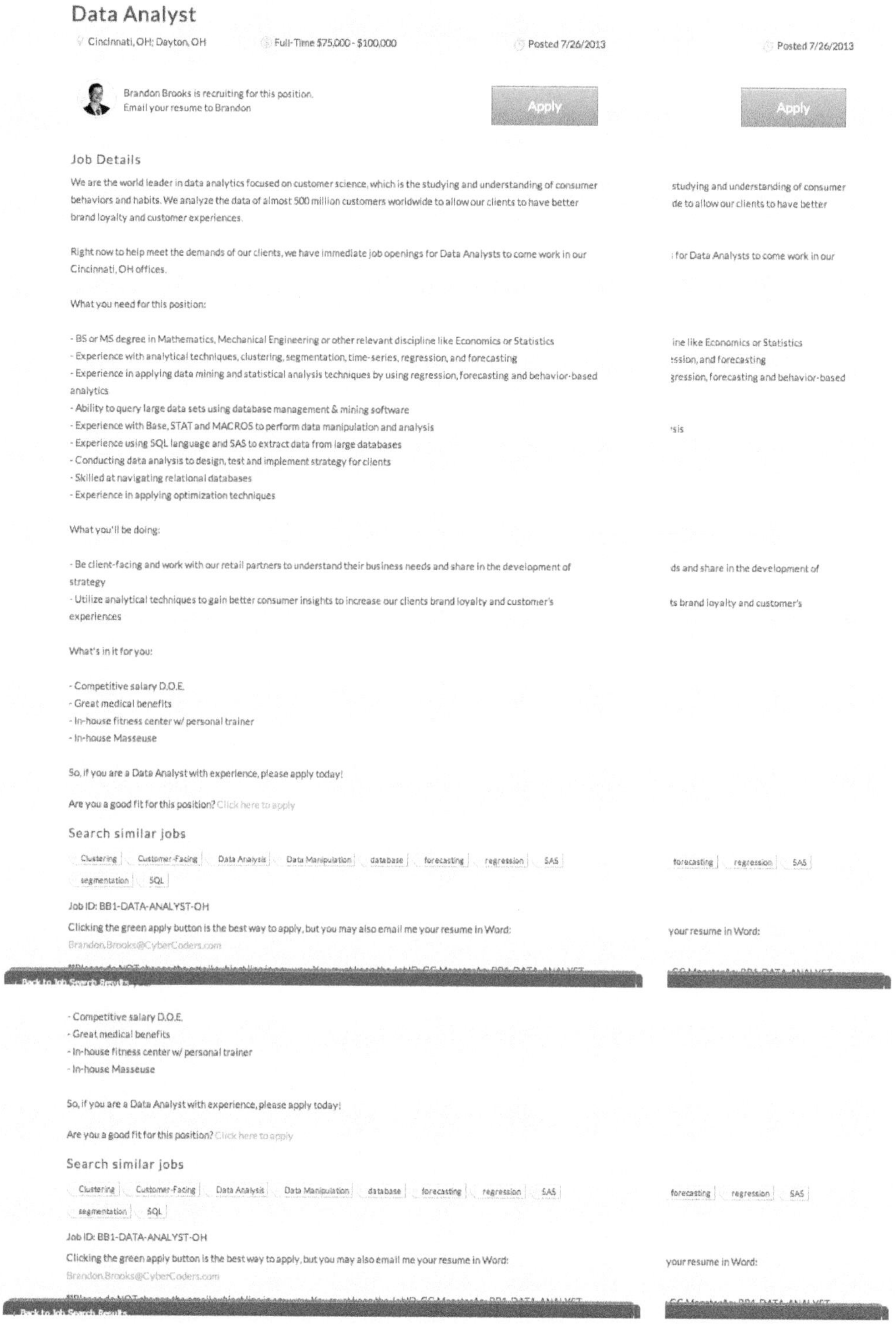

Data Analyst

Cincinnati, OH; Dayton, OH   Full-Time $75,000 - $100,000   Posted 7/26/2013

Brandon Brooks is recruiting for this position.
Email your resume to Brandon

Apply

Job Details

We are the world leader in data analytics focused on customer science, which is the studying and understanding of consumer behaviors and habits. We analyze the data of almost 500 million customers worldwide to allow our clients to have better brand loyalty and customer experiences.

Right now to help meet the demands of our clients, we have immediate job openings for Data Analysts to come work in our Cincinnati, OH offices.

What you need for this position:

- BS or MS degree in Mathematics, Mechanical Engineering or other relevant discipline like Economics or Statistics
- Experience with analytical techniques, clustering, segmentation, time-series, regression, and forecasting
- Experience in applying data mining and statistical analysis techniques by using regression, forecasting and behavior-based analytics
- Ability to query large data sets using database management & mining software
- Experience with Base, STAT and MACROS to perform data manipulation and analysis
- Experience using SQL language and SAS to extract data from large databases
- Conducting data analysis to design, test and implement strategy for clients
- Skilled at navigating relational databases
- Experience in applying optimization techniques

What you'll be doing:

- Be client-facing and work with our retail partners to understand their business needs and share in the development of strategy
- Utilize analytical techniques to gain better consumer insights to increase our clients brand loyalty and customer's experiences

What's in it for you:

- Competitive salary D.O.E.
- Great medical benefits
- In-house fitness center w/ personal trainer
- In-house Masseuse

So, if you are a Data Analyst with experience, please apply today!

Are you a good fit for this position? Click here to apply

Search similar jobs

Clustering | Customer-Facing | Data Analysis | Data Manipulation | database | forecasting | regression | SAS | segmentation | SQL

Job ID: BB1-DATA-ANALYST-OH

Clicking the green apply button is the best way to apply, but you may also email me your resume in Word:
Brandon.Brooks@CyberCoders.com

- Competitive salary D.O.E.
- Great medical benefits
- In-house fitness center w/ personal trainer
- In-house Masseuse

So, if you are a Data Analyst with experience, please apply today!

Are you a good fit for this position? Click here to apply

Search similar jobs

Clustering | Customer-Facing | Data Analysis | Data Manipulation | database | forecasting | regression | SAS | segmentation | SQL

Job ID: BB1-DATA-ANALYST-OH

Clicking the green apply button is the best way to apply, but you may also email me your resume in Word:
Brandon.Brooks@CyberCoders.com

Figure 12.5: Job Post on `monster.com`. *This is a cross-listed job posting on* `monster.com` *that actually comes from the* `cybercoders.com` *Web site. As with the Kaggle posts, much of the post is free-form text. There are several lists that provide metadata about the job. There is also structured content for* similar jobs, *listing the related skills.*

Ideally, we can identify a collection of common fields or variables about each post such as the location, the salary range, the minimum educational requirement, and a collection of keywords about the skills required. We might then keep all of the words as a separate variable. Alternatively, we could have a variable for each possible word and a logical value indicating whether the word was present in the post or not. We might instead store the count of the number of instances of that word in a post. When deciding how to store the data, we want to think about what operations we might want to do with them in order to explore, visualize, summarize, and potentially model them. As a result, we have to pose specific questions we may want to ask. For this, we might choose first to explore the raw data a little before doing this, perhaps working with a small sample to think about questions and how to express them in R. To do this, we'll start scraping and exploring some pages and experiment with different organizations of the data.

12.3 Preliminary/Exploratory Scraping: The Kaggle Job List

Let's start by scraping a post for which we know the URL for the post. We will look at a Kaggle post. We can use the browser to click on a link to any post, say http://www.kaggle.com/forums/t/5153/pandora-media-inc-is-looking-for-a-senior-scientist-growth-hacking, which we displayed above in Figure 12.3. If we quickly read the actual post, we see that it is mostly free-form text in 3 paragraphs, followed by 2 bullet-lists. These lists give requirements for the job and also optional requirements that would help a candidate. There is no salary listed in the text. The location (Oakland, California) is not explicitly identified in a separate part of the post, but is contained in the penultimate line of text of the third paragraph. As a result, we might think about structuring the result of processing this document as a simple list with 3 elements: the raw text, the requirements, and the optional requirements.

To understand the structure of the document, we can look at its *HTML* source, either in the browser or in R. Use the View page source item in your browser, or the R commands

```
library(XML)
doc = htmlParse(u)
doc
```

to parse and print the document in the console. Here, u contains the URL address http://www.kaggle.com/forums/t/5153/pandora-media-inc-is-looking-for-a-senior-scientist-growth-hacking as a string. We can also look at the rendered page in the browser and identify the individual elements in the source and the page using the Developer tools or Inspector. You can move the cursor over part of the page and see the corresponding *HTML* element(s). Familiarize yourself with the basic elements and structure of the *HTML* content, i.e., <head>, <body>, <p> for identifying a paragraph, section header elements <h1>, <h2>, ... <h6>, <a> for hyperlinks, <img> for images, <div> for an abstract container that is often used for layout and appearance. The source for the page also contains *JavaScript* code in <script> elements. Some of these elements are in the <head> element and others are directly in the body. The *JavaScript* code can create dynamic content (i.e., at load and view time), but is also used for adding event handlers. We will ignore the *JavaScript* code and focus on the content in the *HTML* elements. We will attempt to extract the information in the post, making use of the *HTML* elements and attributes to understand the context and meaning of the content.

As we look at the *HTML* document, we see an image (of a goose) on the left with some text underneath it and the main content of the post to its right. These are actually columns in an *HTML* `<table>`. The `<table>` node has a *class* attribute with a value `post`. We can use this to find the post and differentiate between it and any other tables on the page. In this case, there is only one. We will have to look at other postings to see if this is always the case.

Within an *HTML* `<table>` element, there are rows identified by `<tr>` elements. Within a row, there are cells, each identified by a `<td>` (table data) or `<th>` (for the table's header row(s)) elements/nodes. Let's look at the table in our post in *R*. We can fetch the table node with

```
tbls = getNodeSet(doc, "//table[@class = 'post ']")
```

(Note the space at the end of the string "post ".) This uses the *XPath* language to search for all nodes within the document (i.e., all descendants of the root node) that are named `<table>` and which also have an attribute *class* with a value of `post`. The result is a list with 0 or more matching elements. We should check whether there is exactly one element using

```
length(tbls)
```

Indeed, we see there is one, and only one, element matching this *XPath* query in this document.

We can look at the sub-elements within the `<table>`. We should see `<tr>` nodes. We can use names() to get these:

```
table = tbls[[1]]
names(table)
```

```
   tr    tr
 "tr"  "tr"
```

The first row contains the image and poster information and then the post itself. We can access this first `<tr>` element with `row1 = table[[1]]`. The post is in the second column/cell of this row. However, there are other text elements in the row along with the `<td>` elements. We can see the element names of the row's children elements with

```
names(row1)
```

```
     td      text        td      text   comment       text
   "td"    "text"      "td"    "text" "comment"    "text"
```

The post itself is in the third element, i.e, the second `<td>` element. We can access this element and query its attributes with

```
xmlAttrs(row1[[3]])
```

```
     class
"postbox"
```

We see that this has a class named `postbox`. So rather than relying on the presence or absence of text nodes and numerical indices that might change, we could get the post node more robustly by looking for the element named `<td>` that has a class attribute value of `postbox`. We can find this with getNodeSet() and the *XPath* expression `//td[@class = 'postbox']`. Hopefully there will be only one such node in the document, and indeed there is. If there were more than one, we would make the *XPath* expression more specific and restrictive to identify the node we want. However, we can retrieve the node with

```
postNodes = getNodeSet(doc, "//td[@class = 'postbox']")[[1]]
```

The second row of the post table has some extra information that doesn't interest us, but you should verify that.

Our goal above was to search the document to identify the node(s) we wanted and then to create an *XPath* expression to retrieve that node programmatically. To this end, we can explore the structure and content of the document in *R* (or *Python*) or in the Web browser. We typically start in the Web browser and view the *HTML* source. In the source page, we search for text that we see in the rendered page. We then look at the surrounding *HTML* elements and try to identify unique aspects to these, typically a *class* attribute, but also a particular combination of *HTML* elements such as a `<div>` containing a `<table>` with the first element of the first row being a `<img>` element.

Now that we have the `<td>` node containing the content of the post, we can extract its content. We could very simply get all of the text with the xmlValue() function. This combines the content of all of the text nodes within this node and its children and their children into a single string:

```
txt = xmlValue(postNodes)
```

If we just want the words, this is all we need to do. However, we lose context by doing this, which is harder to recover with just the text. For example, in this post, we have a `Requirements` and another `Plus Requirements` section and the content within these are in bullet lists. We would like to be able to associate the text in these sections with necessary requirements and optional requirements. We would also like to be able to process each element of each list separately from the other list elements. If we discard this markup, we lose this context.

By examining the *HTML* document, we see these sections are encoded in *HTML* elements as

```
<p><strong>Requirements:</strong></p>
<ul>
<li>BS/BA in quantitative field or equivalentMinimum 3 years
 professional experience</li>
<li>Experience with large scale growth and retention systems</li>
<li>Experience with the Hadoop technology stack</li>
<li>Experience with R, Matlab, or other scientific computing
language</li>
<li>Experience with SQL databases Familiarity with the modern web
technology stack</li>
<li>Familiarity with software engineering practice</li>
<li>Strong communication skills</li>
<li>A natural sense of curiosity and drive to experiment</li>
</ul>
```

and

```
<p><strong>Plus Requirements:</strong></p>
<ul>
<li>Graduate degree in quantitative field</li>
 .....
```

We have a `<p>` element with the section title within a `<strong>` element. After this `<p>`, we have a `<ul>` element identifying an unordered list (i.e., no numbers). There may be a text node in between these. However, we can find these `<ul>` nodes with *XPath* by searching from the `<td>` post node through all its descendants looking for a paragraph node (`<p>`) that has a child element named `<strong>`. When we find this, we want the next `<ul>` "beside," i.e., a sibling of, the `<p>` node. We express this with the following *XPath* query and evaluate it with getNodeSet():

```
uls = getNodeSet(postNodes,
                 ".//p[strong]/following-sibling::ul")
```

Again, we check if there are 2 of these as we expect, and that they contain the text we expect.

We can process each of these `<ul>` elements and extract the text for each item (`<li>` for list item). Using xpathSApply(), we do this via

```
items = xpathSApply(uls[[1]], ".//li", xmlValue)
```

for the first list (or we can use the readHTMLList() function in the **XML** package). We can process both lists with

```
items = lapply(uls, function(node)
                    xpathSApply(node, ".//li", xmlValue))
```

Before we process the words within each of these items, we want to find the names for these lists, i.e., the titles Requirements and Plus Requirements. We can use getSibling() with each of the nodes in the uls variable to find the previous sibling. Unfortunately, this is a text element containing white space. It is the sibling of this text element we want, i.e.,

```
getSibling(getSibling(uls[[1]], TRUE), TRUE)
```

The value TRUE in this call instructs the getSibling() function to get the preceding sibling, not the following sibling.

Instead of navigating text or any other irrelevant elements with calls to getSibling(), we can use *XPath* again. Given a `<ul>` node, we can find the preceding paragraph containing the `<strong>` node with

```
tmp = getNodeSet(uls[[1]],
                 "./preceding-sibling::p[strong]/strong")
xmlValue(tmp[[1]])
```

We can do this for each of the nodes in uls to get the title/label for the list.

As an alternative approach, we can use the previous *XPath* query we used to find the `<ul>` but get the corresponding `<p>` nodes

```
titles = xpathSApply(doc,
           "//p[strong and following-sibling::ul]/strong",
           xmlValue)
```

Which of these different approaches is better depends on the structure of other documents and how this format is likely to be the same or different across other posts. Indeed, our calculations seem to be quite specific to the particular content and format of this *HTML* document and this post. Do we think it will be similar in other posts? How robust will this approach be? Before we spend too much time refining this, we should look at other posts. Furthermore, it is not clear that all the Kaggle posts have this information or that we will be able to reliably extract it. As we explore more posts, we may decide to get just the collection of words and treat the entire post as free-form text. This doesn't mean we won't be able to extract more structured metadata from posts on other Web sites, however.

Now that we have the text from the semi-structured lists, we want the text and words from the remainder of the post. In order to extract these, we want to process the other elements in the post node. We should avoid processing the `<ul>` and `<li>` nodes we have already processed a second time. We do this by finding all of the `<p>` nodes that are not being used as titles, i.e., that have a `<strong>` element. Again, we can express this succinctly with *XPath*:

```
paras = getNodeSet(postNodes[[1]], ".//p[not(strong)]")
```

This will ignore the `<ul>` and `<li>` nodes. We have to be careful that there aren't any `<p>` nodes within any of the ones that we want to ignore.

12.3.1 Processing the Text

At least for this one Kaggle post, we now have all the text in the different sections. We have the `<p>` nodes containing the free-form text in `paras` and the text from the lists in `items`. We want the words, not the sentences or paragraphs. We can easily get these using the strsplit() function, separating the words by white space or punctuation characters.

```
itemWords = lapply(items, strsplit, "[[:space:][:punct:]]+")
```

Note that the second argument in each call to strsplit() is a regular expression (`[[:space:][:punct:]]+`). This splits the words by space and punctuation. For the `paras`, we get the text and split it into words with

```
words = lapply(paras, function(p)
                        strsplit(xmlValue(p),
                                 "[[:space:][:punct:]]+"))
```

Note that by splitting the text using punctuation as a separator, we remove the opportunity to use this when post-processing the text. For example, we might try to identify the location for the job in the post by looking for text of the form `City, State`. This would have matched `Oakland, California`. However, if we just have the 2 words Oakland and California sequentially in a vector, we do not know if they were separated by a period, comma, or space.

We can combine the different sets of words from the different parts of the post into a list to represent a post:

```
post = list(freeForm = unlist(words),
            requirements = itemWords[[1]],
            optionalSkills = itemWords[[2]])
```

Note that we have given new names to the vectors of words in the 2 Requirement sections. These titles may not be the same in each post and we'll have to manage this and make the correct associations. We are using more informative names in our *R* object to identify the nature of the content.

We are not interested in all of the words in the post. We can discard pronouns, conjunctions, and prepositions such as `a`, `and`, `this`, `that`, `it`, `to` and verbs such as `is`, `are`, `was` and so on. We often call these *stop words*. We can create our own list of these words.

However, collections of stop words are available on the Web. Choosing which collection to use is almost as complex as getting the collections. Of course, we can also combine them. For now, we'll just assume we have the collection of stop words in a character vector in *R*, say the variable `StopWords`. We will see how to fetch them in an exercise at the end of the chapter (Q.23 (page 504)). For now, we can load the object from *StopWords.rda*:

```
load(url("http://rdatasciencecases.github.io/StopWords.rda"))
```

Once we have the vector of stop words, we can discard the words in our post that are also in the stop words. We use setdiff() applied to each *character* vector. We can write a general function that operates on the vector or a *list* and that does the appropriate thing. Our function is

```
removeStopWords =
function(x, stopWords = StopWords)
{
   if(is.character(x))
      setdiff(x, stopWords)
   else if(is.list(x))
      lapply(x, removeStopWords, stopWords)
   else
      x
}
```

We can apply this to our *post list* with

```
post = lapply(post, removeStopWords)
```

or more directly with `removeStopWords(post)`. This preserves the structure of our *list* and its sub-elements but processes only the *character* sub-elements. Note that the function is recursive, i.e., it calls itself.

12.3.2 Generalizing to Other Posts

We have seen all the general steps to process a post. However, we have processed only one post and other posts will have different characteristics, e.g., no lists, different sections. The posts can differ very significantly, even within a single site. However, within Kaggle, there are similarities across posts and many of them are semi-structured text with lists. Let's look at the post at http://www.kaggle.com/forums/t/5192/data-mining-architect-toronto-on, which is also shown in Figure 12.6.

Let's use the same code we used above to try to extract the information from this page. As we do so, we'll probably have to adapt the code. While we are doing this, we will try to generalize the code so that it can work with both pages, and hopefully others. We will build functions so that we don't have to repeat the code. This is a good approach to programming. We have developed code outside of a function for a particular post. Now we are generalizing that code and turning it into functions to handle different content. This is better than starting out writing very general functions without really knowing what the inputs will look like. As we develop the function, we can add parameters to control how it behaves and also program it to adapt to differently structured content.

Our function to process a Kaggle post will take the URL of the page for that job. It parses the *HTML* document and then finds the content of the post in the `<td>` element with a class *postbox*. Then we will process the sections and the free-form text and remove

Data Mining Architect, Toronto, ON.

Start Watching

0 votes

Jr. Data Mining Architect

Duration: 6 month contract position, with subsequent extensions based on performance

Application submission: Please submit resume to hr@objectifi.com

Objectifi Description

Objectifi prides itself on being more than just a Technology and Management Consulting company. We value our relationships with our clients and our teammates primarily and back that up with the perfect technical solution. Our clients are leading North American Fortune 500 clients across diverse industries. Our workplace is open and collaborative, and a great environment to design leading solutions in. Our offices are located centrally in Toronto, with easy access to highways and public transit.

Position Overview

The Data Mining Architect will be a key member on the Objectifi team. The selected candidate will have the opportunity to work directly with and learn from our Professional Services team, and actively on client engagements. The role is also client facing, allowing the candidate to directly work with our Fortune 500 clients.

The selected candidate will be a member of our client engagement teams, providing insights using data mining. The scope of analysis will include large, complex, data sets, for complex marketing and business challenges. Recommendations from this analysis will be included in the solution for the engagement, with the candidate ensuring successful implementation of recommendations.

This position requires approximately 10%-25% (flexible timing) travel to client sites within North America.

Education

- A degree from a Canadian or U.S. University in Computer Science or Computer Engineering
- Short term travel within North America is required. Candidate must be able to qualify for TN Visa

Experience

- 2+ years of data modeling and data analytics experience
- 2+ years of experience with data management and analysis software tools (e.g. R, SQL, SAS)
- 1-2 years of experience in statistics/informatics
- 2+ years of software development experience
- 2+ years of advanced SQL experience (including Stored Procedures, complex queries)
- Strong theoretical understanding data mining and statistical concepts and technologies
- Experience with CRM and Reporting tools is a plus

Non-Technical Skills

- Excellent communicator, including client-facing skills
- Above average attention to detail
- Team player/works well with peers
- Strong presentation skills
- Strong documentation skills

Figure 12.6: Semi-Structured Job Post on Kaggle. *This shows another Kaggle job posting. Again, much of the content is free-form text within different sections that have a title. There are 2 lists with metadata about education and experience qualifications necessary for the job.*

the stop words for each of the different groups of words. We will need to know the stop words, so we add this as a parameter to our function. We give it a default of StopWords. This allows us to override the set of stop words, but we do not have to specify them.

While we want our function to extract all of the information from a post, it is often useful to be able to pre- or post- process the *HTML* document to extract additional information. As a result, it can be useful to be able to download and parse the document once outside of the function and then pass it to our function. So, we also allow the caller to specify a previously parsed *HTML* document rather than the URL. We can do this in different ways, but it is perhaps simplest to have this as a separate parameter for a function, with a default value that parses the URL specified by the caller. The skeleton of our function might look like

```
readKagglePost =
function(u, stopWords = StopWords, doc = htmlParse(u))
{

}
```

If the caller specifies a URL, *doc* will be assigned the parsed document via its default value. However, if the caller specifies the parsed document via the *doc* parameter, she doesn't need to specify the URL and the function does not need to retrieve and repair the document.

As an alternative approach, we could have checked the class of *u* and if it was not a previously parsed *HTML* document, then call htmlParse(). We could implement this via

```
readKagglePost =
function(u, stopWords = StopWords)
{
    if(!is(u, "HTMLInternalDocument"))
       u = htmlParse(u)
}
```

If retrieving the document uses *HTTPS* or requires some additional information such as a password or cookie, we will have to use a different approach than htmlParse() to fetch the document before we parse its content. We will use getURLContent() or getForm() from the **RCurl** [10] package to retrieve the document and then parse it with htmlParse(¬ getURLContent(u), asText = TRUE). This makes the first version of our function more flexible as the caller can control how to fetch the document, perhaps reusing different connections to the Web server, etc., and pass the parsed document to readKagglePost().

Given the parsed document, we start our function by obtaining the node containing the post:

```
post = getNodeSet(doc, "//td[@class = 'postbox']")
```

If that doesn't exist, we want to raise an error, so we can add code such as

```
if(length(postNodes) == 0)
   stop("cannot find <td class='postbox'> element in HTML")
else if(length(postNodes) > 1)
   stop("found more than one <td class='postbox'>
         elements in HTML")
```

The next step is to find all our special sections, i.e., those that follow a

```
<p><strong>....</strong><p>
```

Exploring Data Science Jobs with Web Scraping and Text Mining 473

In the first post we explored, the next element was a `<ul>`. In the second post, these sections are just a sequence of `<p>` nodes. Each `<p>` node starts with a – to indicate a list item. We can try to develop some heuristics for this, but first we have to differentiate between having `<p>` and `<ul>` nodes.

At this point, we see many dissimilarities and special cases in just 2 posts. As we suggested earlier, it may not be worth the effort to try to extract the metadata from lists in the reasonably free-form Kaggle posts. Instead, we will just extract all the words. This will be our generic Kaggle post-processing function. We can do this with the following definition:

```
readKagglePost =
function(u, stopWords = StopWords, doc = htmlParse(u))
{
    post = getNodeSet(doc, "//td[@class = 'postbox']")

    if(length(postNodes) == 0)
       stop("cannot find <td class='postbox'> element in HTML")
    else if(length(postNodes) > 1)
       stop("found more than one <td class='postbox'>
           elements in HTML")

    txt = xmlValue(post[[1]])
    strsplit(txt, "[[:space:][:punct:]]+")[[1]]
}
```

12.3.3 Scraping the Kaggle Post List

We are now ready to scrape all of the posts from the Kaggle job list. This will be a little simpler than for other sites we look at next. This is because a) there are currently about 750 posts and b) we don't have to specify a particular query to identify the posts of interest, e.g., a particular job title or location. Instead, we will read all of the posts so that we can examine the distribution of words and any of the metadata we extract.

We can look at the Kaggle job listing page and see how many pages there are in total. These are listed as 1 2 3 4 5 6 7 8 9 10 .. 39 40. The final number is the last page. We can create the links to the second and subsequent pages with

```
baseURL = "https://www.kaggle.com/forums/f/145/data-science-jobs"
sprintf("%s?page=%d", baseURL, 2:40)
```

We are of course hard-coding the number of pages and not determining this programmatically. That's probably reasonable since we don't have to do this for sub-lists, i.e., by different search term or job classification. However, it will limit us when want to get more recent posts in the future.

We can loop over these URLs giving the successive lists of job postings. For one of these pages, we find the links to each post in that page. We then have the URLs for all of the job posts on the Kaggle forum. We use readKagglePost() to read each of these posts. We can then write our function to read all of the posts as something like

```
readAllKagglePosts =
function(numPages, baseURL =
         "https://www.kaggle.com/forums/f/145/data-science-jobs")
{
```

```
    pageURLs = c(baseURL,
                 sprintf("%s?page=%d", baseURL, 2:numPages))

    postLinks = unlist(lapply(pageURLs, getPostLinks))
    lapply(postLinks, readKagglePost)
}
```

We need to define the getPostLinks() function to get the URLs for each post on a specific page of job listings. We look at one of these pages to determine how the links to the posts are formatted. We can retrieve one of these pages and parse it with

```
baseURL = "https://www.kaggle.com/forums/f/145/data-science-jobs"
txt = getForm(baseURL, page = "2")
doc = htmlParse(txt, asText = TRUE)
```

or we can use one of the URLs we created above.

These pages contain *HTML* elements something like

```
<div class="topiclist-topic">
 <div class="topiclist-topic-name name-col">
  <h3>
   <a href="/forums/t/9663/graduate-data-scientist-edinburgh-uk"
      title="Go to topic 'Graduate Data Scientist - Edinburgh, UK'">
                      Graduate Data Scientist - Edinburgh, UK
   </a>
  </h3>
 </div>
</div>
```

We can explore different *XPath* queries to find the (up to) 20 posts per page. We might use the specific

```
getNodeSet(doc, "//a[starts-with(@href, '/forums/t')]/@href")
```

However, that yields 40 matching links. This is exactly twice the number we expected. This is because we are also matching the links in the third column with the title Last Post. We don't want these, so we have to be more specific in our *XPath* query. We could get the first of these links in each row of the table. However, this is not structured as a table, so this approach won't work. Instead, let's look for the these links directly within a <h3> element:

```
getNodeSet(doc, "//h3/a[starts-with(@href, '/forums/t')]/@href")
```

This indeed yields the 20 links we want. We could also write this as

```
getNodeSet(doc, "//h3/a/@href[starts-with(., '/forums/t')]")
```

This is very marginally shorter!

We can define our function to get the links within a page as

```
getPostLinks =
function(link)
{
   txt = getURLContent(link, followlocation = TRUE)
   doc = htmlParse(txt, asText = TRUE)
   getNodeSet(doc, "//h3/a[starts-with(@href, '/forums/t')]
                          /@href")
}
```

We now have all the pieces of our function

12.4 Scraping `CyberCoders.com`

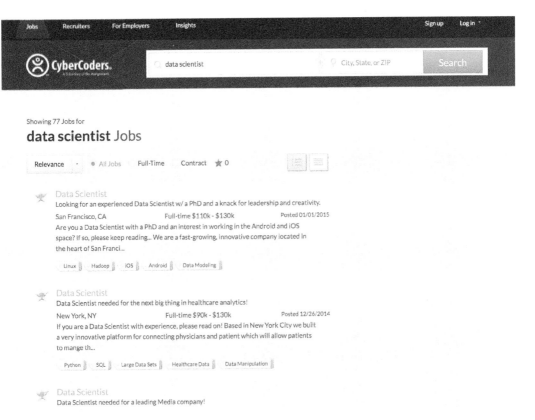

Figure 12.7: Search Results for Data Scientist Jobs on CyberCoders.com. *We enter the search term in the text field at the top of the Web page. The results are displayed below. Note that in addition to the link to each particular job, the results page also shows metadata about each job, e.g., salary range, location, and a list of required skills. If this is all the information we wanted, we would not have to scrape the actual postings.*

Like many of the job posting sites, `cybercoders.com` allows us to specify a search query to find jobs of interest. This is shown Figure 12.7, along with the page of results and links to the matching jobs. Unlike Kaggle, the list of matching job postings contains a lot of important information about each job, e.g., the title, the location, the salary range, and a list of necessary skills. In addition to extracting the content of the post, we also want to fetch this meta-information, if it is easier to extract here, rather than from the full content of each posting.

When we click on a post in the results page, we also see that it has more structure than Kaggle's posts. An example is shown in Figure 12.8. As with Kaggle posts, most of the post is free-form text with some lists. However, because these posts are created by CyberCoders itself, they are much more uniform. This should make it significantly easier for us to extract data across different posts on this site.

Importantly, the posts on `cybercoders.com` contain the location and salary in the top right of the page and an itemized list of `Preferred Skills` at the bottom left of

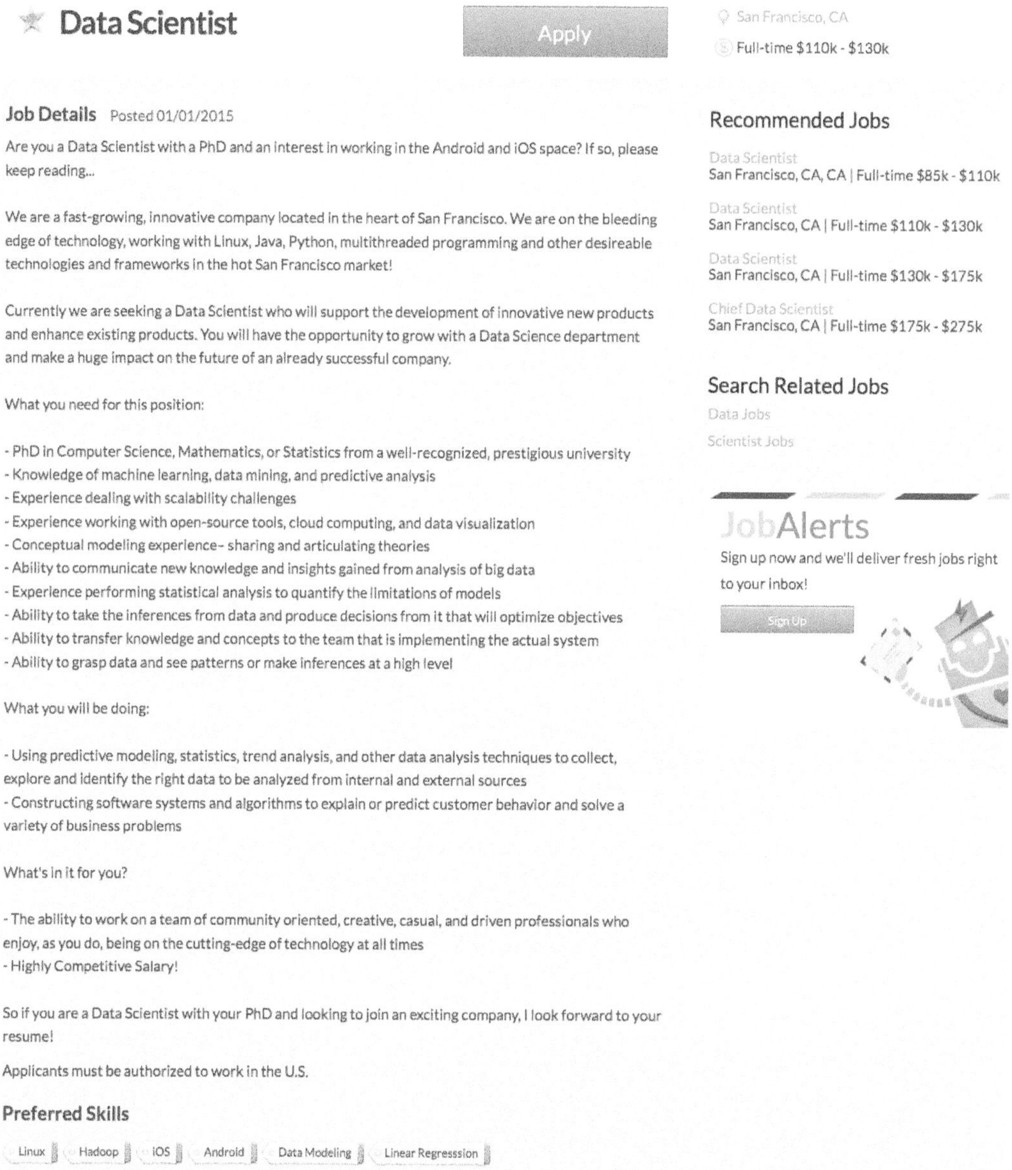

Figure 12.8: Sample Job Post on CyberCoders.com. *This job posting, like others on* cybercoders.com, *has several paragraphs of free-form text, and also some itemized lists including one listing the skills necessary for the position. Additionally, the post has the location and salary separately in the top-right corner. It also displays the preferred skills as a list of phrases.*

the post. Let's focus on getting these 3 pieces of information first. We can get the free-form text in much the same way as we did for the Kaggle posts.

Again, we start by looking at the *HTML* source of two or more posts on the cybercoders.com site. The source for the post shown in Figure 12.8 includes the following *HTML* elements

```
<div class="job-info-main">
 <div class="location">
  <span class="pin"></span>
  <a href="/jobs/san-francisco-ca-area-jobs/">San Francisco, CA</a>
 </div>

 <div class="wage">
  <span class="money"></span>
  <span class="text">Full-time</span> $110k - $130k</div>
 </div>
</div>
```

This is the information about the location and salary that we want. How did we find this? We looked at the rendered page in the Web browser and then searched in the source page for the salary, i.e., $110k. This brought us directly to these *HTML* elements.

Given the parsed *HTML* document for the job post, we can get the job location and salary information using an *XPath* query. We are looking for the *<div>* node with a *class* attribute with the value job-info-main. We do this with

```
node = getNodeSet(cydoc, "//div[@class = 'job-info-main']")[[1]]
```

Note that we only kept the first matching node. There is a second node in the *HTML* document which is identical. However, it is not visible on the page as it is within a *<div>* element that is made invisible by the *CSS style* attribute with the value display: none. We ignore this second node.

We can then get the location information from node with xmlValue(node[[2]]). Why is the location in the second element, not the first? The first element is just blank text between the end of the opening *<div>* and the start of the location *<div>*. Instead of having to know about these "hidden" text elements consisting of white space, we could either make our *XPath* query return the location and wage *<div>* elements or, alternatively, we can access the resulting *<div>* elements directly in *R* via the node variable.

Given either of the *<div>* elements, we get the content of interest with xmlValue(). Therefore, we can use *XPath* to get this information with

```
info = xpathSApply(cydoc, "//div[@class='job-info-main'][1]/div",
                   xmlValue)
```

The xpathSApply() function is a convenience function equivalent to lapply(getNodeSet(doc, query), fun). Note that we are not using node here, and so restrict the query to the first of the job-info-main *<div>* nodes. Alternatively, we can retrieve the *<div>* elements from node via node["div"] to get all child elements of node named div. We can then apply xmlValue() to each of those. That is,

```
sapply(node["div"], xmlValue)
```

gives the same result as the xpathSApply() call above.

12.4.1 Getting the Skill List from a Job Post

Now that we have the location and salary from a post, we can focus on extracting the entries in the Preferred Skills list at the bottom of the post in Figure 12.8. Again, we look at the *HTML* source and search for one of the terms. This term may also be in the regular text of the post, so we have to make certain we find the one in the skills list. The relevant *HTML* elements starts with

```
<div class="skills-section">
 <h4>Preferred Skills</h4>
 <div class="skills">
   <ul class="skill-list">
     <li class="skill-item">
      <a href="/jobs/linux-skills/">
        <span class="left"></span>
        <span class="skill-name">Linux</span>
        <span class="right-off"></span>
      </a>
     </li>
     <li class="skill-item">
      <a href="/jobs/hadoop-skills/">
        <span class="left"></span>
        <span class="skill-name">Hadoop</span>
        <span class="right-off"></span>
      </a>
     </li>
```

The content of each skill is within a *<li>* (list item) element with a *class* attribute with value skill-item. The actual text is within a *<span>* element with skill-name as the *class* attribute. Again, we can fetch all of these elements directly with a single *XPath* expression:

```
lis = getNodeSet(cydoc, "//div[@class = 'skills-section']//
                        li[@class = 'skill-item']//
                        span[@class = 'skill-name']")
```

We get the text for all of the skills with sapply(lis, xmlValue) (or just use xpath-SApply() instead of getNodeSet() and sapply().)

We can also get the date on which the job was posted from the page. This is in the *HTML* content

```
<div class="job-details">
  <div class="jobDetailsHeader">
    <h4>Job Details</h4>
  </div>
  <div class="posted">
    <span class="text">Posted</span> 01/01/2015
  </div>
```

Again, a simple *XPath* query can fetch this value:

```
xmlValue(getNodeSet(doc,
                   "//div[@class = 'job-details']//
                      div[@class='posted']/
                       span/following-sibling::text()")[[1]],
          trim = TRUE)
```

The `following-sibling::text()` expression matches all of the text nodes at the same level as the `<span>` element but after it. We get the first of these from the *R* list.

Both of the *XPath* queries we have used above specify a very specific path. This ensures we don't match other nodes. However, we may be able to use simpler queries if they uniquely match the nodes. For example, when finding the skills, we may be able to use just `//¬span[@class = 'skill-name']` rather than qualifying it with the `<li>` and `<div>` ancestors.

We can combine all of these different steps into a function (see cy.readPost() below) to read a post. Rather than have one function that performs all of the steps, we will make a separate function for each step. This makes each of the functions easier to read, test, maintain, and adjust if the format of the pages changes. We'll define one function which calls these three functions. Again, we'll use the same basic signature (i.e., parameters and their order) as we did for the Kaggle function readKagglePost(). We can define this job-level function as

```
cy.readPost =
function(u, stopWords = StopWords, doc = htmlParse(u))
{
  ans = list(words = cy.getFreeFormWords(doc, stopWords),
             datePosted = cy.getDatePosted(doc),
             skills = cy.getSkillList(doc))
  o = cy.getLocationSalary(doc)
  ans[names(o)] = o

  ans
}
```

We collect the vector of words, date posted, and the skills into a *list*. Then we add the location and salary information as separate elements to this *list*.

Note that we have used the prefix "`cy.`" for each of the three sub-functions. This makes these specific to the `cybercoders.com` site and helps identify the functions as being for this particular site. We might have alternatively collected all of these functions into a list and assigned it to a variable, say, cyFuns, e.g.,

```
cyFuns =
  list(readPost = function(u, stopWords = StopWords,
                           doc = htmlParse(u)) {...},
       datePosted = function(doc) {...},
       skills = function(doc) { ...}
```

Here we avoid assigning the functions individually to top-level *R* variables by defining the functions inline when defining the *list* of functions. We can then loop over these functions to create the initial list ans in cy.readPost().

We also added a *doc* parameter to the cy.readPost() function. The default value for this is to retrieve and parse the *HTML* document from the URL *u*. We allow the caller to specify this for two reasons. Firstly, she might have already parsed this document and so does not want to have to repeat this step, especially if she is off-line. Secondly, we may need to use a more flexible *HTTP* request mechanism to retrieve the page than htmlParse() provides. Having this as a parameter that the caller can provide makes cy.readPost() more flexible.

Our cy.readPost() function refers to cy.getFreeFormWords() and cy.getSkillList(). We have to implement these two. The cy.getFreeFormWords() function has to find the part of the *HTML* document that contains the visible text in the rendered document that describes the job. We want to omit the information about the location, salary, date posted, and skill list, since we have collected these separately from the free-form text. We could collect all of the remaining words together. However, we may want to identify any lists or sections that provide more context. For example, many of the posts have a `What you will be doing` section, and a `What you need for this position` section. We'll explore how to find the different sections of a post. Again, we have to look at two or more postings and try to identify a pattern.

The free-form text of the post is inside the

```
<div class="job-details">
```

node, but so too is the location information. We can find the node containing the free-form text using *XPath* by looking for the node that contains the start of the text, e.g.,

```
a = getNodeSet(cydoc,
         "//*[starts-with(., 'Are you a Data Scientist')]")[[1]]
```

Unfortunately, the result is empty. The issue is that the text starts with white space. We could use `contains()` rather than `starts-with()` in the *XPath* expression. Alternatively, we can remove (or trim) the white space using the `normalize-space()` function:

```
a = getNodeSet(cydoc,
         "//*[normalize-space(., 'Are you a Data Scientist')]")[[1]]
```

Now that we have the node, we can traverse its ancestors with either repeated calls to xmlParent() or a single call to xmlAncestors(). We can then try to identify the node that includes all of the text of the post, but no more. Then we can examine these to identify the pattern to find that node or its sub-nodes of interest in other posts.

For this CyberCoders document, the free-form text is inside several separate sibling `<div>` nodes within the `job-details` `<div>`. This contains several top-level text nodes and `<div>` nodes that we do not want to include. We can see this with

```
names(xmlParent(a))
   text    div   text    div   text      p    text      p    text
 "text"  "div" "text"  "div" "text"    "p" "text"    "p" "text"
```

In other documents, we have numerous `<h4>` elements and comments. All of these top-level text nodes are white space, and any comment nodes are of no interest to us. Any `<h4>` elements are the start of the sections such as `What You Will Be Doing`. We can look at the *XML* attributes of each of the `<div>` elements with

```
sapply(xmlParent(a)["div"], xmlAttrs)
$div
            class
"jobDetailsHeader"

$div
   class
"posted"
```

This document does not have <h4> elements identifying the different sections. These section titles are merely included in the text. In other documents, they are in <h4> elements and have attributes

```
$div
        class     data-section
"section-data"       "1"

$div
                           class          data-section
"section-data section-data-title"            "5"

$div
                           class          data-section
"section-data section-data-title"            "8"

$div
        class     data-section
"section-data"       "9"
```

All of these <div> elements in which we are interested have a *data-section* attribute and also contain the name section-data in the *class* attribute. So we could retrieve these with

```
getNodeSet(doc, "//div[@class='job-details']/
               div[@data-section]")
```

We can then process the text within each of these, for example, using xmlValue() on the entire *HTML* sub-tree.

If the document has no <h4> elements, we could use some heuristic approach to finding the different sections. However, we won't bother. For those documents that do use <h4> elements to identify and separate sections, we can process these separately, if we desire. We could find these nodes with

```
details = getNodeSet(doc, "//div[@class='job-details']")[[1]]
xpathSApply(details, "./h4", getSibling)
```

Note that the second *XPath* query here is searching from the node <div class='job-details'> down that sub-tree. It is not searching the entire document. This is why we preceded the h4 element name with ./, anchoring the search from the node assigned to details. Also note that each <h4> element is followed by a <div> node which contains the content for the section. This is why we use getSibling() to obtain that <div> node.

Once we have these nodes, we can process them as *HTML* lists, or lists with free-form text. These are often also problematic as they may not be <ul> or <ol> nodes with <li> elements identifying the list elements. Instead, they are formatted to look like lists but are actually separated by
 (line breaks). We can attempt to identify and process these, but we have to decide whether it is worth the effort, and if we want this information differently from just words in the post.

After determining where the text is and how it is structured, let's write our cy.getFreeFormWords() function as

```
cy.getFreeFormWords =
function(doc, stopWords = StopWords)
{
  words = xpathApply(doc, "//div[@class='job-details']/
                          div[@data-section]",
                    function(x)
                        asWords(xmlValue(x)))
}
```

This fetches the lists of free-form text in the *HTML* document and then decomposes the text into the words in each element, using spaces and punctuation characters to separate them. We could improve this to add the section (<h4>) titles as names of the elements in the list, when they are available. However, this suffices for our current purposes of looking at the words.

Note that we have used a function asWords() to process the text of each node. We leave this as an exercise for the reader to implement in Q.2 (page 482).

Q.1 Implement the cy.getSkillList() and cy.getLocationSalary() functions. Use the code we explored to extract the skill sets and salary and location from the parsed *HTML* document.

Q.2 Implement the function asWords(). It accepts a *character* vector and separates each element into words, using white space and appropriate punctuation characters as delimiters. It should also remove stop words, and allow the caller to optionally specify the set of stop words. It should also optionally stem each of the words. Stemming is discussed in Chapter 3, specifically Section 3.5.3.

We now have all our functions and we can test them on some of the CyberCoder posts, e.g.,

```
u = "http://www.cybercoders.com/data-scientist-job-140783"
ans1 = cy.readPost(u)
names(ans1)
```

```
[1] "words"      "datePosted" "skills"     "location"   "salary"
```

```
ans1$salary
```

```
[1] "Full-time $90k - $130k"
```

We need to try another job post, and then another, and so on to test the cy.readPost() function and discover any formatting differences. We can find other posts manually on the Web site. However, we can also identify them programmatically, which we'll discuss next.

12.4.2 Finding the Links to Job Postings in the Search Results

Now that we can read an individual post with reasonable generality, we need to be able to programmatically find the posts. Again, recall that we start with the first page returned by a search based on a query string. That page contains links to the first set of results. We have to find out how to programmatically make the initial query to get the first page of results, and then determine how to find all the links to the job postings on that page.

One way to determine how to submit the search query programmatically is to perform a search in the browser on the CyberCoders page and examine the URL to which that brings us. If we type "Data Scientist" into the query textfield, the browser shows the page http://www.cybercoders.com/search/?searchterms="Data+Scientist"&searchlocation=&newsearch=true&sorttype= This indicates that the *HTML* form used *HTTP*'s **GET** operation to submit the request. It also identifies the named parameters as searchterms, searchlocation, newsearch, and sorttype, and shows their values (after each = character). We could also get this information from the Web browser's developer tools by examining the network requests. This may be important if a) the *HTTP* operation is a **POST** rather than a **GET** request, or b) if there are hidden parameters or cookies that are not shown in the request. For our purposes, we have sufficient information to make the request in *R* based on the URL shown in the browser's navigation field.

Rather than using the URL above directly, we can decompose it into a URL and the form parameters and values. Then we can reuse this easily to make different queries. We can use *getFormParams()* to do this or do it manually:

```
u = "http://www.cybercoders.com/search/?searchterms=Data+Science\
&searchlocation=&newsearch=true&sorttype="
p = getFormParams(u)
```

```
   searchterms  searchlocation      newsearch         sorttype
"Data+Science"            ""         "true"               ""
```

We can then submit the query with *getForm()* via

```
txt = getForm("http://www.cybercoders.com/search/",
          searchterms = '"Data Scientist"',
          searchlocation = "",  newsearch = "true",
          sorttype = "")
```

(or use the *.params* parameter and p.) The resulting value in txt is a character vector with a single element containing the *HTML* document as it would be received by the Web browser. Instead of rendering it, we will retrieve the links to the job postings it contains. To do this, we first parse the *HTML* document and then find the relevant <a> elements, i.e., those that have links to actual job postings. The first step is as simple as

```
doc = htmlParse(txt, asText = TRUE)
```

To find the links, we again explore the structure of the *HTML* document. We can search for a job posting that we can see in the Web browser and then look at the surrounding *HTML* elements. After some exploring, we can see that each job posting is contained in a <*div*> element with a *class* attribute value of job-listing-item. An example of this is

```
<div class="job-listing-item">
  <div class="job-status">
    <div class="status-item default" data-jobid="127835"></div>
  </div>
  <div class="job-details-container">
    <div class="job-title">
      <a href="/data-scientist-job-127835">Data Scientist</a>
    </div>
    <div class="details">
```

```
    ...
    </div>
</div>
```

We are looking for the *href* attribute value in the *a* element within the `<div>` with a *class* attribute with a value `job-title`. This is in a `<div>` with a *class* attribute with the value `job-details-container`. We can retrieve the *URL* of interest with

```
links = getNodeSet(doc, "//div[@class = 'job-title']/a/@href")
```

The first of these links looks something like `"/data-scientist-job-127835"`. This is a relative *URL*, relative to the base URL of our query, i.e., http://www.cybercoders.com/search/. We need to merge these links with this URL. We can use getRelativeURL() to do this, e.g.,

```
getRelativeURL("/data-scientist-job-127835",
               "http://www.cybercoders.com/search/")
```

```
[1] "http://www.cybercoders.com/data-scientist-job-127835"
```

We can retrieve and complete all of the *URLs* in a vectorized manner with the getNodeSet() call and

```
links = getRelativeURL(as.character(links),
                       "http://www.cybercoders.com/search/")
```

We are now ready to read the actual job posts contents from the first page of results with the command

```
posts = lapply(links, cy.readPost)
```

However, we want to combine all of these steps into a function. Given a page of search results, we want to find the links and then read each post. We can break these into two separate functions to make them easier to test and reuse. The first finds the links. We can define this function as

```
cy.getPostLinks =
function(doc, baseURL = "http://www.cybercoders.com/search/")
{
    if(is.character(doc))
        doc = htmlParse(doc)

    links = getNodeSet(doc, "//div[@class = 'job-title']/a/@href")
    getRelativeURL(as.character(links), baseURL)
}
```

To make this more flexible, we have allowed the caller to pass either the parsed *HTML* document (*doc*) or a string specifying its URL.

We can then define the second function, cy.readPagePosts(), to process all of the posts on a single page of search results as

```
cy.readPagePosts =
function(doc, links = cy.getPostLinks(doc, baseURL),
         baseURL = "http://www.cybercoders.com/search/")
```

```
{
    if(is.character(doc))
        doc = htmlParse(doc)
    lapply(links, cy.readPost)
}
```

Here, the default value for *links* performs half of the typical calculations. However, the caller can specify a different collection of links. This is useful for testing individual links or a subset of links.

As usual, we need to test this. We have the parsed version of the first page of results in doc. Therefore, we can get the posts with

```
posts = cy.readPagePosts(doc)
```

A quick way to check if these results are reasonable is to examine the collection of salary values in each post:

```
sapply(posts, `[[`, "salary")
```

Similarly, let's count the number of words in each post (after removing the stop words)

```
summary(sapply(posts, function(x) length(unlist(x$words))))
```

At least one of the posts appears to have no words. Let's investigate that.

```
which(sapply(posts, function(x) length(unlist(x$words))) == 0)
```

```
/data-scientist-job-127835
                16
```

We can examine this URL in a Web browser with `browseURL(links[16])`. When we see this in the Web browser, there are clearly many words and not all of them are stop words! There is a problem in our code. Let's look at the result of reading that post:

```
posts[[16]]

$words
list()

$datePosted
[1] "06/23/2014"

$skills
[1] "linux"        "Hadoop"           "iOS"
[4] "Android"      "Data Modeling"    "Linear Regresssion"

$location
[1] "San Jose, CA"

$salary
[1] "Full-time $110k - $130k"
```

This confirms that a) the words element is empty, and b) that the other entries appear to be reasonable and correct. An initial guess about why words is empty is that the *XPath* query to identify the text does not apply here and yielded no nodes. While this is the only post that yielded no words, we may have missed elements in other posts, so we need to check that somehow.

To examine the *HTML* source for post 16, we can again use either *R* or the Web browser. We can again look for the node starting with the text Are you a Data Scientist:

```
nodes = getNodeSet(doc,
                   "//*[starts-with(normalize-space(.),
                        'Are you a Data Scientist with a PhD')]")
```

This removes the leading white space and we also made the query more specific to eliminate matching a generic node at the end of the post. This node appears after the Posted information

```
<p>
  Are you a Data Scientist with a PhD and an ....
</p>
```

This is a paragraph node (<p>), not a <div> and there is no section-data *class* attribute. This is indeed why we missed this text. We can modify our cy.getFreeFormWords() function to include these <p> nodes. We may chose to do it conditionally when there are no regular matching <h4> and <div> nodes, or to match these in all cases. Our updated function is

```
cy.getFreeFormWords =
function(doc, stopWords = StopWords)
{
  nodes = getNodeSet(doc, "//div[@class='job-details']/
                          div[@data-section]")
  if(length(nodes) == 0)
    nodes = getNodeSet(doc, "//div[@class='job-details']//p")

  if(length(nodes) == 0)
    warning("did not find any nodes for the free form text in ",
            docName(doc))

  words = lapply(nodes,
                 function(x)
                   strsplit(xmlValue(x),
                            "[[:space:][:punct:]]+"))

  removeStopWords(words, stopWords)
}
```

Note that we switched from using xpathApply() to getNodeSet(). This is because we have to wait to process the nodes since we may find none and have to find the <p> nodes. We also added a warning to identify a page in which we did not find any free-form text. This will help to identify potential issues for the caller.

We can test this version of the function with w = cy.readPost(links[16])$¬ words. However, it may be more prudent to call cy.getFreeFormWords() directly since we have already retrieved the document. We did this since it took a few iterations to identify

Exploring Data Science Jobs with Web Scraping and Text Mining 487

the precise pattern and there is no point waiting to retrieve the document each time when we already have it. Also, it reduces the number of unnecessary requests to the server, saving its resources.

With the cy.getFreeFormWords() function now apparently working, we can reload the links and test cy.readPost() again, perhaps looking at the post with the next smallest number of words, and also verifying the other information returned in each post object.

Q.3 Test the cy.getFreeFormWords() function on several posts as suggested above.

12.4.3 Finding the Next Page of Job Post Search Results

We have seen how to make the initial query for a particular search term using getForm(), find the links to the job posts in the resulting page, and read the contents of each post. The last step in scraping the job post information from CyberCoders is to find the next page of results after the first one, and generally, the next one after that. Recall that we have the result of the initial query via getForm() in the variable txt. We'll parse this *HTML* document again and look for the link to the next page:

```
doc = htmlParse(txt, asText = TRUE)
```

Again, we can look at the rendered page in the Web browser to see how to identify the "next page" link. This is displayed as a right-facing arrow near the very bottom of the page. The link is http://www.cybercoders.com/search/?page=2&searchterms=Data%20Scientis&searchlocation=&newsearch=true&sorttype=. This is essentially the same URL as for our initial query, but it contains the additional page=2. We could add that as a parameter in our call to getForm(). We could just keep requesting the next page and when we got an error, we would know that we had reached the end. However, let's do this a little more elegantly. Instead, let's find the link to the next page and if there is none, we will stop.

How do we find the link to the next page? If we examine the *HTML* source, we find the following near the end:

```
<li class="lnk-next pager-item ">
  <a class="get-search-results next" rel="next" href="....">»</a>
</li>
```

So, we can look for an *HTML* node named a with a *rel* attribute containing the word next. There are actually two of these, but they have identical *href* attributes, which is what we want. One is at the top of the page and the other is at the bottom to make it convenient for the reader. The call

```
getNodeSet(doc, "//a[@rel='next']/@href")[[1]]
```

gives us the result ./?page=2&searchterms=Data%20Science&searchlocation=&newsearch=true&sorttype=

We'll write a function that takes the parsed *HTML* document as an input and returns the link to the next page of results. It needs to convert this link to an absolute URL rather than a relative link. We'll get the name of the base URL via docName() from the original parsed *HTML* document. However, there may be cases where this is not known, e.g., when we use getForm() to retrieve the initial document as we do here. Therefore, we'll let the caller specify the base URL but calculate it from the parsed *HTML* document if they don't. If that results in an NA value, we'll use the hard-coded URL to http://www.cybercoders.com/search. Our function is

```
cy.getNextPageLink =
function(doc, baseURL = docName(doc))
{
  if(is.na(baseURL))
     baseURL = "http://www.cybercoders.com/search/"
  link = getNodeSet(doc, "//a[@rel='next']/@href")
  if(length(link) == 0)
     return(character())

  getRelativeURL(link[[1]], baseURL)
}
```

Note that if there is no next page link, we know we are on the last page of results and so return an empty character vector. The caller will use this to know not to request any more pages.

As with any function, we need to test this. We can start with the first page in doc:

```
tmp = cy.getNextPageLink(doc,
                  "http://www.cybercoders.com/search/")
```

Q.4 Test the cy.getNextPageLink() function by calling it again with the result of the previous call to cy.getNextPageLink().

12.4.4 Putting It All Together

At this point, we have all of the pieces to retrieve all the job posts on cybercoders.com for a given search query. We now have to put it all together into a top-level function that we can call with a search string for the type of jobs in which we are interested. This function submits the initial query and then reads the posts from each result page. We might define the function as

```
cyberCoders =
function(query)
{
   txt = getForm("http://www.cybercoders.com/search/",
                searchterms = query,  searchlocation = "",
                newsearch = "true",  sorttype = "")
   doc = htmlParse(txt)

   posts = list()
   while(TRUE) {
      posts = c(posts, cy.readPagePosts(doc))
      nextPage = cy.getNextPageLink(doc)
      if(length(nextPage) == 0)
         break

      nextPage = getURLContent(nextPage)
      doc = htmlParse(nextPage, asText = TRUE)
   }
   invisible(posts)
}
```

We can test this function with

```
dataSciPosts = cyberCoders("Data Scientist")
```

This takes some time. For each post and each page query, we have to connect to the Web server and wait for the result. Each of these can take several seconds. Since this is a potentially infinite loop, we don't know if the function is stuck in the loop or waiting for a response from the Web server. Accordingly, we may want add a *verbose* parameter and output a message at the start of each iteration, i.e., page of results.

We can also speed up the requests very slightly and also be better "netizens." Instead of connecting to the same Web server for each request, we can create one connection and reuse this each time. We can create a connection using getCurlHandle(). We can then use this in each request, i.e., calls to getForm() and getURLContent(). We pass the connection via the *curl* parameter in these functions.

Currently, we only call getForm() and getURLContent() to get the search result pages, not the individual posts. We retrieve the job post documents in our cy.getPostLinks() function with a call to the htmlParse() function. That uses its own mechanism to make the HTTP request. We can change this to use getURLContent() and then parse the resulting document. Then we can use `curl = getCurlHandle()` in those calls to getURLContent().

Not only does using a curl handle maintain the connection to the Web server and reduce the overhead of re-connecting for each request, it also allows us to control the HTTP requests in much richer ways, such as supporting cookies and many options, and also handles secure HTTP, i.e., HTTPS.

Now that we have the functionality to get the posts from cybercoders.com, we can do so for any search term. Having retrieved the posts for the query "Data Scientist", we can look at the distribution of the corresponding most common skills with

```
tt = sort(table(unlist(lapply(cy.dataSciPosts, `[[`, "skills"))),
          decreasing = TRUE)
tt[tt >= 2]
```

We'll look at these in Section 12.8.

12.5 A Reusable Generic Framework for Arbitrary Sites

When we look at how we scraped the posts from both Kaggle and CyberCoders, we can see a shared pattern. There are 4 components to get the information from all of the job posts for a given query:

1. A mechanism to submit the search query and get the first page of results.

2. A means to extract the links to the individual job posts from a page of results.

3. A function to read the contents of an individual job post.

4. A way to find the next page of results, relative to the current page.

Each of these steps is quite different for each Web site, but the way we coordinate the different operations is the same across sites. Basically, we submit the query and get the first page of results. Then we apply step 2 to this page and then apply step 3 to each of the resulting links. We then find the next page and repeat steps 2, 3, and 4 on the new page of results. We can use a `while` loop to repeat the entire process until there is no "next" page of results. If we have a collection of Web site-specific functions for each of steps 2, 3, and 4, we can call those functions within our `while` loop.

We can implement this generic mechanism to scrape an arbitrary Web site of job posts with the following function:

```
searchJobs =
  # Given a search query, get the pages listing the jobs.
  # we loop over these pages and harvest the
  # individual jobs in each.
function(firstPage, getNextPage_f, getJobDescriptionLinks_f,
         getJobDescription_f = getJobDescription,
         max = NA, curl = getCurlHandle(followlocation = TRUE))
{
  curPage = firstPage
  jobs = list()

  pageNum = 1L
  while(is.na(max) || length(jobs) < max) {

     doc = htmlParse(curPage, asText = TRUE)

     links = getJobDescriptionLinks_f(doc, curl = curl)
     posts = structure(lapply(links,
                              function(l)
                                try(getJobDescription_f(
                                    getURLContent(l,
                                                  curl = curl),
                                    stem = FALSE,
                                    curl = curl))),
                       names = links)
     jobs = c(jobs, posts)

     curPage = getNextPage_f(doc, curl = curl)
     if(length(curPage) == 0)
       break
     pageNum = pageNum + 1L
  }

  invisible(jobs[!sapply(jobs, inherits, "try-error")])
}
```

This function iterates until there are no more search results pages for this query or until we have collected more than *max* posts. The *max* parameter allows us to limit the number of job postings we scrape. This can be useful for debugging and also to limit our requests to the Web server.

The caller provides the content of the first page of results. We leave it to her to submit the query and retrieve this page of results. This is site-specific. As we saw previously, for Kaggle, this is simply

```
u = "http://www.kaggle.com/forums/f/145/data-science-jobs"
getURLContent(u)
```

For CyberCoders, we use

```
getForm("http://www.cybercoders.com/search/",
        searchterms = queryString, searchlocation = "",
        newsearch = "true", sorttype = "")
```

Our searchJobs() function parses this *HTML* document and then applies the site-specific function getJobDescriptionLinks_f() to extract the links to the individual job posts. It then iterates over these and calls the site-specific function getJobDescription_f() to process each job posting. It appends these to the list of job descriptions and then determines the next page of results using, again, the site-specific function getNextPage_f().

The caller of the function specifies the site-specific functions for steps 2, 3, and 4 above. The searchJobs() function doesn't need to know how they work, just that they expect specific inputs and return specific outputs. This abstraction means that we can scrape new Web sites by implementing these three focused functions and making the initial query. Each of these should be quite short. We can implement a site-specific job post reader function to extract the metadata along with the free-form text. However, if we are content with reading only the words, we can use the generic getJobDescription() function which will work for arbitrary job posts (or any *HTML* page). This is the default value for the *getJobDescription_f* parameter.

We would also probably implement the initial query mechanism as a function so that we can reuse it. However, we envisage people writing a new high-level function for a particular job Web site. This would hide the details of making the initial query to get the first page of job listings, and specifying the different site-specific functions in a call to searchJobs(). For example, we could define a function for searching the CyberCoders Web site as

```
searchCyberCoders =
function(query, ...)
{
   txt = getForm("http://www.cybercoders.com/search/",
             searchterms = query, searchlocation = "",
             newsearch = "true", sorttype = "")

   searchJobs(txt, cy.getNextPageLink,
              cy.getPostLinks,
              cy.readPost, ...)
}
```

This generic searchJobs() function helps us focus on the essential elements of scraping a new site. Furthermore, it requires us to decompose the steps into separate functions. Each of these does one thing and so is relatively simple. We can also test them separately. This abstraction results in significantly more modular code that we (and others) can easily substitute. This framework also supports more complicated Web sites, e.g., if we have to use, and login to, a specific account to access the information.

There are several more detailed aspects of the searchJobs() function that are interesting. We explicitly retrieve the *HTML* document with getURLContent() before passing it to the getJobDescription_f() function provided by the caller. This ensures that we don't rely on htmlParse()'s ability to make the *HTTP* request. This allows us to process some atypical postings on some sites that require *HTTPS* rather than *HTTP* requests. On some sites, "sponsored" job postings required these different requests to access the job, and getURLContent() gives us the flexibility to do this.

We also call the getJobDescription_f() function within a call to try(). This means that if getJobDescription_f() raises an error for any reason for a particular post, we will catch the error here and continue on to the next post. That error will not cause the entire `while` loop to fail. If getJobDescription_f() does raise an error, the resulting element in the list of `jobs` will have class `try-error`. We then remove any of these at the end of the loop.

Q.5 Modify the searchJobs() function to have it optionally give progress messages as it completes pages of results.

Q.6 Suppose we download the first 400 jobs using searchJobs() and later decide to retrieve the next 200. How do we avoid processing the first 400 jobs again?

Q.7 Implement the top-level Kaggle scraping function using searchJobs(). How much did you have to change? Is this an improvement?

Q.8 We could have made our searchJobs() function entirely generic by having the caller specify a function for performing the initial search query. Do this. Do you think this is a worthwhile addition in comparison with having a separate function for each job site, such as cyberCoders() or searchCyberCoders() which we defined earlier?

Q.9 Instead of having the caller specify individual functions, we could define a closure or use a reference class with methods for getting the next page, finding the URLs for the job postings, and reading an individual post. We could then extend this class for each target Web site. Do this. Does this organization improve the code?

12.6 Scraping Career Builder

Now that we have developed a generic way to process a new Web site, we can see how we can use this to implement the functions for scraping the `careerbuilder.com` site. We need to find the first page of results for a given query term, find the links to the posts in that page, and find the next page of results from that page. If we want to provide a customized way to read each post, we can do this. Otherwise, we can simply read all the words in the text with our generic getJobDescription() function.

We can determine how to make the initial query by exploring the site using a Web browser. This amounts to another call to getForm() with several parameters and the query string. This is done with

```
txt = getForm(baseURL,
              IPath= "QH", qb = "1",
              s_rawwords = "queryString", s_freeloc = "",
              s_jobtypes = "ALL",
              sc_cmp2 = "js_findjob_home",
              FindJobHomeButton="hptest_ignore2",
              .opts = list(followlocation = TRUE))
```

where `baseURL` is

`http://www.careerbuilder.com/jobseeker/jobs/jobresults.aspx`

Exploring Data Science Jobs with Web Scraping and Text Mining

The `followlocation = TRUE` is an important option for the query in case this redirects to another URL.

Getting the links to the individual posts in the results page involves a simple *XPath* query. We can retrieve them with a function for processing the parsed *HTML* page as

```
cb.getJobLinks =
function(doc)
   getNodeSet(doc, "//a[@class = 'jt prefTitle']/@href")
```

We discover this, as we did before, by examining sample *HTML* pages and looking for the defining pattern.

The last remaining component is to find the next page. Again, this site displays the text `Next Page` that is a hyperlink to the next page of results. We can retrieve this *URL* from the current results page with the function

```
cb.getNextPage =
function(doc)
{
   nxt = getNodeSet(doc, "//a[. = 'Next Page']/@href")
   if(length(nxt) == 0)
       return(character())

   nxt[[1]]
}
```

This is remarkably similar to other versions of the same function and only differs by the particular *XPath* query. We might abstract this into a function and just specify this query string, however, it is probably not worth this extra effort.

We can now define the simple function to scrape collections of job posts on `careerbuilder.com` using these helper functions. This amounts to getting the text for the first page of results and calling searchJobs(). We can define this as

```
searchCareerBuilders =
function(query, ..., baseURL =
   'http://www.careerbuilder.com/jobseeker/jobs/jobresults.aspx')
{

   txt = getForm(baseURL, IPath= "QH", qb = "1",
              s_rawwords=query,
              s_freeloc="", s_jobtypes="ALL",
              sc_cmp2= "js_findjob_home",
              FindJobHomeButton="hptest_ignore2",
              .opts = list(followlocation = TRUE))

   searchJobs(getNextPage_f = cb.getNextPage,
           getJobDescriptionLinks_f = cb.getJobLinks,
           txt = txt, ...)
}
```

This allows us to collect posts from `careerbuilder.com` for any search term.

Q.10 Enhance this function to use a single curl handle for all of the *HTTP* requests.

12.7 Scraping Monster.com

We now have a simple recipe, and an example, for scraping a new Web site. We can replicate this for the `monster.com` site. We'll use a Web browser to identify the URL for a particular job search query term. When we enter `"Data Scientist"` in the text field, our browser brings us to `http://jobsearch.monster.com/search/Data-Scientist_5?`. This is quite different from previous sites we have scraped. The query string `Data Science` is not a value of a parameter on the right of the ? in the URL. Instead, the query string is in the path in the URL. Where did the _5 come from? This was appended by a mixture of the *HTML* form in the Web browser and redirection on the Web server.

We can use the developer tools in the Web browser to investigate the Web requests when we submit the search query. Unfortunately, among all of the many requests for images, ads, etc., the request for the primary page of results seems to go straight to the URL above. There does not seem to be a regular form request. This could be because it is a **POST**, rather than a **GET**, submission. However, we would see that request in the network developer tools panel. So something is quite different for this site.

We can explore the source of the top-level *HTML* page to understand the form. We can use the **RHTMLForms** package to programmatically query this page to get a description of each of the elements. However, this form is complicated by the presence of *JavaScript*. Indeed, this may be the cause of the different behavior.

We could spend time determining how to mimic the search request from R. However, while this is interesting for understanding how to scrape in general, it distracts us from our goal of actually getting the job postings. Sometimes it is important to pursue these apparent tangents. In this case, we can sidestep it quite easily. Basically, to scrape jobs for a particular search term, we need to determine the first page of the results. Then we find the next page link within that page, and so on. We can by-pass the initial form submission to find the first page and just enter the *URL* ourselves. We saw the *URL* for the search term `Data Scientist`. For the search term `Visualization`, it is `http://jobsearch.monster.com/search/Visualization_5`, and for `Data Science Engineer` it ends in `Data-Science-Engineer_5?`. We see a pattern: spaces in the search string are replaced with '-'; we prepend the URL `http://jobsearch.monster.com/search/` and append _5. For example, `Machine Learning` corresponds to `http://jobsearch.monster.com/search/Machine-Learning_5?`. We can write a function to map a search string to this form:

```
monsterSearchURL =
function(q)
{
  sprintf("http://jobsearch.monster.com/search/%s_5?",
          gsub(" ", "-", q))
}
```

We can write a function monsterSearch() to take the query, construct this URL and then call searchJobs().

```
monsterSearch =
function(q, firstPage = monsterSearchURL(q), ...)
{
  txt = getURLContent(firstPage, followlocation = TRUE)
  searchJobs(txt, monsterNextPage, monsterJobLinks, ...)
}
```

Exploring Data Science Jobs with Web Scraping and Text Mining

The first parameter q is our search term and we use it to construct the URL. However, we have also added the *firstPage* as a parameter with a default value of the URL from the query string. By having this as an explicit parameter, the caller can specify the exact URL for the first page of the search, in case it does not correspond to the scheme we deduced above.

Again, in our monsterSearch() function, we are using the generic word collector to process each post. We need to provide the `monster.com`-specific functions to find the next page and find the links to the posts in a page of results. We can find the next page using the *XPath* query

```
nxt = getNodeSet(doc, "//a[@title='Next']/@href")
```

Similarly, we can find the URLs for each job post with

```
nodes = getNodeSet(doc, "//a[starts-with(@class, 'slJobTitle')]
                         /@href")
```

We can use these expressions to define monsterNextPage() and monsterJobLinks(), respectively.

12.8 Analyzing the Results: The Important Skills

When we scraped the different Web sites, dice.com had 10,363 results for the search term `Data Science`. In contrast, CyberCoders has 35 results for the same term. However, the search term `Data Scientist` returned 70 posts on CyberCoders and 486 on Dice. For a different search term, the query `visualization` returned 979 results on dice, and 113 on CyberCoders. CareerBuilder.com returns over 12,000 results for Data Science! What do these numbers tell us about the number of jobs in this field? About the sites themselves? For one, some sites focus on specific fields more than others, e.g., information technology only, rather than all jobs, such as nursing, accounting, etc. Secondly, some sites match queries quite liberally, e.g., returning a job post title `Food Science Lab Technician` as a match for `Data Science`! What if we enclosed the query term in quotes, i.e., `"Data Science"`?

The job postings we scraped are not a random sample. They are all of the recent posts from these Web sites for the particular queries. Therefore, it is hard to say what larger population they are drawn from and represent. We have to be careful with any statistical inference we might make. For example, if we could compute the average salary for the posts found under Data Scientist and similarly for Statisticians, could we use a t-test to see if they are statistically different? Similarly, can we look at the counts of key words such as *Python*, *R*, and *Hadoop* in job posts from two different sites and use a test to see if the counts/proportions come from similar populations?

In short, there are lots of issues about making inferences from these data. However, the data are very informative and we can explore and summarize them to learn about the nature of different jobs, without using statistical tests.

Let's look at which skills and technologies are important for different types of jobs corresponding to our search queries. Of course, almost all of the jobs include the words data, scientist, statistics, communication, experience, business, and so on. We'll focus on the technologies and programming languages, but these other terms are very important. However, you should take the time to explore these.

While each of the Kaggle posts were structured very differently, we were able to extract all the words for each post. We can look at the frequency of all the words, but of course many are not that interesting. Instead, we'll look for words such as *R*, *Python*, *SAS*, *Hadoop*, *SQL* and so on. Figure 12.9 shows the frequencies of these selected words. We see that *R*, *Python* and *SQL* are the most common. There are two important things to note. Firstly, we have selected the terms based on our knowledge of what is important in Data Science. This helps to compare the important terms. However, it is entirely possible that we have omitted some that appear in many of the posts. We should carefully examine all of the 13 thousand words, or find a good way to identify new important words. Secondly, we should be aware that Kaggle is a specific community. Its focus and members are different from the general business community. Accordingly, the results for Kaggle may be quite different from other job sites. Indeed, each site has a different focus and niche.

For CyberCoders postings, we were able to explicitly extract the set of skill keywords. For the query `Data Science`, we get the count for each skill with

```
cyber.dsSkills = table(unlist(lapply(cyber.DataScientist,
                                     `[[`, "skills")))
```

There are only 35 postings in this collection, and 97 different skills. We should merge this with the postings for `Data Scientist`. This gives us 102 postings in total and 167 different skills. There are 28 skills that occurred in 5 or more of these postings. We might display the counts for the commonly occurring words via a dotplot as shown in Figure 12.10. We create this with:

```
i = (names(cyber.dsSkills) == "Natural Language Processing")
if(any(i))
  names(cyber.dsSkills)[i] = "NLP"
dotchart(sort(cyber.dsSkills[cyber.dsSkills > 4]),
         main = "Skills from Data Science/Scientist on CyberCoders")
```

Note that we have changed the phrase `Natural Language Processing` to its common abbreviation `NLP` in order to make the axis labels more appropriate for the plot.

Q.11 Check that no posting had any skill repeated twice or more.

Alternatively, we might draw a word cloud for the CyberCoders skills as in Figure 12.11. We create this via the `wordcloud` [4] package for *R* and the commands:

```
library(wordcloud)
wordcloud(names(cyber.dsSkills), cyber.dsSkills)
```

Is this a more effective way to visualize this data than the dot plot? Does the frequency correspond to the height of the letters or the area of the phrase? Is this another example of using two dimensions to represent one and mislead the viewer to compare area in contrast to height. (See [11].)

Q.12 Visualize the skills for Data Science job postings from other Web sites, e.g., `dice.com` and `monster.com`. Are they similar across Web sites? If so, combine the words across all posts and visualize them.

The salary is most readily available (and accurate) in the cybercoders and dice.com posts. However, to extract the salary range for each job, we have to perform a some text manipulation on each salary string.

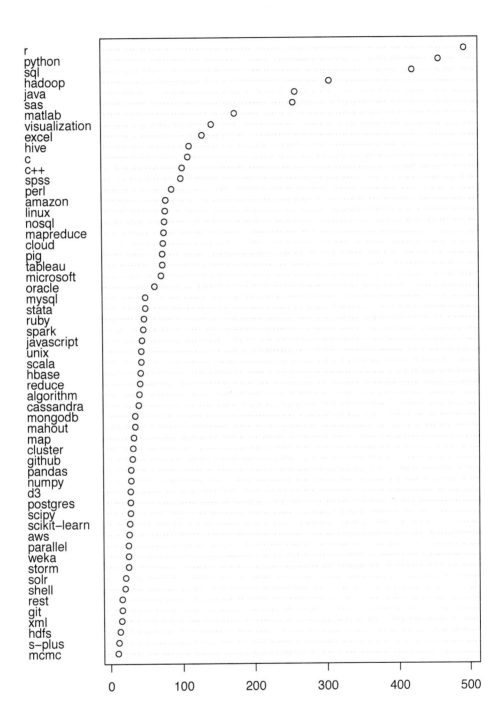

Figure 12.9: Frequency of Selected Terms over all Kaggle Job Posts. *This shows the number of occurrences of each of the selected terms across all 842 job posts on Kaggle by January 2015.*

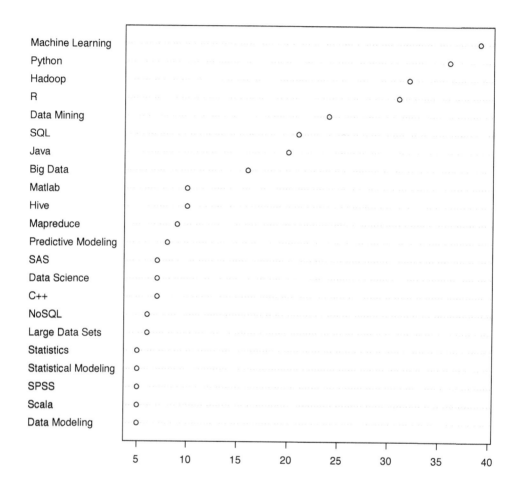

Figure 12.10: Dotplot for Frequency of Skill Phrases across CyberCoder.com Data Science Job Posts. *This shows the number of occurrences of different terms we selected across Data Science job postings on this Web site.*

Figure 12.11: Word Cloud for Frequency of Skill Phrases Across CyberCoder.com Job Posts. This is a different display of the counts of different words across posts on `cybercoders.com`.

For CyberCoder posts, the salary is in the `salary` element for each post and is in the form `"Full-time $90k - $200k"`, for example. Let's convert this to a minimum and maximum. We'll write a function to do this:

```
cy.processSalaries =
function(posts)
{
  tmp = strsplit(gsub(".* \\$([0-9]+)k - \\$([0-9]+)k",
                      "\\1,\\2", sapply(posts, `[[`, "salary")),
                 ",")
  vals = lapply(1:2, function(i)
                      1000*as.integer(sapply(tmp, `[`, i)))
```

```
  ans = as.data.frame(vals)
  names(ans) = c("low", "high")
  ans
}
```

We can combine the different posts together and then call this with

```
dsPosts = c(cyber.DataScientist, cyber.DataScience)
sl = cy.processSalaries(dsPosts)
```

Some of these values were mapped to NA. We need to examine the original value for our salary string in the posts corresponding to these NAs in order to adjust our function to be able handle these values. We can find which posts were problematic with

```
which(is.na(sl$low))
```

```
 [1] 10 15 19 22 26 28 34 40 58 89
```

and the original values with

```
sapply(dsPosts[is.na(sl$low)], `[[`, "salary")
```

Each of these has the value "Full-time Compensation Unspecified", so NA seems appropriate.

Now we can look at the distribution of the differences between the highs and lows within the same post:

```
summary(rowMeans(sl))
```

```
   Min. 1st Qu.  Median    Mean 3rd Qu.    Max.    NA's
      0  113800  130000  130700  150000  225000      10
```

We also have salary information from dice.com posts. The salary information is in the baseSalary element of each post. This is less structured than in the CyberCoder posts. Accordingly, we'll have to deal with more possible specifications to extract the actual values. We first examine the different values

```
unique(unlist(sapply(dice.dsPosts, `[[`, "baseSalary")))
```

and attempt to identify the different possibilities and patterns. There are numerous values that provide no information, e.g., inquiry, Salary DOE (depends on experience), Experience Based, Market, competitive. We'll map these to NA. There are many values that give a range, like CyberCoders. For example, two are "120-150K" and "$90-110K". We also have other values that are single values, e.g., "120000". Others mention stock options, e.g, "80k-120k +Stock Opts". Other salaries are given as a minimum, e.g., "200,000 + Bonus", "$100,000+", "150k+ DOE". It is not obvious how we should map these to both a low and high value. The value can be the low value, but there is no high value. We might map this upper bound on the salary to NA, but we will have to determine how we will use these values.

Some salaries are given as hourly rates, e.g., "60/hr on W2". We'll ignore these and map them to NA.

One strategy for mapping the strings to pairs of (low, high) values is to use a sequence of regular expressions. We start with the first and use it to find matches and convert those values into the form we want. For those values that were not matched by the first regular expression, we use the second regular expression. Generally, we apply the next regular expression in the collection to the subset of values that were not matched by any of the previous expressions. At the end, we hope that all values are matched by one regular expression.

We can write a function to perform the conversion of the general salary strings and then convert the results to a data frame of low and high values. The idea is that we have numerous pairs of regular expressions that we will pass to gsub() on the remaining strings that have not yet been converted. Each call to gsub() attempts to create new strings of the form "low;high". It would take "$110,000 - $140,000" and transform it to "1¬10,000;140,000". Similarly, it would take "$100,000+" and return "100,000;NA". We define the collection of regular expressions pairs as a named character vector:

```
SalaryRegularExpressions =
  c("^([0-9,]+)(\\.00)?( DOE)?$" = "\\1;\\1",
    "^([0-9,]+)[Kk]$" = "\\1;\\1",
    "^\\$?([0-9,]+)[Kk]\\+$" = "\\1;NA",
    "^to \\$?([0-9,]+)[Kk]$" = "NA;\\1",
    "\\$?([0-9,]+)[Kk]?( - |-)\\$?([0-9,]+)[Kk]?" = "\\1;\\3",
    "\\$?([0-9,]+)[Kk]?(\\+DOE)? ?\\+ ?\
(Equity|Bonus|Stock|Stock Opts).*$" = "\\1;\\1",
    "^\\$([0-9,]+)([Kk]|\\.00|\\+)?$" = "\\1;\\1",
    "\\$?([0-9,]+)[Kk]? to \\$?([0-9,]+)[Kk]?.*" = "\\1;\\3",
    "\\$?([0-9,]+)[Kk]?\\+?( DOE)?$" = "\\1;\\1",
    "\\$?([0-9,]+)[Kk]?/ann.*" = "\\1;\\1",
    "\\$?([0-9,]+)[Kk]?/ann.*" = "\\1;\\1",
    "\\$?([0-9,]+)[Kk]? all in" = "NA;\\1")
```

Each element in this vector has a corresponding name. This name is used to match a salary string. For the salary strings that this regular expression matches, we use the corresponding regular expression element to transform those salary strings into the form low;high. These regular expressions look quite complex and are certainly hard to read. Take the time to figure out what each does. Consider some of the sample salary strings we discussed above.

We now define our function to convert all of the salaries to a data.frame containing low and high values:

```
getSalaryRange =
function(values, asDataFrame = TRUE,
         rx = SalaryRegularExpressions)
{
  done = rep(FALSE, length(values))
  ans = rep(NA, length(values))

  for(i in seq(along = rx)) {
     w = grepl(names(rx)[i], values[!done])
     tmp = gsub(names(rx[i]), rx[i], values[!done][w])
     ans[!done][w] = tmp
     done = !is.na(ans)
  }

  ans = structure(gsub(",", "", ans), names = values)
```

```
    if(asDataFrame)
       convertLowHighToDataframe(ans)
    else
       ans
}
```

Q.13 Why do we use a `for` loop in getSalaryRange() rather than sapply() or lapply()?

There are two helper functions used in getSalaryRange() above: convertLowHighToDataframe() and fixNum(). The former splits the resulting `low;high` strings and then creates a data frame with 2 columns. It also converts values that are given in units of 1000 to their full amounts using fixNum(), e.g., 100K or 100, which should be `100000`. We define convertLowHighToDataframe() with

```
convertLowHighToDataframe =
function(ans)
{
    ans = as.data.frame(do.call(rbind, strsplit(ans, ";")),
                         stringsAsFactors = FALSE)
    names(ans) = c("low", "high")
    ans[] = lapply(ans, fixNum)
    ans
}
```

Q.14 Implement the fixNum() function. It takes a number as a salary value and determines whether to multiply it by 1000 to put it on the proper scale. This is used to adjust salaries that are reported as, e.g., $100K.

With these two functions in place, we can look at the dice.com salaries:

```
dice.salaries = getSalaryRange(unlist(sapply(dice.dsPosts,
                                              `[[`, "baseSalary")))
summary(rowMeans(dice.salaries))
   Min. 1st Qu.  Median    Mean 3rd Qu.    Max.    NA's
      0   97500  120000  115300  147500  225000     497
```

Note that there are almost 500 NA values out of 632. However, there are 40 rows that have one missing value, corresponding to the low or the high not being specified. We identify these with

```
w = apply(dice.salaries, 1, function(x) sum(is.na(x))) == 1
avgSalary = rowMeans(dice.salaries)
avgSalary[w] = apply(dice.salaries[w,], 1,
                      function(x) x[!is.na(x)])
summary(avgSalary)
   Min. 1st Qu.  Median    Mean 3rd Qu.    Max.    NA's
      0   82500  117500  110800  140000  250000     457
```

We see that including these shifts the distribution slightly to the left, but increases the maximum!

There are many more aspects to explore. For example, we can look at the data geographically. How many jobs are there in different states? cities? How does salary vary geographically? How do the necessary skill sets differ by salary? Is the set of skills changing over time? Compare the skill sets for different search queries – data analyst, machine learning, statistician, ...

12.9 Note on Web Scraping

Scraping a single page is quite easy. In many cases, helper functions such as readHTMLTable() will extract the data we want. If the data of interest is in a less structured form, we just have to find the particular rendering to identify the data we want and then we can extract them with the corresponding *XPath* queries. If there are two similar pages, we may have to adapt, generalize, or restrict the *XPath* queries. However, if we want to scrape many "similar" pages, we are moving from one-off code to actual software. This is why it is significantly more involved.

Before you scrape data from a Web site, you should first ask two questions, at least. The first is whether there is an easier and more structured way to obtain the data. Most sites have the data we want in some form of a database and they generate the *HTML* pages from this. Sites are increasingly creating Web services and APIs (Application Programming Interfaces) for accessing data. Instead of mimicking a browser and scraping *HTML* content, we can use *HTTP* to send our query request more directly and get the result as *JSON* or *XML*. This is much more reliable and easier to process in robust ways.

Some sites provide the data of interest in "bulk" form and they prefer people to use that rather than making many requests for small pages. If the site doesn't appear to offer the data in bulk form, you can email them and ask if there is a way to access the data. For many companies, the data are the primary assets and also often confidential and so they won't want to share them. However, in other cases, people may be willing to share the data in a much more structured format that will avoid many of the headaches of scraping Web pages.

If you do have to scrape the pages, the second question you need to ask is whether you are legally allowed to do this. Most sites have an explicit terms of service (ToS) that you agree to when you use their site. If this states that you are not to programmatically access the site, then you are not permitted to scrape the pages. If you do, the Web site can easily detect this and can block your IP address or even your entire local network. Other sites will rate-limit you and you are not to exceed that many requests in a fixed period of time.

12.10 Exercises

Q.15 Display all of the jobs from different sites geographically on a Google Earth or Google Maps display. Group the points to allow them to be toggled on and off. For example, group them by the Web sites on which they were found, or by general job category, or by salary range. See Chapter 11 for information about Google Earth and *KML*.

Q.16 Compare job postings for different keywords or job titles and compare their characteristics in various dimensions.

Q.17 Find other job sites and scrape data from those. How do these compare with the sites we have looked at?

Q.18 Compare the salaries for Data Science postings for the two sites Dice and CyberCoders.

Q.19 Compare the data you scraped now to the data we obtained in 2013 (available on the Web site). How have the salaries and skills changed?

Q.20 Determine the date for each job posting and see how the demand for different technologies has changed over time.

Q.21 Use pairs of words, i.e., 2-grams, rather than single words to analyze the posts. What are common pairs of words?

Q.22 Use Natural Language Processing to analyze the words in sentences to identify the metadata about the jobs from the free-form text.

Q.23 We assumed the set of stop words was available. There are many different sets of stop words and we can retrieve others. Extract the stop words from http://www.ranks.nl/resources/stopwords.html. Additionally, there is a collection of stop words for different languages in a zip archive at http://stop-words.googlecode.com/files/stop-words-collection-2011.11.21.zip. Use the **RCurl** and **Rcompression** packages to retrieve and extract the files for your language and create the set of stop words from these.

Bibliography

[1] Apache Software Foundation. Apache Lucene: Open-source search software. http://lucene.apache.org, 2011.

[2] Shay Banon. Elasticsearch: An open source, distributed, *REST*ful search engine. http://www.elasticsearch.org, 2011.

[3] Alex Couture-Beil. rjson: Converts *R* object into *JSON* and vice-versa. http://cran.r-project.org/web/packages/rjson/, 2011. R package version 0.2.6.

[4] Ian Fellows. wordcloud: Pretty word clouds. http://cran.r-project.org/web/packages/wordcloud, 2014. R package version 2.5.

[5] Deborah Nolan and Duncan Temple Lang. *XML and Web Technologies for Data Sciences with R*. Springer, New York, 2013.

[6] R Development Core Team. *R: A Language and Environment for Statistical Computing*. Vienna, Austria, 2012. http://www.r-project.org.

[7] John Simpson. *XPath and XPointer: Locating Content in XML Documents*. O'Reilly Media, Inc., Sebastopol, CA, 2002.

[8] Duncan Temple Lang. **RJSONIO**: Serialize R objects to *JSON* (*JavaScript* Object Notation). http://www.omegahat.org/RJSONIO, 2011. R package version 0.95.

[9] Duncan Temple Lang. **XML**: Tools for parsing and generating *XML* within R and *S-PLUS*. http://www.omegahat.org/RSXML, 2011. R package version 3.4.

[10] Duncan Temple Lang. **RCurl**: General network (HTTP, FTP, etc.) client interface for R. http://www.omegahat.org/RCurl, 2012. R package version 1.95-3.

[11] Edward Tufte. *The Visual Display of Quantitative Information*. Graphics Press, Connecticut, 1983.

Index

.C(), 272, 347–349, 354–356
.Call(), 347, 348, 356
[(), 175
[[.data.frame(), 335, 336
%in%(), 148

abline(), 248, 252
access point, 3, 4
addChildren(), 446
addPlacemark(), 444–446, 452
addPlacemarks.fast(), 447
addPlacemarks.slow(), 447
addRunners(), 86
aggregate(), 190
all(), 148
all.equal(), 339
AND, 404
any_count(), 389, 390, 393
API, 457, 501
AS, 404
as.data.frame(), 175
as.numeric(), 53
asWords(), 480
ATT, 240, 246, 253, 257
attach.big.matrix(), 226, 231, 232
attributes(), 80

bag of words, 127–130, 134, 164
Bayes' Rule, 111
bet(), 390, 391, 393
big.matrix(), 232
big.matrix, 225, 227–229, 232–234
biganalytics, 235
biglm(), 235
biglm, 235
bigmatrix, 225–227, 234, 235
bigmemory, 217, 224, 225, 227, 228, 231–233
bigmemory, 224
birth-and-assassination process, 281
BMLGrid, 319, 324, 325, 354
boundary string, 105, 108, 115, 117, 139

boxplot, 14, 16, 21, 22, 24, 54, 55, 63, 131, 132, 156, 405
bwplot(), 21
by(), 17, 190, 401, 404

C, iii, v, vi, 243, 272, 273, 308, 326, 346–350, 352–357, 360, 505
C++, 272, 346, 356
calcError(), 41
call frame, 94, 317, 384
call stack, 94, 317
careerbuilder.com, 457, 490, 491, 493
cbind(), 175, 339
Central Intelligence Agency, 418
Cherry Blossom Ten Mile Run, 43
circle.fit(), 202, 203
class(), 80
classification tree, iv, 105, 136, 160, 165
cleanUp(), 98
cmpFun(), 272, 360, 362
codetools, 20, 41
col(), 326
colnames(), 80
color, 61–63, 184, 185, 301, 303, 323, 362, 418, 419, 422, 446, 449, 452
  transparent, 62, 64, 65
colorRamp(), 185
combine2Stocks(), 247
compile, 271
compiler, 272, 360
computeFreqs(), 128, 164
computeMsgOdds(), 130, 164
computer experiment, 243, 261, 262, 266, 268, 297, 357, 362
confusion matrix, 207
convertLowHighToDataframe(), 500
convertTime(), 58, 99
CoPrimeBMLGrid, 324
COUNT(), 223, 401, 402, 412
count(), 385
count_payoff(), 391, 393
Counter, 384, 385
CRAWDAD, 4, 42

507

createDerivedDF(), 152, 153
createDF(), 58–60, 99
createGrid(), 312, 316, 317, 319, 322–324, 332, 344
cross-sectional, 44, 72, 76, 86, 91
cross-validation, 4, 37–40, 42, 105, 133, 166, 210, 211
crunBML(), 355, 356
CSS, 475
CSV, 172, 216, 218, 219, 222, 225, 245, 246, 418–420, 422, 427, 448
cumsum(), 40, 263
cumulative distribution function, 278
cut(), 63, 423
cy.getFreeFormWords(), 478, 479, 484, 485
cy.getLocationSalary(), 480
cy.getNextPageLink(), 486
cy.getPostLinks(), 487
cy.getSkillList(), 478, 480
cy.readPagePosts(), 482
cy.readPost(), 477, 478, 480, 485
cyberCoders(), 490
cybercoders.com, 462, 473–475, 477, 478, 480, 481, 485–487, 489, 493, 494, 497, 498

data structures, 4, 45, 53, 82, 105, 108, 138, 146, 174, 217, 221, 224, 279, 291, 308, 398, 419, 421
data.frame(), 341
data.frame, 224, 233, 342, 499
data.table, 342
Date, 245, 252, 258
dbGetQuery(), 223, 411
dbHasCompleted(), 413
DBI, 399
dbSendQuery(), 411
DCF, 140
dd(), 371, 373, 393
dealer_cards(), 375
debugging, 4, 45, 105, 126, 138, 144, 153, 154, 176, 177, 227, 248, 254, 278, 279, 290, 317, 371, 377, 379, 381, 386
decks_left(), 386, 393
density curve, 74, 75, 99, 269, 381, 382
densityplot(), 21, 74
deparse(), 247, 288, 305
dice.com, 493, 494, 500
dim(), 40, 80, 225, 334, 335, 408
display.brewer.all(), 63

DISTINCT, 403, 405
distribution
    Exponential, 277–281, 292
    Normal, 76, 262
    Poisson, 279, 285, 287, 292
    Uniform, 279, 286, 287, 305
do.call(), 10, 60, 61
DocBook, 505
docName(), 485
domain-specific language, 456
doMC, 231
doMPI, 231
doRedis, 231
doSMP, 231
doSNOW, 231
Dow Jones, 239, 241, 242
download.file(), 244
draw.circle(), 446, 452
draw_n(), 386, 388, 393
dropAttach(), 114, 122, 124, 144, 162, 163
DSL, 456
duplicated(), 190

efficiency, iv, 105, 218, 221, 227, 230, 271, 279, 308, 340, 346, 356, 360, 408, 446
ElasticSearch, 456
error, 94, 317
evalSegment(), 208
evalSegments(), 208, 209
expand.grid(), 274, 300
exploratory data analysis, 4, 14, 15, 17, 45, 105, 154, 170, 173–175, 177, 191, 457
exptOne(), 297, 305
ExtendedBMLGrid, 325
extractResTable(), 94, 97, 98, 100
extractRestTable(), 94
extractVariables(), 52, 57, 98, 99

f(), 383, 384
familyTree(), 293, 295–297, 305, 306
fetch(), 411, 412
fields, 26
file.info(), 171
filled.contour(), 301, 303
findColLocs(), 50–52
findGlobals(), 20, 41, 329
findMsg(), 163
findMsgWords(), 114, 124, 164
findNextPosition(), 251, 254, 270, 271, 273

Index

findNN(), 34, 35, 40, 41
fitOne(), 88
fixNum(), 500
force(), 288, 305
foreach(), 229–232, 234
`foreach`, 227, 228, 231
`foreach`, 229
FROM, 402, 404, 407, 409

gambling strategy, 366, 381
    card counting, 382, 388, 389, 391, 392
    optimal, 377–379, 382, 391
    simple, 374, 376, 382
gc(), 219
genBirth(), 283, 284
genKids(), 282–288, 304, 305
genKidsB(), 304
genKidsR(), 284, 305
genKidsU(), 286–289, 305
genKidsV(), 289–292, 305, 306
GET, 244, 481, 492
getBestK(), 267, 268, 274
getBoundary(), 119, 124, 139, 146, 162, 163
getCarLocations(), 329, 330, 332, 336, 340–343, 354, 362
getCurlHandle(), 487
getForm(), 244, 470, 481, 485, 487, 490
getFormParams(), 481
getJobDescription(), 489, 490
getJobDescription_f(), 489, 490
getJobDescriptionLinks_f(), 489
getNextPage_f(), 489
getNextPosition(), 328, 329, 334, 335
getNextPositions(), 330
getNodeSet(), 92, 438, 439, 451, 464, 466, 476, 482, 484
getPositions(), 255, 267
getPostLinks(), 472
getProfit.K(), 267, 268, 274
getRangeErrors(), 190
getRelativeURL(), 482
getSalaryRange(), 500
getSegments(), 194–197, 206
getSibling(), 466, 479
getURLContent(), 470, 487, 489
getWrappedSegments(), 197, 198, 201, 206, 207
`ggplot2`, 232
global variable, 271
globalenv(), 20
Google, 240

Google Earth™, iv, 417–419, 422, 427–430, 432, 433, 440, 442, 444, 446, 452
Google Earth™, 418, 428, 429
GPU, ix
gregexpr(), 49
grep(), 48, 115, 116, 174, 176
grepl(), 116, 174
grid search, 199, 243
gridToIntegerGrid(), 355
GROUP BY, 223, 398, 403–405, 410
GROUP BY, 404
gsub(), 78, 79, 118, 148, 499

Hadoop, 493, 494
hand, 371
handValue(), 369
HAVING, 398, 404, 405
HAVING, 404
head(), 225, 403
heat map, 26–28, 41, 303, 304
hi_low(), 389, 391, 393
hit(), 371, 373, 393
HTML, iv, 91, 92, 96, 97, 100, 106, 107, 136, 145, 163, 244, 430, 432, 442, 446, 448, 451, 456, 457, 463–466, 468, 470, 472, 475–482, 484, 485, 489, 491, 492, 501
HTML, 149, 448, 451
htmlParse(), 91, 97, 100, 451, 470, 478, 487, 489
HTTP, v, 244, 419, 456, 457, 478, 481, 487, 489, 491, 501
HTTPS, 470, 487, 489

identical(), 20, 339
ifelse(), 336, 340
image(), 320–322
image map, 28, 303, 304
increment(), 384, 385
indeed.com, 457
infant mortality, 418, 421
initialize(), 384, 385
interactive maps, 428
interactive visualization, 419, 429
International Organization for Standardization, 420
invisible(), 87, 254
isRe(), 151, 152, 165
isYelling(), 150

JavaScript, 463, 492

JSON, iv, 456, 501–503

Kaggle, 457–459, 462, 463, 466–469, 471, 487, 488, 490, 494
KML, iv, v, 429–433, 440–443, 446, 452, 453, 501
kmlLegend(), 432, 433
kmlPoints(), 430–432, 440

Lahman's baseball archive, 397, 399
lapply(), 11, 20, 45, 94, 176, 327, 500
LaTeX, 505
lattice, 21, 74, 180
least squares, 45, 65, 71, 89, 90, 345
    linear, 66
    non-linear, 170, 199
LEFT OUTER JOIN, 398, 409
legend(), 432
length(), 7, 78, 109, 213, 403
likelihood, 104, 112, 113, 129, 130
    ratio, 112, 113, 124, 126–131, 133, 134, 164, 166
LIMIT, 403, 405
line plot, 240, 248, 249
linear model, 67, 216, 235
    piecewise, 45, 69, 86, 99
    simple, 65, 67, 68, 70
lines(), 67
linkedin.com, 457
list.dirs(), 108
list.files(), 108–110, 171
lm(), 61, 65–68, 87, 235
lm, 67, 68, 71, 72, 235
local area networks, 3
loess(), 24, 61, 67–69, 91, 99
loess, 68, 72
longitudinal, 45, 87, 100
ls(), 11

MAC address, 6
machine learning, v, 457
makeBaseDocument(), 443, 446
makeCIAPlot(), 452
makeStyleNode(), 445, 446, 452
makeStyleNodes(), 446
map(), 424
mapply(), 27, 45, 97, 289, 444
mapproj, 449
MapReduce, v
maps, 424
mashup, 417, 427

match(), 114, 288, 320, 322, 334, 355
MATLAB®, 347, 353, 356
matrix(), 37
matrix, 224, 225, 233, 320, 324, 326
matrix subsetting, 308
MAX(), 402
max(), 79
MaxMind, 418
MCBA(), 300
mean(), 273
merge(), 408, 409, 421, 449
methods(), 325
MIME type, 105, 165
MIN(), 402, 403, 405
mode(), 80
monster.com, 457–459, 461
monsterJobLinks(), 493
monsterNextPage(), 493
monsterSearch(), 492, 493
Monte Carlo, 277–279
mosaic plot, 159
moveCars(), 327–335, 338, 339, 341–344, 346, 348, 362

naïve Bayes, iv, 104, 111–113, 131
names(), 464
nchar(), 163
nearest neighbor, iv, 4, 29, 32, 34–38, 41, 100, 166
new(), 385
new_hand(), 371, 374
new_shoe(), 387
newXMLDoc(), 440
newXMLNode(), 440, 441, 444–447
NextMethod(), 325
nlm(), 199–203, 211, 213
non-parametric curve, 45, 69, 100
nonlinear, 90, 99, 199
NOT, 404
numerical accuracy, 135
numerical optimization, 170, 199

object-oriented programming, 308, 323, 325, 371
object.size(), 40
ON, 410
on.exit(), 186
open(), 212
options(), 94
OR, 404
order(), 247

Index

ORDER BY, 405
outer(), 190
over plotting, 61, 62, 65, 67, 86

pairs trading, 239, 240, 243, 248
palette, 62, 304, 422, 452
par(), 186
parallel, 346, 362
parallel computing, 216, 225, 227–231, 362
parseXMLAndAdd(), 446
paste(), 163
payoff(), 381, 388
PDF, 106, 107, 505
perCaps(), 150, 151, 154
perCaps2(), 154
play_hand(), 375–377, 380, 381, 391, 393
played(), 386, 391
Player Project, 172
plot(), 61, 62, 64, 74, 99, 184, 186, 188, 247, 248, 323, 324, 354, 390, 430, 446, 452
plot.BMLGrid(), 321–325, 334, 354, 355
plot.default(), 324
plot.ExtendedBMLGrid(), 325
plot.RobotLog(), 184
plot.rpart(), 160
plot.surface(), 27
plotLook(), 186–188, 196, 200
plotLook2(), 188, 194, 195
plotRatio, 252
pmax(), 70, 425
PNG, 433, 446, 452
png(), 452
points(), 184, 186, 430
Poisson process, 279, 280, 282, 285
positionProfit(), 255, 257, 267
POSIXt, 12
POST, 481, 492
predict(), 68, 71, 72, 99, 160
predict.lm(), 68
predict.loess(), 68
predictSurface(), 27
predXY(), 35, 38–40
print(), 160, 320, 371
print.BMLGrid(), 320, 321, 324, 325
print.hand(), 371
processAllEmail(), 146, 147
processAllWords(), 125, 146
processAttach(), 138, 144, 146, 165
processHeader(), 138, 141, 146, 166
processLine(), 10–12, 20

profiling, 243, 269, 270, 273, 279, 287, 288, 305, 308, 356, 390, 394

queryFcn(), 409
quote(), 271

R2HTML, 432
random number generator, 265, 279, 290, 305, 373
rbind(), 60, 61
RColorBrewer, 62, 422
Rcompression, 502
Rcpp, 346, 356
RCurl, 244, 456, 470, 502
rdyncall, 346
read.big.matrix(), 225, 226
read.csv(), 173, 218, 225, 244, 245, 419, 420
read.dcf(), 140, 165
read.fwf(), 46, 98
read.table(), 8, 9, 45, 173, 218, 419, 452
readData(), 19, 20, 31, 40, 245–247
readEmail(), 146
readHTMLList(), 466
readHTMLTable(), 244, 501
readKagglePost(), 470, 471, 477
readLines(), 7, 9, 46, 100, 110, 115, 174, 212, 213
readLog(), 176, 177, 190, 212
recover(), 45, 94, 317
rect(), 320
recursive, 279, 284, 468
recursive partitioning, 136, 137, 162, 166
recycling rule, 255
reference class, 366, 385
reference class, 325, 366, 384, 385
regexpr(), 49
regression, 199, 235
regular expression, iii, iv, 8, 9, 45, 48, 55, 57, 98, 104, 116, 117, 119, 123, 140, 163, 165, 171, 177, 457, 467, 498, 499
relational database, iv, 216, 217, 221, 223, 224, 397–399, 401, 407
rep(), 53, 127, 248, 313, 314, 318
rep_len(), 318
replicate(), 268, 380
reshapeSS(), 33, 37, 41
rexp(), 281, 288, 289
Rffi, 346
rgb(), 184
RHTMLForms, 492

RITA, 216
rjson, 456
RJSONIO, 456
RKML, 419, 429, 430, 444
rle(), 192–194, 196
Rllvm, 272
RLLVMCompile, 272, 360
rm(), 219
RMySQL, 399, 400
rnorm(), 263–265
robot.evaluation(), 201–203, 205, 208, 211, 213
ROracle, 399
roundOrientation(), 14
row(), 326
rpart(), 158, 160, 162, 166
rpart, 160
rpart, 158
rpart.control(), 161, 166
rpart.plot, 160
rpois(), 289
RPostgreSQL, 399
Rprof(), 40, 287, 305, 333
RSpamData, 108, 110, 153
RSQLite, 222
RSQLite, 221
runBML(), 333–335, 343, 345, 346, 349, 350, 355
runGrid(), 357
runif(), 286, 289
runSim(), 267, 268, 274

S, ix
S3, 308, 323, 325, 371
S4, 325
S&P 500, 239–242
sample(), 126, 166, 315
sapply(), 45, 50, 87, 116, 132, 174–176, 218, 255, 268, 327, 476, 500
SAS, 494
saveXML(), 433, 444, 446
scale(), 190
scatter plot, 25, 29, 30, 41, 61, 62, 157, 300, 302, 320, 410
    3d, 301
    smooth, 26, 63
scatterplot3d, 301
searchCyberCoders(), 490
searchJobs(), 489–492
seed, 265, 290, 291, 293, 295, 296, 318, 373, 376, 381

SELECT, 223, 398, 401, 402, 404, 405, 408–410
SELECT, 402
selectCols(), 51, 52, 55
selectTrain(), 33, 34, 41
set.seed(), 265, 373
setdiff(), 468
setRefClass(), 385
setupRpart(), 159
Shoe, 390, 393
shoe(), 371, 376, 384, 388
shoe, 385
showPosition(), 252–254
shuffle(), 386, 393
shuffle_decks(), 386
signal strength, 3, 4, 7, 14, 20
simple_strategy(), 374, 376
simplyhired.com, 457
simProfitDist(), 266, 268–270, 274
simulation, iv, 243, 262, 263, 279, 292, 308, 365, 381, 391, 393
smoothScatter(), 24, 63, 67
sort(), 411
source(), 318
SpamAssassin, 104, 108, 109
spline
    loess, 67
    thin plate, 26
split(), 229
splitMessage(), 114, 115, 124, 138, 146
splitPair(), 371–373, 393
SQL, iv, vi, 216, 217, 221–225, 397–399, 401, 402, 404, 407–410, 413, 459, 494, 505
SquareBMLGrid, 324
stand(), 371–373
statscareers.com, 457
stem, 480
stemming, 123, 127, 164, 457
stochastic process, 243, 276, 281, 288
stockSim(), 265–267, 270–273
stockSim.c(), 273
stop(), 317
stop words, 123, 164, 457
stopifnot(), 318, 379
stopwords(), 164
strategy_optimal(), 379
streaming data, 170
strsplit(), 8, 9, 53, 58, 148, 163, 174, 467
sub(), 117, 118, 140
substitute(), 247

Index

substr(), 7, 48–50, 163
summary(), 15, 66, 67, 71, 324
summary.BMLGrid(), 324
summary.default(), 324
summary.matrix(), 324
summaryRProf(), 335
summaryRprof(), 287, 333
surfaceSS(), 27, 41
SVG, 362
symbols(), 252
system(), 221
system.file(), 108
system.time(), 40, 164, 287, 447
system2(), 221

table(), 15, 78, 316
table, 80
tapply(), 45, 99, 190, 401, 404, 409
text mining, 104, 113, 122, 159, 163, 164
text processing, iv, 170, 457
textConnection(), 244
textConnection, 177
time, 57, 99
time series, 240, 242, 248, 250, 252–254, 262
    autoregression, 261
tm, 123, 164
tolower(), 48
Tps(), 26
trace(), 271, 323, 334
trainSelect(), 35
tree structure, 276, 288, 291, 419, 433–438, 446
trimBlanks(), 78
try(), 490
try-error, 490
type I and II error, 105, 132–134, 136, 161–164, 166
typeIErrorRate(), 132, 166
typeIErrorRates(), 164
typeIIErrorRates(), 164

unique(), 128, 411
UNIX, iii, vi, 216, 217, 219–221, 227
unlist(), 128
usajobs.gov, 457
UseMethod(), 324

vectorized, 127, 132, 133, 181, 190, 243, 287, 289, 308, 311, 313, 326, 332, 335, 336, 338–340, 345, 394, 446
Verizon, 246, 247, 253, 257

version control, v
visualization, 4, 45, 63, 74, 170, 182, 185, 248, 252, 254, 279, 293, 301, 305, 417–419, 422, 457

Web scraping, 44, 45, 91, 448, 451, 457, 459, 461, 463, 471, 501
WHERE, 224, 398, 403–405, 407, 408, 410
WHERE, 404
which(), 80, 250
winnings(), 368, 369, 374, 392, 393
word cloud, 497
word vector, 126, 133, 136, 162, 166
wordcloud, 494
writeLines(), 98, 100

XDocTools, 505
XDynDocs, 505
XML, 91, 419, 433–435, 440, 441, 444, 446, 456, 466
xmlAncestors(), 478
xmlChildren(), 434, 437
xmlGetAttr(), 437, 439
xmlName(), 436
xmlParent(), 478
xmlParse(), 434, 435
xmlParseAndAdd(), 452
xmlRoot(), 434, 436, 451
xmlSize(), 436
xmlValue(), 92, 452, 465, 475, 479
XPath, 92, 97, 100, 419, 435, 437–439, 450, 451, 456, 457, 464–467, 472, 475–479, 484, 491, 493, 501, 505
xpathApply(), 484
xpathSApply(), 466, 475, 476
XPointer, 453, 502
XSL, 505

YAHOO, 240

Colophon

The content of this book was authored using *DocBook*, an *XML* vocabulary for technical documents. We introduced numerous extensions to the vocabulary for R code and plots, other languages and concepts such as *SQL*, *C*, *XPath*, and the UNIX shell. The content was edited primarily using the Emacs text editor, using our extended version of nxml-mode (which is available on the book's Web site http://rdatasciencecases.org).

The typesetting for the book was performed by transforming the *XML* to LaTeX using *xsltproc*. We used the *dblatex XSL* style sheets that build on top of the docbook-xsl distribution (version 1.74-0). Again, we implemented extensions to these *XSL* style sheets to process our extensions to the *DocBook* vocabulary and also to customize the appearance to conform with the Chapman & Hall format. The resulting LaTeX markup is processed by *pdflatex* to create the final *PDF* file, using the Chapman & Hall *krants.cls* LaTeX class file.

The authoring tools for Emacs and the *XSL* stylesheets are available on the book's Web site and also in the **XDynDocs** package available via Github and the Omegahat Web site (http://www.omegahat.org/XDynDocs). Similarly, tools for programmatically querying, validating, and updating the document are available in the **XDocTools** package (https://github.com/omegahat/XDocTools). Code from the book and supplementary materials are available at http://rdatasciencecases.org